THE MANY FACETS OF GEOMETRY

The Many Facets of Geometry

A tribute to Nigel Hitchin

Edited by

Oscar García-Prada
Jean Pierre Bourguignon
Simon Salamon

OXFORD
UNIVERSITY PRESS

OXFORD
UNIVERSITY PRESS

Great Clarendon Street, Oxford OX2 6DP

Oxford University Press is a department of the University of Oxford.
It furthers the University's objective of excellence in research, scholarship,
and education by publishing worldwide in

Oxford New York

Auckland Cape Town Dar es Salaam Hong Kong Karachi
Kuala Lumpur Madrid Melbourne Mexico City Nairobi
New Delhi Shanghai Taipei Toronto

With offices in

Argentina Austria Brazil Chile Czech Republic France Greece
Guatemala Hungary Italy Japan Poland Portugal Singapore
South Korea Switzerland Thailand Turkey Ukraine Vietnam

Oxford is a registered trade mark of Oxford University Press
in the UK and in certain other countries

Published in the United States
by Oxford University Press Inc., New York

First published 2010

British Library Cataloguing in Publication Data
Data available

Library of Congress Cataloging in Publication Data

Data available

Typeset by SPI Publisher Services, Pondicherry, India
Printed in Great Britain
on acid-free paper by
CPI Antony Rowe

ISBN 978–0–19–953492–0

1 3 5 7 9 10 8 6 4 2

To Nigel Hitchin

SHORT CONTENTS

FULL CONTENTS

PREFACE

A conference to honour Professor Nigel Hitchin on the occasion of his 60th birthday took place at the Consejo Superior de Investigaciones Científicas in Madrid during the week of 4–8 September 2006. The meeting was devoted to the many research topics covered by Professor Hitchin, and was a satellite of the International Congress of Mathematicians, that had closed in the same city a few days earlier.

This book was conceived at the time of the September conference, though it took longer to acquire its present unity. It incorporates chapters from most of the speakers, as well as chapters by Simon Donaldson, Michael Murray, and Edward Witten, all of whom attended the conference 'in spirit'. We are delighted too that Nigel Hitchin himself agreed to write a personal view on 'Geometry and physics' that places all the contributions in context.

Nigel Hitchin's mathematical influence has been enormous, and his work is frequently cited in widely different branches of mathematics. In accordance with the book's title, his work concerns not just differential geometry, but strikes at the heart of algebraic geometry, complex analysis, Lie groups, as well as abstract algebra and topology. The present contributions reflect his work on four-manifolds, monopoles and instantons, Nahm's equations, hyperkähler manifolds, Higgs bundles, integrable systems, the geometrical structure of moduli spaces, gerbes, harmonic forms, reduced holonomy, special Lagrangian submanifolds, and generalized complex structures. Still this list is not exhaustive, and a number of chapters address aspects of theoretical physics that underlie some of Hitchin's most significant results.

Seven of the contributors either obtained their doctorates under Nigel Hitchin's supervision (as did two of the editors), or studied actively with him during their time as graduate students. It is likely that each of the remaining authors would claim to have been a 'student of Nigel' at some point in their professional career. Many of Nigel's students have inherited the clear presentational style of announcing deep results that we have become accustomed to in his own lectures and seminars. This aspect was all too evident in the conference, and we feel that it is well represented in this book.

The editors would like to extend special thanks to all the contributors and, above all, to Nigel Hitchin himself.

Oscar García-Prada
Jean Pierre Bourguignon
Simon Salamon

June 2009

LIST OF EDITORS AND CONTRIBUTORS

Editors

Oscar García-Prada (CSIC, Madrid, Spain)
Jean Pierre Bourguignon (IHES, France)
Simon Salamon (Politecnico di Torino, Italy)

Contributors

Jørgen Ellegaard Andersen (Aarhus Universitet, Denmark): Toeplitz operators and Hitchin's projectively flat connection.

Sir Michael Atiyah (University of Edinburgh, UK): Mathematical work of Nigel Hitchin.

Philip Boalch (ENS, Paris, France): Towards a non-linear Schwarz's list.

Robert L. Bryant (MSRI, USA): Non-embedding and non-extension results in special holonomy.

Benoit Charbonneau (Duke University, Durham, USA) and **Jacques Hurtubise** (McGill University, Canada): The Nahm transform for calorons.

Simon K. Donaldson (Imperial College, London, UK): Nahm's equations and free-boundary problems.

Daniel S. Freed (University of Texas at Austin, USA), **Michael J. Hopkins** (Harvard University), and **Constantin Teleman** (UC Berkeley, USA): Consistent orientation of moduli spaces.

William M. Goldman (University of Maryland, USA): Higgs bundles and geometric structures on surfaces.

Marco Gualtieri (MIT, USA): Branes on Poisson varieties.

F. Reese Harvey (Rice University, USA) and **H. Blaine Lawson, Jr.** (SUNY at Stony Brook, USA): Projective linking and boundaries of positive holomorphic chains in projective manifolds, part I.

Tamás Hausel (University of Oxford, UK): S-duality in hyperkähler Hodge theory.

Nigel Hitchin (University of Oxford, UK): Geometry and Physics: a personal view.

Claude LeBrun (SUNY at Stony Brook, USA): The Einstein–Maxwell equations, extremal Kähler metrics, and Seiberg–Witten theory.

Naichung Conan Leung (Chinese University of Hong Kong, Hong Kong) and **Shing-Tung Yau** (Harvard University, USA): Mirror symmetry of Fourier–Mukai transformation for elliptic Calabi–Yau manifolds.

Nicholas S. Manton (University of Cambridge, UK): Skyrmions and nuclei.

Michael K. Murray (University of Adelaide, Australia): An introduction to bundle gerbes.

S. Ramanan (Chennai Mathematical Institute, India): Some aspects of the theory of Higgs pairs.

Graeme Segal (University of Oxford, UK): Locality of holomorphic bundles, and locality in quantum field theory.

Edward Witten (IAS, Princeton, USA): Mirror symmetry, Hitchin's equations, and Langlands duality.

I

GEOMETRY AND PHYSICS: A PERSONAL VIEW

Nigel Hitchin

This excellent collection of chapters prompts me to recall here how their various themes emerged in my own mathematical development. It may help the reader to navigate a passage through them and also perhaps explain how the links between physics and geometry which seem to underlie much of my own work came about. I hope these background comments do not seem too self-indulgent, but they might shed some light on the curious process of doing mathematical research – a process which is largely hidden in the chapters we finally write.

I was never a serious student of physics. As a mathematics undergraduate in Oxford from 1965 to 1968 I avoided the courses in relativity and quantum mechanics that my peers were taking. Instead I would browse in the Jesus College library trying to make sense of Weyl's *The Classical Groups* or in particular Harley Flanders' *Differential Forms with Applications to the Physical Sciences*, one which in a way defined my interests right from the start.

Although my DPhil thesis was about the Dirac operator, this was largely in the context of the Atiyah–Singer index theorem and associated vanishing theorems and it hardly occurred to me that physicists would be interested in this Riemannian version. When I went to the Institute for Advanced Study in 1971–3 my neighbour in the next apartment was the physicist Andy Hanson but we rarely spoke about our work. He once asked me what was so special about 26 dimensions and all I could think of was that it was of the form $8k+2$ (where the mod 2 index exists). My main thoughts at that time were on the two big unsolved problems in differential geometry: the Yamabe conjecture and the Calabi conjecture. Yau's presence at the Institute in the second year was a clear influence here. My first paper, on four-dimensional Einstein manifolds (Hitchin 1974), was heavily influenced by the Calabi conjecture and the K3 surface. There was a great lack of understanding of the Einstein equations at the time amongst pure mathematicians, concerning both global and local existence.

It was the conformal invariance of the Dirac operator which brought me closer to physics. I spent the year 1973–4 in New York at the Courant Institute and I began looking for a systematic way of finding all first-order conformally invariant differential operators. I started to read papers of Penrose on the zero rest-mass field equations; the conformal invariance of Maxwell's equations on a four-manifold gave an elliptic complex

$$\Omega^2_{\pm} \to \Omega^3 \to \Omega^4$$

which intrigued me at the time. I remember rather nervously giving a talk one cold December day at MIT to an audience of Atiyah, Kostant, Singer, and Irving Segal about Szegö kernels for the Dirac operator and I became aware then of Segal's interest in conformal invariance, causality, and Mach's principle.

I returned to Oxford in 1974 as Michael Atiyah's research assistant. Roger Penrose had also recently been appointed as the Rouse Ball Professor and I began to learn from him and his students about twistor theory, in particular the non-linear graviton construction. At the time everything was performed over the complex numbers, or specialized to Minkowski space, and my interest was still driven by the Calabi conjecture and whether a K3 surface, or some sort of complexification, could be found using twistors. The first I heard of the Yang–Mills equations was overhearing, while in the queue for tea in the Mathematical Institute, Penrose describing his student Richard Ward's new twistorial construction: 'You just turn the handle and out it comes ...' However it was Singer's visit in early 1977 that introduced us all to the Euclidean version of the Yang–Mills equations.

For me this was a real eye-opener. First of all physicists like 't Hooft and Jackiw had very concrete constructions of connections in Euclidean space which satisfied the self-duality equations – the Yang–Mills instantons. At the same time there were gravitational analogues – concrete four-dimensional Einstein metrics produced by Hawking and Gibbons, and Eguchi and Hanson (my former neighbour in Princeton). Their insight and intuition had achieved just the sort of thing that pure mathematicians should have been able to do. But twistor theory gave a link to algebraic geometry, so there was a chance that we could contribute in reverse. While Atiyah, Ward, and Penrose were developing a version of twistor theory to deal with the Euclidean case, I was recalling how my earlier interests were appropriate for this particular problem. The inequality in my paper (Hitchin 1974) could now be seen simply as the statement that the bundle Λ^2_+ on an Einstein manifold is self-dual; the conformal invariance of the Yang–Mills equations meant that the four-sphere and vanishing theorems could be used; and by turning around the elliptic complex which was my earlier interest, one could get the deformation complex for self-dual connections

$$\Omega^0(\mathfrak{g}) \to \Omega^1(\mathfrak{g}) \to \Omega^2_-(\mathfrak{g}).$$

The index theorem then showed that there were more solutions than the physicists had yet found, which was one up for the mathematicians. Quite soon, using differential geometry and recent results on the algebraic geometry of vector bundles, Atiyah and I, and independently Manin and Drinfeld, found a way to construct all of them. This experience was not only satisfying, but also formative – precisely stated problems, originating in physics and unthought of by pure mathematicians, yielding to the latest pure mathematical techniques.

Chapter 3 by Claude LeBrun represents well this general area – he takes the Einstein–Maxwell equation in four dimensions and links it with Kähler geometry and topological obstructions. The work draws also on the more recent Seiberg–Witten equations, one of the most important contributions to four-dimensional geometry from the physicists. This of course followed on from Donaldson's own spectacular use of gauge theory in four-dimensional topology. I have to admit that, when writing the 1978 paper with Atiyah and Singer, I thought that self-dual connections on self-dual spaces was a cosy self-contained theory – a sort of quaternionic version of Riemann surfaces. But then Taubes showed how to construct self-dual connections in general and there was also clearly a story to be told for algebraic surfaces. When Simon Donaldson became my research student in 1980, I asked him to look into these questions, and the rest is history.

Following on from the instantons, I began to look at related equations, and in particular magnetic monopoles in Euclidean space. Here the physical intuition of Nick Manton had acquired solid mathematical support in the existence proof of Jaffe and Taubes, but the Bogomolny equations which describe the monopoles were also clear cousins of the instanton equations and I wanted to use twistor methods to describe the solutions more concretely. I began by asking what was the natural background geometry in three dimensions to support these – an analogue of the self-dual conformal structures in four dimensions. These turned out to be what are known as Einstein–Weyl manifolds (after Weyl's early attempt to introduce the electromagnetic field into general relativity). The important feature of these is that the space of geodesics has the structure of a two-dimensional complex manifold, giving a version of twistor theory. I then abandoned the general picture and concentrated on the Euclidean case and came up with the spectral curve description of $SU(2)$ monopoles. The solution to the equations was determined by an algebraic curve which satisfied some transcendental constraints. This was in some respects a geometrical version of Corrigan and Goddard's application of Ward's method. I was about to move on to something else when a preprint of Nahm fell on my desk (1982).

Nahm's construction of $SU(2)$ monopoles of charge k involved a system of ordinary differential equations for three $k \times k$ matrices T_1, T_2, T_3: $T_1' = [T_2, T_3]$, etc. For $k = 2$ he showed that this could be solved with elliptic functions. Since my spectral curve for charge 2 was elliptic there was a clear challenge to relate directly the two approaches, which led to Hitchin (1983). In writing this, I learned more about non-linear systems which can be linearized on the Jacobian of a curve. Nahm's equations – generalizations of the spinning top equations – are examples of this type and the curve in question turns out to be the spectral curve of the monopole as defined by the twistor approach. The Nahm approach, however, has the specific advantage that one can see directly that the solutions to the Bogomolny equations are nonsingular. One should note that singularity issues dogged the early development of the theory – Ward's charge 2 solution had to be checked numerically and Forgács, Horváth, and Palla consigned essentially the same solution to the waste bin because they thought it was singular.

The Nahm method is the subject of Chapter 4 by Benoit Charbonneau and Jacques Hurtubise, though not for monopoles – instead for the more sophisticated *calorons*, which are instantons on $S^1 \times \mathbb{R}^3$. This is a more elaborate closing of a circle of ideas than in Hitchin (1983). I am particularly pleased to see it here since I began to make headway on the problem in 1983 and had every intention of spending a sabbatical in Stony Brook working out the details, but only got as far as giving a talk to Jacques and his colleagues in Montreal when other issues intervened, of which more later. One of the most beautiful applications of Nahm's equations to monopoles was Simon Donaldson's proof (1984) that a circle bundle over the moduli space is naturally diffeomorphic to the space of based rational maps from the projective line to itself. In his contribution to this collection, Simon returns to Nahm's equations but where the matrices T_i are replaced by vector fields. He pointed out to me long ago that Ashtekar's Hamiltonian formulation of general relativity could be viewed this way (I gave a minor extension to this in Hitchin (1998) in the context of hypercomplex geometry). Donaldson's chapter links Nahm's equations to the physical area of free boundary problems and their analogues on compact Riemannian manifolds.

Arriving in Stony Brook in August 1983, I made contact with Martin Roček in the Theoretical Physics Department. Earlier in the year Blaine Lawson had visited Oxford and he was puzzled by a construction of hyperkähler manifolds by Martin and his co-workers. 'They take a group representation, turn a handle, and you get a hyperkähler metric.' My own interest at the time in hyperkähler geometry was dominated by the twistor approach, so I was curious to learn what was going on. We soon discovered that this was a straightforward generalization of the symplectic quotient construction, which was very much in vogue in Oxford at the time, for example, in the work of Atiyah and Bott on the Yang–Mills equations on Riemann surfaces, and in Frances Kirwan's work in algebraic geometry. I worked hard contributing to the joint paper Hitchin *et al.* (1987*a*) in trying to reconcile the physicists' language of supersymmetric fields with the twistor approach, but I think the reader can easily discern the borderline between mathematics and physics. In fact Martin tried at the time to convince me to work on another piece of geometry related to the supersymmetric sigma model, but I wasn't listening. It finally surfaced 25 years later in Marco Gualtieri's 2003 thesis as a *generalized Kähler manifold.*

On my return to Oxford in 1984, Atiyah and I set to work in evaluating the hyperkähler metric on the moduli space of charge 2 monopoles – the hyperkähler quotient construction when applied to a certain infinite-dimensional affine space showed that every monopole moduli space has such a metric. I then began looking at other moduli spaces with this property, where the gauge-theoretic equations are hyperkähler moment maps. One of these could be defined on a Riemann surface and became the study of Higgs bundles (the terminology is due to Carlos Simpson). At first (as in Hitchin 1986) I thought I was just getting a hyperkähler metric on the cotangent bundle of the moduli space of stable bundles, and I began explaining this on the blackboard to Simon Donaldson but got stuck. Having decided that there must be a joint stability condition it was easy to see what should happen because one knew the equations to be satisfied and the type of vanishing theorem that would ensue. From then on, virtually every day a new facet of the subject seemed to appear. The paper Hitchin (1987*b*) by no means wrote itself but it certainly had an inner momentum. There was one aspect however – the integrable system – which came a little later, because it was not really part of the hyperkähler story. I was visiting Jacques Hurtubise in 1985 and we went up to his family cottage in the Laurentian mountains. While he was doing some maintenance work I sat on the porch with a pen and paper and suddenly realized that we had a completely integrable Hamiltonian system where the fibres were recognizable abelian varieties. Until then, my experience with such objects was confined to Nahm's equations but here was a vast natural source of such things.

Chapter 6 by Ramanan gives a good survey of the subject from the point of view of algebraic geometry, including the Hecke correspondence which links up with Chapter 7 by Witten on the very active recent work on the geometric Langlands programme from a physics point of view. This makes essential use of both the gauge-theoretic equations and the integrable system. Chapter 8 by Bill Goldman is again about Higgs bundles, but focuses on the interpretation as flat bundles and in particular on representations of the fundamental group of the surface into a real form of the group. One of the current problems in this area is to give an understanding of the generalized Teichmüller spaces of Hitchin (1982) as structures on the surface modulo diffeomorphism and Goldman discusses some recent results in this direction.

From 1979 to 1989, I was a Fellow of St Catherine's College, and apart from my teaching duties there I also had multiple opportunities over lunch to talk to Graeme Segal and Michael Atiyah, who were also Fellows. I never wrote a joint paper with Graeme, but I valued greatly the questions and remarks which came in our discussions. Our approaches were different – I would always want to talk about stable bundles and a finite-dimensional moduli space whereas for him stability was not an issue and, as he says in Chapter 9, the consideration of the category or stack of bundles is actually closer to quantum field theory which was always his prime motivation. In Autumn 1988 we had a seminar in Oxford run by Atiyah, Segal, and Ruth Lawrence on quantum field theory and the Jones polynomial. During that summer, and in particular at the International Congress of Mathematical Physicists in Swansea, Witten, aided by Atiyah and Segal, had seen how these invariants could be viewed via Chern–Simons theory. One aspect which I liked very much was the geometric quantization of the moduli space of flat unitary connections on a surface. As a research student I had learned a bit about quantization from another student John Rawnsley, but I never understood pairings between different polarizations. The interpretation here as a projectively flat connection on a bundle over Teichmüller space was more appealing, and I set to work with the aim of giving a talk at Michael Atiyah's 60th birthday meeting in 1989. For some reason, I chanced upon the paper of Welters (1983) using a heat equation in the abelian case, and before long I saw that the Poisson-commuting functions of my integrable system were instrumental in defining this connection for non-abelian theta functions. My construction was a mixture of algebraic geometry and differential geometry, but Chapter 10 by Jørgen Andersen shows how much further one can go by a deeper analysis of the non-abelian heat equation.

In January 1990 I left Oxford for Warwick, and began to learn about the Painlevé equations. It was really the work of Paul Tod in Oxford on self-dual four-manifolds with $SO(3)$ symmetry that spurred my interest. The two-monopole metric which Atiyah and I had calculated using complete elliptic integrals was one such example, but one knew that the general solution of the Painlevé equation involved genuinely new transcendental functions, so there was a question about how far one could go in hoping to get closed-form expressions for such metrics. On the other hand twistor theory said that these could all be described by a complex three-manifold, so I tried to approach the problem that way by looking at threefolds with an action of the complex group $SO(3,\mathbb{C})$. This led me to some algebraic solutions which I spoke about for the 60th birthdays of Narasimhan and Seshadri (Hitchin 1996). It also led into the classical geometry of Poncelet's theorem, which I had learned about earlier from Atiyah in the context of hyperbolic monopoles. The twistor spaces in these cases were equivariant compactifications of $SO(3,\mathbb{C})$ modulo a dihedral group and the obvious question was to find solutions of Painlevé's sixth equation for any finite subgroup. Chapter 11 by Philip Boalch describes this subject and its general context well. He has given a classification of icosahedral solutions, the most challenging case. I am only now beginning to understand the twistorial interpretation of any of these.

It was a visit of Jean-Luc Brylinski to Warwick that got me interested in the subject of gerbes, an interest that was later reinforced by conversations with Dan Freed. Chapter 12 by Michael Murray recalls the enthusiasm that I conveyed, though it was more in terms of the potential of the subject rather than any theorems. I felt that at the right time and place I would see the correct role of gerbes and that I should look out for that in my research. Some years later, after I moved to Cambridge, I noticed that there was a twistor transform analogous to the Atiyah–Ward approach to the self-dual Yang–Mills equations and I had my student David Chatterjee write his thesis on various aspects of gerbes, including this one. Murray's chapter carefully leads the reader through his bundle gerbe approach, connecting it with my own viewpoint. One of Chatterjee's contributions was to use the harmonic theory of the current defined by a codimension three submanifold to define a gerbe with connection – the direct analogue of a divisor of a holomorphic line bundle. This distributional approach is central to Chapter 13 by Harvey and Lawson on holomorphic linking numbers, a concept which is also present in twistor theory.

In October 1994 I took up the Rouse Ball Chair in Cambridge, and started talking to Nick Manton again about monopoles. He and Michael Murray had been working on monopoles which were symmetric under the isometric action of a finite group. I joined in, with the aim of using Nahm's equations. It was clear that the quotient of the spectral curve by the finite group was elliptic and so one expected to solve the Nahm equations with elliptic functions. This we could do (and learned some important lessons about nonsingularity in doing so), but we also recruited Paul Sutcliffe, who was a postdoc at the time, to do some numerical and analytical work to plot energy densities of the resulting monopoles. Manton's Chapter 14 discusses his other interest in the area – skyrmions. His pictures of energy densities reveal the remarkable similarities in the qualitative behaviour of symmetric skyrmions and monopoles – as if the actual equations matter less than their ability to encompass the non-linearities of the fields. I should say that this experience taught me to appreciate more the problems of even asking for explicit solutions. Working on charge 3 monopoles using genus two theta functions, it became clear that it was far more efficient to numerically solve the Nahm equations than evaluate the constrained functions analytically.

The geometry seminar in Cambridge was always held in the Department of Applied Mathematics and Theoretical Physics (in particular because Stephen Hawking would usually attend). It also meant that we would choose topics of common interest to discuss and one of these was the Strominger–Yau–Zaslow (SYZ) approach to mirror symmetry. I found this a very attractive idea. It led me to write a couple of papers (Hitchin 1997, 1999) about moduli spaces of special Lagrangian submanifolds and to learn from Dan Freed about special Kähler geometry on moduli spaces. It also led to my first application of the theory of gerbes in lectures I gave at a meeting in Harvard in 1999 (Hitchin 2001*a*). Chapter 15 by Leung and Yau discusses the consequences of the SYZ approach for an elliptically fibred Calabi–Yau manifold, and the mirror symmetry in Witten's chapter is again of this type. Tamás Hausel also writes about this approach in Chapter 16.

Tamás became my student in Cambridge at the time of yet another question about monopoles. (It seemed as if this subject would shadow me for the rest of my life.) These were the remarkable conjectures of the physicist Ashoke Sen about the $\mathcal{L}^2$ harmonic forms on the moduli space. A lot of progress had been made by Graeme Segal and his student Alex Selby on this, and I asked Tamás to look at the corresponding question on the hyperkähler moduli space of Higgs bundles. His results, obtained by a number of different methods, form part of his chapter.

In 1997 I returned to Oxford as the Savilian Professor of Geometry. My interests were still directed towards trying to understand the special Kähler geometry on the moduli spaces coming from physics. The moduli space of complex structures on a Calabi–Yau threefold led me to the study of invariant functionals on differential forms (Hitchin 2001*b*). This seemed to give a natural setting for the central objects studied in string theory and M-theory, but also a practical one when considering the associated flow equations which describe the geometry in terms of an evolving hypersurface geometry. Chapter 17 by Robert Bryant addresses the initial value problems associated with such flows, using the methods of exterior differential systems. My approach to Calabi–Yau complex structures was based on the fact that the action of $GL(6,\mathbb{R})$ on $\Lambda^3\mathbb{R}^6$ has an open dense orbit; G_2 structures appeared from a similar remark about the group $GL(7,\mathbb{R})$. In both instances I learned a lot about the background to these facts from Bryant. I began to notice the occurrence of this type of situation in a number of rather different places, most notably after a lecture in London by Merkulov on the holonomy of affine connections, and I began wondering about a geometrical use of $SO(6,6)$ in the spin representation.

In the Spring of 2001 I spent a sabbatical term in Madrid, visiting Oscar García-Prada at the Universidad Autónoma. I went with the intention of finishing a book on hyperkähler manifolds (which has still not been written) but I got distracted by thinking again about the common features of $SL(6,\mathbb{R})$ and $SO(6,6)$. The first week in May seemed like a succession of public holidays in Madrid and it rained all the time that year but at the end of it I had the concept of a generalized complex structure – a hybrid object that interpolates between symplectic and complex geometry and should, I thought, have something to contribute to mirror symmetry. Moreover the role of two-forms seemed to parallel the physicists' B-field. Even more, the natural setting seemed to require a gerbe in the background. Marco Gualtieri, who wrote his thesis on the subject, addresses this in Chapter 18, discussing the objects within this generalized area which replace holomorphic vector bundles and submanifolds, for which he adopts the physicists' term *brane*. There are still many questions unanswered in this theory, not least a tantalizing connection with noncommutative geometry. In this area, the twisted de Rham complex – replacing the exterior derivative d by $d+H$ where H is a closed three-form – is a natural process. The K-theoretic analogue of this uses the gerbe idea much more closely, and Chapter 19 by Dan Freed, a continuation of a series with Mike Hopkins and Constantin Teleman, focuses again on the equivariant twisted K-theory of a Lie group. I have much to learn about twisted K-theory and the cycles that represent it, but it clearly fits into the framework of the generalized geometry around which much of my current research revolves.

If any conclusion can be drawn from this personal timeline, it is that it is specific problems that have engaged me in research projects, and that theoretical physics is perhaps the richest source of these. I have generally sought to learn about an area of mathematics by seeing how it impinges on a particular problem, which by itself might seem unimportant (the 'Miss Marple' approach). So progress this way requires an ever-changing source of mathematical challenges, of the sort that string theory seems to provide in abundance. This is surely one of the reasons for the currently active interface between geometry and physics.

Acknowledgements

I would like to wholeheartedly thank the contributors to this book, including some who were unable to be at the Madrid conference. It is a wonderfully rich cross-section of articles which appeal to my own interests. I would like to thank Oxford University Press for agreeing to publish the book, and the editors for their hard work in assembling it. Above all I would like to express my immense thanks to Oscar García-Prada for instigating the whole project – the original 2006 conference and this book based on it.

References

Donaldson, S. K. (1984). Nahm's equations and the classification of monopoles. *Commun. Math. Phys.* **96**, 387–407.

Hitchin, N. J. (1974). Compact four-dimensional Einstein manifolds. *J. Differential Geometry* **9**, 435–41.

Hitchin, N. J. with Atiyah, M. F. and Singer, I. M. (1978). Self-duality in four-dimensional Riemannian geometry. *Proc. R. Soc. London Ser A* **362**, 425–61.

Hitchin, N. J. (1983). On the construction of monopoles. *Commun. Math. Phys.* **89**, 145–90.

Hitchin, N. J. (1986). Metrics on moduli spaces, The Lefschetz centennial conference, Part I (Mexico City, 1984). *Contemp. Math.* **58**, American Mathematical Society, Providence, RI, 157–78.

Hitchin, N. J. with Karlhede, A. Lindström, U. and Roček, M. (1987*a*). Hyperkähler metrics and supersymmetry. *Commun. Math. Phys.* **108**, 535–89.

Hitchin, N. J. (1987*b*). The self-duality equations on a Riemann surface. *Proc. London Math. Soc.* **55**(3), 59–126.

Hitchin, N. J. (1992). Lie groups and Teichmüller space. *Topology* **31**, 449–73.

Hitchin, N. J. (1996). Poncelet polygons and the Painlevé equations, in *Geometry and Analysis*. (Bombay, 1992), Oxford University Press, Bombay, India, pp. 151–85.

Hitchin, N. J. (1997). The moduli space of special Lagrangian submanifolds. *Ann. Scuola Norm. Sup. Pisa Cl. Sci.* **25**, 503–15.

Hitchin, N. J. (1998). Hypercomplex manifolds and the space of framings, in *The Geometric Universe: Science, Geometry and the Work of Roger Penrose.* S. A. Huggett *et al.* (eds.), Oxford University Press, Oxford, pp. 9–30.

Hitchin, N. J. (1999). The moduli space of complex Lagrangian submanifolds. *Asian J. Math.* **3**, 77–92.

Hitchin, N. J. (2001*a*). Lectures on special Lagrangian submanifolds, in *Winter School on Mirror Symmetry, Vector Bundles and Lagrangian Submanifolds.* C. Vafa and S. T. Yau (eds.), Studies in Advanced Mathematics **23**, AMS/International Press, Providence, RI, pp. 151–82.

Hitchin, N. J. (2001*b*). Stable forms and special metrics, in *Global Differential Geometry: The Mathematical Legacy of Alfred Gray.* M. Fernández and J.A. Wolf (eds.),

Contemporary Mathematics **288**, American Mathematical Society, Providence, RI, pp. 70–89.

Nahm, W. (1982). The construction of all self-dual monopoles by the ADHM method, in *Monopoles in Quantum Field Theory (Trieste, 1981)*. World Scientific Publishing, Singapore, pp. 87–94.

Welters, G. E. (1983). Polarized abelian varieties and the heat equations. *Compositio Math.* **49**, 173–94.

II

MATHEMATICAL WORK OF NIGEL HITCHIN

Sir Michael Atiyah

Dedicated to Nigel Hitchin

It was a great pleasure for me to take part in the 60th birthday celebration for Nigel Hitchin. I have known him ever since he was a graduate student in Oxford and subsequently as my assistant at the institute in Princeton and then as a colleague and collaborator. I have watched him mature mathematically over the years and gradually establish a unique niche for himself on the frontier between differential geometry and theoretical physics. This is now a very popular and active field but Nigel occupies a singular place in it by virtue of the many important and beautiful ideas he has introduced.

It has been said that the real measure of a mathematician's contribution is how long it would have taken the community to discover these results in his absence. In Nigel's case it is clear that he would score highly by this criterion, since many of his results have been somewhat neglected on their first appearance and their significance has only become apparent some years later.

Nigel has a large output with many beautiful papers which deserve to be read and re-read. Because of the constraints of time I will limit myself to half-a-dozen topics which I particularly like, and which show Nigel's real knack of finding natural and elegant ideas.

I will omit reference to the papers we have written jointly, except tangentially, but I would like to mention our work on instantons which became known as the ADHM construction (Atiyah *et al.* 1978). One morning, after a long struggle, we finally saw the light. We adjourned to lunch in St Catherine's College euphoric at our success, though post-prandial analysis often punctures premature celebration. On this occasion there was no unseen error but instead there was a letter from Manin informing us that he and Drinfeld had just reached the same conclusion! This was the genesis of the four-author paper and it was some years later before Nigel or I met Drinfeld.

My first topic, which dates back to 1982, is Nigel's introduction of the spectral curve of a magnetic monopole (Hitchin 1982). I remember his letter to me at the time saying he had found something which he felt I would like. He was completely right. The result was really simple, beautiful, and important. It was at the basis of much subsequent work on monopoles by many people, as I will recall.

Let me first review the details.

The Bogomolny equations

$$D_A\phi = *F_A$$

are equations in $\mathbb{R}^3$ for a gauge-field (or G-connection) A, where D_A is the covariant derivative, ϕ (the Higgs field) is a section of the adjoint bundle, F_A is the curvature, and $*$ is the Hodge dual. Under appropriate decay conditions at infinity the solutions describe magnetic monopoles. Up to gauge equivalence the solutions are parametrized by a finite-dimensional moduli space. The case most studied is when $G = SU(2)$. Nigel introduced the notion of a *spectral line* of a monopole. This is by definition an oriented line L for which the linear differential equation on L

$$(D_A - i\phi)s = 0$$

(for a section s of the $\mathbb{C}^2$-bundle) has *square-integrable* solutions. Since solutions s have exponential behaviour at $\pm\infty$ this means that we have a solution exponentially decaying at both ends.

Nigel's brilliant observation is that the spectral lines of any monopole are parametrized by a compact Riemann surface in the complex surface which parametrizes all oriented lines in $\mathbb{R}^3$. Moreover, the spectral curve completely determines the monopole and so is a very effective tool in the study of monopoles.

Algebraic curves had previously appeared in the theory of integrable systems, such as the KdV equation, but these were equations in one space and one time dimension. The novelty in Hitchin's case is that the Bogomolny equation is a PDE in three dimensions. Moreover, the spectral curve has a beautifully simple geometric origin.

Among the many applications of the spectral curve I will single out just two:

1. The study of symmetric monopoles by Manton and collaborators (Manton and Sutcliffe 2004). These correspond to symmetric spectral curves which are easy to study.
2. The metric on the (relative) moduli space for magnetic charge 2. This metric was explicitly described in (Atiyah and Hitchin 1988) and turned out to be asymptotic to a Taub-NUT metric with negative mass-parameter. A purely formal observation, this was many years later given an interesting M-theory interpretation in terms of 'anti-branes' by Seiberg and Witten (1996).

My second example is the introduction (Hitchin 1987) by Nigel of 'Higgs bundles' in 1987. For $SU(2)$ these are solutions of the equations, on a compact Riemann surface X,

$$F_A + [\phi, \phi^*] = 0$$
$$d''_A\phi = 0$$

where A is a connection on a rank 2 vector bundle E and ϕ is a section of End $(E) \otimes K$, with K the canonical line-bundle. This leads to a moduli space M with an extremely rich structure. Here are some of its key properties:

1. M is hyperkähler of dimension $4(3g-3)$ where g is the genus of X.
2. The determinant gives a map
$$M \longrightarrow H^0(X, K^2)$$
whose generic fibre is a complex torus (Prym variety).
3. M contains as a dense open set the cotangent bundle of stable $SU(2)$ bundles on X.
4. For all complex structures except two $(\pm I)$ M can be identified with the moduli space of flat $SL(2,\mathbb{C})$ bundles on X.
5. M contains a copy of Teichmüller space given by pairs (A, ϕ) with $E = K^{1/2} \oplus K^{-1/2}$
$$\phi = \begin{pmatrix} 0 & a \\ 1 & 0 \end{pmatrix}, \quad a \in H^0(X, K^2).$$

When I first saw this large and meaty paper I expected that it would have significant applications, but it appeared not to have had much impact at the time. An exception is the use Witten (1991) made of it in his study of $SL(2,\mathbb{C})$ Chern–Simons theory. But, very recently, after nearly 20 years, the moduli space of Higgs bundles plays a central role in the geometric Langlands program (Kapustin and Witten 2007).

In this the theory on a Riemann surface X is obtained by dimensional reduction from a super-symmetric Yang–Mills theory on the four-manifold $\mathbb{R}^2 \times X$. A key fact is that this theory is a 'twisted' version of more physical theories, corresponding to the twisting (tensoring with K) introduced by Nigel. Although such twistings have become familiar over the past 20 years it appears that Nigel was the first to introduce one (in the two-dimensional version).

One potential application of Nigel's work would be the abelianization programme put foward in Atiyah (1990) where the aim would be to identify the space of 'non-abelian theta-functions' as the subspace of a space of ordinary abelian theta-functions invariant under an appropriate discrete group.

My third example is the work of Nigel and his collaborators (Hitchin *et al.* 1987) on the hyperkähler quotient construction. This arose from the work of physicists in super-symmetry, but it fits well into conventional symplectic and complex geometry. What it does is to produce a hyperkähler manifold of dimension $4(n-d)$ from the action of a compact Lie group G of dimension d on a hyperkähler manifold of dimension $4n$.

In particular one can start with a linear action of G on a quaternion vector space H^n. This produces a host of interesting hyperkähler manifolds as quotients. Even better we can start with an infinite-dimensional linear space over H and an action of an infinite-dimensional Lie group. Suitable examples occur naturally in the physics of gauge theories and the formal quotienting process can be rigorously justified.

Cases where this programme has been successfully used include the work of Kronheimer (1989) on ALE spaces, moduli spaces of instantons on $\mathbb{R}^4$ and monopoles or Higgs bundles on $\mathbb{R}^3$.

A more recent application has been (Atiyah and Bielawski 2002) the construction of hyperkähler metrics on manifolds associated with nilpotent conjugacy classes in Lie groups. The sub-regular class leads to the four-dimensional manifolds studied by Kronheimer (extending the work of Brieskorn), while other classes lead to higher dimensional generalizations.

All of this has totally transformed the subject of hyperkähler manifolds. While compact manifolds are rare, non-compact ones are numerous and of great interest. Before this general theory was developed Nigel wrote an elegant little paper (Hitchin 1979) which has always been one of my favourites. It deals with the Gibbons–Hawking gravitational instantons (Gibbons and Hawking 1978). These are the special cases of ALE manifolds associated to the A_n singularities, that is, $\mathbb{C}^2/\Gamma$ where Γ is cyclic. What Nigel did was to show how the Gibbons–Hawking metric could be simply derived by twistor methods. These connected naturally to the algebraic geometry used by Brieskorn and the outcome is the hyperkähler extension of the Brieskorn theory, which includes the metric as well as the complex structure.

Two things should be noted about this paper of Nigel's. First it eventually led to Kronheimer's complete treatment of all the ADE singularities. Second it showed in an explicit way how twistor theory could be harnessed in such problems. In fact twistor theory lies behind much of Nigel's work and he was one of the first to successfully add twistor theory to other techniques in the differential geometric study of moduli spaces. This was illustrated again in our joint work on monopoles (Atiyah and Hitchin 1988) as well, of course, as in the better known work on instantons (Atiyah *et al.* 1978). It is interesting that a new link between twistor theory and gauge theory has been initiated by Witten (2004).

Many mathematicians reaching the age of 60 may be slowing down, at least in terms of producing really novel ideas. This is not the case with Nigel, so let me end with two examples of more recent work, each of which illustrates Nigel's knack of identifying fruitful new geometric ideas. In both cases physicists have been quick to latch on to Nigel's work and exploit it.

The first is his use of stable three forms in connection with special metrics (Hitchin 2001). A three-form ρ on a linear space V is said to be *stable* if it is in an open orbit of $GL(V)$. There are two interesting special cases:

1. $\dim V = 6$ with stabilizer $SL(3) \times SL(3)$
2. $\dim V = 7$ with stabilizer G_2

Both preserve a volume element $\phi(\rho)$.

If now M is a compact-oriented manifold of dimension 6 or 7 with an everywhere stable three-form ρ then the total volume

$$V(\rho) = \int \phi(\rho)$$

is a functional of ρ. If we consider the variational problem for $V(\rho)$ restricted to closed forms ρ in a given cohomology class in $H^3(M,\mathbb{R})$, then critical points of $V(\rho)$ give, in the two cases:

1. dim 6 is a complex three-manifold with trivial canonical bundle (i.e. a Calabi–Yau manifold).
2. dim 7 is a Riemannian manifold with holonomy G_2.

These are of interest to physicists for string theory in dimension 6 and for M-theory in dimension 7.

One of the merits of Hitchin's approach in dimension 6 is that explains why the local moduli space is given by $H^3(M,\mathbb{R})$. The Hitchin functional in dimension 7 turns out to be related to 11-dimensional supergravity, which is the large volume limit of M-theory.

My final example (Hitchin 2003), is Nigel's introduction of the notion of a generalized complex structure. This helps to unify symplectic and complex geometry and should be helpful in understanding mirror symmetry.

The definition is quite simple. We consider a manifold X with tangent bundle TX and a linear automorphism τ of

$$TX \oplus T^*X$$

with $\tau^2 = -1$, and satisfying an integrability condition. Two special cases are

1. A complex structure J leads to a generalized complex structure with
$$\tau = \begin{pmatrix} J & 0 \\ 0 & -J^* \end{pmatrix}$$
2. A symplectic structure $\omega : TX \longrightarrow T^*X$ gives a generalized complex structure with
$$\tau = \begin{pmatrix} 0 & -\omega^{-1} \\ \omega & 0 \end{pmatrix}$$

Already this notion of generalized complex structure is being used by physicists as in (Kapustin and Witten 2007), where it fits in very naturally.

If we look back over these examples of Nigel's work we see that he always seems able to identify some natural geometric concepts, which have a good motivation and elegant definitions. It is not surprising that these lead to fruitful applications in geometry and physics.

Finally, I should emphasize how clear and readable all Nigel's papers are. Based on simple novel ideas the papers almost seem to write themselves, but this is the hallmark of real mathematics.

References

Atiyah, M. F. (1990). *The Geometry and Physics of Knots.* Cambridge University Press.

Atiyah M. F. and Bielawski, R. (2002). Nahm's equations, configuration spaces and flag manifolds. *Bull. Braz. Math. Soc., New Ser.* **33**(2), 157–76.

Atiyah, M. F., Drinfeld, V. G., Hitchin, N.J. and Manin, Yu. I. (1978). Construction of instantons. *Phys. Lett. A* **65**, 185–7.

Atiyah, M. F. and Hitchin, N. J. (1988). *The Geometry and Dynamics of Magnetic Monopoles.* Princeton University Press.

Gibbons, G. W. and Hawking, S. W. (1978). Grativational multi-instantons. *Phys. Lett. B* **78**, 430.

Hitchin, N. J. (1979). Polygons and gravitons. *Math. Proc. Camb. Phil. Soc.* **85**, 465–76.

Hitchin, N. J. (1982). Monopoles and geodesics. *Commun. Math. Phys.* **83**, 579–602.

Hitchin, N. J. (1987). The self-duality equations on a Riemann surface. *Proc. London Math. Soc.* **55**(3), 59–126.

Hitchin, N. J. (2001). Stable forms and special metrics. *Contemp. Math.* **288**, 70–89.

Hitchin, N. J. (2003). Generalized Calabi-Yau manifolds. *Q. J. Math.* **54**, 281–308.

Hitchin, N. J., Karlhede, A., Lindström, U. and Roček, M. (1987). Hyperkähler metrics and supersymmetry. *Commun. Math. Phys.* **108**, 535–89.

Kapustin, A. and Witten, E. (2007). Electro-magnetic duality and the geometric Langlands program, hep-th/0604151.

Kronheimer, P. B. (1989). The construction of ALE spaces as hyper-Kahler quotients. *J. Diff. Geom.* **29**, 665–83.

Manton, N. J. and Sutcliffe, P. M. (2004). *Topological Solitons.* Cambridge University Press, Cambridge.

Seiberg, N. and Witten, E. (1996). Gauge Dynamics and Compactification to Three-dimensions in The Mathematical Beauty of Physics: A Memorial Volume for Claude Itzykson Saclay, France, June 5–7 (Advanced Series in Mathematical Physics, Vol. 24) by Claude Itzykson, Jean Bernard Zuber, and Jean-Michel Drouffe, pp. 333–66.

Witten, E. (1991). Quantization of Chern-Simons gauge theory with complex gauge group. *Commun. Math. Phys.* **137**, 29–66.

Witten, E. (2004). Perturbative gauge theory as a string theory in twistor space. *Commun. Math. Phys.* **252**, 189–258.

III

THE EINSTEIN–MAXWELL EQUATIONS, EXTREMAL KÄHLER METRICS, AND SEIBERG–WITTEN THEORY

Claude LeBrun

Dedicated to Nigel Hitchin on the occasion of his 60th birthday

Nigel Hitchin has played a key role in the exploration of four-dimensional Riemannian geometry, and in particular has made foundational contributions to the theory of self-dual manifolds, four-dimensional Einstein manifolds, spin^c structures, and Kähler geometry (Atiyah, Hitchin, and Singer 1978; Hitchin 1974*a*; 1974*b*; 1975; 1979; 1981; Hitchin, Karlhede, Lindström, and Roček 1987). In the process, he has often alerted the rest of us to the profound mathematical interest of beautiful geometric problems that had previously only been considered by physicists. I would therefore like to use the occasion of Nigel's 60th birthday as an opportunity to draw the attention of an audience of geometers and physicists to an interesting relationship between the four-dimensional Einstein–Maxwell equations and Kähler geometry, and point out some fascinating open problems that directly impinge on this relationship.

Let us begin by recalling that a two-form F on an oriented pseudo-Riemannian n-manifold (M, g) is said to satisfy *Maxwell's equations* (in vacuo) iff

$$dF = 0$$
$$d \star F = 0$$

where $\star$ is the Hodge star operator. If M is compact and g is Riemannian, these equations of course just mean that F is a harmonic two-form, and Hodge theory thus asserts that there is in fact exactly one solution in each de Rham cohomology class $[F] \in H^2(M, \mathbb{R})$. This solution may be found by minimizing the action

$$F \longmapsto \int_M |F|_g^2 d\mu_g$$

among all closed forms $F \in [F]$. In this context, dimension 4 enjoys a somewhat privileged status, because it is precisely when $n = 4$ that both the action and the solutions themselves become conformally invariant, in the sense that they are unaltered by replacing g with any conformally related metric $\tilde{g} = u^2 g$.

When $n > 2$, coupling these equations to the gravitational field (Hawking and Ellis 1973; Misner, Thorne, and Wheeler 1973) gives rise to the so-called

Einstein–Maxwell equations (with cosmological constant)

$$dF = 0$$

$$d \star F = 0$$

$$\left[r + F \circ F \right]_0 = 0$$

where r is the Ricci tensor of g, $(F \circ F)_{jk} = F_j{}^\ell F_{\ell k}$ is obtained by composing F with itself as an endomorphism of TM, and $[\ ]_0$ indicates the trace-free part with respect to g. In the compact Riemannian setting, these equations may be understood as the Euler–Lagrange equations of the functional

$$(g, F) \longmapsto \int_M (s_g + |F|_g^2) d\mu_g$$

where F is again allowed to vary over all closed two-forms in a given de Rham class, and g is allowed to vary over all Riemannian metrics of some fixed total volume V. The privileged status of dimension 4 becomes more pronounced in this context, for it is only when $n = 4$ that the Einstein–Maxwell equations imply that the scalar curvature s of g is *constant.* Indeed, this just reflects Yamabe's observation (Yamabe 1960) that a Riemannian metric has constant scalar curvature iff it is a critical point of the restriction of the Einstein–Hilbert action $\int s d\mu$ to the space of volume V metrics in a fixed conformal class. When $n = 4$, the conformal invariance of $\int |F|^2 d\mu$ thus implies that critical points of the above functional must have constant scalar curvature; but when $n \neq 4$, by contrast, the scalar curvature turns out to be constant only when F has constant norm.

We have just observed that the Einstein–Maxwell equations on a four-manifold imply that the scalar curvature is constant. But what happens in the converse direction is far more remarkable: namely, *any constant-scalar-curvature Kähler metric on a four-manifold may be interpreted as a solution of the Einstein–Maxwell equations.* Indeed, suppose that (M^4, g, J) is a Kähler surface, with Kähler form $\omega = g(J\cdot, \cdot)$ and Ricci form $\rho = r(J\cdot, \cdot)$. Let

$$\mathring{\rho} = \rho - \frac{s}{4}\omega$$

denote the primitive part of the Ricci form, corresponding to the trace-free Ricci tensor

$$\mathring{r} := [r]_0 = r - \frac{s}{4} g.$$

Suppose that the scalar curvature s of (M, g) is constant, and set

$$F = \omega + \frac{\mathring{\rho}}{2}.$$

Then (g, F) automatically solves the Einstein–Maxwell equations. This generalizes an observation due to Flaherty (1978) concerning the scalar-flat case.

On a purely calculational level, this fact is certainly easy enough to check. Indeed, on any oriented Riemannian four-manifold, the two-forms canonically decompose

$$\Lambda^2 = \Lambda^+ \oplus \Lambda^-$$

into self-dual and anti-self-dual parts, and it is then easily shown that for any two-form

$$F = F^+ + F^-$$

one has

$$\left[F \circ F\right]_0 = 2F^+ \circ F^- .$$

Since in our special case we have $F^+ = \omega$ and $F^- = \mathring{\rho}/2$, it therefore follows that

$$\left[F \circ F\right]_0 = -\mathring{r},$$

so that

$$\left[r + F \circ F\right]_0 = 0$$

as required. Moreover, since ρ is automatically closed, and its self-dual part $s\omega/4$ is closed if s is assumed to be constant, we conclude that F is indeed harmonic on a constant-scalar-curvature Kähler surface, exactly as claimed.

But there is actually a great deal more going on here. Recall that Calabi (1982) defined an *extremal Kähler metric* on a compact complex manifold (M^{2m}, J) to be a Kähler metric which is a critical point of the Riemannian functional

$$g \longmapsto \int_M s^2 d\mu$$

restricted to a fixed Kähler class $[\omega] \in H^2(M)$. The associated Euler–Lagrange equations then amount to requiring that the gradient ∇s of the scalar curvature be the real part of a holomorphic vector field. In particular, any constant-scalar-curvature Kähler metric is extremal in this sense. In fact, as was recently proved by Chen (2006), extremal Kähler metrics actually always *minimize* $\int s^2 d\mu$ in their Kähler classes. In the constant-scalar-curvature case, this was long ago shown by Calabi, using a simple but elegant argument. Indeed, if (M^{2m}, g, J) is a compact Kähler manifold of complex dimension m, then

$$\int_M s_g d\mu_g = \int_M \rho \wedge \star\omega = \frac{4\pi}{(m-1)!} c_1 \cdot [\omega]^{m-1}$$

and

$$\int_M 1 d\mu = \int_M \frac{\omega^{\wedge m}}{m!} = \frac{1}{m!}[\omega]^m$$

so that the Cauchy–Schwarz inequality tells us that

$$\int_M s^2 d\mu \geq \frac{(\int s d\mu)^2}{\int 1 d\mu} = \frac{16\pi^2 m}{(m-1)!} \frac{(c_1 \cdot [\omega]^{m-1})^2}{[\omega]^m} \tag{3.1}$$

with equality iff s is constant.

Calabi (1982) also considered the Riemannian functionals

$$g \longmapsto \int_M |r|_g^2 \, d\mu_g$$

and

$$g \longmapsto \int_M |\mathcal{R}|_g^2 \, d\mu_g$$

obtained by squaring the L^2 norms of the Ricci curvature r and the full Riemann curvature $\mathcal{R}$. Here, his observation was that the restriction of either of these functionals to the space of Kähler metrics can be rewritten in the form

$$g \longmapsto a + b \int_M s^2 d\mu$$

where a and $b > 0$ are constants depending only on the Kähler class. For example,

$$\int_M |r|_g^2 d\mu_g = \frac{1}{2} \int_M s_g^2 \, d\mu_g - \frac{8\pi^2}{(m-2)!} c_1^2 \cdot [\omega]^{m-2},$$

so that

$$\int_M |r|_g^2 d\mu_g \geq \frac{8\pi^2}{(m-2)!} \left[\frac{m}{m-1} \frac{(c_1 \cdot [\omega]^{m-1})^2}{[\omega]^m} - c_1^2 \cdot [\omega]^{m-2} \right], \tag{3.2}$$

with equality iff s is constant. Thus extremal Kähler metrics turn out to minimize these functionals too.

I would now like to point out an interesting Riemannian analog of Calabi's variational problem that leads to the Einstein–Maxwell equations on a smooth compact four-manifold. To this end, first notice that the Kähler form of a Kähler surface is self-dual and harmonic. Let us therefore introduce the following notion:

Definition 3.1 *Let M be smooth compact oriented four-manifold, and let $[\omega] \in H^2(M, \mathbb{R})$ be a deRham class with $[\omega]^2 > 0$. We will then say that a Riemannian metric g is* adapted *to $[\omega]$ if the harmonic form ω representing $[\omega]$ with respect to g is* self-dual.

This allows us to introduce the Riemannian analog of a Kähler class:

Definition 3.2 *Let M be smooth compact-oriented four-manifold, and let $[\omega] \in H^2(M, \mathbb{R})$ be a deRham class with $[\omega]^2 > 0$. We then set*

$$\mathcal{G}_{[\omega]} = \Big\{ \textit{smooth metrics } g \textit{ on } M \textit{ which are adapted to } [\omega] \Big\}.$$

In particular, if ω is the Kähler form of a metric g on M which is Kähler with respect to some complex structure J, then $\mathcal{G}_{[\omega]}$ contains the entire Kähler class of ω on (M, J). Of course, however, $\mathcal{G}_{[\omega]}$ is vastly larger than a Kähler class. In particular, if g belongs to $\mathcal{G}_{[\omega]}$, so does every conformally related metric $\tilde{g} = u^2 g$. It is also worth noticing that

$$\mathcal{G}_{\lambda[\omega]} = \mathcal{G}_{[\omega]}$$

for any non-zero real constant λ.

It is now germane to ask precisely how large $\mathcal{G}_{[\omega]}$ really is relative to

$$\mathcal{G} = \Big\{ \text{smooth metrics } g \text{ on } M \Big\},$$

and to ponder the dependence of $\mathcal{G}_{[\omega]} \subset \mathcal{G}$ on $[\omega]$ as we allow this cohomology class to vary through the open cone

$$\mathcal{C} = \Big\{ [\omega] \in H^2(M, \mathbb{R}) \;\Big|\; [\omega]^2 > 0 \Big\}.$$

Proposition 3.1 (Donaldson/Gay–Kirby) *Let M be any smooth compact four-manifold with $b_+(M) \neq 0$ For any $[\omega] \in \mathcal{C}$, $\mathcal{G}_{[\omega]} \subset \mathcal{G}$ is a Fréchet submanifold of finite codimension $b_-(M)$. Moreover, $\mathcal{G}_{[\omega]} \neq \emptyset$ for all $[\omega]$ belonging to an open dense subset of $\mathcal{C}$.*

Indeed, if $g \in \mathcal{G}_{[\omega]}$ and if $\omega \in [\omega]$ is the harmonic representative, then Donaldson (1986, p. 336) has shown that $T_g\mathcal{G}_{[\omega]}$ is precisely the L^2 orthogonal of the $b_-(M)$-dimensional subspace

$$\{\omega \circ \varphi \mid \varphi \in \mathcal{H}_g^-\} \subset \Gamma(\odot^2 T^* M),$$

where $\mathcal{H}_g^-$ is the space of anti-self-dual harmonic two-forms with respect to g; moreover, his proof also shows that the subset of $[\omega] \in \mathcal{C}$ for which $\mathcal{G}_{[\omega]} \neq \emptyset$ is necessarily open. On the other hand, Gay and Kirby (2004) found an essentially explicit way of constructing a metric g adapted to any $[\omega] \in \mathcal{C} \cap H^2(M, \mathbb{Z})$, so that $\mathcal{G}_{[\omega]} \neq \emptyset$ for any $[\omega]$ in the dense subset $[\mathcal{C} \cap H^2(M, \mathbb{Q})] \subset \mathcal{C}$.

We now consider the natural generalization of Calabi's variational problem to this broader context.

Proposition 3.2 *An $[\omega]$-adapted metric g is a critical point of the Riemannian functional*

$$g \longmapsto \int_M s_g^2 d\mu_g$$

restricted to $\mathcal{G}_{[\omega]}$ iff either

- *g is a solution of the Einstein–Maxwell equations, in conjunction with a unique harmonic form F with $F^+ = \omega$; or else*
- *g is scalar-flat ($s \equiv 0$).*

Proof. Consider a one-parameter family of metrics

$$g_t := g + th + O(t^2)$$

in $\mathcal{G}_{[\omega]}$. By Donaldson's result, we know that h can be taken to be any smooth symmetric tensor field which satisfies

$$\int_M \langle h, \omega \circ \varphi \rangle d\mu = 0$$

for all harmonic forms $\varphi \in \Gamma(\Lambda^-)$, where ω is the g-harmonic representative of $[\omega]$. On the other hand, a standard calculation Besse (1987) shows that

$$\left.\frac{d}{dt}s\right|_{t=0} = \Delta h_a^a + \nabla^a\nabla^b h_{ab} - h^{ab}r_{ab},$$

and

$$\left.\frac{d}{dt}[d\mu]\right|_{t=0} = \frac{1}{2}h_a^a d\mu,$$

so that

$$\begin{aligned}\left.\frac{d}{dt}\left[\int_M s^2 d\mu\right]\right|_{t=0} &= \int_M 2s\dot{s}d\mu + \int_M s^2 \dot{d\mu} \\ &= \int 2s\left(\Delta h_a^a + \nabla^a\nabla^b h_{ab} - h^{ab}\mathring{r}_{ab}\right)d\mu\end{aligned}$$

where $\mathring{r}$ again denotes the trace-free part of the Ricci tensor.

Let us now ask when a metric is critical within its conformal class. This corresponds to setting $h = vg$ for some smooth function v. We then have

$$\frac{d}{dt}\int_M s^2 d\mu = \int 2s(3\Delta v)d\mu = 6\int_M \langle ds, dv\rangle d\mu,$$

so the derivative is zero for all such variations iff s is constant.

We may thus assume henceforth that s is constant. We then have

$$\begin{aligned}\frac{d}{dt}\int_M s^2 d\mu &= 2s\int(\Delta h_a^a + \nabla^a\nabla^b h_{ab} - h^{ab}\mathring{r}_{ab})d\mu \\ &= -2s\int_M \langle h, \mathring{r}\rangle d\mu.\end{aligned}$$

If $s \equiv 0$, this obviously vanishes for every h, and g is a critical point. Otherwise, g will be critical iff $\mathring{r}$ belongs to the L^2 orthogonal complement of $T_g\mathcal{G}_{[\omega]}$. But we already have seen that this orthogonal complement precisely consists of tensors of the form $\omega \circ \varphi$, $\varphi \in \mathcal{H}_g^-$. Thus, when $s \not\equiv 0$, g is a critical point iff s is constant and $\mathring{r} = \omega \circ \varphi$ for some $\varphi \in \mathcal{H}_g^-$. But, setting

$$F = \omega + \frac{\varphi}{2},$$

this is in turn equivalent to saying that (g, F) solves the Einstein–Maxwell equations, as claimed. □

So, why are constant-scalar-curvature Kähler metrics critical points of $\int s^2 d\mu$ restricted to $\mathcal{G}_{[\omega]}$? Well, we will now see that they typically turn out not only to be critical points, but actually to be *minima*. Indeed, the following result (LeBrun 1995, 2001, 2003) may be thought of as a Riemannian generalization of Calabi's inequalities (3.1 and 3.2):

Theorem 3.1 *Let (M^4, J) be a compact complex surface, and suppose that $[\omega]$ is a Kähler class with $c_1 \cdot [\omega] \leq 0$. Then any* Riemannian *metric $g \in \mathcal{G}_{[\omega]}$ satisfies the inequalities*

$$\int s^2 d\mu \geq 32\pi^2 \frac{(c_1 \cdot [\omega])^2}{[\omega]^2} \tag{3.3}$$

$$\int |r|^2 d\mu \geq 8\pi^2 \Big[2\frac{(c_1 \cdot [\omega])^2}{[\omega]^2} - c_1^2\Big] \tag{3.4}$$

with equality iff g is constant-scalar-curvature Kähler.

In the equality case, the complex structure $\tilde{J}$ with respect to which g is Kähler will typically be different from J, but must have the same first Chern class c_1, while its Kähler class must be a positive multiple of $[\omega]$.

We also remark that if (M, J) is not rational or ruled, the hypothesis that $c_1 \cdot [\omega] \leq 0$ holds automatically, and that in this setting a Kähler metric is extremal iff it has constant scalar curvature. In this context, the relevant constant is of course necessarily non-positive.

By contrast, if (M, J) is rational or ruled, there will always be Kähler classes for which $c_1 \cdot [\omega] > 0$. When this happens, the above generalization (3.3) of (3.1) turns out definitely *not* to hold for arbitrary Riemannian metrics. Instead, the correct generalization (LeBrun 1997) is that

$$Y_{[g]} \leq \frac{4\pi \; c_1 \cdot [\omega]}{\sqrt{[\omega]^2/2}}, \tag{3.5}$$

where the Yamabe constant $Y_{[g]}$ is obtained by minimizing the Einstein–Hilbert action $\int s d\mu$ over all unit-volume metrics $\tilde{g} = u^2 g$ conformal to g. Moreover, the inequality is strict unless the Yamabe minimizer is a constant-scalar-curvature Kähler metric, so that (3.3) is in fact violated by an appropriate conformal rescaling of any generic Riemannian metric of positive scalar curvature.

It is also worth remarking that no sharp lower bound in the spirit of Theorem 3.1 is currently known for the square-norm $\int |\mathcal{R}|^2 d\mu$ of the Riemann curvature tensor. Deriving one would be extremely interesting and potentially very useful, but, for reasons I will now explain, the technical obstacles to doing so seem formidable.

Recall that, by raising an index, the Riemann curvature tensor may be reinterpreted as a linear map $\Lambda^2 \to \Lambda^2$, called the *curvature operator*. The decomposition $\Lambda^2 = \Lambda^+ \oplus \Lambda^-$ thus allows one to view this linear map as consisting of four blocks:

$$\mathcal{R} = \left(\begin{array}{c|c} W_+ + \frac{s}{12} & \mathring{r} \\ \hline \mathring{r} & W_- + \frac{s}{12} \end{array} \right). \tag{3.6}$$

Here $W_\pm$ are the trace-free pieces of the appropriate blocks, and are called the self-dual and anti-self-dual Weyl curvatures, respectively. The scalar curvature s is understood to act by scalar multiplication, whereas the trace-free Ricci curvature $\mathring{r} = r - \frac{s}{4}g$ acts on two-forms by $\varphi_{ab} \mapsto 2\varphi_{[a}{}^c\mathring{r}_{b]c}$.

When (M, g) happens to be Kähler, $\Lambda^{2,0} \subset \ker \mathcal{R}$, and the entire upper-left-hand block is therefore entirely determined by the scalar curvature s. For Kähler metrics, one thus obtains the identity

$$|W_+|^2 \equiv \frac{s^2}{24},$$

and Gauss–Bonnet-type formulae like

$$(2\chi + 3\tau)(M) = \frac{1}{4\pi^2} \int_M \left(\frac{s^2}{24} + 2|W_+|^2 - \frac{|\mathring{r}|^2}{2} \right) d\mu$$

reduce many questions about square norms of curvature to questions about the scalar curvature alone. But for general Riemannian metrics, the norms of s and W_+ are utterly independent quantities, so if one wants to use the identity

$$\int |r|^2 d\mu = -8\pi^2(2\chi + 3\tau)(M) + 8 \int \left(\frac{s^2}{24} + \frac{1}{2}|W_+|^2 \right) d\mu \tag{3.7}$$

to prove a generalization of (3.2) for Riemannian metrics, information must be obtained concerning not only the scalar curvature, but also concerning the self-dual Weyl curvature as well.

The curvature estimates of Theorem 3.1 are derived by means of Seiberg–Witten theory (Witten 1994), making it clear that this really is an essentially four-dimensional story. The complex structure J determines a spin^c structure on M with twisted spin bundles $\mathbb{S}_\pm \otimes L^{1/2}$, where L^{-1} is the canonical line bundle $\Lambda^{2,0}$ of (M, J). For simplicity, suppose that $c_1 \cdot [\omega] < 0$. For each metric $g \in \mathcal{G}_{[\omega]}$, one then considers the Seiberg–Witten equations

$$D_A \Phi = 0$$

$$F_A^+ = -\frac{1}{2}\Phi \odot \bar{\Phi}$$

where the unknowns are a unitary connection A on the line-bundle $L \to M$ and a twisted spinor $\Phi \in \Gamma(\mathbb{S}_+ \otimes L^{1/2})$; here $D_A : \Gamma(\mathbb{S}_+ \otimes L^{1/2}) \to \Gamma(\mathbb{S}_- \otimes L^{1/2})$ denotes the twisted Dirac operator associated with A, and F_A^+ is the self-dual part of the curvature of A. One then shows that there must be at least one solution for each $g \in \mathcal{G}_{[\omega]}$ by establishing a count of solutions modulo gauge equivalence which is independent of the metric and which is obviously non-zero for a Kähler metric.

However, the Seiberg–Witten equations can be shown to imply various curvature estimates via Weitzenböck formulae. In particular, the existence of at least one solution for each metric $\tilde{g} = u^2 g$ conformal to g is enough to guarantee that the curvature of g satisfies

$$\int_M s^2 d\mu_g \geq 32\pi^2 [c_1(L)^+]^2$$

$$\int_M (s - \sqrt{6}|W_+|)^2 d\mu_g \geq 72\pi^2 [c_1(L)^+]^2$$

where $[c_1(L)]^+$ is the orthogonal projection of $c_1(L) \in H^2(M, \mathbb{R}) = \mathcal{H}_g^+ \oplus \mathcal{H}_g^-$ into the space $\mathcal{H}_g^+$ of harmonic self-dual two-forms, defined with respect to g. Since ω is assumed to be self-dual with respect to g, we therefore have

$$[c_1(L)^+]^2 \geq \frac{(c_1 \cdot [\omega])^2}{[\omega]^2}$$

by the Cauchy–Schwarz inequality. Inequality (3.3) follows. Since yet another Cauchy–Schwarz argument shows that

$$\int \left(\frac{s^2}{24} + \frac{1}{2}|W_+|^2 \right) d\mu_g \geq \frac{1}{36} \int (s - \sqrt{6}|W_+|)^2 d\mu_g,$$

the second inequality and (3.7) together imply (3.4). The fact that only Kähler metrics can saturate (3.3) or (3.4) is then deduced by examining the relevant Weitzenböck formulae.

One might be tempted to expect the story to be similar for the norm of the full Riemann tensor. After all, the identity

$$\int_M |\mathcal{R}|^2 d\mu = -8\pi^2(\chi + 3\tau)(M) + 2\int_M \left(\frac{s^2}{24} + 2|W_+|^2 \right) d\mu_g$$

certainly provides a good analog of (3.7). In the Kähler case, one has

$$\frac{s^2}{24} = |W_+|^2,$$

so this simplifies to become

$$\int_M |\mathcal{R}|^2 d\mu = 8\pi^2(c_2 - c_1^2) + \frac{1}{4}\int_M s^2 \, d\mu_g,$$

and applying (3.1) we therefore obtain Calabi's inequality

$$\int_M |\mathcal{R}|^2 d\mu \geq 8\pi^2 \left[\frac{(c_1\cdot[\omega])^2}{[\omega]^2} + c_2 - c_1^2\right] \tag{3.8}$$

for any Kähler metric. In light of Theorem 3.1, it might therefore seem reasonable to hope that one could simply extend this inequality to general Riemannian metrics by means of Seiberg–Witten theory. However, we will now show that this cannot work. The key idea is to examine certain extremal Kähler metrics from the vantage point of their *reversed orientations.*

Theorem 3.2 *Calabi's inequality (3.8) cannot possibly be extended to general Riemannian metrics by means of Seiberg–Witten theory. Indeed, there actually exist smooth compact-oriented Riemannian four-manifolds (M, g) which admit a spin^c structure of almost-complex type with non-zero Seiberg–Witten invariant, but such that*

$$\int_M |\mathcal{R}|^2 d\mu < 8\pi^2 \left[\frac{(c_1\cdot[\omega])^2}{[\omega]^2} - (\chi + 3\tau)(M)\right]$$

for some self-dual harmonic two-form ω on (M, g).

Proof. A *Kodaira-fibered complex surface* is by definition a compact complex surface X equipped with a holomorphic submersion $\varpi : X \to \mathcal{B}$ onto a compact complex curve, such that the base $\mathcal{B}$ and fiber $\mathcal{F}_z = \varpi^{-1}(z)$ both have genus ≥ 2. The product $\mathcal{B}\times\mathcal{F}$ of two complex curves of genus ≥ 2 is certainly Kodaira fibered, but such a product signature $\tau = 0$. However, one can also construct examples (Atiyah 1969; Kodaira 1967) with $\tau > 0$ by taking *branched covers* of products.

Let X be any such Kodaira-fibered surface with $\tau(X) > 0$, and let $\varpi : X \to \mathcal{B}$ be its Kodaira fibration. Let p denote the the genus of the base $\mathcal{B}$, and let q denote the genus of some fiber $\mathcal{F}$ of ϖ. A beautiful result of Fine 2004 then asserts that X actually admits a family of extremal Kähler metrics; namely, *for any sufficiently small $\epsilon > 0$,*

$$[\omega_\epsilon] = 2(p-1)\mathcal{F} - \epsilon c_1$$

is a Kähler class on X which is represented by a Kähler metric g_ϵ of constant scalar curvature.

These metrics, being Kähler, have total scalar curvature

$$\int s_{g_\epsilon} d\mu_{g_\epsilon} = 4\pi c_1\cdot[\omega_\epsilon] = -4\pi(\chi + \epsilon c_1^2)(X)$$

and total volume

$$\int d\mu_{g_\epsilon} = \frac{[\omega_\epsilon]^2}{2} = \frac{\epsilon}{2}(2\chi + \epsilon c_1^2)(X).$$

Since s_{g_ϵ} is constant, it follows that

$$\int s_{g_\epsilon}^2 \, d\mu_{g_\epsilon} = \frac{32\pi^2}{\epsilon} \frac{(\chi + \epsilon c_1^2)^2}{2\chi + \epsilon c_1^2}.$$

These metrics therefore satisfy

$$\int_X |\mathcal{R}|_{g_\epsilon}^2 \, d\mu_{g_\epsilon} = 8\pi^2 (c_2 - c_1^2) + \frac{1}{4} \int_X s^2 \, d\mu_g$$

$$= 8\pi^2 \left[-(\chi + 3\tau)(X) + \frac{(\chi + \epsilon c_1^2)^2}{\epsilon(2\chi + \epsilon c_1^2)} \right]$$

On the other hand, there are symplectic forms on X which are compatible with the *non-standard* orientation of X; for example, the cohomology class $\mathcal{F} + \varepsilon c_1$ is represented by such forms if ε is sufficiently small. A celebrated theorem of Taubes (1994) therefore tells us that the reverse-oriented version $M = \overline{X}$ of X has a non-trivial Seiberg–Witten invariant (Kotschick 1998; Leung 1996). The relevant spin^c structure on $\overline{X}$ is of almost-complex type, and its first Chern class, which we will denote by $\bar{c}_1$, is given by

$$\bar{c}_1 = c_1 + 4(p-1)\mathcal{F}.$$

Since this a $(1,1)$ class, one has

$$(\bar{c}_1)^+ = \frac{\bar{c}_1 \cdot [\omega_\epsilon]}{[\omega_\epsilon]^2} \omega_\epsilon = -\frac{(\chi + 3\epsilon\tau)}{[\omega_\epsilon]^2} \omega_\epsilon,$$

relative to the Kähler metric g_ϵ, so that

$$|(\bar{c}_1)^+|^2 = \frac{(\chi + 3\epsilon\tau)^2}{[\omega_\epsilon]^2} = \frac{(\chi + 3\epsilon\tau)^2}{\epsilon(2\chi + \epsilon c_1^2)}.$$

Now since $\bar{c}_1$ arises from an almost-complex structure on $\overline{X}$, we have

$$|(\bar{c}_1)^-|^2 - |(\bar{c}_1)^+|^2 = 2\chi - 3\tau,$$

so that

$$|(\bar{c}_1)^-|^2 = 2\chi - 3\tau + \frac{(\chi + 3\epsilon\tau)^2}{\epsilon(2\chi + \epsilon c_1^2)},$$

and

$$|(\bar{c}_1)^-|^2 - (\chi - 3\tau)(X) = \chi(X) + \frac{(\chi + 3\epsilon\tau)^2}{\epsilon(2\chi + \epsilon c_1^2)},$$

where, for example, τ indicates $\tau(X)$. But it therefore follows that

$$\begin{aligned}\frac{1}{8\pi^2}\int_X |\mathcal{R}|^2_{g_\epsilon} d\mu_{g_\epsilon} - \left[|(\bar{c}_1)^-|^2 - (\chi - 3\tau)\right] &= -(2\chi + 3\tau) \\ &\quad + \frac{(\chi + \epsilon c_1^2)^2 - (\chi + 3\epsilon\tau)^2}{\epsilon(2\chi + \epsilon c_1^2)} \\ &= -(2\chi + 3\tau) + \frac{4\chi\epsilon(\chi + 3\epsilon\tau) + 4\chi^2\epsilon^2}{\epsilon(2\chi + \epsilon c_1^2)} \\ &= -(2\chi + 3\tau) + 2\chi\frac{2\chi + \epsilon c_1^2 + 3\epsilon\tau}{2\chi + \epsilon c_1^2} \\ &= -3\tau(X)\left[1 - \frac{2\chi\epsilon}{2\chi + \epsilon c_1^2}\right],\end{aligned}$$

which is negative for any sufficiently small ϵ. The result therefore follows once we take "ω" to be the anti-self-dual harmonic form $(\bar{c}_1)^-$, which becomes self-dual on $M = \overline{X}$. □

Similarly, careful examination of these examples also shows that, for any constant $t > 1$, the Seiberg–Witten equations cannot imply an estimate of

$$\int (s - t\sqrt{6}|W_+|)^2 d\mu$$

which is saturated by constant-scalar-curvature Kähler metrics. Of course, the Seiberg–Witten equations still imply lower bounds for such quantities, but they are simply never as sharp as those obtained for $t \in [0, 1]$.

In this chapter, we have seen that constant-scalar-curvature Kähler metrics occupy a privileged position in four-dimensional Riemannian geometry. I would therefore like to conclude this discussion by indicating a bit of what we now know concerning their existence.

There are several ways to phrase the problem. From the Riemannian point of view, one might want to fix a smooth compact-oriented four-manifold M, and simply ask whether there exists an extremal Kähler metric g, where the associated complex structure is not specified as part of the problem. Since M must in particular admit a Kähler metric, two necessary conditions are that M must admit a complex structure and have even first Betti number. Provided these desiderata are fulfilled, Shu (2006) has then shown that an extremal metric g always exists. For all but two diffeotypes, moreover, one can actually arrange for the extremal Kähler metric g to have constant scalar curvature. However, these two exceptional diffeotypes are $\mathbb{CP}_2\#\overline{\mathbb{CP}}_2$, and $\mathbb{CP}_2\#2\overline{\mathbb{CP}}_2$, and it is now known (Chen, LeBrun and Weber 2008) that both these manifolds carry Einstein metrics—indeed, even Einstein metrics which are conformal rescalings of extremal Kähler metrics! Since, in conjunction with $F = 0$, any Einstein metric of course satisfies the Einstein–Maxwell equations, we thus immediately deduce the following:

Theorem 3.3 *Let M be the underlying four-manifold of any compact complex surface of Kähler type. Then M admits a Riemannian solution (g, F) of the Einstein–Maxwell equations.*

While the above formulation of Shu's result certainly suffices to imply Theorem 3.3, it unfortunately also obfuscates the nature of the proof, which involves constructing extremal Kähler metrics compatible with some *fixed* complex structure in each possible deformation class. The key tool used for this purpose is due to Arezzo and Pacard (2006, 2009) who have shown that constant-scalar-curvature Kähler metrics can be constructed on blow-ups and desingularizations of constant-scalar-curvature Kähler orbifolds, under only very mild assumptions on the complex automorphism group; similar results moreover have even been proved concerning the strictly extremal case (Arezzo, Pacard, and Singer 2007). These gluing results represents a vast generalization of earlier work by the present author and his collaborators (Kim, LeBrun and Pontecorvo 1997; LeBrun 1991; LeBrun and Singer 1993) regarding the limited realm of scalar-flat Kähler surfaces. In fact, by invoking the theory of Kähler–Einstein metrics (Aubin 1976; Yau 1977), Arezzo and Pacard (2006) had already shown that every Kähler-type complex surface of Kodaira dimension 0 or 2 admits compatible constant-scalar-curvature Kähler metrics. Shu's results concerning the remaining cases of Kodaira dimensions $-\infty$ and 1 are much less robust, but still easily produce enough examples to imply most of Theorem 3.3.

Of course, one ultimately does not want to settle for mere existence statements; we would really like to completely understand the moduli space of solutions! From this point of view, the first question to ask is whether there can only be one solution for any given complex structure and Kähler class. Modulo complex automorphisms' uniqueness always holds in this setting, as was proved in a series of a fundamental papers by Donaldson (2001), Mabuchi (2004), and Chen and Tian (2005). For a fixed complex structure, one also knows that the Kähler classes of extremal Kähler metrics sweep out an open subset of the Kähler cone (LeBrun and Simanca 1993), and somewhat weaker results are also available regarding deformations of complex structure (Fujiki and Schumacher 1990; LeBrun and Simanca 1994). However, it turns out that the set of Kähler classes which are representable by extremal Kähler metrics may sometimes be a *proper* non-empty open subset of the Kähler cone (Apostolov and Tønnesen-Friedman 2006; Ross 2006). The latter phenomenon is related to algebro-geometric stability problems (Mabuchi 2005; Ross and Thomas 2006) in a manner which is still only partly understood, but there is reason to hope that a definitive understanding of such issues may result from the incredible ferment of research currently being carried out the field.

As we saw in Theorem 3.3, Kähler geometry supplies a natural and beautiful way of constructing solutions of the Einstein–Maxwell equations on many compact four-manifolds. In the opposite direction, we also have the following easy but rather suggestive result:

Proposition 3.3 *Let M be the underlying four-manifold of any compact complex surface of non-Kähler type with vanishing geometric genus. Then M does not carry any Riemannian solution of the Einstein–Maxwell equations.*

Proof. Let us begin by remembering the remarkable fact (Barth, Peters, and Van de Ven 1984; Buchdahl 1999; Siu 1983) that a compact complex surface is of Kähler type iff it has b_1 even. Consequently, for any non-Kähler-type complex surface M, b_1 is odd, and $b_+ = 2p_{\mathbf{g}}$, where $p_{\mathbf{g}} = h^{2,0}$ is the geometric genus (Barth, Peters, and Van de Ven, 1984, Theorem IV.2.6). Since the latter is assumed to vanish, M then has negative-definite intersection form, and Hodge theory tells us that there are no non-trivial self-dual harmonic two-forms for any metric g on M.

Now suppose that (g, F) is a Riemannian solution of the Einstein–Maxwell equations on M. Then the harmonic two-form F satisfies $F^+ = 0$, and hence

$$\mathring{r} = -[F \circ F]_0 = -2F^+ \circ F^- = 0.$$

The metric g must therefore be Einstein. But the negative-definiteness of the intersection form also tells us that $(2\chi + 3\tau)(M) = c_1^2(M) \leq 0$. The Hitchin–Thorpe inequality for Einstein manifolds (Hitchin 1974*b*) therefore guarantees that M has a finite normal cover $\tilde{M}$ which is diffeomorphic to either $K3$ or T^4, and so, in particular, has b_1 even. Pulling back the complex structure J of M to this cover, we therefore obtain a complex surface $(\tilde{M}, \tilde{J})$ of Kähler type. Averaging an arbitrary Kähler metric h on $(\tilde{M}, \tilde{J})$ over the finite group of deck transformation of $\tilde{M} \to M$ then gives us a Kähler metric which descends to (M, J). Thus (M, J) is of Kähler type, in contradiction to our hypotheses. Our supposition was therefore false, and M thus cannot carry any Riemannian solution of the Einstein–Maxwell equations. □

This chapter has endeavored to convince the reader that the four-dimensional Einstein–Maxwell equations represent a beautiful and natural generalization of the constant-scalar-curvature Kähler condition. However, it still remains to be seen whether solutions exist on many compact four-manifolds other than complex surfaces of Kähler type. In this direction, my guess is that Proposition 3.3 will actually prove to be rather misleading. For example, there are ALE Riemannian solutions of the Einstein–Maxwell equations, constructed (Yuille 1987) via the Israel–Wilson ansatz, on manifolds very unlike any complex surface. It thus seems reasonable to conjecture that there are plenty of compact solutions that in no sense arise from Kähler geometry. Perhaps some interested reader will feel inspired to go out and find some!

Acknowledgments

The author would like to thank Maciej Dunajski, Caner Koca, and Galliano Valent for their interesting comments and queries regarding an earlier version of this chapter, as some of these resulted in important improvements to this chapter. This chapter was supported in part by NSF grant DMS-0604735.

References

Apostolov, V. and Tønnesen-Friedman, C. (2006). A remark on Kähler metrics of constant scalar curvature on ruled complex surfaces. *Bull. London Math. Soc.* **38**(3), 494–500.

Arezzo, C. and Pacard, F. (2006). Blowing up and desingularizing constant scalar curvature Kähler manifolds. *Acta Math.* **196**(2), 179–228.

Arezzo, C. and Pacard, F. (2009). Blowing up and desingularizing Kähler orbifolds of constant scalar curvature II. *Ann. Math.* **170**, 685–738. E-print math.DG/0504115.

Arezzo, C., Pacard, F. and Singer, M. (2007). Extremal metrics on blow ups. E-print math.DG/0701028.

Atiyah, M. F. (1969). The signature of fibre-bundles, in *Global Analysis* (Papers in Honor of K. Kodaira), University Tokyo Press, Tokyo, pp. 73–84.

Atiyah, M. F., Hitchin, N. J. and Singer, I. M. (1978). Self-duality in four-dimensional Riemannian geometry. *Proc. R. Soc. London Ser. A* **362**, 425–61.

Aubin, T. (1976). Equations du type Monge-Ampère sur les variétés kählériennes compactes. *C. R. Acad. Sci. Paris A* **283**, 119–21.

Barth, W., Peters, C. and Van de Ven, A. (1984). *Compact Complex Surfaces.* Springer-Verlag, Berlin, Heidelberg, New York, Tokyo.

Besse, A. L. (1987). *Einstein manifolds*, Vol. 10 of *Ergebnisse der Mathematik und ihrer Grenzgebiete (3)* [*Results in Mathematics and Related Areas (3)*], Springer-Verlag, Berlin.

Buchdahl, N. (1999). On compact Kähler surfaces. *Ann. Inst. Fourier (Grenoble)* **49**(1), 287–302.

Calabi, E. (1982). Extremal Kähler metrics, in Seminar on Differential Geometry, Vol. 102 of *Ann. of Math. Stud.*, Princeton University Press, Princeton, NJ, pp. 259–90.

Chen, X. X. (2006). Space of Kähler metrics III on the lower bound of the Calabi energy and geodesic distance. E-print math.DG/0606228.

Chen, X. X., LeBrun, C. and Weber, B. (2008). On conformally Kähler, Einstein manifolds. *J. Am. Math. Soc.* **21**(4), 1137–1168. E-print arXiv:0705.0710.

Chen, X. X. and Tian, G. (2005). Uniqueness of extremal Kähler metrics. *C. R. Math. Acad. Sci. Paris* **340**(4), 287–90.

Donaldson, S. K. (1986). Connections, cohomology and the intersection forms of 4-manifolds. *J. Differential Geom.* **24**(3), 275–341.

Donaldson, S. K. (2001). Scalar curvature and projective embeddings. I. *J. Differential Geom.* **59**(3), 479–522.

Fine, J. (2004). Constant scalar curvature Kähler metrics on fibred complex surfaces. *J. Differential Geom.* **68**(3), 397–432.

Flaherty, E. J., Jr. (1978). The nonlinear graviton in interaction with a photon. *Gen. Relativity Gravit.* **9**(11), 961–78.

Fujiki, A. and Schumacher, G. (1990). The moduli space of extremal compact Kähler manifolds and generalized Weil-Petersson metrics. *Publ. Res. Inst. Math. Sci.* **26**(1), 101–83.

Gay, D. T. and Kirby, R. (2004). Constructing symplectic forms on 4-manifolds which vanish on circles. *Geom. Topol.* **8**, 743–77 (electronic).

Hawking, S. W. and Ellis, G. F. R. (1973). *The Large Scale Structure of Space-Time*, Cambridge University Press, London.

Hitchin, N. J. (1974*a*). Harmonic spinors. *Adv. Math.* **14**, 1–55.

Hitchin, N. J. (1974*b*). On compact four-dimensional Einstein manifolds. *J. Differential Geom.* **9**, 435–42.

Hitchin, N. J. (1975). On the curvature of rational surfaces, in *Differential Geometry (Proceedings of the Symposium on Pure Mathematics*, Vol. XXVII, Part 2, Stanford University, Stanford, CA, 1973), American Mathematical Society, Providence, RI, pp. 65–80.

Hitchin, N. J. (1979). Polygons and gravitons. *Math. Proc. Cambridge Philos. Soc.* **85**(3), 465–76.

Hitchin, N. J. (1981). Kählerian twistor spaces. *Proc. London Math. Soc. (3)* **43**(1), 133–50.

Hitchin, N. J., Karlhede, A., Lindström, U. and Roček, M. (1987). Hyper-Kähler metrics and supersymmetry. *Commun. Math. Phys.* **108**(4), 535–89.

Kim, J., LeBrun, C. and Pontecorvo, M. (1997). Scalar-flat Kähler surfaces of all genera. *J. Reine Angew. Math.* **486**, 69–95.

Kodaira, K. (1967). A certain type of irregular algebraic surfaces. *J. Analyse Math.* **19**, 207–15.

Kotschick, D. (1998). Signatures, monopoles and mapping class groups. *Math. Res. Lett.* **5**(1–2), 227–34.

LeBrun, C. (1991). Scalar-flat Kähler metrics on blown-up ruled surfaces. *J. Reine Angew. Math.* **420**, 161–77.

LeBrun, C. (1995). Polarized 4-manifolds, extremal Kähler metrics, and Seiberg-Witten theory. *Math. Res. Lett.* **2**, 653–62.

LeBrun, C. (1997). Yamabe constants and the perturbed Seiberg-Witten equations. *Commun. Anal. Geom.* **5**, 535–53.

LeBrun, C. (2001). Ricci curvature, minimal volumes, and Seiberg-Witten theory. *Invent. Math.* **145**, 279–316.

LeBrun, C. (2003). Einstein metrics, four-manifolds, and differential topology, in *Surveys in Differential Geometry*, Vol. VIII (Boston, MA, 2002), Surveys in Differential Geometry, VIII, International Press, Somerville, MA, pp. 235–55.

LeBrun, C. and Simanca, S. R. (1993). On the Kähler classes of extremal metrics, in *Geometry and Global Analysis* (Sendai, Japan, 1993), Tohoku University, Sendai, Japan, pp. 255–71.

LeBrun, C. and Simanca, S. R. (1994). Extremal Kähler metrics and complex deformation theory. *Geom. Funct. Anal.* **4**(3), 298–336.

LeBrun, C. and Singer, M. (1993). Existence and deformation theory for scalar-flat Kähler metrics on compact complex surfaces. *Invent. Math.* **112**(2), 273–313.

Leung, N. C. (1996). Seiberg Witten invariants and uniformizations. *Math. Ann.* **306**(1), 31–46.

Mabuchi, T. (2004). Uniqueness of extremal Kähler metrics for an integral Kähler class. *Int. J. Math.* **15**(6), 531–46.

Mabuchi, T. (2005). An energy-theoretic approach to the Hitchin-Kobayashi correspondence for manifolds. I. *Invent. Math.* **159**(2), 225–43.

Misner, C. W., Thorne, K. S. and Wheeler, J. A. (1973). *Gravitation*, W. H. Freeman and Co., San Francisco, CA.

Ross, J. (2006). Unstable products of smooth curves. *Invent. Math.* **165**(1), 153–62.

Ross, J. and Thomas, R. (2006). An obstruction to the existence of constant scalar curvature Kähler metrics. *J. Differential Geom.* **72**(3), 429–66.

Shu, Y. (2006). Compact complex surfaces and constant scalar curvature Kähler metrics. E-print math.DG/0612013.

Siu, Y. (1983). Every K3 surface is Kähler. *Invent. Math.* **73**, 139–50.

Taubes, C. H. (1994). The Seiberg-Witten invariants and symplectic forms. *Math. Res. Lett.* **1**, 809–22.

Witten, E. (1994). Monopoles and four-manifolds. *Math. Res. Lett.* **1**, 809–22.

Yamabe, H. (1960). On a deformation of Riemannian structures on compact manifolds. *Osaka Math. J.* **12**, 21–37.

Yau, S. T. (1977). Calabi's conjecture and some new results in algebraic geometry. *Proc. Natl. Acad. USA* **74**, 1789–99.

Yuille, A. L. (1987). Israel-Wilson metrics in the Euclidean regime. *Classical Quantum Gravity* **4**(5), 1409–26.

IV

THE NAHM TRANSFORM FOR CALORONS

Benoit Charbonneau and Jacques Hurtubise

Dedicated to Nigel Hitchin on the occasion of his 60th birthday

4.1 Introduction

One rather mysterious feature of the self-duality equations on $\mathbb{R}^4$ is the existence of a quite remarkable non-linear transform, the Nahm transform. It maps solutions to the self-duality equations on $\mathbb{R}^4$ invariant under a closed translation group $G \subset \mathbb{R}^4$ to solutions to the self-duality equations on $(\mathbb{R}^4)^*$ invariant under the dual group G^*. This transform uses spaces of solutions to the Dirac equation, it is quite sensitive to boundary conditions, which must be defined with care, and it is not straightforward: for example, it tends to interchange rank and degree.

The transform was introduced by Nahm as an adaptation of the original ADHM construction of instantons (Atiyah *et al.* 1978) having the advantage that it can be generalized. The series of papers (Corrigan and Goddard 1984; Nahm 1983, 1984) details Nahm's original transform. In the monopole case, $G = \mathbb{R}$, and the transform takes the monopole to a solution to some non-linear matrix valued ordinary differential equations, Nahm's equations on an interval, and there is an inverse transform giving back the monopole. Much of the mathematical development of this transform is due to Nigel Hitchin (1983), who showed for the $SU(2)$ monopole how the monopole and the corresponding solution to Nahm's equations are both encoded, quite remarkably, in the same algebraic curve, the spectral curve of the monopole. This work was extended to the cases of monopoles for the other classical groups by Hurtubise and Murray (1989). This extension illustrates just how odd the transform can be: one gets solutions to the Nahm equations on a chain of intervals, but the size of the matrices jump from interval to interval.

The Nahm transform for other cases of G invariance has been studied by various authors (among others, Braam and van Baal 1989; Charbonneau 2004; Cherkis and Kapustin 2001; Jardim 2001, 2002*a*, 2002*b*; Nye 2001; Schenk 1988; Szabó 2005); a nice survey can be found in Jardim (2004). Here, we study the case of 'minimal' invariance, under $\mathbb{Z}$; the fields are referred to as calorons, and the Nahm transform sends them to solutions of Nahm's equations over the circle. This case is very close to the monopole one, and indeed calorons can be considered as Kač–Moody monopoles (Garland and Murray 1988).

Calorons have been studied from different angles by various authors, including Bruckmann and van Baal (2002); Bruckmann *et al.* (2003, 2004), Chakrabarti (1987), Kraan (2000), Kraan and van Baal (1998*a*; 1998*b*; 1998*c*), Lee (1998), Lee and Lu (1998), Lee and Yi (2003), Nógrádi (2005), Norbury (2000), and Ward (2004). In particular, many explicit solutions have been found.

This project started while we were studying the works of Nye (2001) and Nye and Singer (2000) on calorons; they consider the Nahm transform directly, and do most of the work required to show that the transform is involutive. The missing ingredients turn out to lie in complex geometry, in precisely the same way Hitchin's extra ingredient of a spectral curve complements Nahm's. The complex geometry allows us to complete the equivalence, which can then be used to compute the moduli. It thus seems to us quite appropriate to consider this problem in a volume dedicated to Nigel Hitchin, as it allows us to revisit some of his beautiful mathematics, including some on calorons which has remained unpublished.

In Section 4.2, we summarize the work of Nye and Singer towards showing that the Nahm transform is an equivalence between calorons and appropriate solutions to Nahm's equations. In Section 4.3, following in large part on the work of Garland and Murray (1988), we describe the complex geometry ('spectral data') that encodes a caloron. In Section 4.4, we study the process by which spectral data also correspond to solutions to Nahm's equations. In Section 4.5, we close the circle, showing the two Nahm transforms are inverses. In Section 4.6, we give a description of moduli, expounded in Charbonneau and Hurtubise (2007).

4.2 The work of Nye and Singer

4.2.1 *Two types of invariant self-dual gauge fields on* $\mathbb{R}^4$

Nye and Singer study the Nahm transform between the following two self-dual gauge fields (we restrict ourselves to $SU(2)$, though they study the more general case of $SU(N)$):

4.2.1.1 $SU(2)$ *Calorons of charge* (k, j)

$SU(2)$ calorons are *self-dual* $SU(2)$ connections on $S^1 \times \mathbb{R}^3$, satisfying appropriate boundary conditions. We view $S^1 \times \mathbb{R}^3$ as the quotient of the standard Euclidean $\mathbb{R}^4$ by the time translation $(t, x) \mapsto (t + 2\pi/\mu_0, x)$. Let A be such a connection, defined over a rank 2 vector bundle V equipped with a unitary structure; we write it in coordinates over $\mathbb{R}^4$ as

$$A = \phi dt + \sum_{i=1,2,3} A_i dx_i,$$

with associated covariant derivatives

$$\nabla = \left(\frac{\partial}{\partial t} + \phi\right) dt + \sum_{i=1,2,3} \left(\frac{\partial}{\partial x_i} + A_i\right) dx_i \equiv \nabla_t dt + \sum_{i=1,2,3} \nabla_i dx_i.$$

We require that the L^2 norm of the curvature of A be finite, and that in suitable gauges, the functions A_i be $O(|x|^{-2})$ as $|x| \to \infty$, and that ϕ be conjugate to $\mathrm{diag}(i(\mu_1 - j/2|x|), i(-\mu_1 + j/2|x|)) + O(|x|^{-2})$ for a positive real constant μ_1 and a positive integer j (the monopole charge). We also have bounds on the derivatives of these fields.

The boundary conditions tell us in essence that in a suitable way the connection extends to the two-sphere at infinity times S^1; furthermore, one can show that the extension is to a fixed connection, which involves fixing a trivialization at infinity; there is thus a second invariant we can define, the relative second Chern class, which we represent by a (positive) integer k. There are then two integer charges for our caloron, k and j, the instanton and monopole charges, respectively.

There is a suitable group of gauge transformations acting on these fields (Nye 2001): Nye's approach is to compactify to $S^1 \times \bar{B}^3$ with a fixed trivilization over the boundary $S^1 \times S^2$. The gauge transformations are those extending smoothly to the identity on the boundary. Two solutions are considered to be equivalent if they are in the same orbit under this group.

4.2.1.2 *Solutions to Nahm's equations on the circle*

The second class of objects we consider are skew adjoint matrix-valued functions $T_i(z), i = 0, \ldots, 3$, of size $(k+j) \times (k+j)$ over the interval $(-\mu_1, \mu_1)$, and of size $(k) \times (k)$ over the interval $(\mu_1, \mu_0 - \mu_1)$ (hence $\mu_1 < \mu_0 - \mu_1$, so we impose that condition) that are solutions to Nahm's equations

$$\frac{dT_{\sigma(1)}}{dz} + [T_0, T_{\sigma(1)}] = [T_{\sigma(2)}, T_{\sigma(3)}], \qquad \text{for } \sigma \text{ even permutations of } (123), \quad (4.1)$$

on the circle $\mathbb{R}/(z \mapsto z + \mu_0)$. These equations are reductions of the self-duality equations to one dimension, and are invariant under a group of gauge transformations under which the $\frac{d}{dz} + T_0$ transforms as a connection.

We think of the T_i as sections of the endomorphisms of a vector bundle K whose rank jumps at the two boundary points; at these points $\pm\mu_1$ we need boundary conditions.

Case 1: $j \neq 0$. At each of the boundary points, there is a large side, with a rank $(k+j)$ bundle, and a small side, with a rank k bundle. We attach the two at the boundary point using an injection ι from the small side into the large side, and a surjection π going the other way, with $\pi \cdot \iota$ the identity. One asks, from the small side, that the T_i have well-defined limits at the boundary point. From the large side, one has a decomposition near the boundary points of the bundle $\mathbb{C}^{k+j} \times [-\mu_1, \mu_1]$ into an orthogonal sum of subbundles of rank k and j invariant with respect to the connection $\frac{d}{dz} + T_0$ and compatible with the maps ι, π. With respect to this decomposition the T_i have the form

$$T_i = \begin{pmatrix} \hat{T}_i & O\left(\hat{z}^{(j-1)/2}\right) \\ O(\hat{z}^{(j-1)/2}) & R_i \end{pmatrix}, \quad (4.2)$$

in a basis where T_0 is gauged to zero and for a choice of coordinate $\hat{z} := z - (\pm\mu_1)$. At the boundary point $\hat{z} = 0$, each $\hat{T}_i$ has a well-defined limit that coincides using π, ι with the limit from the small side. The R_i have simple poles, and we ask that their residues form an irreducible representation of $su(2)$.

Case 2: $j = 0$. Here the boundary conditions are different. We have at the boundary, identification of the fibres from both sides. With this identification, we ask that, in a gauge where T_0 is zero, the T_i have well-defined limits $(T_i)_\pm$ from both sides, and that, setting

$$A(\zeta) = (T_1 + iT_2) - 2iT_3\zeta + (T_1 - iT_2)\zeta^2, \tag{4.3}$$

one has that taking the limits from both sides,

$$A(\zeta)_+ - A(\zeta)_- = (\alpha_0 + \alpha_1\zeta)(\bar{\alpha}_1 - \bar{\alpha}_0\zeta)^{\mathrm{T}} \tag{4.4}$$

for vectors α_0, α_1 in $\mathbb{C}^k$. In particular, $A(\zeta)_+ - A(\zeta)_-$ is of rank at most one for all ζ.

In both cases, there is a symmetry condition, that, in a suitable gauge,

$$T_i(0) = T_i(0)^{\mathrm{T}}$$

and, finally, an irreducibility condition, that there be no covariant constant sections of the bundle that are left invariant under the matrices T_i. Solutions for which there are such covariant constant sections should correspond to calorons where instanton charge has bubbled off, leaving behind a caloron of lower charge.

4.2.2 *The Nahm transform*

From caloron to a solution to Nahm's equations.

Let z be a real parameter. Using the Pauli matrices e_i, we define the Dirac operators D_z acting on sections of the tensor product of the vector bundle V with the spin bundles $S_\pm \simeq \mathbb{C}^2 \times S^1 \times \mathbb{R}^3$ by

$$D_z \colon \Gamma(V \otimes S_+) \to \Gamma(V \otimes S_-)$$
$$s \mapsto (\nabla_t + iz)s + \sum_i (e_i \nabla_i)s.$$

The Weitzenbock formula $D_z D_z^* = \nabla^*\nabla + \rho\left(F_{A_z}^-\right)$ guarantees that D_z^* is injective. Nye and Singer (2000) show that for z not in the set of lifts $\pm\mu_1 + n\mu_0$ $(n \in \mathbb{Z})$ of the boundary point $\pm\mu_1$ from the circle to $\mathbb{R}$, and with suitable choices of function spaces, the operator D_z is Fredholm, and its index away from these points is $k + j$ for z lying in the intervals $(-\mu_1 + n\mu_0, \mu_1 + n\mu_0)$ and is k in the intervals $(\mu_1 + n\mu_0, -\mu_1 + (n+1)\mu_0)$. There is thus a bundle over $\mathbb{R}$ whose rank jumps at $\pm\mu_1 + n\mu_0$ $(n \in \mathbb{Z})$, with fibre $\ker(D_z)$ at z. There is a natural way of shifting by μ_0, identifying $\ker(D_z)$ with $\ker(D_{z+\mu_0})$, giving a bundle K over the circle.

Over each interval, this bundle sits inside the trivial bundle whose fibre is the space of L^2 sections of $V \otimes S_-$. Let P be the orthogonal projection from this trivial bundle onto K. As elements of K decay exponentially, the operation X_i of multiplying by the coordinate x_i can be used to define operators on sections of K by

$$\frac{d}{dz} + T_0 = P \cdot \frac{d}{dz},$$
$$T_i = P \cdot X_i.$$

Theorem 4.1 (Nye 2001, section 4.1.2, p. 108) *The operators defined in this way satisfy Nahm's equations.*

This theorem and Theorem 4.2 below fall in line with the general Nahm transform heuristic philosophy: the curvature of the transformed object, seen as an invariant connection on $\mathbb{R}^4$, is always composed of a self-dual piece and another piece depending on the behaviour at infinity of the harmonic spinors. Since the latter are decaying exponentially, that other piece is zero and the transformed object is self-dual, or equivalently once we reduce, it satisfies Nahm's equations (see for instance Charbonneau 2006, section 3).

From a solution to Nahm's equations to a caloron.
For the inverse transform, we proceed in the same way: from a solution to Nahm's equations, we define a family of auxiliary operators parameterized by points of $\mathbb{R}^4$ acting on sections of $K \otimes \mathbb{C}^2$ by

$$D_{x,t} = i\left(\frac{d}{dz} + T_0 - it + \sum_{j=1,2,3} e_j(T_j - ix_j)\right).$$

One must again distinguish two cases, $j > 0$ and $j = 0$.

Case 1: $j > 0$. We define the space W of L^2_1 sections of $K \otimes \mathbb{C}^2$ such that at $\pm\mu_1$ the values of the sections coincide: for the sections s_1 on the small side and s_2 on the large side, we need $\iota(s_1) = s_2$. Let X be the space of L^2 sections of $K \otimes \mathbb{C}^2$ over the circle. Set

$$\hat{D}_{x,t} := D_{x,t} \colon W \to X \tag{4.5}$$

Case 2: $j = 0$. We have as above the space W of sections. We define in addition a two-dimensional space U associated with the end points $\pm\mu_1$. The jump condition given by (4.4) at μ_1 imply that there is a vector $u_+ \in K_{\mu_1} \otimes \mathbb{C}^2$ such that, as elements of $\operatorname{End}(K_{\mu_1} \times \mathbb{C}^2)$,

$$\sum_{j=1}^{3}\big((T_j)_+ - (T_j)_-\big) \otimes e_j = \big(u_+ \otimes u_+^*\big)_0 .$$

Here the subscript '0' signifies taking the trace-free $\mathrm{End}(K) \otimes Sl(2, \mathbb{C})$ component inside $\mathrm{End}(K) \otimes \mathrm{End}(\mathbb{C}^2) \simeq \mathrm{End}(K \otimes \mathbb{C}^2)$.

One has a similar vector u_- at $-\mu_1$. Let U be the vector space spanned by u_+, u_-. Let $\Pi\colon K_{\mu_1} \oplus K_{-\mu_1} \to U$ be the orthogonal projection, let X be the sum of the space of L^2 sections of K with the space U, and set

$$\hat{D}_{x,t} := (D_{x,t}, \Pi)\colon W \to X \tag{4.6}$$

The kernel of this operator consists of sections s in the kernel of $D_{x,t}$, with values at $\pm\mu_1$ lying in $U^\perp$; the cokernel consists of triples $(s, c_{\mu_1}u_{\mu_1}, c_{-\mu_1}u_{-\mu_1})$, where s is a section of $K \otimes \mathbb{C}^2$, lying in the kernel of $D^*_{x,t}$, with jump discontinuity c_+u_+ at μ_1 and c_-u_- at $-\mu_1$.

Theorem 4.2 (Nye 2001, p. 104) *The operator $\hat{D}_{x,t}$ has trivial kernel for all (x, t), and has a cokernel of rank 2, defining a rank 2 vector bundle over $\mathbb{R}^4$, with natural time periodicity which allows one to build a vector bundle V over $S^1 \times \mathbb{R}^3$, defined locally as a subbundle of the infinite-dimensional bundle $X \times S^1 \times \mathbb{R}^3$. Let P denote the orthogonal projection from X to V. Setting, on sections of V,*

$$\nabla_i = P \cdot \frac{\partial}{\partial x_i}, \qquad i = 1, 2, 3$$

$$\nabla_t = P \cdot \frac{\partial}{\partial t},$$

defines an $SU(2)$ caloron.

4.2.3 *Involutivity of the transforms*

We would like these two transforms to be inverses of each other, so that they define a one-to-one correspondence between gauge equivalence classes of calorons and gauge equivalence classes of solutions of Nahm's equations on the circle. The results of Nye and Singer get us most of the way there. What is missing is a proof that the solutions to Nahm's equations one obtains from a caloron satisfy the correct boundary and irreducibility conditions and then that the two constructions are inverses to one another.

The easiest way to do so, as Hitchin did in his original paper on $SU(2)$ monopoles, is to exploit the regularity given by a third equivalent set of data, which involves complex geometry. To do this, we examine some work of Garland and Murray, building on unpublished work of Hitchin.

4.3 Twistor transform for calorons/Kač–Moody monopoles

4.3.1 *Upstairs: twistor transform for calorons*

Like all self-dual gauge fields on $\mathbb{R}^4$ or its quotients $\mathbb{R}^4/G$, calorons admit a twistor transform, translating the gauge fields into holomorphic vector bundles on an auxiliary space, the twistor space associated to $\mathbb{R}^4/G$. Following Garland

and Murray (1988, section 2), we summarize the construction, again for $SU(2)$ only.

It is convenient first to recall from Hitchin (1982, section 3) the twistor space for $\mathbb{R}^3$. It can be interpreted as the space of oriented lines in $\mathbb{R}^3$ and it is isomorphic to $T\mathbb{P}^1$, the tangent bundle of the Riemann sphere. Let ζ be the standard affine coordinate on $\mathbb{P}^1$, and η be the fibre coordinate in $T\mathbb{P}^1$ associated to the basis vector $\partial/\partial\zeta$. The incidence relation between the standard coordinates on $\mathbb{R}^3$ and (ζ, η) is given by

$$\eta = (ix_1 - x_2) + 2x_3\zeta + (ix_1 + x_2)\zeta^2.$$

The space $T\mathbb{P}^1$ comes equipped with standard line bundles $\mathcal{O}(k)$, lifted from $\mathbb{P}^1$, and line bundles L^t, parameterized by $t \in \mathbb{C}$, with transition function $\exp(t\eta/\zeta)$ from the open set $\zeta \neq \infty$ to the open set $\zeta \neq 0$. For $t \in \mathbb{R}$, these line bundles correspond to the standard $U(1)$ monopoles on $\mathbb{R}^3$ given by a flat connection and constant Higgs field it. Let $L^t(k) := L^t \otimes \mathcal{O}(k)$. These bundles are in some sense the building blocks for monopoles for higher gauge groups.

According to Garland and Murray, the twistor space $\mathcal{T}$ for $S^1 \times \mathbb{R}^3$ parametrizes pairs consisting of a point in $S^1 \times \mathbb{R}^3$ and a unit vector in $\mathbb{R}^3$. Such a pair gives a cylinder along which to integrate, and that projects to an oriented line in $\mathbb{R}^3$. There is then a $\mathbb{C}^*$ fibration $\pi\colon \mathcal{T} \to T\mathbb{P}^1$, which is in fact holomorphic and the complement of the zero section in L^{μ_0}. It has a natural fibrewise compactification $\tilde{\mathcal{T}} = \mathbb{P}(\mathcal{O} \oplus L^{\mu_0})$ compactifying the cylinders into spheres. The complement $\tilde{\mathcal{T}} \backslash \mathcal{T}$ is a sum of two disjoint divisors $\mathcal{T}^0$ and $\mathcal{T}^\infty$ mapping isomorphically to $T\mathbb{P}^1$.

In the monopole case, the holomorphic vector bundle was obtained by integrating $\nabla_s - i\phi$ along real lines using a metric coordinate s along the line, and the Higgs field ϕ. Here, the analogous operation is integrating $\nabla_s - i\nabla_t$ over the cylinders to obtain a vector bundle E over $\mathcal{T}$. The boundary conditions allow us to extend the bundle to the compactification $\tilde{\mathcal{T}}$; we denote this extension again by E. The boundary conditions also give line subbundles over these divisors, $E_1^0 = L^{-\mu_1}(-j)$ over $\mathcal{T}^0$, corresponding to solutions of $\nabla_s - i\phi$ which decay as one goes to plus infinity on the cylinder, and $E_1^\infty = L^{\mu_1}(-j)$ over $\mathcal{T}^\infty$, corresponding to solutions which decay as one goes to minus infinity.

The twistor space has a real structure τ, which acts on the bundle by $\tau^*(E) = E^*$, and maps the line subbundle $L^{-\mu_1}(-j)$ over $\mathcal{T}^0$ to the annihilator of the subbundle $L^{\mu_1}(-j)$ over $\mathcal{T}^\infty$.

4.3.2 *Downstairs: caloron as a Kač–Moody monopole*

As Garland and Murray (1988, section 6) show, there is a very nice way of thinking of the caloron as a monopole over $\mathbb{R}^3$, with values in a Kač–Moody algebra, extending a loop algebra: the fourth variable t becomes the internal loop algebra variable. This way of thinking goes over to the twistor space picture.

Indeed, the twistor transform for the caloron gives us a bundle E over $\mathcal{T}$. Taking a direct image (restricting to sections with poles of finite order along $0, \infty$) gives one an infinite-dimensional vector bundle F over $T\mathbb{P}^1$. If w is a standard fibre coordinate on $\mathcal{T}$ vanishing over $\mathcal{T}^0$ and with a simple pole at $\mathcal{T}^\infty$, it induces an endomorphism W of F, and quotienting F by the $\mathcal{O}$-module generated by the image of $W - w_0$ gives us the restriction of E to the section $w = w_0$, so that E and (F, W) are equivalent. One has more, however; the fact that the bundle E extends to $\tilde{\mathcal{T}}$ gives a subbundle F^0 of sections in the direct image which extend over $\mathcal{T}^0$, and a subbundle F^∞ of sections in the direct image which extend over $\mathcal{T}^\infty$. One can go further and use the flags $0 = E_0^0 \subset E_1^0 \subset E_2^0 = E$ over $\mathcal{T}^0$, $0 = E_0^\infty \subset E_1^\infty \subset E_2^\infty = E$ over $\mathcal{T}^\infty$ to define for $p \in \mathbb{Z}$ and $q = 0, 1$ subbundles $F_{p,q}^0, F_{p,q}^\infty$ of F as

$$F_{-p,q}^0 = \left\{ s \in F \mid w^{-p} s \text{ finite at } \mathcal{T}^0 \text{ with value in } E_q^0 \right\},$$

$$F_{p,q}^\infty = \left\{ s \in F \mid w^{-p} s \text{ finite at } \mathcal{T}^\infty \text{ with value in } E_q^\infty \right\}.$$

We now have infinite flags

$$\begin{gathered} \cdots \subset F_{-1,0}^0 \subset F_{-1,1}^0 \subset F_{0,0}^0 \subset F_{0,1}^0 \subset F_{1,0}^0 \subset F_{1,1}^0 \subset \cdots, \\ \cdots \supset F_{2,0}^\infty \supset F_{1,1}^\infty \supset F_{1,0}^\infty \supset F_{0,1}^\infty \supset F_{0,0}^\infty \supset F_{-1,1}^\infty \supset \cdots. \end{gathered} \tag{4.7}$$

Note that $W \cdot F_{p,q}^0 = F_{p+1,q}^0$ and $W \cdot F_{p,q}^\infty = F_{p-1,q}^\infty$. Garland and Murray show

- $F_{p,0}^0$ and $F_{-p+1,0}^\infty$ have zero intersection and sum to F away from a compact curve S_0 lying in the linear system $|\mathcal{O}(2k)|$.
- $F_{p,1}^0$ and $F_{-p,1}^\infty$ have zero intersection and sum to F away from a compact curve S_1 lying in the linear system $|\mathcal{O}(2k + 2j)|$.
- Quotients $F_{p,0}^0 / F_{p-1,1}^0$ are line bundles isomorphic to $L^{(p-1)\mu_0 + \mu_1}(j)$.
- Quotients $F_{p,1}^0 / F_{p,0}^0$ are line bundles isomorphic to $L^{p\mu_0 - \mu_1}(-j)$.
- Quotients $F_{p,0}^\infty / F_{p-1,1}^\infty$ are line bundles isomorphic to $L^{-(p-1)\mu_0 - \mu_1}(j)$.
- Quotients $F_{p,1}^\infty / F_{p,0}^\infty$ are line bundles isomorphic to $L^{-p\mu_0 + \mu_1}(-j)$.

It is worthwhile stepping back now and seeing what we have from a group theoretic point of view. On $T\mathbb{P}^1$, we have a bundle with structure group $G = \widetilde{Gl}(2, \mathbb{C})$ of $Gl(2, \mathbb{C})$-valued loops. The flags that we have found give two reductions R_0, R_∞ to the opposite Borel subgroups B_0, B_∞ (of loops extending to $0, \infty$, respectively, in $\mathbb{P}^1$, and preserving flags over these points) in G. As for finite-dimensional groups, we have exact sequences relating the Borel subgroups, their unipotent subgroups, and the maximal torus:

$$\begin{gathered} 0 \to U_0 \to B_0 \to T \to 0, \\ 0 \to U_\infty \to B_\infty \to T \to 0. \end{gathered} \tag{4.8}$$

In a suitable basis, B_0 and B_∞ consist, respectively, of upper and lower triangular matrices, and T of diagonal matrices.

The two reductions are generically transverse, and fail to be transverse over the spectral curves of the caloron. This failure of transversality gives us geometrical data which encodes the bundle and hence the caloron. This approach is expounded in Hurtubise and Murray (1990) for monopoles, but can be extended to calorons in a straightforward fashion. We summarize the construction here. The vector bundle F can be thought of as defining an element f of the cohomology set $H^1(T\mathbb{P}^1, G)$, and so a principal bundle P_f; one simply thinks of f as the transition functions for P_f, in terms of Čech cohomology. The reductions to B_0, B_∞ can also be thought of as elements f_0, f_∞ of $H^1(T\mathbb{P}^1, B_0)$, $H^1(T\mathbb{P}^1, B_\infty)$, respectively, or as principal bundles P_{f_0}, P_{f_∞}, or, since they are reductions of P_f, as sections R_0 of $P_f \times_G G/B_0 = P_{f_\infty} \times_{B_\infty} G/B_0$ and R_∞ of $P \times_G G/B_\infty = P_{f_0} \times_{B_0} G/B_\infty$.

We note also that the elements f_0, f_∞ when projected to $H^1(T\mathbb{P}^1, T)$ give fixed elements α_0, α_∞, in the sense that they are independent of the caloron, or rather depend only on the charges and the asymptotics, which we presume fixed. This is a consequence of the fact that the successive quotients in sequence (4.7) depend only on the μ_i and the charges. Let A_0, A_∞ denote the principal T bundles associated to α_0, α_∞.

Let $A_0(U_0)$ be the sheaf of sections of the U_0 bundle associated to A_0 by the action of T on U_0. The cohomology set $H^1(T\mathbb{P}^1, A_0(U_0))$ describes (Hurtubise and Murray 1990, section 3) the set of B_0 bundles mapping to α_0. To find the bundle f_0, and so f, then, one simply needs to have the appropriate class in $H^1(T\mathbb{P}^1, A_0(U_0))$.

Note that U_0 also serves as the big cell in the homogeneous space G/B_∞. Following Gravesen (1989), we think of G/B_∞ as adding an infinity to U_0, so if $\mathcal{U}_0$ denotes the sheaf of holomorphic maps into U_0, the sheaf $\mathcal{M}$ of holomorphic maps into G/B_∞ can be thought of as meromorphic maps into U_0. Hence there is an sequence of sheaves of pointed sets (base points are chosen compatibly in both U_0 and G/B_∞)

$$0 \to \mathcal{U}_0 \to \mathcal{M} \to \mathcal{P}r \to 0$$

defining the sheaf $\mathcal{P}r$ of principal parts. As B_0, T act on these sheaves, we can build the twisted versions

$$P_{f_0}(\mathcal{U}_0) \to P_{f_0}(\mathcal{M}) \xrightarrow{\phi} P_{f_0}(\mathcal{P}r),$$

$$A_0(\mathcal{U}_0) \to A_0(\mathcal{M}) \longrightarrow A_0(\mathcal{P}r).$$

As we are quotienting out the action of U_0, we have $A_0(\mathcal{P}r) \simeq P_{f_0}(\mathcal{P}r)$.

We are now ready to define the *principal part data* of the caloron. The principal part data of the caloron is the image under ϕ of the class of the reduction given as a section R_∞ of $P_{f_0} \times_{B_0} G/B_\infty = P_{f_0}(\mathcal{M})$:

$$\phi(R_\infty) \in H^0(A_0(\mathcal{P}r)).$$

To get back the caloron bundle from the principal part data, one takes the coboundary $\delta(\phi(R_\infty))$, in the obvious Čech sense, for the second sequence, obtaining a class in $H^1(T\mathbb{P}^1, A_0(\mathcal{U}_0))$. One checks that this is precisely the class corresponding to the bundle P_0; while this seems a bit surprising, the construction is fairly tautological, and is done in detail in (Hurtubise and Murray 1990, section 3). In our infinite-dimensional context, the bundles are quite special, as they, and their reductions, are invariant under the shift operator W.

Of course, this construction does not tell us what the mysterious class $\phi(R_\infty)$ actually corresponds to, but it turns out that, over a generic set of calorons, it is quite tractable. Indeed, the principal part data, as a sheaf, is supported over the spectral curves, and, if the curves have no common components and no multiple components, then the principal part data amounts to the following *spectral data* (Garland and Murray 1988):

- The two spectral curves, S_0 and S_1
- The ideal $\mathcal{I}_{S_0 \cap S_1}$, decomposing as $\mathcal{I}_{S_{01}} \cdot \mathcal{I}_{S_{10}}$; the real structure interchanges the two factors
- An isomorphism of line bundles $\mathcal{O}[-S_{10}] \otimes L^{\mu_0 - 2\mu_1} \simeq \mathcal{O}[-S_{01}]$ over S_0
- An isomorphism of line bundles $\mathcal{O}[-S_{01}] \otimes L^{2\mu_1} \simeq \mathcal{O}[-S_{10}]$ over S_1
- A real structure on the line bundle $\mathcal{O}(2k + j - 1)[-S_{01}] \otimes L^{\mu_1}|S_1$, lifting the real involution τ on $T\mathbb{P}^1$

The structure of this data can be understood in terms of the Schubert structure of G/B_0. The sheaf of principal parts, by definition, lives in the complement of U_0, where the two flags cease to be transverse. In the case that concerns us here (remember the invariance under W), there are essentially two codimension one varieties at infinity to U_0 that we consider, whose pull-backs give the two spectral curves. The decomposition of $S_0 \cap S_1$ into two pieces is simply a pull-back of the Schubert structure from G/B_0. For the line bundles, the basic idea is that in codimension one, the principal parts can be understood using embeddings of $\mathbb{P}^1$ into G/B_0 given by principal $Sl(2, \mathbb{C})$'s in G (see for instance Boyer *et al.* 1994, section 4), reducing the question at least locally to understanding principal parts for maps into $\mathbb{P}^1$, a classical subject. For simple poles, the principal part of a map into $\mathbb{P}^1$ is encoded by its residue. Here, the poles are over the spectral curves, and globally, the principal part is then encoded as a section of a line bundle over each curve; we identify these below. A way of seeing that the spectral data splits into components localized over each curve is that we can choose appropriately two parabolic subgroups Par_0, Par_1 and project from G/B_0 to G/Par_0 and G/Par_1.

To see what the 'residues' correspond to here, we exploit a description of the bundle first given in Hurtubise and Murray (1989, proposition 1.12), the short exact sequence

$$0 \longrightarrow F \longrightarrow \begin{array}{c} \vdots \\ F/\left(F^0_{p-1,1}+F^\infty_{-p+1,0}\right) \\ \oplus \\ F/\left(F^0_{p,0}+F^\infty_{-p,1}\right) \\ \oplus \\ F/\left(F^0_{p,1}+F^\infty_{-p,0}\right) \\ \vdots \end{array} \longrightarrow \begin{array}{c} \vdots \\ \oplus \\ F/\left(F^0_{p,0}+F^\infty_{-(p-1),0}\right) \\ \oplus \\ F/\left(F^0_{p,1}+F^\infty_{-p,1}\right) \\ \oplus \\ \vdots \end{array} \longrightarrow 0. \tag{4.9}$$

The quotients at the right are supported over the spectral curves ($F/(F^0_{p,0}+F^\infty_{-(p-1),0})$ over S_0, and $F/(F^0_{p,1}+F^\infty_{-p,1})$ over S_1) and those in the middle are generically line bundles.

The shift operator W moves the quotients in the middle and last columns two steps down. The quotients are in fact direct image sheaves $R^1\tilde{\pi}_*$ of sheaves on $\tilde{\mathcal{T}}$ derived from E; for example, let $E_{p,0,-(p-1),\infty}$ be the sheaf of sections of E over $\mathcal{T}$ with poles of order p at $\mathcal{T}_0$ and a zero of order $p-1$ at $\mathcal{T}_\infty$. The sheaf $F/\left(F^0_{p,0}+F^\infty_{-p,1}\right)$ can be identified with $R^1\tilde{\pi}_*(E_{p,0,-(p-1),\infty})$. One can make similar identifications for the other sheaves. In a way that parallels Hitchin (1983), we have the following result:

Proposition 4.3

1. *The spectral curves S_0 and S_1 are real (τ-invariant); the quotient $F/(F^0_{p,1}+F^\infty_{-p,1})$ over S_1 inherits from E a quaternionic structure, lifting τ.*
2. *The sheaves $F/(F^0_{p,0}+F^\infty_{-(p-1),0})$ and $F/(F^0_{p,1}+F^\infty_{-p,1})$ satisfy a vanishing theorem:*

$$H^0\left(T\mathbb{P}^1, F/\left(F^0_{p,0}+F^\infty_{-(p-1),0}\right)\otimes L^{-z}(-2)\right)=0,$$

$$\textit{for } z\in[(p-1)\mu_0+\mu_1, p\mu_0-\mu_1], \textit{ and}$$

$$H^0\left(T\mathbb{P}^1, F/\left(F^0_{p,1}+F^\infty_{-p,1}\right)\otimes L^{-z}(-2)\right)=0,$$

$$\textit{for } z\in(p\mu_0-\mu_1, p\mu_0+\mu_1).$$

The vanishing holds because these cohomology groups encode L^2 solutions to the Laplace equation in the caloron background, which must be zero (see Hitchin 1983, theorem 3.7, and Hurtubise and Murray 1989, theorem 1.17).

For generic spectral curves, following a line of argument of Hurtubise and Murray (1989), Garland and Murray show directly that the spectral data determines the caloron. To this end, they identify in Garland and Murray (1988, section 6) the quotients

$$\begin{aligned}
F/\left(F^0_{p,0}+F^\infty_{-(p-1),0}\right) &= L^{p\mu_0-\mu_1}(2k+j)[-S_{10}]|_{S_0}\\
&= L^{(p-1)\mu_0+\mu_1}(2k+j)[-S_{01}]|_{S_0},\\
F/\left(F^0_{p,1}+F^\infty_{-p,1}\right) &= L^{p\mu_0+\mu_1}(2k+j)[-S_{01}]|_{S_1}\\
&= L^{p\mu_0-\mu_1}(2k+j)[-S_{10}]|_{S_1},\\
F/\left(F^0_{p,0}+F^\infty_{-p,1}\right) &= L^{p\mu_0-\mu_1}(2k+j)\otimes\mathcal{I}_{S_{10}},\\
F/\left(F^0_{p,1}+F^\infty_{-p,0}\right) &= L^{p\mu_0+\mu_1}(2k+j)\otimes\mathcal{I}_{S_{01}}.
\end{aligned}$$

These identifications realized in exact sequence (4.9) give

$$F \hookrightarrow \begin{matrix}\vdots\\ L^{(p-1)\mu_0+\mu_1}(2k+j)\otimes\mathcal{I}_{S_{01}}\\ \oplus\\ L^{p\mu_0-\mu_1}(2k+j)\otimes\mathcal{I}_{S_{10}}\\ \oplus\\ L^{p\mu_0+\mu_1}(2k+j)\otimes\mathcal{I}_{S_{01}}\\ \vdots\end{matrix} \longrightarrow \begin{matrix}\vdots\\ L^{(p-1)\mu_0+\mu_1}(2k+j)[-S_{01}]|_{S_0}\\ = L^{p\mu_0-\mu_1}(2k+j)[-S_{10}]|_{S_0}\\ \oplus\\ L^{p\mu_0+\mu_1}(2k+j)[-S_{01}]|_{S_1}\\ = L^{p\mu_0-\mu_1}(2k+j)[-S_{10}]|_{S_1}\\ \vdots\end{matrix} \longrightarrow 0. \tag{4.10}$$

The 'residues' are then the sheaves on the right, supported over the spectral curves; the maps from the middle to the right-hand sheaves give the various isomorphisms. This diagram shows that F, and hence the caloron, is encoded in the spectral data.

More generally, we can define the *spectral data* composed of the curves S_0, S_1, the pull-backs to $T\mathbb{P}^1$ of the Schubert varieties in G/B_0, and the sheaves on the right-hand side of the sequence (4.9), with isomorphisms similar to the ones for generic spectral data given by maps from the sheaves in the middle column.

To summarize, we have that the caloron determines principal part data, a section of a sheaf of principal parts supported over the spectral curve; this data determines the caloron in turn. In the generic case, the principal part data equivalent to spectral data can be described in terms of two curves, their intersections, and sections of line bundles over these curves. We note that these generic calorons exist; indeed, we already know that a caloron is determined by a solution to Nahm's equations, and it is easy to see that generic solutions to Nahm's equations exist, as we shall see in Section 4.6.

4.4 From Nahm's equations to spectral data, and back

4.4.1 *Flows of sheaves*

The solutions to Nahm's equations we consider also determine equivalent spectral data, as we shall see. To begin, note that by setting $A(\zeta, z)$ as in (4.3), and

$$A_+(\zeta, z) = -iT_3(z) + (T_1 - iT_2)(z)\zeta,$$

Nahm's equations (4.1) are equivalent to the Lax form

$$\frac{dA}{dz} + [A_+, A] = 0. \tag{4.11}$$

The evolution of A is by conjugation, so the spectral curve given by

$$\det\,(A(\zeta, z) - \eta\mathbf{I}) = 0 \tag{4.12}$$

in $T\mathbb{P}^1$ is an obvious invariant of the flow (4.11); if the matrices are $k \times k$, the curve is a k-fold branched cover of $\mathbb{P}^1$. We define for each z a sheaf $\mathcal{L}_z$ over the curve via the exact sequence

$$0 \to \mathcal{O}(-2)^k \xrightarrow{A(\zeta,z)-\eta\mathbf{I}} \mathcal{O}^k \to \mathcal{L}_z \to 0. \tag{4.13}$$

This correspondence taking a solution to Nahm's equations to a curve and a flow of line bundles over the curve is fundamental to the theory of Lax equations (see Adams *et al.* 1990). In our case, we have an equivalence:

Proposition 4.4 *There is an equivalence between*

1. *Solutions to Nahm's equations on an interval (a, b), given by $k \times k$ matrices with reality condition built from skew-Hermitian matrices T_i as in (4.3)*
2. *Spectral curves S that are compact and lie in the linear system $|\mathcal{O}(2k)|$ lying in $T\mathbb{P}^1$, and flows $\mathcal{L}_z$, for $z \in (a, b)$, of sheaves supported on S, such that*
   a. $H^0(T\mathbb{P}^1, \mathcal{L}_z(-1)) = H^1(T\mathbb{P}^1, \mathcal{L}_z(-1)) = 0.$
   b. $\mathcal{L}_z = \mathcal{L}_{z'} \otimes L^{z-z'}$, *for* $z, z' \in (a, b)$.
   c. *The curve S is real, that is, invariant under the antiholomorphic involution $\tau(\eta, \zeta) = (-\bar{\eta}/\bar{\zeta}^2, -1/\bar{\zeta})$ corresponding to reversal of lines in $T\mathbb{P}^1$.*
   d. *There is a linear form μ on $H^0(S, \mathcal{L}_z \otimes \tau^*(\mathcal{L}_z))$ vanishing on $H^0(S, \mathcal{L}_z \otimes \tau^*(\mathcal{L}_z)(-C))$ for all fibres C of $T\mathbb{P}^1 \to \mathbb{P}^1$ and inducing a positive definite Hermitian metric $(\sigma_1, \sigma_2) \mapsto \mu(\sigma_1\tau^*(\sigma_2))$ on $H^0(S \cap C, \mathcal{L}_z)$.*

We start with the passage from (1) to (2). For the relation $H^0(T\mathbb{P}^1, \mathcal{L}_z(-1)) = H^1(T\mathbb{P}^1, \mathcal{L}_z(-1)) = 0$ to hold, we need $\mathcal{O}(-3)^k \xrightarrow{A(\zeta,z)-\eta\mathbf{I}} \mathcal{O}(-1)^k$ to induce an isomorphism on H^1. As shown in Hurtubise and Murray (1989, lemma 1.2), the groups $H^1(T\mathbb{P}^1, \mathcal{O}(p))$ are infinite-dimensional spaces nicely filtered by finite-dimensional pieces, corresponding to powers of η in the cocycle. A basis for $H^1(T\mathbb{P}^1, \mathcal{O}(-3))$ is $1/\zeta, 1/\zeta^2$, $\eta/\zeta, \ldots, \eta/\zeta^4, \eta^2/\zeta, \ldots, \eta^2/\zeta^6, \ldots$, and a basis for

$H^1(T\mathbb{P}^1, \mathcal{O}(-1))$ is obtained from it by multiplying by η. Multiplication by η then induces an isomorphism, and so multiplication by $A(\zeta, z) - \eta\mathbf{I}$ gives an isomorphism $H^1(T\mathbb{P}^1, \mathcal{O}(-3)^k) \to H^1(T\mathbb{P}^1, \mathcal{O}(-1)^k)$. For the relation $\mathcal{L}_z = \mathcal{L}_{z'} \otimes L^{z-z'}$, see Hurtubise and Murray (1989, section 2.6) or Adams *et al.* (1990) for the general theory of the Lax flows. The reality of the curve follows from the fact that the T_i are skew Hermitian. For the final property, the positive definite inner product on $\mathbb{C}^k$ with respect to which the T_i are skew Hermitian induces one on $H^0(T\mathbb{P}^1, \mathcal{L}_z)$, by passing to global sections in sequence (4.13). The inner product is a linear map $H^0(S, \mathcal{L}_z) \otimes H^0(S, \tau^*(\mathcal{L}_z)) \to \mathbb{C}$. As $\mathcal{L}_z$ represents, in essence, the dual to the eigenspaces of the $A(\zeta)^{\mathrm{T}}$, this linear map factors through $H^0(S, \mathcal{L}_z \otimes \tau^*(\mathcal{L}_z))$, and it must be zero on sections that vanish on fibres.

Now that the passage from (1) to (2) is established, note that since $\tau^* L^{z-z'} = L^{z'-z}$, the product $\mathcal{L}_z \otimes \tau^*(\mathcal{L}_z)$ is constant. The last condition of (2) imposes severe constraints on $\mathcal{L}_z$: when S is smooth, and $\mathcal{L}_z$ a line bundle, it tells us that

$$\mathcal{L}_z \otimes \tau^*(\mathcal{L}_z) \simeq K_S(2C). \tag{4.14}$$

Indeed, in that case the vanishing of the cohomology of $\mathcal{L}_z(-1)$ tells us $\deg(\mathcal{L}_z) = g + k - 1$ and $\deg(\mathcal{L}_z \otimes \tau^*(\mathcal{L}_z)) = 2g + 2k - 2$. The space of sections $H^0(S, \mathcal{L}_z \otimes \tau^*(\mathcal{L}_z))$ is of dimension $g + 2k - 1$, while $H^0(S, \mathcal{L}_z \otimes \tau^*(\mathcal{L}_z)(-C))$ and $H^0(S, \mathcal{L}_z \otimes \tau^*(\mathcal{L}_z)(-C'))$ are of dimension $g + k - 1$, and $\dim H^0(S, \mathcal{L}_z \otimes \tau^*(\mathcal{L}_z)(-C - C')) = g - 1$, unless it is the canonical bundle, in which case the dimension is g. There cannot be a non-zero form on $H^0(S, \mathcal{L}_z \otimes \tau^*(\mathcal{L}_z))$ vanishing on $H^0(S, \mathcal{L}_z \otimes \tau^*(\mathcal{L}_z)(-C)) + H^0(S, \mathcal{L}_z \otimes \tau^*(\mathcal{L}_z)(-C'))$ unless the intersection $H^0(S, \mathcal{L}_z \otimes \tau^*(\mathcal{L}_z)(-C - C'))$ is of dimension g, in which case $\mathcal{L}_z \otimes \tau^*(\mathcal{L}_z)(-C - C')$ is the canonical bundle.

Now we deal with the passage (2) to (1). In essence, it has been treated by Hitchin (1983), but we give a more invariant construction of A suitable for our purposes. The fibre product FP of $T\mathbb{P}^1$ with itself over $\mathbb{P}^1$ is the vector bundle $\mathcal{O}(2) \oplus \mathcal{O}(2)$, with fibre coordinates η, η', and projections π, π' to $T\mathbb{P}^1$; the lift of $\mathcal{O}(2)$ to this vector bundle has global sections $1, \zeta, \zeta^2, \eta, \eta'$, and the diagonal Δ is cut out by $\eta - \eta'$. Hence over FP we have the exact sequence

$$0 \to \mathcal{O}(-2) \xrightarrow{\eta-\eta'} \mathcal{O} \to \mathcal{O}_\Delta \to 0. \tag{4.15}$$

Lifting $\mathcal{L}_z$ via π, tensoring it with this sequence, and pushing it down via π' gives

$$0 \to V(-2) \xrightarrow{A(\zeta)-\eta'} V \to \mathcal{L}_z \to 0 \tag{4.16}$$

for some rank k vector bundle V, which over $(\zeta, \eta) = (\zeta_0, \eta_0)$ is the space of sections of $\mathcal{L}$ in the fibre of $T\mathbb{P}^1$ over $\zeta = \zeta_0$, and so is independent of η: V is lifted from $\mathbb{P}^1$. We then identify V. The fibre product $FP = \mathcal{O}(2) \oplus \mathcal{O}(2)$ lives in the product $T\mathbb{P}^1 \times T\mathbb{P}^1$, and is cut out by $\zeta - \zeta'$, and one has the exact sequence

$$0 \to \mathcal{O}(-1,-1) \xrightarrow{\zeta-\zeta'} \mathcal{O} \to \mathcal{O}_{FP} \to 0. \tag{4.17}$$

Lifting $\mathcal{L}_z$ to $T\mathbb{P}^1 \times T\mathbb{P}^1$, tensoring it with this sequence and taking direct image on the other factor gives the long exact sequence

$$\begin{aligned} 0 &\to H^0(T\mathbb{P}^1, \mathcal{L}_z(-1)) \otimes \mathcal{O}(-1) \to H^0(T\mathbb{P}^1, \mathcal{L}_z) \otimes \mathcal{O} \to V \\ &\to H^1(T\mathbb{P}^1, \mathcal{L}_z(-1)) \otimes \mathcal{O}(-1) \cdots . \end{aligned}$$

The vanishing of the cohomolology of $\mathcal{L}_z(-1)$ shows that $V \simeq \mathcal{O}^k$.

We then have given $\underline{A}(\zeta, z)$ as a map of bundles; the process of turning it into a matrix-valued function $A(\zeta, z)$, and of showing that it evolves according to Nahm's equations as one tensors it by L^t, is given in Hitchin (1983, proposition 4.16). Similarly, the definition of an inner product on $H^0(T\mathbb{P}^1, \mathcal{L}_z)$ and the proof that the T_i are skew adjoint with respect to the inner product follow the line given in Hitchin (1983, section 6).

4.4.2 *Boundary conditions*

Now suppose that we have a solution to Nahm's equations satisfying the boundary conditions for an $SU(2)$ caloron, as given above. We shall show that the boundary behaviour allows us to identify the initial conditions for the flow, in other words to say what the sheaves $\mathcal{L}_z$ are. As there are two intervals, there are two spectral curves: S_0, which is a k-fold cover of $\mathbb{P}^1$, and S_1, which is $k+j$-fold. We suppose the curves are generic, in that they have no common components and no multiple components. A solution to Nahm's equations is called generic if its curves are.

Proposition 4.5 *There is an equivalence between*

1. *Generic solutions $A(\zeta, z)$ to Nahm's equations over the circle, denoted $A^0(\zeta, z)$ on $(\mu_1, -\mu_1+\mu_0)$ and $A^1(\zeta, z)$ on $(-\mu_1, \mu_1)$, satisfying the conditions of Section 4.1.2.2*
2. *Generic spectral curves S_0 in $T\mathbb{P}^1$ of degree k over $\mathbb{P}^1$ that are the support of sheaves $\mathcal{L}^0_z$, $z \in [\mu_1, \mu_0 - \mu_1]$, and spectral curves S_1 in $T\mathbb{P}^1$ of degree $k+j$ over $\mathbb{P}^1$ that are the support of sheaves $\mathcal{L}^1_z$, $z \in [-\mu_1, \mu_1]$, with the following properties:*
   - *$\mathcal{L}^0_z$, for $z \in [\mu_1, \mu_0 - \mu_1]$, $\mathcal{L}^1_z$, for $z \in (-\mu_1, \mu_1)$, S_0, and S_1 have the properties of Proposition 4.4.*
   - *The intersection $S_0 \cap S_1$ decomposes as a sum of two divisors S_{01} and S_{10}, interchanged by τ.*
   - *At $-\mu_1 = \mu_0 - \mu_1$ (on the circle), $\mathcal{L}^0_{-\mu_1} = \mathcal{O}(2k+j-1)[-S_{10}]|_{S_0}$ and $\mathcal{L}^1_{-\mu_1} = \mathcal{O}(2k+j-1)[-S_{10}]|_{S_1}$.*
   - *At μ_1, $\mathcal{L}^0_{\mu_1} = \mathcal{O}(2k+j-1)[-S_{01}]|_{S_0}$ and $\mathcal{L}^1_{\mu_1} = \mathcal{O}(2k+j-1)[-S_{01}]|_{S_1}$.*
   - *There is a real structure on $\mathcal{L}^1_0$ lifting τ.*

As a consequence there are isomorphisms of line bundles $\mathcal{O}[-S_{01}] \otimes L^{\mu_0 - 2\mu_1} \simeq \mathcal{O}[-S_{10}]$ over S_0 and $\mathcal{O}[-S_{10}] \otimes L^{2\mu_1} \simeq \mathcal{O}[-S_{01}]$ over S_1.

Let us begin with a discussion of the case $k = 0$, studied in Hitchin (1983). He showed that the solution to Nahm's equations for an $SU(2)$ monopole corresponded to the flow of line bundles $L^{t+\mu_1}(j-1)$, $t \in [-\mu_1, \mu_1]$ on the monopole's spectral curve, with the flow being regular in the middle of the interval, and having simple poles at the ends of the interval, with residues giving an irreducible representation of $SU(2)$.

In the construction given in Proposition 4.4, we obtain for this flow a bundle V over $\mathbb{P}^1 \times [-\mu_1, \mu_1]$, and a regular section of $\mathrm{Hom}(V(-2), V)$ over this interval. For $t \in (-\mu_1, \mu_1)$ the bundle V is trivial on $\mathbb{P}^1 \times \{t\}$. The singularity at the end of the interval is caused by a jump in the holomorphic structure in V at the ends. (Both ends are identical, because of the identification $\mathcal{O} = L^{2\mu_1}$ over the spectral curve.)

To understand the structure of V at $\mathcal{L}_{-\mu_1} = \mathcal{O}(j-1)$, we note that in the natural trivializations lifted from $\mathbb{P}^1$, local sections of $\mathcal{O}(j-1)$ over $T\mathbb{P}_1$ are filtered by the order of vanishing along the zero section:

$$\mathcal{O}(j-1) \supset \eta\mathcal{O}(j-3) \supset \eta^2\mathcal{O}(j-5) \supset \cdots;$$

this filtration can be turned onto a sum, as we have the subsheaves L_i of sections $\eta^i s$, with s lifted from $\mathbb{P}^1$. The construction, by limiting to a curve which is a j-sheeted cover of $\mathbb{P}^1$, essentially says that the degrees of vanishing of interest are at most $j-1$, as one is taking the remainder by division by the equation of the curve. The bundle V over $\mathbb{P}^1 \times \{\mu_1\}$ decomposes as a sum

$$V = \mathcal{O}(j-1) \oplus \mathcal{O}(j-3) \oplus \mathcal{O}(j-5) \oplus \cdots \oplus \mathcal{O}(-j+1).$$

More generally, for later use, set

$$V_{k,k-2\ell} \stackrel{\mathrm{def}}{=} \mathcal{O}(k) \oplus \mathcal{O}(k-2) \oplus \mathcal{O}(k-4) \oplus \cdots \oplus \mathcal{O}(k-2\ell),$$

so $V = V_{j-1,-j+1}$. Writing $0 = \eta^j + \eta^{j-1}p_1(\zeta) + \cdots + p_j(\zeta)$ for the equation of the spectral curve, with polynomials p_i of degree $2i$, we can write the induced automorphism $A(\zeta) - \eta\mathbf{I}$ in this decomposition as

$$\begin{pmatrix} -\eta & 0 & 0 & \dots & 0 & -p_j(\zeta) \\ 1 & -\eta & 0 & \dots & 0 & -p_{j-1}(\zeta) \\ 0 & 1 & -\eta & \dots & 0 & -p_{j-2}(\zeta) \\ \vdots & \vdots & \vdots & \vdots & \vdots & \vdots \\ 0 & 0 & 0 & \dots & 1 & -\eta - p_1(\zeta) \end{pmatrix}.$$

There is another way of understanding how the pole of Nahm's equations generates a non-trivial V. We first restrict to $\zeta = 0$, and vary z. For a moment, suppose by translation that $z = 0$ is the point where the structure jumps. From the boundary conditions, the residues at 0 are given in a suitable basis by

$$\mathrm{Res}(A_+) = \mathrm{diag}\left(\frac{-(j-1)}{2}, \frac{2-(j-1)}{2}, \ldots, \frac{(j-1)}{2}\right), \tag{4.18}$$

$$\mathrm{Res}(A) = \begin{pmatrix} 0 & 0 & \ddots & 0 & 0 \\ 1 & 0 & \ddots & 0 & 0 \\ 0 & 1 & \ddots & 0 & 0 \\ \vdots & \ddots & \ddots & \ddots & \vdots \\ 0 & 0 & 0 & 1 & 0 \end{pmatrix}. \tag{4.19}$$

As in Hurtubise (1989, proposition 1.15), we solve $\frac{ds}{dz} + A_+ s = 0$ along $\zeta = 0$, for s of the form $z^{\frac{j-1}{2}}((1,0,\ldots,0) + z \cdot \text{holomorphic})$. The sections $A^j s$ also solve the equation, and, using these sections as a basis, one conjugates $A - \eta\mathbf{I}$ to

$$\begin{pmatrix} -\eta & 0 & 0 & \ldots & 0 & -p_j(0) \\ 1 & -\eta & 0 & \ldots & 0 & -p_{j-1}(0) \\ 0 & 1 & -\eta & \ldots & 0 & -p_{j-2}(0) \\ \vdots & \vdots & \ddots & & \ddots & \vdots \\ 0 & 0 & 0 & \ldots & 1 & -\eta - p_1(0) \end{pmatrix} \tag{4.20}$$

using a matrix of the form $M(z) = (\text{Holomorphic in } z)N(z)$, with

$$N(z) := \mathrm{diag}\left(z^{\frac{-(j-1)}{2}}, z^{\frac{2-(j-1)}{2}}, \ldots, z^{\frac{(j-1)}{2}}\right). \tag{4.21}$$

This process can be applied over any point ζ, as the matrices $A(\zeta)$, $A_+(\zeta)$ have residues (in z) conjugate to (4.18) and (4.19). Indeed, at $\zeta = 0$, they represent standard generators of a representation of $Sl(2)$, and moving away from $\zeta = 0$ simply amounts to a change of basis. Explicitly, conjugating the residues of $A(\zeta)$, $A_+(\zeta)$ by $N(\zeta)$ takes them to the residues of $\zeta A(1)$, $A_+(1)$. Conjugating again by a matrix T takes them to the residues of $\zeta A(0), A_+(0)$, and then by $N(\zeta)^{-1}$ to the residues of $A(0), A_+(0)$. Thus, for a suitable

$$M(\zeta, z) = (\text{Holomorphic in } z, \zeta) \cdot N(\zeta^{-1}z)TN(\zeta) \tag{4.22}$$

one has

$$M(\zeta,z)(A(\zeta,z) - \eta\mathbf{I})M(\zeta,z)^{-1} = \begin{pmatrix} -\eta & 0 & 0 & \ldots & 0 & -p_j(\zeta) \\ 1 & -\eta & 0 & \ldots & 0 & -p_{j-1}(\zeta) \\ 0 & 1 & -\eta & \ldots & 0 & -p_{j-2}(\zeta) \\ \vdots & \vdots & \vdots & \vdots & \vdots & \vdots \\ 0 & 0 & 0 & \ldots & 1 & -\eta - p_1(\zeta) \end{pmatrix}. \tag{4.23}$$

At $z = 0$, it is a holomorphic section of $\mathrm{Hom}(V(-2), V)$ in standard trivializations.

The same procedures work when $k \neq 0$. Indeed, the singular solution to $\frac{ds}{dz} + A_+ s$ has the same behaviour. Starting from a solution to Nahm's equations, and integrating the connection as in Hurtubise (1989), we find that, near μ_1 on the interval $(-\mu_1, \mu_1)$, a change of basis of the form

$$M(\zeta, z) = (\text{Holomorphic in } z, \zeta) \cdot \operatorname{diag}(\mathbf{I}_{k\times k}, N(\zeta^{-1}z)TN(\zeta)) \tag{4.24}$$

conjugates $A^1(\zeta, z) - \eta\mathbf{I}$ to the constant (in z) matrix

$$\begin{pmatrix} a_{11}(\zeta)-\eta & a_{12}(\zeta) & \dots & a_{1k}(\zeta) & 0 & 0 & 0 & \dots 0 & g_1(\zeta) \\ a_{21}(\zeta) & a_{22}(\zeta)-\eta & \dots & a_{1k}(\zeta) & 0 & 0 & 0 & \dots 0 & g_2(\zeta) \\ \vdots & \vdots & \vdots & \vdots & \vdots & \vdots & \vdots & \vdots\ \vdots & \vdots \\ a_{k1}(\zeta) & a_{k2}(\zeta) & \dots & a_{kk}(\zeta)-\eta & 0 & 0 & 0 & \dots 0 & g_k(\zeta) \\ f_1(\zeta) & f_2(\zeta) & \dots & f_k(\zeta) & -\eta & 0 & 0 & \dots 0 & -p_j(\zeta) \\ 0 & 0 & \dots & 0 & 1 & -\eta & 0 & \dots 0 & -p_{j-1}(\zeta) \\ 0 & 0 & \dots & 0 & 0 & 1 & -\eta & \dots 0 & -p_{j-2}(\zeta) \\ \vdots & \vdots & \vdots & \vdots & \vdots & \vdots & \vdots & \vdots\ \vdots & \vdots \\ 0 & 0 & \dots & 0 & 0 & 0 & 0 & \dots 1 & -\eta - p_1(\zeta) \end{pmatrix}. \tag{4.25}$$

Setting $\mathcal{C}(p(\zeta))$ to denote the companion matrix of $p(\zeta)$, we write this matrix schematically as

$$A^1(\zeta) - \eta\mathbf{I} = \begin{pmatrix} A^0(\zeta) - \eta\mathbf{I} & \begin{pmatrix} 0 & G(\zeta) \end{pmatrix} \\ \begin{pmatrix} F(\zeta) \\ 0 \end{pmatrix} & \mathcal{C}(p(\zeta)) - \eta\mathbf{I} \end{pmatrix}. \tag{4.26}$$

Let M_{adj} denote the classical adjoint of M, so that $M_{\mathrm{adj}}M = \det(M)\mathbf{I}$. Then

$$\begin{aligned} \det(A^1(\zeta) - \eta\mathbf{I}) &= \det(A^0(\zeta) - \eta\mathbf{I})\Big(\eta^j + \sum \eta^{j-i} p_i(\zeta)\Big) \\ &\quad + (-1)^j F\big(A^0(\zeta) - \eta\mathbf{I}\big)_{\mathrm{adj}} G. \end{aligned} \tag{4.27}$$

The matrix $A^0(\zeta)$ is equal to the limit $A^0(\zeta, \mu_1)$. At the boundary point μ_1, the bundle $V^0_{\mu_1}$ for S_0 is trivial, since the solution on S_0 is smooth at that point; one has

$$0 \to \mathcal{O}(-2)^k \xrightarrow{A^0(\zeta)-\eta\mathbf{I}} \mathcal{O}^k \to \mathcal{L}^0_{\mu_1} \to 0. \tag{4.28}$$

The limit bundle $V^1_{\mu_1}$ for S_1 is

$$V^1_{\mu_1} = \mathcal{O}^k \oplus V_{j-1,-j+1} \tag{4.29}$$

with

$$0 \to V^1_{\mu_1}(-2) \xrightarrow{A^1(\zeta)-\eta\mathbf{I}} V^1_{\mu_1} \to \mathcal{L}^1_{\mu_1} \to 0. \tag{4.30}$$

We now want to identify the limit bundles $\mathcal{L}^0_{\mu_1}$, $\mathcal{L}^1_{\mu_1}$, and the gluing between them. We want to show that there is a divisor D contained in the intersection $S_0 \cap$

S_1 such that $\mathcal{L}^0_{\mu_1} \simeq \mathcal{O}(2k+j-1)[-D]|_{S_0}$ and $\mathcal{L}^1_{\mu_1} \simeq \mathcal{O}(2k+j-1)[-D]|_{S_1}$, and that the correspondence between the matrices is mediated by the maps

$$\mathcal{O}(2k+j-1)[-D]|_{S_0} \leftarrow \mathcal{O}(2k+j-1) \otimes \mathcal{I}_D \rightarrow \mathcal{O}(2k+j-1)[-D]|_{S_1}.$$

To prove its existence, we need to understand how to lift and push down $\mathcal{O}(2k+j-1) \otimes \mathcal{I}_D$ (here $\mathcal{I}_D$ is the sheaf of ideals of D on $T\mathbb{P}^1$) and so $\mathcal{O}(2k+j-1)$ through the sequence of (4.15). To do so, we compactify $T\mathbb{P}^1$ by embedding it into $\mathbf{T} = \mathbb{P}(\mathcal{O}(2) \oplus \mathcal{O})$, adding a divisor at infinity P_∞. Let C be the fibre of the projection $\mathbf{T} \rightarrow \mathbb{P}^1$, then $P_\infty + 2C$ is linearly equivalent to the zero section P_0 of $T\mathbb{P}^1$. Our fibre product now compactifies to a $\mathbb{P}^1 \times \mathbb{P}^1$ bundle over $\mathbb{P}^1$. Similarly, the sequence of (4.15) compactifies to

$$0 \rightarrow \mathcal{O}\left(-2C - P_\infty - P'_\infty\right) \xrightarrow{\eta-\eta'} \mathcal{O} \rightarrow \mathcal{O}_\Delta \rightarrow 0. \tag{4.31}$$

The line bundle $\mathcal{O}(mC + nP_\infty)$ over $\mathbf{T}$ has over each fibre C a $(n+1)$-dimensional space of sections, and these sections are graded by the order in η, as before. Define an $\ell \times (\ell-1)$ matrix $S(\ell,\eta)$ by

$$S(\ell,\eta) = \begin{pmatrix} -\eta & 0 & \cdots & \cdots & 0 \\ 1 & -\eta & \ddots & & \vdots \\ 0 & 1 & \ddots & \ddots & \vdots \\ \vdots & \ddots & \ddots & -\eta & 0 \\ \vdots & & \ddots & 1 & -\eta \\ 0 & \cdots & \cdots & 0 & 1 \end{pmatrix}. \tag{4.32}$$

Lemma 4.6 *Let $m, n > 0$. Lifting $\mathcal{O}(mC + nP_\infty)$ to the fibre product and tensoring with the sequence (4.28), then pushing down, we obtain an exact sequence*

$$0 \longrightarrow \bigoplus_{i=0}^{n-1} \mathcal{O}\big((m-2i-2)\,C + (n-i-1)P_\infty\big)$$

$$\downarrow S(n,\eta)$$

$$\bigoplus_{i=0}^{n} \mathcal{O}\big((m-2i)C + (n-i)P_\infty\big) \xrightarrow{(1,\eta,\eta^2,\ldots,\eta^n)} \mathcal{O}(mC + nP_\infty) \rightarrow 0.$$

Over $T\mathbb{P}^1$, it becomes

$$0 \rightarrow V_{m-2,m-2n} \xrightarrow{S(n,\eta)} V_{m,m-2n} \xrightarrow{(1,\eta,\eta^2,\ldots,\eta^n)} \mathcal{O}(m) \rightarrow 0.$$

The proof is straightforward. We now want to define a subsheaf of $\mathcal{O}(2k+j)$. Set

$$R(\zeta,\eta) = \begin{pmatrix} A^0(\zeta) - \eta\mathbf{I} & 0 \\ \begin{pmatrix} F \\ 0 \end{pmatrix} & S(j,\eta) \end{pmatrix}.$$

We define a vector of polynomial functions in (ζ, η)

$$(\phi_1, \ldots, \phi_k) = -(-1)^k F(\zeta)(A^0(\zeta) - \eta\mathbf{I})_{\text{adj}}.$$

We have

$$(\phi_1, \ldots, \phi_k)\big(A^0(\zeta) - \eta\mathbf{I}\big) + F(\zeta)(-1)^k \det\big(A^0(\zeta) - \eta\mathbf{I}\big) = 0. \tag{4.33}$$

Write $\phi_i = \sum_{j=0}^{k-1} \phi_{ji}(\zeta)\eta^j$, and $(-1)^k \det(A^0(\zeta) - \eta\mathbf{I}) = \eta^k + \sum_{j=0}^{k-1} h_j(\zeta)\eta^j$. Decomposing (4.33) into different powers of η, we obtain

$$-\begin{pmatrix} 0 & \cdots & 0 \\ \phi_{01} & \cdots & \phi_{0k} \\ \vdots & \vdots & \vdots \\ \phi_{k-1,1} & \cdots & \phi_{k-1,k} \end{pmatrix} + \begin{pmatrix} \phi_{01} & \cdots & \phi_{0k} \\ \vdots & \vdots & \vdots \\ \phi_{k-1,1} & \cdots & \phi_{k-1,k} \\ 0 & \cdots & 0 \end{pmatrix} A^0(\zeta) + \begin{pmatrix} h_0 \\ \vdots \\ h_{k-1} \\ 1 \end{pmatrix} F = 0. \tag{4.34}$$

Let M_n be the $(k+n) \times (k+n)$ matrix:

$$M_n = \begin{pmatrix} \phi_{01} & \cdots & \phi_{0k} & h_0 & 0 & 0 \\ \vdots & \vdots & \vdots & \vdots & \ddots & 0 \\ \phi_{k-1,1} & \cdots & \phi_{k-1,k} & h_{k-1} & & h_0 \\ 0 & \cdots & 0 & 1 & \ddots & \vdots \\ \vdots & \vdots & \vdots & 0 & \ddots & h_{k-1} \\ 0 & \cdots & 0 & 0 & 0 & 1 \end{pmatrix}. \tag{4.35}$$

Using (4.34), we have the commuting diagram

$$\begin{array}{ccccc} \mathcal{O}(-2)^{\oplus k} \oplus V_{j-3,-j+1} & \xrightarrow{R(\zeta,\eta)} & \mathcal{O}^{\oplus k} \oplus V_{j-1,-j+1} & \longrightarrow & \mathcal{R} \\ \downarrow{\scriptstyle M_{j-1}} & & \downarrow{\scriptstyle M_j} & & \downarrow \\ V_{2k+j-3,-j+1} & \xrightarrow{S(k+j,\eta)} & V_{2k+j-1,-j+1} & \xrightarrow{(1,\eta,\ldots,\eta^{k+j-1})} & \mathcal{O}(2k+j-1). \end{array}$$

It defines a rank 1 sheaf $\mathcal{R}$, and embeds it in $\mathcal{O}(2k+j-1)$. This embedding fails to be surjective when $(1, \eta, \ldots, \eta^{k+j-1})M_j = 0$, or equivalently when

$$F(A^0(\zeta) - \eta\mathbf{I})_{\text{adj}} = 0 \text{ and } \det\big(A^0(\zeta) - \eta\mathbf{I}\big) = 0.$$

These conditions can only be met on S_0, and, because of (4.27), on S_1. In short, there is a subvariety D of $S_0 \cap S_1$, with $\mathcal{R} = \mathcal{O}(2k+j-1) \otimes \mathcal{I}_D$. There are natural surjective maps from $\mathcal{R}$ to $\mathcal{L}^0_{\mu_1}$, $\mathcal{L}^1_{\mu_1}$: for the projection $\Pi_{m+n,n}$ on a sum with $m+n$ summands onto the first n summands and the injection $I_{n,n+m}$ into the first n summands, we have that the diagram of exact sequences

$$\begin{array}{ccccc}
\mathcal{O}(-2)^{\oplus k} & \xrightarrow{A^0(\zeta)-\eta\mathbf{I}} & \mathcal{O}^{\oplus k} & \longrightarrow & \mathcal{L}^0_{\mu_1} \\
\Big\uparrow{\scriptstyle \Pi_{k+j-1,k}} & & \Big\uparrow{\scriptstyle \Pi_{k+j,k}} & & \Big\uparrow \\
\mathcal{O}(-2)^{\oplus k} \oplus V_{j-3,-j+1} & \xrightarrow{R(\zeta,\eta)} & \mathcal{O}^{\oplus k} \oplus V_{j-1,-j+1} & \longrightarrow & \mathcal{R} \\
\Big\downarrow{\scriptstyle I_{k+j-1,k+j}} & & \Big\downarrow{\scriptstyle I} & & \Big\downarrow \\
\mathcal{O}(-2)^{\oplus k} \oplus V_{j-3,-j-1} & \xrightarrow{A^1(\zeta)-\eta\mathbf{I}} & \mathcal{O}^{\oplus k} \oplus V_{j-1,-j+1} & \longrightarrow & \mathcal{L}^1_{\mu_1}
\end{array}$$

commutes. This diagram identifies $\mathcal{L}^0_{\mu_1}$, $\mathcal{L}^1_{\mu_1}$ as $\mathcal{O}(2k+j-1)[-D]$ over their respective curves.

A similar analysis for $-\mu_1$ shows that $\mathcal{L}^0_{-\mu_1}$, $\mathcal{L}^1_{-\mu_1}$ are $\mathcal{O}(2k+j-1)[-D']$, for some D'. To relate D to D', note that as in Hitchin (1983, section 6), there is a real structure on $\mathcal{L}_0$ lifting the real involution τ. Hence $T_i(0) = T_i(0)^{\mathrm{T}}$. On the other hand, for a solution $A(z)$ to Nahm's equation (4.11),

$$\frac{dA(-z)^{\mathrm{T}}}{dz} + [A_+(-z)^{\mathrm{T}}, A(-z)^{\mathrm{T}}] = 0. \tag{4.36}$$

As solutions to the same differential equation with the same initial condition, all the way around the circle,

$$A(-z) = A^{\mathrm{T}}(z). \tag{4.37}$$

This symmetry tells us that in terms of the matrix $A^1(\zeta), A^0(\zeta)$ at μ_1, the equation for D' is

$$0 = (A^0(\zeta) - \eta\mathbf{I})_{\mathrm{adj}} G. \tag{4.38}$$

Using (4.27), we see that along S_0 where $\det\big(A^0(\zeta) - \eta\mathbf{I}\big) = 0$, the intersection $S_0 \cap S_1$ is cut out by $0 = F\big(A^0(\zeta) - \eta\mathbf{I}\big)_{\mathrm{adj}} G$. Generically along S_0, the matrix $(A^0(\zeta) - \eta\mathbf{I})$ has corank 1, and so $(A^0(\zeta) - \eta\mathbf{I})_{\mathrm{adj}}$ has rank 1. We can thus write it as the product UV of a column vector and a row vector. The equation for the intersection is then $0 = (FU)(VG)$, the product of the defining relations for D and D'. If the two curves are smooth, intersecting transversally, then $S_0 \cap S_1 = D + D'$. This property holds independently of whether the matrices arise from calorons or not. As one has the result for generic intersections, varying continuously gives $S_0 \cap S_1 = D + D'$ even for non-generic curves.

The skew Hermitian property of the T_i implies that $A(\zeta, z) = -\zeta^2 \bar{A}(-1/\bar{\zeta}, z)^{\mathrm{T}}$. This symmetry transforms the equation $0 = F(A^0(-1/\bar{\zeta}) - \eta\mathbf{I})_{\mathrm{adj}}$ into $0 = (A^0(\zeta) - \eta\mathbf{I})_{\mathrm{adj}} G$, showing that $\tau(D) = D'$.

We have now obtained our complete generic spectral data from our generic solution to Nahm's equations. The converse, starting with the spectral data, is essentially done above, apart from the boundary behaviour. This last piece is dealt with in Hurtubise and Murray (1989, section 2); the conditions on the intersections of the spectral curves used here are more general, but the proof goes through unchanged.

4.5 Closing the circle

We have two different types of gauge fields, with a transform relating them; both have complex data associated to them, which encodes them; this data, we have seen (at least in the generic case), satisfies the same conditions, with, for example, the sheaves $\left(F/(F^0_{p,0} + F^\infty_{-(p-1),0})\right) \otimes L^z(-1)$ and $\left(F/(F^0_{p,1} + F^\infty_{-p,1})\right) \otimes L^z(-1)$ obtained from the caloron satisfying the same conditions as the $\mathcal{L}^t$ obtained from the solution to Nahm's equation for appropriate values of z and t. We now want to check that the complex data associated to the object is the same as that associated to the object's transform.

4.5.1 *Starting with a caloron*

Starting with a caloron, one can define, as above, the spectral curves and sheaves over them:

- The curve S_0 is the locus where $F^0_{p,0}$ and $F^\infty_{-p+1,0}$ have non-zero intersection.
- The curve S_1 is the locus where $F^0_{p,1}$ and $F^\infty_{-p,1}$ have non-zero intersection.
- The quotient $F/(F^0_{p,0} + F^\infty_{-(p-1),0})$, supported over S_0, is generically isomorphic to the line bundle $L^{p\mu_0-\mu_1}(2k+j)[-S_{10}]|_{S_0} = L^{(p-1)\mu_0+\mu_1}(2k+j)[-S_{01}]|_{S_0}$.
- The quotient $F/(F^0_{p,1} + F^\infty_{-p,1})$, supported over S_1, is generically isomorphic to the line bundle $L^{p\mu_0+\mu_1}(2k+j)[-S_{01}]|_{S_1} = L^{p\mu_0-\mu_1}(2k+j)[-S_{10}]|_{S_1}$.

We can 'shift' the caloron by the $U(1)$ monopole with constant Higgs field iz. Let us consider the direction in $\mathbb{R}^3$, corresponding to $\zeta = 0$: the positive x_3 direction. If one does this, for $z \in ((p-1)\mu_0 + \mu_1, p\mu_0 - \mu_1)$, $F^0_{p,0}$ represents the sections in the kernel of $\nabla_3 - i\nabla_0 + z$ along each cylinder that decay at infinity in the positive direction; similarly, $F^\infty_{-(p-1),0}$ represents the sections along each cylinder that decay at infinity in the negative direction. For $z \in (p\mu_0 - \mu_1, p\mu_0 + \mu_1)$, $F^0_{p,1}$ represents the sections along each cylinder that decay at infinity in the positive direction; similarly, $F^\infty_{-p,1}$ represents the sections along each cylinder that decay at infinity in the negative direction. With these shifts, the spectral curves represent the lines for which the solutions that decay at one of the ends do not sum to the whole bundle. The quotient $F/(F^0_{p,0} + F^\infty_{-(p-1),0})$ represents this failure and therefore the existence of a solution decaying at both ends. Suppose that $z \in (p\mu_0 - \mu_1, p\mu_0 + \mu_1)$. One has the same for $F/(F^0_{p,1} + F^\infty_{-p,1})$.

Following the Nahm transform heuristic, starting with the caloron, we solve the Dirac equation for the family of connections shifted by z, and obtain in doing so a bundle over an interval and a solution to Nahm's equations on this bundle. The spectral curve, from this point of view, over $\zeta = 0$, consist of the eigenvalues of $A(0,z) = T_1 + iT_2(z)$; there is a sheaf over the curve whose fibre over the point η in the spectral curve given by the cokernel of $A(0,z) - \eta\mathbf{I}$.

A priori, it is not evident what link there is between eigenvalues on a space of solutions to an equation defined over all of space, and the behaviour of solutions to an equation along a single line. The link is provided by a remarkable formulation of the Dirac equation (see for instance Donaldson and Kronheimer 1990 or Cherkis and Kapustin 2001. The idea, roughly, is to write the Dirac equation as the equations for the harmonic elements of the complex

$$L^2(V) \xrightarrow{D_1 = \begin{pmatrix} \nabla_3 + i\nabla_0 + z \\ -\nabla_1 - i\nabla_2 \end{pmatrix}} L^2(V)^{\oplus 2} \xrightarrow{D_2 = \begin{pmatrix} \nabla_1 + i\nabla_2 & \nabla_3 + i\nabla_0 + z \end{pmatrix}} L^2(V). \tag{4.39}$$

Proposition 4.7 *The operators D_1, D_2 commute with multiplication by $w = x_1 + ix_2$, and for self-dual connections, $D_2 D_1 = 0$. The kernel $K_z = \ker(D_2) \cap \ker(D_1^*)$ of the Dirac operator D_z^* is naturally isomorphic to the cohomology $\ker(D_2)/\mathrm{Im}(D_1)$ of the complex.*

On a compact manifold, this equivalence is part of standard elliptic theory. In the non-compact case, the analysis must be done with care. The simplest way, for us, is to adapt Nye and Singer (2000, section 4). Using the techniques developed by Mazzeo and Melrose (1998), they show that the Dirac operator D_z^* is Fredholm if and only if an 'operator on fibres along the boundary' $P = (\nabla_0 - z)_\infty + i(\eta_1 e_1 + \eta_2 e_2 + \eta_3 e_3)$ is invertible along all the circles S^1 in $S^1 \times S^2_\infty = \partial(S^1 \times \mathbb{R}^3)$. Again, the e_1 are the Pauli matrices, and the constraint must hold for all choices of constants (η_1, η_2, η_3). Nye and Singer show that it does as long as z does not have values $\pm\mu_1 + n\mu_0$. In the elliptic complex, the constraint gets replaced by the essentially equivalent condition that the complex

$$L^2(V|_{S^1}) \xrightarrow{\begin{pmatrix} \eta_3 + i(\nabla_0)_\infty + z \\ -\eta_1 - i\eta_2 \end{pmatrix}} L^2(V|_{S^1})^{\oplus 2} \xrightarrow{\begin{pmatrix} \eta_1 + i\eta_2, \eta_3 + i(\nabla_0)_\infty + z \end{pmatrix}} L^2(V|_{S^1})$$

defined over the circle be exact. The equivalence between the Euler characteristic of the complex and the index of the Dirac operator then goes through establishing the isomorphism.

We now turn to analysing the way a solution to Nahm's equations is extracted from this complex. The operator $A(0,z)$ on the kernel of D_z^x is defined by multiplication by $x_1 + ix_2$ followed by the projection on that kernel. On the cohomology, it is simply multiplication by $w = x_1 + ix_2$, simplifying matters considerably.

Suppose η is an eigenvalue of $A(0,z)$; hence there is a representative v of the cohomology class such that $(x_1 + ix_2 - \eta)(v) = D_1(s)$ for some s in L^2. In particular, along $w = \eta$ in $S^1 \times \mathbb{R}^3$, we have $(\nabla_3 + i\nabla_0 - z)s = 0$, and so $(\eta, \zeta) = (w, 0)$ belongs to the spectral curve. Conversely, if $(\nabla_3 + i\nabla_0 - z)s = 0$ along

$w = \eta$, we can build a cohomology class v by extending s to a neighbourhood so that it is compactly supported in the x_1, x_2 directions and satisfies $(\nabla_1 + i\nabla_2)s = 0$ along $w = \eta$; one then sets $v = \frac{D_1 s}{w-\eta}$. The spectral curves are thus identified.

We can further identify sections of the quotients $F/(F^0_{p,1} + F^\infty_{-p,1})$ (for the caloron twisted by z) over $S_1 \cap \{\zeta = 0\}$ with sections of the cokernel of $A^1(0,z) - \eta\mathbf{I}$ as both are supported over $S_1 \cap \{\zeta = 0\}$. Indeed, let $\sigma(w)$ and $\rho(x_3)$ be smooth functions such that

$$\sigma(w) = \begin{cases} 1, & |w| < \epsilon; \\ 0, & |w| > \epsilon; \end{cases} \qquad \text{and} \qquad \rho(x_3) = \begin{cases} 0, & x_3 < 0; \\ 1, & x_3 > 1. \end{cases}$$

Notice that the derivative $(\nabla_1 + i\nabla_2)(\sigma(w))$ is supported on the annulus $\epsilon < |w| < 2\epsilon$.

Consider a ball B centred at η in the fibre above $\zeta = 0$ of radius 2ϵ chosen such that $B \cap S_1 = \{\eta\}$. The ball parameterizes cylinders $S^1 \times \{w\} \times \mathbb{R}$, for $w \in B$, and thus a section $\phi \in H^0(B, F)$ is in fact a section of V on $S^1 \times B \times \mathbb{R}$ satisfying $(\nabla_3 + i\nabla_0 + z)\phi = 0$. The holomorphicity condition is then $(\nabla_1 + i\nabla_2)\phi = 0$.

Let $\phi_0 \in H^0(B, F^0_{p,1}|_{(0,\eta)})$. Then $\sigma(w-\eta)\rho(x_3)\phi_0$ lies in $L^2(S^1 \times \mathbb{R}^3, V)$. Thus the section $D_1(\sigma(w-\eta)\rho(x_3)\phi_0)$ is a coboundary for the complex (4.39). Similarly, if ϕ_∞ lies in $F^\infty_{-p,1}$, then $\sigma(w-\eta)(\rho(x_3)-1)\phi_\infty$ also lies in L^2, and its image by D_1 is also a coboundary.

Suppose for simplicity $S_1 \cap \{\zeta = 0\}$ contains only points of multiplicity 1. Consider now a general section $\phi \in H^0(B, F)$. Away from the spectral curve, ϕ decomposes into a sum $\phi_0 + \phi_\infty$, and combining the two constructions above gives a coboundary $K(\phi)$ for the complex (4.39). This decomposition has a pole at η if $\phi(\eta)$ represents a non-trivial element in the quotient $F/(F^0_{p,1} + F^\infty_{-p,1})$. However, the section

$$\begin{aligned} K(\phi) &= D_1\big(\sigma_\epsilon(w-\eta)\rho(x_3)\phi - \rho_\infty\big) \\ &= \begin{pmatrix} \sigma_\epsilon(w-\eta)\big(\nabla_3 + i\nabla_0 + z\big)\big(\rho(x_3)\phi - \rho_\infty\big) \\ -\big(\nabla_1 + i\nabla_2\big)\big(\sigma_\epsilon(w-\eta)\rho(x_3)\phi - \rho_\infty\big) \end{pmatrix} \end{aligned}$$

is L^2. Because of the pole of $\phi_0 + \phi_\infty$, it is not in $D_1(L^2(V))$ but by construction it is definitely in the kernel of D_2. It thus represents a non-trivial cohomology class for the complex (4.39).

Doing this for each point of $S_1 \cap \{\zeta = 0\}$ expresses K_z as a sum of classes localized along the lines in $S^1 \times \mathbb{R}^3$ corresponding to the intersection of the spectral curve with $\zeta = 0$, identifying K_z with sections of $F/(F^0_{p,1} + F^\infty_{-p,1})$ over $\zeta = 0$. Let $K_z(\phi)$ correspond to a non-zero element ϕ of $F/(F^0_{p,1} + F^\infty_{-p,1})$ over $(\eta, 0)$.

This class projects non-trivially to the cokernel of $A(0,z) - \eta\mathbf{I}$. Indeed, the action of $A(0,z)$ on K_z is multiplication by $w = x_1 + ix_2$, so if $K_z(\phi) = (A(0,z) - \eta\mathbf{I})K_z(\psi) = (w-\eta)K_z(\psi)$ for some ψ with $\phi = (w-\eta)\psi$, then the decomposition $\phi = \phi_0 + \phi_1$ is holomorphic and so $K_z(\phi) = 0$.

For intersections of higher multiplicity, the case presents only notational difficulties, but is conceptually identical. The choice of ζ made above just simplified the analysis. Varying it we obtain the following result:

Proposition 4.8 *For $z \in (p\mu_0 - \mu_1, p\mu_0 + \mu_1)$, the curves S_1 associated to the caloron and S_1' associated to its Nahm transform are identical; there is a natural isomorphism between the sheaf $F/(F^0_{p,1} + F^\infty_{-p,1})$ for the caloron shifted by z and the cokernel sheaf of $A^1(\zeta, z) - \eta$.*

Similarly, for $z \in ((p-1)\mu_0 + \mu_1, p\mu_0 - \mu_1)$, the curves S_0 associated to the caloron and S_0' associated to its Nahm transform are identical; there is a natural isomorphism between the sheaf $F/(F^0_{p,0} + F^\infty_{-(p-1),0})$ for the caloron shifted by z and the cokernel sheaf of $A^0(\zeta, z) - \eta$.

We remark, in particular, for generic solutions, that flowing to a boundary point gives the divisor D in the intersection of the two curves, and so an identification of the spectral data.

4.5.2 *Starting with a solution to Nahm's equation*

We now want to identify the spectral data for a solution to Nahm's equations and the caloron it produces via the Nahm transform.

Case 1: $j > 0$. Again, as for the opposite direction, it is more convenient to give a cohomological interpretation of the Dirac equation. This interpretation is governed by choosing a particular direction in $\mathbb{R}^3$, and so a particular equivalence of $S^1 \times \mathbb{R}^3$ with $\mathbb{C}^* \times \mathbb{C}$. We choose the direction corresponding to $\zeta = 0$, so set $\beta = T_1 + iT_2$, $w = x_1 + ix_2$, $\alpha = T_0 - iT_3$, $y = \mu_0 x_0 - ix_3$, and consider the complex

$$\tilde{L}^2_2 \xrightarrow{D_1 = \begin{pmatrix} i(d_z + \alpha - iy) \\ -\beta + iw \end{pmatrix}} (\tilde{L}^2_1)^{\oplus 2} \xrightarrow{D_2 = \left(\beta - iw\, i(d_z + \alpha - iy)\right)} \tilde{L}^2. \tag{4.40}$$

For this complex to be defined, we must display a bit of care in our choice of function spaces:

- $\tilde{L}^2_2$ is the space of sections s of K that are L^2_2 over each interval and such that, (subscripts denoting the limits on the appropriates sides of the jump points) $\pi(s_{\text{large}}) = s_{\text{small}}$, $\pi(d_z s_{\text{large}}) = d_z s_{\text{small}}$ when $j = 1$ and $s_{\text{large}} = \iota(s_{\text{small}})$, $d_z s_{\text{large}} = \iota(d_z s_{\text{small}})$ when $j > 1$.
- $(\tilde{L}^2_1)^{\oplus 2}$ is the space of sections (s_1, s_2) of $K \oplus K$ that are in L^2_1 over each interval, with $\pi(s_{\text{large}}) = s_{\text{small}}$ at the jump points when $j = 1$ and $s_{\text{large}} = \iota(s_{\text{small}})$ when $j > 1$.
- $\tilde{L}^2$ is the space of L^2 sections of K.

Proposition 4.9 *For solutions to Nahm's equations, $D_2 D_1 = 0$, and*

$$D_1^* D_1 = -\big(d_z + (T_0 - i\mu_0 x_0)\big)^2 - \sum_{j=1}^{3} (T_j - ix_j)^2. \tag{4.41}$$

So for $v \neq 0$ satisfying $D_1^* D_1(v) = 0$, there is a point-wise relation

$$d_z^2 \|v\|^2 = 2\|\big(d_z + (T_0 - i\mu_0 x_0)\big)v\|^2 + 2\sum_{j=1}^{3} \|(T_j - ix_j)v\|^2 \geqslant 0. \tag{4.42}$$

Hence if v is a regular solution vanishing at two points, $v = 0$.

We look at solutions to the equations $D_1^* D_1(v) = 0$ near a boundary point $\pm\mu_1$; one has on the interval $(-\mu_1, \mu_1)$ near $\pm\mu_1$, a decomposition $\mathbb{C}^{k+j} = \mathbb{C}^k \oplus \mathbb{C}^j$: the $2k$-dimensional space of solutions with boundary values (and derivatives) in $\mathbb{C}^k$ continue outside the interval, and solutions with boundary values in $\mathbb{C}^j$ are governed by the theory of regular singular points for o.d.e.s. As the sum of the squares of the residues at the boundary points of the T_i is half of the Casimir element in the enveloping algebra of $su(2)$, its value is $-(j-1)^2/4$. The theory of regular singular o.d.e. then gives, for ρ one of $(1 \pm j)/2$, j-dimensional spaces of solutions of the form $\hat{z}^\rho f(\hat{z})$ to $D_1^* D_1(v) = 0$, where $f(\hat{z})$ is analytic, at each of the boundary points. Here $\hat{z}$ is a coordinate whose origin is at the boundary point.

We can use this knowledge to build a Green's function. This operation is somewhat complicated by the poles of the T_i, as we now see in a series of lemmas.

Lemma 4.10 *Let $x \in (-\mu_1, \mu_1)$, and $u \in \mathbb{C}^{k+j}$. There is a unique solution to $D_1^* D_1(v) = \delta_x u$ on the circle.*

Proof. Such a solution must be continuous, smooth on the circle but for a single jump in the derivative at x, of value u. Let U be the $2k$-dimensional space of solutions on the small side, outside $(-\mu_1, \mu_1)$. Those solutions propagate into the interval from both ends, and for $f \in U$, let f_- be the continuation into the interval $(-\mu_1, \mu_1)$ from the $-\mu_1$ side, and f_+ from the μ_1 side. Let V_+ and V_- be the j-dimensional spaces of solutions of the form $\hat{z}^{(1+j)/2} f(\hat{z})$ born at μ_1 and $-\mu_1$, respectively. Consider the map

$$R_x \colon U \oplus V_+ \oplus V_- \to \mathbb{C}^{k+j} \oplus \mathbb{C}^{k+j}$$

$$(f, g_+, g_-) \mapsto \begin{bmatrix} f_-(x) + g_-(x) - f_+(x) - g_+(x) \\ f'_-(x) + g'_-(x) - f'_+(x) - g'_+(x) \end{bmatrix}.$$

This map must be injective (thus bijective) because of the convexity property of solutions given by (4.42). If $R_x^{-1}(0, u) = (f, g_+, g_-)$, the desired Green's function v can be chosen by taking f outside of $(-\mu_1, \mu_1)$, $f + g_-$ on $(-\mu_1, x)$, and $f + g_+$ on (x, μ_1). □

Lemma 4.11 *Let $u_+, u_- \in \mathbb{C}^k$. There is a unique solution to $D_1^* D_1(v) = 0$ on $(-\mu_1, \mu_1)$ with values u_+, u_- at $\mu_1, -\mu_1$, respectively.*

Proof. Let W_- and W_+ be the $(k+j)$-dimensional affine spaces of solutions with boundary value u_- at $-\mu_1$, and u_+ at μ_1, respectively. Consider at an intermediate point x the affine map

$$\begin{aligned} R_x \colon W_+ \oplus W_- &\to \mathbb{C}^{k+j} \oplus \mathbb{C}^{k+j} \\ (g_+, g_-) &\mapsto (g_-(x) - g_+(x), g_-'(x) - g_+'(x)) \,. \end{aligned}$$

It is injective, otherwise one has, taking differences, an element of $\ker(D_1^* D_1)$ vanishing at both ends; it is thus surjective, and the inverse image of zero gives solutions whose values and derivatives match at x. □

A similar result holds on the other interval:

Lemma 4.12 *Let $x \in (\mu_1, \mu_0 - \mu_1)$, $u_+, u_- \in \mathbb{C}^k$, and $u \in \mathbb{C}^k$. There is a unique solution to $D_1^* D_1(v) = \delta_x u$ on the interval, with value u_+ at μ_1, and value u_- at $\mu_0 - \mu_1$.*

Combining the two previous lemmas, we obtain the Green's function for the small side.

Lemma 4.13 *Let $x \in (\mu_1, \mu_0 - \mu_1)$ and $u \in \mathbb{C}^k$. There is a unique solution to $D_1^* D_1(v) = \delta_x u$ on the circle.*

Proof. Again, one has, varying the boundary values at $\pm\mu_1$, an affine $2k$-dimensional space of continuous solutions, with the right jump in derivatives at x, and possibly extra jumps in the derivatives at $\pm\mu_1$. One considers the affine map from this space to $\mathbb{C}^k \oplus \mathbb{C}^k$ taking the jumps in the derivatives at $\pm\mu_1$; it has to be injective (convexity) and so is surjective, allowing us to match the derivatives. □

By the usual ellipticity argument, the Green's function solution to $D_1^* D_1(v) = \delta_x u$ is smooth away from x. By convexity, $d_z \|v\|$ is increasing everywhere away from x. Since we are on a circle, both $\|v\|$ and $d_z \|v\|$ must attain their maximum at x. In addition, integrating over the circle, we get

$$\begin{aligned} |u|^2 &\geqslant d_z <v, v> (x)_- - d_z <v, v> (x)_+ \\ &= \int d_z^2 <v, v> \\ &= \int 2\|(\nabla_0 - i\mu_0 x_0)v\|^2 + 2\sum_j \|(T_j - ix_j)v\|^2 \\ &\geqslant C \int \|(\nabla_0 - i\mu_0 x_0)v\|^2 + \|v\|^2. \end{aligned}$$

The last inequality follows from the fact that the solution to Nahm's equation is irreducible. This L^2_1 bound on v ensures continuity, with $\|v\|_{L^\infty} \leqslant C\|v\|_{L^2_1}$, and hence the L^∞ norm of the Green's function is bounded by a constant times the norm of u.

Case 2: $j = 0$. We can again use a cohomological version of the Dirac operator, but we must modify the function spaces a little bit to account for the jump at μ_1 in the solution to Nahm's equations given by $\Delta_+(A(\zeta)) = (\alpha_{+,0} + \alpha_{+,1}\zeta)\left(\bar{\alpha}^{\mathrm{T}}_{+,1} - \bar{\alpha}^{\mathrm{T}}_{+,0}\zeta\right)$ for suitable column vectors $\alpha_{+,i}$, and the similar jump $\Delta_-(A(\zeta)) = (\alpha_{-,0} + \alpha_{-,1}\zeta)\left(\bar{\alpha}^{\mathrm{T}}_{-,1} - \bar{\alpha}^{\mathrm{T}}_{-,0}\zeta\right)$ at $-\mu_1$. The corresponding jumps for the matrices T_i are

$$\Delta_\pm(T_3) = \frac{i}{2}\left(\alpha_{\pm,1}\bar{\alpha}^{\mathrm{T}}_{\pm,1} - \alpha_{\pm,0}\bar{\alpha}^{\mathrm{T}}_{\pm,0}\right),$$
$$\Delta_\pm(T_1 + iT_2) = \alpha_{\pm,0}\bar{\alpha}^{\mathrm{T}}_{\pm,1}.$$

We want solutions to the Dirac equations $D_1^*(s_1, s_2) = D_2(s_1, s_2) = 0$ with jumps that are multiples of $v_\pm := (i\alpha_{\pm,1}, \alpha_{\pm,0})$ at $\pm\mu_1$, so we modify the function spaces for the complex (4.40):

- $\tilde{L}^2_2$ is the space of sections s of K that are L^2_2 over each interval, continuous at the jumping points, and with a jump discontinuity $\Delta_\pm(d_z s) = \frac{1}{2}(\alpha_{\pm,1}\bar{\alpha}^{\mathrm{T}}_{\pm,1} + \alpha_{\pm,0}\bar{\alpha}^{\mathrm{T}}_{\pm,0})s(\pm\mu_1)$ in derivatives at $\pm\mu_1$.
- $(\tilde{L}^2_1)^{\oplus 2}$ is the space of sections (s_1, s_2) of $K \oplus K$ that are L^2_1 over the intervals, but with a jump discontinuity $\Delta_\pm(s_1, s_2) = c_\pm v_\pm$ at $\pm\mu_1$.
- $\tilde{L}^2$ is the space of L^2 sections of K.

Proposition 4.14 *For solutions to Nahm's equations and for the complex defined using those function spaces just defined, Proposition 4.9 holds, and furthermore, at the jump points,*

$$\begin{aligned}\Delta(d_z(<v, v>)) &= <\Delta(d_z v), v> + <v, \Delta(d_z v)> \\ &= \left(<\bar{\alpha}^{\mathrm{T}}_{\pm,0}v, \bar{\alpha}^{\mathrm{T}}_{\pm,0}v> + <\bar{\alpha}^{\mathrm{T}}_{\pm,1}v, \bar{\alpha}^{\mathrm{T}}_{\pm,1}v>\right) \geqslant 0\end{aligned} \tag{4.43}$$

for $v \neq 0$ satisfying $D_1^* D_1(v) = 0$; hence, if v is a regular solution vanishing at two points, then $v = 0$.

Proceeding as for $j > 0$, one can then build a Green's function. Using the Green's function in both cases $j > 0$ and $j = 0$, we have the analog of Proposition 4.7.

Proposition 4.15 *The cohomology $\ker(D_2)/\mathrm{Im}(D_1)$ of the complex (4.40) is isomorphic to the space of L^2 solutions to the Dirac equation $D_2(v) = 0 = D_1^*(v)$, which are the harmonic representatives in the cohomology classes. As such, they have minimal norm in the class.*

Proof. The cohomology class of a solution to the Dirac equation is non-zero as the convexity property ensures that if $D_1^*D_1(v) = 0$ over the whole circle then $v = 0$. On the other hand, from a solution to $D_2(u) = 0$, one uses the Green's function to solve $D_1^*D_1(v) = D_1^*(u)$ and gets a harmonic representative $u + D_1(v)$. □

We can now turn to identifying the spectral data of a solution to Nahm's equations with the data of the caloron it induces, beginning with the spectral curves. Suppose we are in the generic situation of spectral curves that are reduced and with no common components. Let us work over $\zeta = 0$, and suppose that the intersection of $\zeta = 0$ with the curves is generic, so distinct points of multiplicity 1.

We turn first to analysing the kernel of the matrices $\beta(z) - iw = T_1(z) + iT_2(z) - ix_1 + x_2$, at points w where $\det(\beta(z) - iw) = 0$ so that we are on the Nahm spectral curve, let us say for the interval $(-\mu_1, \mu_1)$. We note that since $[\beta, d_z + \alpha] = 0$, the spectrum is constant along the intervals $(-\mu_1, \mu_1)$ or $(\mu_1, \mu_0 - \mu_1)$, and indeed, if $(\beta(z_0) - iw)f_0 = 0$, solving $(d_z + \alpha)f = 0$, $f(z_0) = f_0$ gives $(\beta(z) - iw)f = 0$ over the interval.

For such an f and a bump function ρ, setting

$$(s_1, s_2) = (\rho f, 0) \tag{4.44}$$

defines a cohomology class of the complex (4.40). This class is zero only when the integral of ρ is zero over the interval as then for some σ, we have $(d_z + \alpha)(\sigma f) = ((d_z\sigma)f, 0) = (\rho f, 0)$. Note that we can modify by a boundary and do $\rho \mapsto \rho + d_z\sigma$ to place the bump anywhere on the interval. This fact is quite useful.

Given this class at a fixed $y = 0$, one can extend it to other y by taking

$$(s_1, s_2) = (e^{iyz}\rho f, 0). \tag{4.45}$$

We can think of this family of cohomology classes as a section over $S^1 \times \mathbb{R}^3$ of the bundle that Nahm's construction produces from the solution to Nahm's equations. This section has exponential decay as $x_3 \to \pm\infty$. To see this, we exploit the fact that we can move the bump function ρ around to give different representatives supported on $(-\mu_1, \mu_1)$ for the class $(s_1, s_2) = (e^{iyz}\rho f, 0) = (e^{-x_3 z}e^{i\mu_0 x_0}\rho f)$. Recall that the norm of any representative in the cohomology class is at least the norm of the harmonic (Dirac) representative. As we take x_3 to $+\infty$, we move the bump towards μ_1, giving up to a polynomial the bound $\exp(-x_3\mu_1)$ on the L^2 norm of the harmonic representative; similarly, as we go to $-\infty$, we move the bump towards $-\mu_1$, and get a bound of the form $\exp(x_3\mu_1)$.

The condition on a section of E over $\mathbb{C}^* = \{e^{i\mu_0 x_0 + x_3}\}$ to extend over $0, \infty$ is given in Garland and Murray (1988, section 3), and amounts to a growth condition. More generally, there are a whole family of growth conditions, giving us not only our bundle E but also a family of associated sheaves $E_{(p,q,0)(p',q',\infty)}$, the sheaf of sections of E over $\mathcal{T}$ with poles of order p at $\mathcal{T}_0$, with leading term in E_q^0, and poles of order p' at $\mathcal{T}_\infty$, with leading term in $E_{q'}^\infty$. The growth condition that $E_{(0,1,0)(0,1,\infty)}$ corresponds to is that of growth bounded (up to polynomial factors) by $\exp(-x_3\mu_1)$ as $x_3 \to \infty$, and by $\exp(x_3\mu_1)$ as $x_3 \to -\infty$; for $E_{(0,0,0)(1,0,\infty)}$, it is growth bounded (up to polynomial factors) by $\exp(x_3\mu_1)$ as $x_3 \to \infty$, and by $\exp(x_3(\mu_0 - \mu_1))$ as $x_3 \to -\infty$. Let us consider the case of $E_{(0,1,0)(0,1,\infty)}$. Our bounds for the section defined by (4.45) tell us that it is a global (holomorphic) section of $E_{(0,1,0)(0,1,\infty)}$ on the $\mathbb{P}^1$ above $(\eta, \zeta) = (w, 0)$.

The sheaves $E_{(p,q,0)(p',q',\infty)}$ have degree $2(p + p' - 2) + q + q'$ on the fibres of the projections to $T\mathbb{P}1$, from our identifications of them given above. In particular, the degree of $E_{(0,1,0)(0,1,\infty)}$ is -2. Now, starting from a point of the Nahm spectral curve, we have produced a global section of $E_{(0,1,0)(0,1,\infty)}$ over the corresponding Riemann sphere; the Riemann–Roch theorem then tells us that the first cohomology is also non-zero, so that we are in the support of $R^1\tilde{\pi}_*(E_{(0,1,0)(0,1,\infty)})$, which is exactly the caloron spectral curve S_1. As both S_1's are of the same degree and unreduced, they are identical.

Proceeding similarly for the spectral Nahm spectral curve S_0, we obtain a global section of $E_{(0,0,0)(1,0,\infty)}$ over the corresponding twistor line, and so show that the line lies in the support of $R^1\tilde{\pi}_*(E_{(0,0,0)(1,0,\infty)})$, which is exactly the caloron spectral curve S_0.

We have now identified the spectral curves for the solutions to Nahm's equations and the caloron that it produces. We now note that, in addition to the bundle V over $S^1 \times \mathbb{R}^3$ that Nahm's construction produces, there is another bundle, $\hat{V}$, obtained by taking the spectral data associated to the solution of Nahm's equations, and feeding it in to the reconstruction of a caloron from its spectral data. Indeed, from a solution to Nahm's equations, we have seen that we can define sheaves $\mathcal{L}_z$; the fact that they satisfy the boundary conditions by Proposition 4.5 tells us we can reconstruct a bundle F of infinite rank over $T\mathbb{P}^1$ using the sequence (4.10). We now identify the bundles V and $\hat{V}$.

Starting from a caloron, we obtained F as the push-down from the twistor space $\mathcal{T}$ of a rank 2 bundle E. If w is a fibre coordinate on $\mathcal{T} \to T\mathbb{P}^1$, we saw that it induced an automorphism W of F, such that, at $w = w_0$, $E_{w_0} \simeq F/\mathrm{Im}(W - w_0)$. In the sequence (4.10), the shift map W identifies the nth entry in the middle column with the $(n+2)$th (e.g. $L^{(p-1)\mu_0+\mu_1}(2k+j) \otimes \mathcal{I}_{S_{01}}$ with $L^{p\mu_0-\mu_1}(2k+j) \otimes \mathcal{I}_{S_{10}}$) and similarly in the right-hand column. If $w = w(\zeta)$ is the equation of a twistor line L in $\mathcal{T}$, then quotienting by $W - w(\zeta)$ in the sequence (4.10) gives for E over the line (recalling that L_0^μ is trivial over the line)

$$E \hookrightarrow \begin{matrix} L^{\mu_1}(2k+j) \otimes \mathcal{I}_{S_{01}} \\ \oplus \\ L^{-\mu_1}(2k+j) \otimes \mathcal{I}_{S_{10}} \end{matrix} \quad \to \quad \begin{matrix} L^{\mu_1}(2k+j)[-S_{01}]|_{S_0} \\ \oplus \\ L^{-\mu_1}(2k+j)[-S_{10}]|_{S_1} \end{matrix} \to 0. \tag{4.46}$$

From the twistor point of view, the space of global sections of E along the real line L_x in $\mathcal{T}$ correspond to the fibre of the caloron bundle $\hat{V}$ over the corresponding $x \in S^1 \times \mathbb{R}^3$. On the other hand, from the Nahm point of view,

V_x correspond to $\ker D_x^*$, and so to the cohomology $\ker(D_2)/\mathrm{Im}(D_1)$ in the complex (4.40).

To identify V and $\hat{V}$, we follow closely Hurtubise and Murray (1989, pp. 80–84), so we simply summarize the ideas. The identification is made for the x whose lines L_x which do not intersect $S_0 \cap S_1$; to do it on this dense set is sufficient, as one is also identifying the connections.

Let us denote the sequence of sections corresponding to (4.46) by

$$0 \to H^0(L_x, E) \to U_{\mu_1} \oplus U_{-\mu_1} \to W_0 \oplus W_1. \tag{4.47}$$

What Hurtubise and Murray (1989) do is to identify W_0 and W_1 with solutions to $D_x^*(s) = 0$ on the interval $(\mu_1, \mu_0 - \mu_1)$ and $(-\mu_1, \mu_1)$, respectively, and U_{μ_1} and $U_{-\mu_1}$ with L^2 solutions on a neighborhood of μ_1 and $-\mu_1$, respectively. The kernel $H^0(L_x, E)$ then gets identified with global L^2 solutions on the circle, which are elements of V_x. Since from the twistor point of view $\hat{V}_x \simeq H^0(L_x, E)$, we are done. The case $j = 0$ is similar.

Having identified the bundles, one then wants to ensure that the connections defined in both case are the same, and we do so again according to Hurtubise and Murray (1989). It suffices to do this identification along one null plane through the point x, as changing coordinates will do the rest, and we can suppose that x is the origin, so $\eta = 0$ in $T\mathbb{P}^1$. Let us choose the plane corresponding to $\zeta = 0$. From the twistor point of view, parallel sections along this null plane correspond to sections on the corresponding family of lines with fixed values at $\zeta = 0$. From the Nahm point of view, a parallel section along this null plane is a family of cohomology classes represented by cocycles $s = (s_1, s_2)$ in $\ker(D_2)$ that satisfy $(\frac{\partial}{\partial x_1} + i\frac{\partial}{\partial x_2})s = 0, (\frac{\partial}{\partial x_0} - i\frac{\partial}{\partial x_3})s = 0$, modulo coboundaries. In particular, the derivatives of s_2 are of the form $(T_1 + iT_2)\phi = A_0\phi$ for a suitable ϕ depending on x. Following through the identifications of Hurtubise and Murray (1989 pp. 80–84) tells us that the derivatives $\nabla\hat{s}$ in suitable trivializations of the corresponding sections $\hat{s}$ on the family of lines are of the form $\nabla\hat{s} = A_0\psi$ for the corresponding section ψ. However, the defining relation for $A(\zeta)$ is $(\eta\mathbf{I} - A_0 - \zeta A_1 - \zeta^2 A_2)\phi = 0$ over the spectral curve. At the boundaries of the inverval, it extends to all of $\mathcal{T}$. In particular, at $(\zeta, \eta) = 0$, $A_0\phi$ vanishes as a function, and so $\hat{s}$ is parallel, as desired.

4.5.3 *From Nahm to caloron to Nahm to caloron*

Let us first place ourselves in the generic situation. We have seen first that the family of generic calorons maps continuously and injectively into the family of generic spectral data; secondly, that generic spectral data and generic solutions to Nahm's equations are equivalent; thirdly, that starting from a caloron and producing a solution to Nahm's equations then taking the spectral data of both objects gives the same result, and fourthly, that the generic solution to Nahm's equation gives the same caloron, whether you pass through the Nahm transform or through the spectral data via the twistor construction. Taken together, those facts tell us that the map from caloron to spectral data is a bijection, that all three sets of data are equivalent, and that the six transforms between them are pairwise inverses of each other.

Let us now leave the generic set. If we now take an arbitrary solution T to Nahm's equations, again satisfying all the conditions, we can fit it into a continuous family $T(t)$ with $T(0) = T$ with $T(t)$ generic for $t \neq 0$; that it can be done follows from our description of moduli given in Section 4.6. The Nahm transform of this family is a continuous family $C(t)$ of calorons, which in turn produces a continuous family of solutions $\tilde{T}(t)$ to Nahm's equations. We do not know a priori whether the boundary and symmetry conditions are satisfied at $t = 0$. However, for $t \neq 0$, $\tilde{T}(t) = T(t)$, and so $\tilde{T}(0) = T(0)$, and we are done; the transform Nahm to caloron to Nahm is the identity.

If the caloron C lies in the closure of the set of generic calorons (presumably all calorons do, but it needs to be proved, perhaps using the methods of Taubes (1984) to show that the moduli space is connected), we again fit it as $C(0)$ into a family $C(t)$ of calorons, with $C(t)$ generic for $t \neq 0$, all of same charge. The Nahm transform produces a family $T(t)$ of solutions to Nahm's equations, with $T(t)$ satisfying all the conditions for $t \neq 0$. We also get a family $S(t)$ of spectral data, which for all t corresponds to both $C(t)$ and $T(t)$. We need to show that $T(0)$ satisfy all the conditions.

Taking a limit, it is fairly clear that the symmetry condition is satisfied also for $t = 0$. The boundary conditions are less obvious. From the small side, there is no problem, as the vanishing theorem (proposition 4.3) holds at $C(0)$, and so the solutions to Nahm's equations are continuous at the boundary of the interval. From the large side, what saves us is the rigidity of representations into $SU(2)$; indeed, the polar parts of the solutions are given by representations, and so are fixed, in a suitable family of unitary gauges, hence preserved in a limit. The process, given in Section 4.4.2, of passing to a constant gauge by solving $ds/dz + A_+(z)s = 0$ applies in the limit near the singular points. We have in the limit the same type of transformations relating the basis of K in which one has the solution to Nahm's equations to the continuous basis obtained from lifting up and pushing down as to obtain the sequence (4.16), with the same process of producing endomorphisms $\underline{A}(z,\zeta)(t)$, giving us again in the limit a continuous endomorphism of a bundle V over $T\mathbb{P}^1 \times S^1$, with V in fact lifted from $\mathbb{P}^1 \times S^1$. The restrictions $V(\pm\mu_1)$ (on the large side) are of fixed type $V^1_{\mu_1} = \mathcal{O}^k \oplus V_{j-1,-j+1}$ for all t. The summand $V_{j-1,-j+1}$ corresponds to the subsheaf $\mathcal{O}(j)$ of the sheaf $F/(F^0_{p,1} + F^\infty_{-p,1}) \otimes L^{\pm\mu_1}$, which exists in the limit. The polar part $\mathcal{C}(p(\zeta,t))$ of (4.26), mapping $V_{j-1,-j+1}$ to itself, has limit $\mathcal{C}(p(\zeta,0))$. By (4.27), this latter limit is determined by the spectral curves, and so is well defined. The other summand $\mathcal{O}^k$ is also well defined in the limit, as it is the piece transferred from the small side, which is still $\mathcal{O}^k$ in the limit because of the vanishing theorem. The off-diagonal vanishing as one goes back to the trivialization for Nahm's equations is simply a consequence of the polar behaviour of solutions to $ds/dz + A_+(z)s = 0$, and of the normal form for $\underline{A}(z,\zeta)(t)$ as in

(4.25). In short, the limit has exactly the same normal form, and so the same boundary behaviour.

Finally, there is the question of irreducibility. The reducible solutions correspond to calorons for which charge has bubbled off; as our family has constant charge, this is precluded. The limit solutions satisfies all the conditions necessary to produce a caloron by the opposite Nahm transform; as the transform is involutive on the generic member of the family, it is also involutive in the limit, and so the circle closes.

In short one has the following theorem:

Theorem 4.16

1. *There is an equivalence between*
   a. *Generic calorons of charge* (k, j)
   b. *Generic solutions to Nahm's equations on the circle satisfying the conditions of Section 4.2.1.2*
   c. *Generic spectral data*

   The equivalences of (a) and (b) are given in both directions by the Nahm transform.
2. *The Nahm transforms give equivalences between*
   a. *Calorons of charge* (k, j)*, in the closure of the generic set*
   b. *Solutions to Nahm's equations on the circle satisfying the conditions of Section 4.2.1*

4.6 Moduli

The equivalence exhibited above allows us to classify calorons by classifying appropriate solutions to Nahm's equations. One has to guide us through the example of monopoles, as classified in Donaldson (1984) and Hurtubise (1989); the (framed) monopoles for gauge group G, with symmetry breaking to a torus at infinity, and of charge k are classified by the space $\mathrm{Rat}_k(\mathbb{P}^1, G_{\mathbb{C}}/B)$ of based degree k rational maps from the Riemann sphere into the flag manifold $G_{\mathbb{C}}/B$. As our calorons are Kač–Moody monopoles, we should have the same theorem, with framed calorons equivalent to rational maps from $\mathbb{P}^1$ into the loop group. Following an idea developed in Atiyah (1984), one thinks of these as bundles on $\mathbb{P}^1 \times \mathbb{P}^1$, with some extra data of a flag along a line, and some framing.

The rank 2 bundle corresponding to an $SU(2)$ caloron, again following the example of monopoles, should be the restriction of the bundle E on $\tilde{T}$ corresponding to the caloron, to the inverse image of a point, say $\zeta = 0$, in $\mathbb{P}^1$. This inverse image is $\mathbb{P}^1 \times \mathbb{C}$; along the divisor $\{0\} \times \mathbb{C}$, the bundle has the flag E_0, and along $\{\infty\} \times \mathbb{C}$, the bundle has the flag E_∞. We extend this bundle to infinity, using a framing, in such a way that E_∞ is a trivial subbundle, and E_0 has degree $-j$. This bundle will be corresponding to the caloron.

The theorem, however, is proven in terms of solutions to Nahm's equations; indeed, we show in Charbonneau and Hurtubise (2007) that both bundles and solutions to Nahm's equations satisfying the appropriate conditions are describable in terms of a geometric quotient of a family of matrices.

Theorem 4.17 *Let $k \geqslant 1, j \geqslant 0$ be integers. There is an equivalence between*

1. *Vector bundles E of rank 2 on $\mathbb{P}^1 \times \mathbb{P}^1$, with $c_1(E) = 0, c_2(E) = k$, trivialized along $\mathbb{P}^1 \times \{\infty\} \cup \{\infty\} \times \mathbb{P}^1$, with a flag $\phi\colon \mathcal{O}(-j) \hookrightarrow E$ of degree j along $\mathbb{P}^1 \times \{0\}$ such that $\phi(\infty)(\mathcal{O}(-j)) = \mathrm{span}(0, 1)$.*
2. *Framed, irreducible solutions to Nahm's equations on the circle, with rank k over $(\mu_1, \mu_0 - \mu_1)$, rank $k + j$ over $(-\mu_1, \mu_1)$, with the boundary conditions defined above, modulo the action of the unitary gauge group.*
3. *Complex matrices A, B ($k \times k$), C ($k \times 2$), D ($2 \times k$), A' ($j \times k$), B' ($1 \times k$), C' ($j \times 2$), satisfying the monad equations*

$$[A, B] + CD = 0,$$

$$\begin{pmatrix} B' \\ 0 \end{pmatrix} A + S(j,0)A' - A'B - C'D = 0,$$

$$-e_+A' + \begin{pmatrix} 1 & 0 \end{pmatrix} D = 0,$$

and the genericity conditions

$$\begin{pmatrix} A - y \\ B - x \\ D \end{pmatrix} \text{ is injective for all } x, y \in \mathbb{C},$$

$$\begin{pmatrix} x - B & A - y & C \end{pmatrix} \text{ is surjective for all } x, y \in \mathbb{C},$$

$$\begin{pmatrix} x - B & A & C & 0 \\ \begin{pmatrix} -B' \\ 0 \end{pmatrix} & A' & C' & x - S(j,0) \\ 0 & 0 & \begin{pmatrix} 1 & 0 \end{pmatrix} & -e_+ \end{pmatrix} \text{ is surjective for all } x \in \mathbb{C},$$

$$\left(\begin{pmatrix} A \\ A' \end{pmatrix} \begin{pmatrix} C_2 \\ C_2' \end{pmatrix} M \begin{pmatrix} C_2 \\ C_2' \end{pmatrix} \cdots M^{j-1} \begin{pmatrix} C_2 \\ C_2' \end{pmatrix} \right) \text{ is an isomorphism,}$$

modulo the action of $Gl(k, \mathbb{C})$ given by

$$(A, B, C, D, A', B', C') \mapsto (gAg^{-1}, gBg^{-1}, gC, Dg^{-1}, A'g^{-1}, B'g^{-1}, C').$$

Here C_i, C_i' are the ith column of C, C', and D_i the ith row of D, (4.32) gives $S(j,0)$, and $e_+ := \begin{pmatrix} 0 & \cdots & 0 & 1 \end{pmatrix}$,

$$M := \begin{pmatrix} B & -C_1e_+ \\ \begin{pmatrix} B' \\ 0 \end{pmatrix} & S(j,0) - C_1'e_+ \end{pmatrix}.$$

We see in Charbonneau and Hurtubise (2007) that B and M are conjugate to the matrix $T_1 + iT_2$ on the intervals $(\mu_1, \mu_0 - \mu_1)$ and $(-\mu_1, \mu_1)$, respectively. Choosing B diagonal, and B and M with distinct and disjoint eigenvalues, it is not difficult to build solutions to the various matrix constraints. Hence, the spectral curves for the obtained solutions to Nahm's equations are reduced and have no common components, as the spectral curve over $\zeta = 0$ is the spectrum of $T_1 + iT_2$. Thus, generic generic solutions to Nahm's equations, and so generic calorons, exist.

Acknowledgement

This research was conducted while the first author was at McGill University and was supported by an NSERC PDF. The second author is supported by NSERC and FQRNT grants. The diagrams in this chapter were created using Paul Taylor's Commutative Diagram package.

References

Adams, M. R., Harnad, J., and Hurtubise, J. (1990). Isospectral Hamiltonian flows in finite and infinite dimensions. II. Integration of flows. *Commun. Math. Phys.* **134**(3), 555–85.

Atiyah, M. F. (1984). Instantons in two and four dimensions. *Commun. Math. Phys.* **93**(4), 437–51.

Atiyah, M. F., Drinfel′d, V. G., Hitchin, N. J., and Manin, Yu. I. (1978). Construction of instantons. *Phys. Lett. A* **65**(3), 185–7.

Boyer, C. P., Mann, B. M., Hurtubise, J. C., and Milgram, R. J. (1994). The topology of the space of rational maps into generalized flag manifolds. *Acta Math.* **173**(1), 61–101.

Braam, P. J. and van Baal, P. (1989). Nahm's transformation for instantons. *Commun. Math. Phys.* **122**(2), 267–80.

Bruckmann, F. and van Baal, P. (2002). Multi-caloron solutions. *Nuclear Phys. B* **645**(1–2), 105–33.

Bruckmann, F., Nógrádi, D., and van Baal, P. (2003). Constituent monopoles through the eyes of fermion zero-modes. *Nuclear Phys. B* **666**(1–2), 197–229.

Bruckmann, F., Nógrádi, D., and van Baal, P. (2004). Higher charge calorons with non-trivial holonomy. *Nuclear Phys. B* **698**(1–2), 233–54.

Chakrabarti, A. (1987). Periodic generalizations of static, self-dual SU(2) gauge fields. *Phys. Rev. D (3)* **35**(2), 696–706.

Charbonneau, B. (2004). Analytic aspects of periodic instantons. PhD thesis, September, Massachusetts Institute of Technology, Cambridge, MA. Available at http://hdl.handle.net/1721.1/26746.

Charbonneau, B. (2006). From spatially periodic instantons to singular monopoles. *Commun. Anal. Geom.* **14**(1), 183–214.

Charbonneau, B. and Hurtubise, J. (2008).Calorons, Nahm's equations on S^1 and bundles over $\mathbb{P}^1 \times \mathbb{P}^1$. *Commun. Math. Phys.* **280**(2), 315–49.

Cherkis, S. and Kapustin, A. (2001). Nahm transform for periodic monopoles and $\mathcal{N} = 2$ super Yang–Mills theory. *Commun. Math. Phys.* **218**(2), 333–71.

Corrigan, E. and Goddard, P. (1984). Construction of instanton and monopole solutions and reciprocity. *Ann. Phys.* **154**(1), 253–79.

Donaldson, S. K. (1984). Nahm's equations and the classification of monopoles. *Commun. Math. Phys.* **96**(3), 387–407.

Donaldson, S. K. and Kronheimer, P. B. (1990). The geometry of four-manifolds. *Oxford Mathematical Monographs.* The Clarendon Press Oxford University Press, New York.

Garland, H. and Murray, M. K. (1988). Kac–Moody monopoles and periodic instantons. *Commun. Math. Phys.* **120**(2), 335–51.

Gravesen, J. (1989). On the topology of spaces of holomorphic maps. *Acta Math.* **162**(3–4), 247–86.

Hitchin, N. J. (1982). Monopoles and geodesics. *Commun. Math. Phys.* **83**(4), 579–602.

Hitchin, N. J. (1983). On the construction of monopoles. *Commun. Math. Phys.* **89**(2), 145–90.

Hurtubise, J. (1989). The classification of monopoles for the classical groups. *Commun. Math. Phys.* **120**(4), 613–41.

Hurtubise, J. and Murray, M. K. (1989). On the construction of monopoles for the classical groups. *Commun. Math. Phys.* **122**(1), 35–89.

Hurtubise, J. and Murray, M. K. (1990). Monopoles and their spectral data. *Commun. Math. Phys.* **133**(3), 487–508.

Jardim, M. (2001). Construction of doubly-periodic instantons. *Commun. Math. Phys.* **216**(1), 1–15.

Jardim, M. (2002*a*). Classification and existence of doubly-periodic instantons. *Q. J. Math.* **53**(4), 431–42.

Jardim, M. (2002*b*). Nahm transform and spectral curves for doubly-periodic instantons. *Commun. Math. Phys.* **225**(3), 639–68.

Jardim, M. (2004). A survey on Nahm transform. *J. Geom. Phys.* **52**(3), 313–27.

Kraan, T. C. (2000). Instantons, monopoles and toric hyperKähler manifolds. *Commun. Math. Phys.* **212**(3), 503–33.

Kraan, T. C. and van Baal, P. (1998*a*). Exact T-duality between calorons and Taub-NUT spaces. *Phys. Lett. B* **428**(3–4), 268–76.

Kraan, T. C. and van Baal, P. (1998*b*). Monopole constituents inside $SU(n)$ calorons. *Phys. Lett. B* **435**, 389–95.

Kraan, T. C. and van Baal, P. (1998*c*). Periodic instantons with non-trivial holonomy. *Nuclear Phys. B* **533**(1–3), 627–59.

Lee, K. (1998). Instantons and magnetic monopoles on $\mathbf{R}^3 \times S^1$ with arbitrary simple gauge groups. *Phys. Lett. B* **426**(3–4), 323–8.

Lee, K. and Lu, C. (1998). Su(2) calorons and magnetic monopoles. *Phys. Rev. D (2)* **58**, 025011.

Lee, K. and Yi, S.-H. (2003). 1/4 BPS dyonic calorons. *Phys. Rev. D (3)* **67**(2), 025012, 9.

Mazzeo, R. and Melrose, R. B. (1998). Pseudodifferential operators on manifolds with fibred boundaries. *Asian J. Math.* **2**(4), 833–66.

Nahm, W. (1983). All self-dual multimonopoles for arbitrary gauge groups. In *Structural Elements in Particle Physics and Statistical Mechanics (Freiburg, 1981)*, Vol. 82 of *NATO Advanced Study Institute Series B: Physics*, pp. 301–10. Plenum, New York.

Nahm, W. (1984). Self-dual monopoles and calorons. In *Group Theoretical Methods in Physics (Trieste, 1983)*, Vol. 201 of *Lecture Notes in Physics*, pp. 189–200. Springer, Berlin.

Nógrádi, D. (2005). Multi-calorons and their moduli. PhD thesis, Institute Lorentz for Theoretical Physics, University of Leiden, Leiden, the Netherlands. arXiv:hep-th/0511125.

Norbury, P. (2000). Periodic instantons and the loop group. *Commun. Math. Phys.* **212**(3), 557–69.

Nye, T. M. W. (2001). The geometry of calorons. PhD thesis, University of Edinburgh, Edinburgh, Scotland. arXiv:hep-th/0311215.

Nye, T. M. W. and Singer, M. A. (2000). An L^2-index theorem for Dirac operators on $S^1 \times \mathbf{R}^3$. *J. Funct. Anal.* **177**(1), 203–18.

Schenk, H. (1988). On a generalised Fourier transform of instantons over flat tori. *Commun. Math. Phys.* **116**(2), 177–83.

Szabó, S. (2007). Nahm transform for integrable connections on the Riemann sphere. *Mém. Soc. Math. Fr. (N.S.)* No. 110.

Taubes, C. H. (1984). Path-connected Yang–Mills moduli spaces. *J. Differential Geom.* **19**(2), 337–92.

Ward, R. S. (2004). Symmetric calorons. *Phys. Lett. B* **582**(3–4), 203–10.

V

NAHM'S EQUATIONS AND FREE-BOUNDARY PROBLEMS

Simon K. Donaldson

Dedicated to Nigel Hitchin, with gratitude and affection

5.1 Introduction

In Donaldson (1984), following up work of Hitchin (1983), the author found it useful to express *Nahm's equations*, for a matrix group, in terms of the motion of a particle in a Riemannian symmetric space, subject to a potential field. This point of view lead readily to an elementary existence theorem for solutions of Nahm's equation, corresponding to particle paths with prescribed end points. The original motivation for this chapter is the question of formulating an analogous theory for the Nahm equations associated to the infinite-dimensional Lie group of area-preserving diffeomorphisms of a surface – in the spirit of Donaldson (1993). We will see that this can be done, and that a form of the appropriate existence theorem holds – essentially a special case of a result of Chen. However the main focus of the chapter is not on existence proofs but on the various formulations of the problem, and connections between them. In these developments, one finds that the natural context is rather more general than the original question, so we will start out of a different tack, and return to Nahm's equations in Section 5.5.

Consider the following set-up in Euclidean space $\mathbb{R}^3$, in which we take coordinates (x_1, x_2, z) – thinking of z as the vertical direction. (We will use the notation $\frac{\partial}{\partial x_i} = \partial_i$, $\frac{\partial}{\partial z} = \partial_z$.) Suppose we have a strictly positive function $H(x_1, x_2)$. This defines a domain

$$\Omega_H = \{(x_1, x_2, z) \; : \; 0 < z < H(x_1, x_2)\},$$

whose boundary has two components $\{z = 0\}$ and $\{z = H\}$. We consider the Dirichlet problem for the standard Laplacian: to find a harmonic function θ on Ω_H with $\theta = 0$ on $\{z = 0\}$ and $\theta = 1$ on $\{z = H\}$. To set up this problem precisely, let us assume that the data H is $\mathbb{Z}^2$ periodic on $\mathbb{R}^2$, and seek a $\mathbb{Z}^2$-periodic solution θ. Now we have a unique solution θ to our Dirichlet problem. Consider the flux of the gradient of θ through the boundary $\{z = H\}$. This defines another function ρ on $\mathbb{R}^2$. To be precise, if ι_H is the obvious map from $\mathbb{R}^2$ to the boundary $\{z = H\}$ then the flux is defined by

$$\iota_H^*(*d\theta) = \rho dx_1 dx_2. \tag{5.1}$$

Explicitly

$$\rho = \partial_z\theta - (\partial_1\theta\partial_1 H + \partial_2\theta\partial_2 H),$$

with the right-hand side evaluated at $(x_1, x_2, H(x_2, x_2))$. By the maximum principle, ρ is a positive function, since the normal derivative of θ is positive in the positive z direction on $\{z = H\}$. We consider the following free-boundary problem: Given a positive periodic function ρ does it arise from some periodic H, and is H unique?

One can gain some physical intuition for this question by supposing that the lower half-space $\{z \leq 0\}$ represents a body with an infinite specific heat capacity fixed at temperature 0 and Ω_H corresponds to a layer of ice covering this body. We choose units so that the melting temperature of the ice is 1. Sunlight shines vertically downwards onto the upper surface $\{z = H\}$ of the ice, but with a variable intensity so that heat is transmitted to the surface according to the density function ρ. We suppose that the surface of the ice is sprinkled by rain, which will instantly freeze if the surface temperature of the ice is less than 1. We also suppose that a wind blows across the surface, instantly removing any surface water. Then we see that the solution to our free-boundary problem represents a static physical state, in which the upper surface of the ice is just at freezing point, the lower surface is at the imposed sub-freezing temperature, and the heat generated by the given sunlight flows through the ice without changing the temperature. Physical intuition suggests that there should indeed be a unique solution.

We can express the free-boundary problem considered above as a special case of another question. Suppose now that we have a pair of periodic functions H_0, H_1 on $\mathbb{R}^2$ with $H_0 < H_1$. Then we have a domain $\Omega_{H_0,H_1} = \{H_0(x_1, x_2) < z < H_1(x_1, x_2)\}$, with two boundary components. Let θ be the harmonic function in this domain equal to $0, 1$ on $\{z = H_0\}, \{z = H_1\}$, respectively. Then we obtain a pair of flux functions ρ_0, ρ_1 as before. By Gauss' theorem, these satisfy a constraint

$$\int_{[0,1]^2} \rho_0 \, d\underline{x} = \int_{[0,1]^2} \rho_1 \, d\underline{x}, \tag{5.2}$$

since $[0, 1]^2$ is a fundamental domain for the $\mathbb{Z}^2$ action. Obviously, if we replace H_0, H_1 by $H_0 + c, H_1 + c$ for any constant c we get the same pair ρ_0, ρ_1. We ask: given ρ_0, ρ_1 satisfying the integral constraint (5.2), is there a corresponding pair (H_0, H_1), and if so is the solution unique up to the addition of a constant? A positive answer to this question implies a positive answer to the previous one, by a simple reflection argument. (Given ρ, as in the first problem, take $\rho_0 = \rho_1 = \rho/2$. Then uniqueness implies that the solution has reflection symmetry about $\theta = 1/2$ and we get a solution to the first problem by changing θ to $2\theta - 1$.)

Of course we can also imagine a physical problem corresponding to this second question: for example, a layer of ice in the region Ω_{H_0,H_1}. We can now vary the problem by supposing that in place of ice we have a horizontally stratified material in which heat can only flow in the horizontal directions. Thus the steady-state condition, for a temperature distribution $\theta(x_1, x_2, z)$, is

$$\left(\partial_1^2 + \partial_2^2\right)\theta = 0. \tag{5.3}$$

We define flux functions ρ_0, ρ_1 by pulling back the two-form

$$\partial_1\theta \; dx_2 dz - \partial_2\theta \; dx_1 dz,$$

and the integral constraint (5.2) still holds. So we ask: given ρ_0, ρ_1 satisfying (5.2), is there a pair H_0, H_1 and a function θ on Ω_{H_0,H_1}, equal to $0, 1$ on the two boundary components, which has these fluxes, and is the solution essentially unique? (In this case, one has to relax the condition on the domain to $H_0 \leq H_1$.)

It is natural to extend these questions to a general compact-oriented Riemannian manifold X (which would be the flat torus $\mathbb{R}^2/\mathbb{Z}^2$ in the discussion above). Write $d\mu$ for the Riemannian volume form on X. We fix a real parameter $\epsilon \geq 0$ and define a map $*_\epsilon$ from $T^*(X \times \mathbb{R})$ to $\Lambda^n T^*(X \times \mathbb{R})$ by

$$*_\epsilon dz = \epsilon d\mu, \quad *_\epsilon \alpha = (*_X \alpha) dz,$$

for $\alpha \in T^*X$, where $*_X$ is the usual Hodge $*$ operator on X. Then for a function θ on a domain in $X \times \mathbb{R}$

$$d *_\epsilon d\theta = (\Delta_\epsilon \theta) dz d\mu,$$

where

$$\Delta_\epsilon \theta = \left(-\epsilon \partial_z^2 + \Delta_X\right)\theta,$$

with Δ_X the standard Laplace operator on X. (We use the sign convention that Δ_X is a positive operator, so when $\epsilon = 1$ our Δ_ϵ is the standard Laplace operator on $X \times \mathbb{R}$.) If θ is defined on a domain Ω_{H_0,H_1}, as above, we define the flux ρ_i on the boundary $\{z = H_i\}$ by pulling back $*_\epsilon d\theta$ just as before. We consider a pair of functions $\rho_0, \rho_1 > 0$ with

$$\int_X \rho_0 \; d\mu = \int_X \rho_1 \; d\mu = \int_X \; d\mu$$

and we ask

Question 5.1 *Is there a pair $H_0 \leq H_1$ and a function θ on the set $\Omega_{H_0,H_1} \subset X \times \mathbb{R}$ with $\theta = 0, 1$ on the hypersurfaces $\{z = H_0\}, \{z = H_1\}$, with fluxes ρ_i and with $\Delta_\epsilon \theta = 0$? If so, is the solution essentially unique?*

For any $\epsilon > 0$ the equation $\Delta_\epsilon \theta = 0$ can be transformed into the standard Laplace equation on the product, by rescaling the z variable. When $\epsilon = 0$ the equation has a very different character: it is not elliptic and we obviously do not have automatic interior regularity with respect to z.

5.2 An infinite-dimensional Riemannian manifold

We now start in a different direction. Given our compact Riemannian manifold X we let $\mathcal{H}$ be the set of functions ϕ on X such that $1 - \Delta_X \phi > 0$. We make $\mathcal{H}$ into a Riemannian manifold, defining the norm of a tangent vector $\delta\phi$ at a point ϕ by

$$\|\delta\phi\|_\phi^2 = \int_X (\delta\phi)^2 \, (1 - \Delta_X \phi) d\mu.$$

Thus a path $\phi(t)$ in $\mathcal{H}$, parametrized by $t \in [0, 1]$ say, is simply a function on $X \times [0, 1]$ and the "energy" of the path is

$$\frac{1}{2} \int_0^1 \int_X \left(\frac{\partial \phi}{\partial t}\right)^2 (1 - \Delta_X \phi) d\mu dt. \tag{5.4}$$

When X is two-dimensional and orientable, this definition coincides with the metric on the space of "Kähler potentials" discussed by Mabuchi, Semmes, and the author (Mabuchi 1987; Semmes 1992; Donaldson 1999). The general context in those references is a compact Kähler manifold: here we are considering a different extension of the two-dimensional case, and we will see that some new features emerge. The account below follows the approach in Donaldson (1999) closely.

It is straightforward to find the Euler–Lagrange equations associated to the energy (5.4). These are

$$\ddot{\phi} = \frac{|\nabla_X \dot{\phi}|^2}{1 - \Delta_X \phi}.$$

These equations define the geodesics in $\mathcal{H}$. We can read off the Levi-Civita connection of the metric from this geodesic equation, as follows. Let $\phi(t)$ be any path in $\mathcal{H}$ and $\psi(t)$ be another function on $X \times [0, 1]$, which we regard as a vector field along the path $\phi(t)$. Then the covariant derivative of ϕ along the path is given by

$$D_t \phi = \frac{d\phi}{dt} + (W_t, \nabla_X \psi), \tag{5.5}$$

where

$$W_t = \frac{-1}{1 - \Delta_X \phi} \nabla_X \dot{\phi}$$

and (,) is the Riemannian inner product on tangent vectors to X. (We write ∇_X, or sometimes just ∇, for the gradient operator on X, so W_t is a vector field on X.) This has an important consequence for the *holonomy group* of the manifold $\mathcal{H}$. Observe that the tangent space to $\mathcal{H}$ at a point ϕ is the space of functions on X endowed with the standard L^2 inner product associated to the measure

$$d\mu_\phi = (1 - \Delta\phi) d\mu_0.$$

So, in a general way, the parallel transport along a path from ϕ_0 to ϕ_1 should be an isometry from $L^2(X, d\mu_{\phi_0})$ to $L^2(X, d\mu_{\phi_1})$. (Here we are ignoring the distinction between, e.g., smooth functions and L^2 functions.) What we see from (5.5) is that this isometry is induced by a diffeomorphism $f : X \to X$ with

$$f^*(d\mu_{\phi_1}) = d\mu_{\phi_0}. \tag{5.6}$$

The diffeomorphism is obtained by integrating the time-dependent vector field W_t and (5.6) follows from the identity

$$\mathcal{L}_{W_t} d\mu_\phi = d * \left(\frac{1}{1 - \Delta_X} d\dot{\phi} *_X d\mu_\phi \right) = \Delta\dot{\phi} = -\frac{d}{dt}\mu_\phi.$$

(Here $\mathcal{L}$ denotes the Lie derivative on X.) We conclude that the holonomy group of $\mathcal{H}$ is contained in the group $\mathcal{G}$ of volume-preserving diffeomorphisms of $(X, d\mu_0)$, regarded as a subgroup of the orthogonal group of $L^2(X, d\mu_0)$. (This can also be expressed by saying that there is an obvious principal $\mathcal{G}$ bundle over $\mathcal{H}$ with the tangent bundle as an associated vector bundle, and the Levi-Civita connection is induced by a connection on this $\mathcal{G}$ bundle.)

We now move on to discuss the curvature tensor of $\mathcal{H}$. Let ϕ be a point of $\mathcal{H}$ and let α, β be tangent vectors to $\mathcal{H}$ at ϕ—so α and β are just functions on X. The curvature $R_{\alpha,\beta}$ should be a linear map from tangent vectors to tangent vectors: that is, from functions on X to functions on X. The discussion of the holonomy above tells us that this map must have the form

$$R_{\alpha,\beta}(\psi) = (\nu_{\alpha,\beta}, \nabla\psi), \tag{5.7}$$

for some vector field $\nu_{\alpha,\beta}$ on X, determined by ϕ, α, and β. Moreover we know that we must have

$$\mathcal{L}_{\nu_{\alpha,\beta}}(d\mu_\phi) = 0.$$

To identify this vector field we introduce some notation. For vector fields v, w on X we write $v \times w$ for the exterior product: a section of the bundle $\Lambda^2 TX$. We define a differential operator

$$\text{curl } : \Gamma(\Lambda^2 TX) \to \Gamma(TX)$$

to be the composite of the standard identification:

$$\Lambda^2 TX \cong \Lambda^{n-2} T^* X$$

(using the Riemannian volume form $d\mu$), the exterior derivative

$$d : \Gamma(\Lambda^{n-2} T^* X) \to \Gamma(\Lambda^{n-1} T^* X),$$

and the standard identification

$$\Lambda^{n-1}T^*X \cong TX$$

(using the volume form $d\mu$ again). Then we have

Theorem 5.1 *The curvature of $\mathcal{H}$ is given by (5.7) and the vector field*

$$\nu_{\alpha,\beta} = \frac{1}{1-\Delta\phi}\text{curl}\left(\frac{1}{1-\Delta\phi}\nabla\alpha \times \nabla\beta\right).$$

Corollary 5.1 *The manifold $\mathcal{H}$ has non-positive sectional curvature.*

The sectional curvature corresponding to a pair of tangent vectors α, β at a point ϕ is

$$K_{\alpha,\beta} = \langle R_{\alpha,\beta}(\alpha), \beta\rangle.$$

In our case this is

$$K_{\alpha,\beta} = \int_X (\nu_{\alpha,\beta}, \nabla\alpha)\beta(1-\Delta_X\phi)d\mu.$$

Unwinding the algebraic identifications we used above, the integrand can be written in terms of differential forms as

$$\frac{1}{1-\Delta_X\phi}d\alpha \wedge d\left(\frac{1}{1-\Delta_X\phi} * (d\alpha \wedge d\beta)\right)\beta(1-\Delta_X\phi).$$

So

$$K_{\alpha,\beta} = \int_X d\alpha \wedge d\left(\frac{1}{1-\Delta_X\phi} * (d\alpha \wedge d\beta)\right)\beta.$$

Applying Stokes' theorem this is

$$K_{\alpha,\beta} = -\int_X \frac{1}{1-\Delta_X\phi}d\alpha \wedge d\beta \wedge *(d\alpha \wedge d\beta) = -\int_X \frac{1}{1-\Delta_X\phi}|d\alpha \wedge d\beta|^2\, d\mu \leq 0.$$

In the proof of Theorem 5.1 we will make use of two identities. For any pair of vector fields v, w and function f

$$\text{curl}\ (v \times w) = [v, w] + (\text{div}\ v)w - (\text{div}\ w)v \tag{5.8}$$

$$\text{curl}\ (f(v \times w)) = f\text{curl}\ (v \times w) + (v, \nabla f)w - (w, \nabla f)v. \tag{5.9}$$

We leave the verification as an exercise. (Considering geodesic coordinates we see that it suffices to treat the case of Euclidean space. Our notation has been chosen to agree with standard notation in the case of vector fields in $\mathbb{R}^3$.)

To calculate the curvature we consider a two-parameter family $\phi(s,t)$ in $\mathcal{H}$, with a corresponding vector field $\psi(s,t)$ along the family. Then we will compute the commutator $(D_tD_s - D_sD_t)\psi(s,t)$. Evaluating at $\phi = \phi(0,0)$ this is $R_{\alpha,\beta}(\psi)$ where $\psi = \psi(0,0), \alpha = \partial_s\phi, \beta = \partial_t\phi$.

Now we write

$$D_s = \frac{\partial}{\partial s} + W_s \ , \ D_t = \frac{\partial}{\partial t} + W_t,$$

where the vector fields W_s, W_t are regarded as operators on the functions on X. So $D_t D_s - D_s D_t$ is the operator given by the vector field

$$\nu = \frac{\partial W_s}{\partial t} - \frac{\partial W_t}{\partial s} - [W_s, W_t],$$

and ν is exactly the vector field $\nu_{\alpha,\beta}$ we need to identify. Recall that

$$W_s = \frac{-\nabla \partial_s \phi}{1 - \Delta\phi} \ , \ W_t = \frac{-\nabla \partial_t \phi}{1 - \Delta\phi}.$$

So

$$\frac{\partial W_s}{\partial t} = \frac{-1}{1-\Delta\phi} \nabla \left(\frac{\partial^2 \phi}{\partial s \partial t} \right) + \frac{1}{(1-\Delta\phi)^2} \Delta \partial_s \phi \nabla \partial_t \phi.$$

Evaluating at $s = t = 0$ where $\partial_s \phi = \alpha, \partial_t \phi = \beta$ we have

$$\frac{\partial W_s}{\partial t} - \frac{\partial W_t}{\partial s} = \frac{1}{(1-\Delta\phi)^2} (\Delta\alpha \nabla\beta - \Delta\beta \nabla\alpha).$$

Write g for the function $(1 - \Delta\phi)^{-1}$. Combining with the Lie bracket term we obtain

$$\nu_{\alpha,\beta} = [g\nabla\alpha, g\nabla\beta] + g^2(\Delta\alpha\nabla\beta - \Delta\beta\nabla\alpha).$$

Now applying (5.8) we have

$$[g\nabla\alpha, g\nabla\beta] = \text{curl } (g^2 \nabla\alpha \times \nabla\beta) - \text{div } (g\nabla\alpha) g\nabla\beta + \text{div } (g\nabla\beta) g\nabla\alpha.$$

Applying (5.9) we have

$$\text{curl } (g^2 \nabla\alpha \times \nabla\beta) = g \text{ curl } (g\nabla\alpha \times \nabla\beta) + g((\nabla g, \nabla\alpha)\nabla\beta - (\nabla g, \nabla\beta)\nabla\alpha).$$

Since

$$\text{div } (g\nabla\alpha) = g\Delta\alpha + (\nabla g, \nabla\alpha), \quad \text{div } (g\nabla\beta) = g\Delta\beta + (\nabla g, \nabla\beta)$$

we see that

$$\nu_{\alpha,\beta} = g \, \text{curl } (g\nabla\alpha \times \nabla\beta),$$

as required.

In the case when X has dimension 2—as discussed in Donaldson (1999), Mabuchi (1987), and Semmes (1992)—the space $\mathcal{H}$ is formally a symmetric space. This is not true in general, since the curvature tensor is not preserved by the action of the group $\mathcal{G}$.

We define a functional on $\mathcal{H}$ by

$$V(\phi) = \int_X \phi \, d\mu.$$

This function is *convex* along geodesics in $\mathcal{H}$, since the geodesic equation implies $\ddot{\phi} \geq 0$. Now introduce a real parameter $\epsilon \geq 0$ and consider the functional on paths in $\mathcal{H}$:

$$E = \int \frac{1}{2}|\dot{\phi}|_\phi^2 + \epsilon V(\phi) \, dt, \tag{5.10}$$

corresponding to the motion of a particle in the potential $-\epsilon V$. The Euler–Lagrange equations are

$$\ddot{\phi} = \frac{|\nabla_X \dot{\phi}|^2 + \epsilon}{1 - \Delta_X \phi}. \tag{5.11}$$

5.3 Three equivalent problems

In this section we will explain that there are three equivalent formulations of the same PDE problem associated to a compact Riemannian manifold X. We have essentially encountered two of these already.

- *θ equation.* This is the problem we set up in Section 5.1. We are given positive functions ρ_0, ρ_1 on X, with

$$\int_X \rho_i \, d\mu = \int_X d\mu. \tag{5.12}$$

  We seek a domain $\Omega_{H_0,H_1} \subset X \times \mathbb{R}$ defined by $H_0, H_1 : X \to \mathbb{R}$ and a function θ on Ω_{H_0,H_1}, equal to $0, 1$ on the two boundary components, with fluxes ρ_0, ρ_1 and satisfying the equation

$$\Delta_\epsilon \theta = 0.$$

- *Φ equation.* Here we are given ϕ_0, ϕ_1 on X, with $1 - \Delta\phi_i > 0$. We seek a function Φ on $X \times [0, 1]$, equal to ϕ_0, ϕ_1 on the two boundary components, with $1 - \Delta\Phi > 0$ for all t and satisfying the non-linear equation

$$\frac{\partial^2 \Phi}{\partial t^2}(1 - \Delta_X \Phi) - |\nabla \left(\frac{\partial \Phi}{\partial t}\right)|^2 = \epsilon. \tag{5.13}$$

  As we have explained in Section 5.2, this is the same as finding a path in the space $\mathcal{H}$, with end points ϕ_0, ϕ_1, corresponding to the motion of a particle in the potential $-\epsilon V$.

  Now we introduce the third problem.

- *U equation.* We are given positive functions ϕ_0, ϕ_1, with $1 - \Delta\phi_i > 0$, as above. Define a function L on $X \times \mathbb{R}$ by

$$L(x, z) = \max(\phi_0(x) - \phi_1(x) + z, 0).$$

We seek a C^1 function $U(x, z)$ on $X \times \mathbb{R}$ with $U \geq L$ everywhere and satisfying the equation

$$\Delta_\epsilon U = (1 - \Delta\phi_0) \tag{5.14}$$

on the open set Ω where $U > L$.

The equivalence of these three problems (assuming suitable regularity for the solutions in each case) arises from elementary, but not completely obvious, transformations. We describe these now.

5.3.1 θ *equation* $\Longrightarrow$ ϕ *equation*

Suppose we have a solution θ on a domain Ω_{H_0,H_1}. Then $\partial_z\theta = \frac{\partial\theta}{\partial z}$ is positive on the boundary components of Ω_{H_0,H_1}. The function $\partial_z\theta$ satisfies the equation $\Delta_\epsilon(\partial_z\theta) = 0$ and it follows from this that $\partial_z\theta$ is positive throughout the domain. This implies that, for any $t \in [0, 1]$, the set $\theta^{-1}(t)$ is the graph of a smooth function h_t on X. By definition $h_0 = H_0$ and $h_1 = H_1$. We also write this function as $h(t, x)$ where convenient. For each fixed t we can define a function ρ_t on X by the flux of $*_\epsilon d\theta$, just as before.

We claim that

$$\frac{\partial\rho_t}{\partial t} = \Delta_X h_t \tag{5.15}$$

We show this by direct calculation (there are more conceptual, geometric arguments). For simplicity we treat the case when the metric on X is locally Euclidean, so $\Delta_X = -\sum \partial_i^2$ where $\partial_i = \frac{\partial}{\partial x_i}$, for local coordinates x_i. The identity

$$\theta(x, h_t(x)) = t$$

implies that

$$\partial_i\theta + \partial_z\theta\ \partial_i h = 0 \tag{5.16}$$

and

$$\partial_z\theta\ \partial_t h = 1. \tag{5.17}$$

Now

$$\Delta_X h_t = -\sum_i (\partial_i + (\partial_i h)\partial_z)\partial_i h,$$

and this is

$$\Delta_X h_t = -\sum_i \left(\partial_i - \frac{\partial_i\theta}{\partial_z\theta}\partial_z\right)\left(-\frac{\partial_i\theta}{\partial_z\theta}\right)$$

which is

$$-\sum_i \left(\frac{\partial_i^2 \theta}{\partial_z \theta} - 2\frac{\partial_i \theta \partial_i \partial_z \theta}{(\partial_z \theta)^2} + \frac{\partial_i \theta \partial_i \theta \partial_z \partial_z \theta}{(\partial_z \theta)^3} \right).$$

On the other hand the flux ρ_t is given by pulling back the differential form $*_\epsilon d\theta$ on the product by the map $x \mapsto (x, h_t(x))$ and this gives

$$\rho_t = \epsilon \partial_z \theta + \frac{1}{\partial_z \theta} \sum_i (\partial_i \theta)^2.$$

So

$$\frac{\partial \rho}{\partial t} = \frac{1}{\partial_z \theta} \partial_z \left(\epsilon \partial_z \theta + \frac{\sum_i (\partial_i \theta)^2}{\partial_z \theta} \right).$$

This is

$$\partial_t \rho_t = \epsilon \frac{\partial_z \partial_z \theta}{\partial_z \theta} + 2 \frac{\sum_i \partial_i \theta \partial_i \partial_z \theta}{(\partial_z \theta)^2} - \frac{\sum_i (\partial_i \theta)^2 \partial_z \partial_z \theta}{(\partial_z \theta)^3}.$$

So we see that $\partial_t \rho_t = -\Delta_X h_t$, since $\epsilon \partial_z \partial_z \theta = -\sum_i \partial_i \partial_i \theta$.

Now the normalization (5.13) implies that there is a function ϕ_0 on X such that $\rho_0 = 1 - \Delta_X \phi_0$. For $t > 0$ we define ϕ_t by

$$\phi_t = \phi_0 + \int_0^t h_\tau d\tau.$$

We can also regard this family of functions as a single function Φ on $X \times [0, 1]$, Then (5.15) implies that $\rho_t = 1 - \Delta_X \phi_t$ for each t. We have

$$\frac{\partial^2 \Phi}{\partial t^2} = \partial_t h = \frac{1}{\partial_z \theta}$$

and

$$1 - \Delta_X \Phi = \epsilon \partial_z \theta + \frac{1}{\partial_z \theta} \sum_i (\partial_i \theta)^2.$$

So

$$\frac{\partial^2 \Phi}{\partial t^2} (1 - \Delta_X \Phi) = \epsilon + \sum_i \left(\frac{\partial_i \theta}{\partial_z \theta} \right)^2.$$

Now since

$$\partial_i \partial_t \Phi = \partial_i h_t = \frac{-1}{\partial_z \theta} \partial_i \theta$$

we can write the above as

$$\frac{\partial^2 \Phi}{\partial t^2} (1 - \Delta_X \Phi) = \epsilon + \left| \nabla_X \frac{\partial}{\partial t} \Phi \right|^2,$$

as required.

5.3.2 Φ *equation* $\Longrightarrow$ *U equation*

Here we suppose we have a solution $\Phi(x,t)$ of the Φ equation and we write $\Phi(x,0) = \phi_0, \Phi(x,1) = \phi_1$. We essentially take the Legendre transform in the t-variable. The discussion is slightly more complicated when $\epsilon = 0$, so for simplicity we treat the case when $\epsilon > 0$ and $\partial_t^2\Phi$ is strictly positive. Write $H_1(x), H_2(x)$ for the derivatives $\partial_t\Phi$ evaluated at $(x,0), (x,1)$, respectively, so $H_0 < H_1$. We calculate first in the open set Ω_{H_0,H_1}. For each fixed $x \in X$ and each z in the interval $(H_0(x), H_1(x))$ there is a $t = t(x,z)$ such that $z = \partial_t\Phi$. We set

$$U(x,z) = \Phi(x,0) - \Phi(x,t) + zt.$$

This defines a function U in Ω_{H_0,H_1}. We define U outside this set by setting $U(x,z) = 0$ if $z \leq H_0(x)$ and $U(x,z) = L(x,z) = \phi_0 - \phi_1 - z$ if $z \geq H_1(x)$. It follows from the definitions that U is C^1, that $U \geq L$, and that the set where $U > L$ is exactly Ω_{H_0,H_1}. We calculate on this set. Then $\partial_z U = t$ and

$$\partial_z^2 U = (\partial_t^2\Phi)^{-1}. \tag{5.18}$$

Differentiating with respect to the parameters x_i we have

$$\partial_i U = \partial_i\phi_0 - \partial_i\Phi,$$

and

$$\partial_i^2 U = \partial_i^2\phi_0 - \partial_i^2\Phi - \frac{\partial t}{\partial x_i}\frac{\partial^2\Phi}{\partial t\partial x_i}.$$

Differentiating the identity $z = \partial_t\Phi$ gives

$$0 = \frac{\partial^2\Phi}{\partial t\partial x_i} + \frac{\partial^2\Phi}{\partial t^2}\frac{\partial t}{\partial x_i},$$

so we can write

$$\partial_i^2 U = \partial_i^2\phi_0 - \partial_i^2\Phi + \frac{1}{\partial_t^2\Phi}(\partial_t\partial_i\Phi)^2.$$

Summing over i and using (5.18) for $\partial_z^2 U$ we obtain

$$\epsilon\partial_z^2 U - \Delta_X U = \frac{1}{\partial_t^2\Phi}\left(\epsilon + |\nabla_{Xt}\Phi|^2\right) - \Delta_X\phi_0 + 1,$$

and so

$$\Delta_\epsilon U = 1 - \Delta_X\phi_0.$$

5.3.3 *U equation* $\Longrightarrow$ θ *equation*

Now suppose we have a solution U of $\Delta_\epsilon U = \rho_0$ in a domain Ω_{H_0,H_1}, satisfying the appropriate boundary conditions, where $\rho_0 = 1 - \Delta_X\phi_0$. We set

$$\theta = \frac{\partial U}{\partial z}.$$

Then $\Delta_\epsilon \theta = 0$ and $\theta = 0, 1$ on the two boundary components. We have to check that the fluxes of $*_\epsilon d\theta$ on the boundary components are $\rho_i = 1 - \Delta_X \phi_i$. Consider first the boundary component where $z = H_0$. The flux is

$$\epsilon \partial_z \theta + \frac{|\nabla_X \theta|^2}{\partial_z \theta} = \epsilon \partial_z^2 F + \frac{1}{\partial_z^2 F} \sum (\partial_z \partial_i F)^2.$$

Now we have identities

$$(\partial_i F)(x, H_0(x)) = 0, \ (\partial_z F)(x, H_0(x)) = 0.$$

Differentiating the first of these with respect to x_i we get

$$\partial_i^2 F + \partial_i H_0 \partial_i \partial_z F = 0$$

on the boundary. Differentiating the second gives

$$\partial_i \partial_z F + \partial_i H_0 \partial_z^2 F = 0$$

on the boundary. Combining these we have

$$(\partial_z \partial_i F)^2 = (\partial_z^2 F)(\partial_i^2 F).$$

Hence the flux is

$$\epsilon \partial_z^2 F + \sum_i \partial_i^2 F = \rho_0.$$

The argument for the other boundary component $\{z = H_1(x)\}$ is similar.

5.4 Existence results and discussion

We have set up a class of PDE problems associated to any compact Riemannian manifold, and seen that these have three equivalent formulations. In this section we will make some remarks about existence results, and comparison with the free-boundary literature. This discussion is unfortunately rather incomplete, mainly due to the author's limited grasp of the background.

5.4.1 *Monge–Ampère and the results of Chen*

For a function Φ on $X \times (0, 1)$ write $q(\Phi)$ for the non-linear differential operator

$$q(\Phi) = \partial_t^2 \Phi (1 - \Delta_X \Phi) - |\nabla_X \frac{\partial}{\partial t} \Phi|^2.$$

So our "Φ equation" is $q(\Phi) = \epsilon$. When X has dimension 1—a circle with local coordinate x—we can write $\Delta_X = -\partial_x^2$ and the equation is the real Monge–Ampère operator

$$q(\Phi) = \det \begin{pmatrix} \partial_t^2 \Phi & \partial_x \partial_t \Phi \\ \partial_x \partial_t \Phi & 1 + \partial_x^2 \Phi \end{pmatrix}$$

When X has dimension 2 the operator can be expressed as a *complex* Monge–Ampère operator. That is, we regard X as a Riemann surface and identify the Laplace operator on X with $i\bar{\partial}\partial$. We take the product with a circle, with coordinate α, and let $\tau = t + i\alpha$ be a complex coordinate on the Riemannn surface $S^1 \times (0,1)$. Then, in differential form notation, our non-linear operator is given by

$$(\omega_0 + i\bar{\partial}\partial\Phi)^2 = q(\Phi)\omega_0 d\tau d\bar{\tau},$$

where ω_0 is the Riemannian area form of X lifted to $X \times S^1 \times (0,1)$. Our Dirichlet problem becomes a Dirichlet problem for S^1-invariant solutions of this complex Monge–Ampère equation on $X \times S^1 \times (0,1)$. This was studied by Chen (2000), and it follows from his results that, for any $\epsilon > 0$ there is a unique solution to our problem, and hence an affirmative answer to Question 5.1 in this case. (Chen does not state this result explicitly, but it follows from the continuity method developed in Chen 2000, section 3, that for any strictly positive smooth function ν on $X \times [0,1]$ there is a solution of the equation $q(\Phi) = \nu$ with prescribed boundary values ϕ_0, ϕ_1.)

It seems quite likely that the techniques used by Chen can be extended to the higher dimensional case. The foundation for this should be provided by a convexity property of the non-linear operator which we will now derive. Let A be a symmetric $(n+1) \times (n+1)$ matrix with entries A_{ij} $0 \le i, j \le n$. Define

$$Q(A) = A_{00} \sum_{i=1}^{n} A_{ii} - \sum_{i=1}^{n} A_{i0}^2.$$

Thus Q is a quadratic function on the vector space of symmetric $(n+1) \times (n+1)$ matrices.

Lemma 5.1

1. *If $A > 0$ then $Q(A) > 0$ and if $A \ge 0$ then $Q(A) \ge 0$.*
2. *If A, B are matrices with $Q(A) = Q(B) > 0$ and if the entries A_{00}, B_{00} are positive then for each $s \in [0,1]$*

$$Q(sA + (1-s)B) \ge Q(A), \; Q(A - B) < 0.$$

Moreover, if $A \neq B$ then strict inequality holds.

To see the first item, observe that we can change basis in $\mathbb{R}^n \subset \mathbb{R}^{n+1}$ to reduce to the case when the block $A_{ij}, 1 \le i, j \le n$ is diagonal, say, with entries b_i. Then if $A \ge 0$ we have $A_{00} b_i \ge A_{0i}^2$ and so

$$Q(A) = \sum A_{00} \sum b_i - \sum A_{0i}^2 \ge 0,$$

with strict inequlality if $A > 0$.

For the second item, we just have to observe that Q is induced from a quadratic form of Lorentzian signature on $\mathbb{R}^{n+2}$ by the linear map

$$\pi : A \mapsto (A_{00}, \sum_{i=1}^{n} A_{ii}, A_{0i}).$$

The hypotheses imply that $\pi(A)$ and $\pi(B)$ are in the same component of a hyperboloid defined by this Lorentzian form and the statements follow immediately from elementary geometry of Lorentz space.

Using this lemma we can deduce the uniqueness of the solution to our Dirichlet problem, in any dimension.

Proposition 5.1 *If $\phi_0, \phi_1 \in \mathcal{H}$ then there is at most one solution Φ of the equation $Q(\Phi) = \epsilon$ on $X \times [0,1]$ with $1 - \Delta_X \Phi > 0$ for all t and with $\Phi(x,0) = \phi_0(x), \Phi(x,1) = \phi_1(x)$.*

We show that the functional $E(\Phi)$ given by (5.10) is convex with respect to the obvious linear structure. Thus we consider a one-parameter family $\Phi_s = \Phi + s\psi$, with the fixed end points. We have

$$\frac{d}{ds} E(\Phi_s) = \int_0^1 \int_X 2\dot{\Phi}_s \dot{\psi}(1 - \Delta_X \Phi_s) - \dot{\Phi}^2 \Delta_X \psi.$$

Integrating by parts (just as in the derivation of the geodesic equation) we obtain

$$\frac{d}{ds} E(\Phi_s) = \int_0^1 \int_X (q(\Phi_s) - \epsilon)\ \psi\ d\mu.$$

Suppose that Φ_0, Φ_1 are two different solutions, so when $s = 0, 1$ the term $q(\Phi_s) - \epsilon$ in the above expression vanishes pointwise. Item (2) in Lemma 5.1 implies that for $s \in (0,1)$ we have $q(\Phi_s) - \epsilon \geq 0$, with strict inequality somewhere. This means that $E(\Phi_1) > E(\Phi_0)$. Interchanging the roles of Φ_0, Φ_1 we obtain the reverse inequality, and hence a contradiction.

One can also prove this uniqueness using the maximum principle. Note too that the uniqueness is what one would expect, formally, from the negative curvature of the space $\mathcal{H}$ and the convexity of the functional V.

5.4.2 *Comparison with the free-boundary literature*

The author is not at all competent to make this comparison properly. Suffice it to say, first, that the problem we are considering is very close to those which have been studied extensively in the applied literature. For example, in the θ-formulation, the condition of prescribing the pull-back of the flux on the free boundary is the same as that in the classical problem of the "porous dam" (Elliot and Ockendon 1982, chapter 4.4; Baiocchi and Capelo 1984, chapter 8), but with the difference that in that case ρ is constant and there are additional boundary conditions on other boundary components. Second, the constructions we have introduced in Section 5.3 all appear in this literature. The transformation from θ to Φ taking the harmonic function θ as a new independent variable is called in Crank (1984, chapter 5) the "isothermal migration method." The transformation from the formulation in terms of θ to that in terms of U is known as the Baiocchi transformation (Elliot and Ockendon 1982; Baiocchi and Capelo 1984; Crank 1984). The transformation of the free-boundary problem for a linear equation to a non-linear Dirichlet problem is used in Kinderlehrer and Nirenberg (1979) to derive fundamental regularity results.

An important feature of the U-formulation is that it admits a variational description. Recall that we are given a function $L = \max(\phi_0 - \phi_1 + z, 0)$ on $X \times \mathbb{R}$ and we seek a C^1 function U with $U \geq L$ satisfying the equation $\Delta_\epsilon U = \rho_0$ on the set where $U > L$. This can be formulated as follows. We fix a large positive M and consider the functional

$$\mathcal{E}_M(U) = \int \frac{1}{2}|\nabla_X U|^2 + \epsilon|\partial_z U|^2 - \rho_0 U \; d\mu dz$$

over the space of functions satisfying the constraint $U \geq L$, where the integral is taken over $X \times [-M, M]$ in $X \times \mathbb{R}$ (which, *a posteriori*, should contain the set Ω_{H_0,H_1} on which $U > L$). Then the solution minimizes $\mathcal{E}_M$ over all functions $U \geq L$. This can be used to give another proof of the uniqueness of the solution to our problem. It seems likely that it could also be made the basis of an existence proof, following standard techniques in the free-boundary literature. Now recall that our Φ-formulation was based on a variational principle, with Lagrangian (5.10). To relate the two, we consider any function Φ on $X \times [0, 1]$ with $\partial_t \partial_t \Phi \geq 0$ and define U by the recipe of Section 5.3. We suppose that $-M < \partial_t \Phi(x, 0)$ and $\partial_t \Phi(x, 1) < M$ for all $x \in X$. Then we have

Proposition 5.2 *The functional $\mathcal{E}_M(U)$ is*

$$E(\Phi) + M \int_X (1 - \Delta_X \phi_0)(\phi_0 - \phi_1) + \frac{1}{2}|\nabla(\phi_1 - \phi_0)|^2 d\mu$$

$$+ \left(\frac{M^2}{2} + \epsilon M\right) \int_X d\mu - \epsilon \int_X \phi_1 d\mu.$$

Thus if we fix M and the end points ϕ_0, ϕ_1 the two functionals differ by a constant. The central step in the proof is the fact that the integrals

$$\int_0^1 \int_X \partial_t^2 \Phi |\nabla_X \Phi|^2 \; d\mu \; dt,$$

$$\int_0^1 \int_X \Delta_X \Phi (\partial_t \Phi)^2 \; d\mu \; dt$$

are equal modulo boundary terms. We leave the full calculation as an exercise for the reader.

5.4.3 *Degenerate case*

So far, in this section, we have discussed the case when $\epsilon > 0$. In that case the equations we are studying are elliptic. The degenerate case, when $\epsilon = 0$, is much more delicate. In fact Chen's main concern in Chen (2000) was to obtain results about this case, taking the limit as ϵ tends to 0. Chen shows that the Dirichlet problem for Φ, with $\epsilon = 0$, has a $C^{1,1}$ solution but the question of smoothness is open. The formulation of the problem in terms of the function U has particular advantages here, because the problem is set-up as a family of elliptic problems, and the issue becomes one of smooth dependence on parameters. (This is related to another approach, involving families of holomorphic maps, discussed in Semmes 1992 and Donaldson 1999.) We can express the central question as follows. Suppose we have a smooth function λ on a compact Riemannian manifold X and fix a smooth positive function ρ. Let J be the functional

$$J(u) = \int_X \frac{1}{2}|\nabla u|^2 - \rho u.$$

For each $z \in \mathbb{R}$ we set $\lambda_z = \max(\lambda, z)$ and minimize the functional J over the set of functions $u \geq \lambda_z$. Suppose we know that there is a minimizer u_z which is smooth on the open set $\Omega_z \subset X$ where $u_z > \lambda_z$. Let $\Omega = \{(x, z) : x \in \Omega_z\} \subset X \times \mathbb{R}$.

Question 5.2 *In this situation, does u_z vary smoothly with z in Ω?*

The interesting case here seems to be when z is a critical value of g.

5.5 Relation with Nahm's equations

We recall that Nahm's equations are a system of ODE for three functions T_1, T_2, and T_3 taking values in a fixed Lie algebra:

$$\frac{dT_i}{dt} = [T_j, T_k], \tag{5.19}$$

where i, j, k run over cyclic permutations of $1, 2, 3$. To simplify notation, let us fix on the Lie algebra $u(n)$. It is equivalent (at least in the finite-dimensional case) to introduce a fourth function T_0 and consider the equations

$$\frac{dT_i}{dt} + [T_0, T_i] = [T_j, T_k], \tag{5.20}$$

with the action of the "gauge group" of $U(n)$-valued functions $u(t)$:

$$T_i \mapsto uT_iu^{-1}, \quad T_0 \mapsto uT_0u^{-1} - \frac{du}{dt}u^{-1}$$

which preserved solutions to (5.20) (i.e. using the gauge group we can transform T_0 to 0). The equations imply that

$$\frac{d}{dt}(T_2 + iT_3) = [T_0 + iT_1, T_2 + iT_3], \tag{5.21}$$

so $T_2 + iT_3$ moves in a single adjoint orbit in the Lie algebra of $GL(n, \mathbb{C})$. Conversely if we fix some B in this complex Lie algebra, introduce a function $g(t)$ taking values in $GL(n, \mathbb{C})$ and define skew-Hermitian matrices $T_i(t)$ by

$$T_0 + iT_1 = \frac{dg}{ds}g^{-1},$$

$$T_2 + iT_3 = gBg^{-1},$$

then two of the three Nahm equations are satisfied identically. The remaining equation can be expressed in terms of the function $h(t) = g^*(t)g(t)$, taking values in the space $\mathcal{H}$ of positive definite Hermitian matrices, which we can also regard as the quotient space $GL(n, \mathbb{C})/U(n)$. This equation for $h(t)$ is a second-order ODE which is the Euler–Langrange equation for the Lagrangian

$$E(h) = \int |\frac{dh}{dt}|^2_{\mathcal{H}} + V_B(h)dt.$$

Here $|\ |_{\mathcal{H}}$ denotes the standard Riemannian metric on $\mathcal{H}$. The function V on $\mathcal{H}$ is

$$V_B(h) = \mathrm{Tr}(hBh^{-1}B^*).$$

If g is any element of $GL(n, \mathbb{C})$ with $g^* = h$ then

$$V_B(h) = |gBg^{-1}|^2,$$

so V_B is determined by the norm of matrices in the adjoint orbit of B. (See Donaldson (1984) for details of the manipulations involved in all the above.) The result in Donaldson (1984), mentioned in the introduction of this chapter, is that for any two points $h_0, h_1 \in \mathcal{H}$ there is a unique solution $h(t)$ to the Euler–Lagrange equations for $t \in [0, 1]$ with $h(0) = h_0, h(1) = h_1$.

These constructions go over immediately to the case when $U(n)$ is replaced by any compact Lie group and $GL(n, \mathbb{C})$ by the complexified group. We want to extend them to the situation where $U(n)$ is replaced by the group $\mathcal{G}$ of Hamiltionian diffeomorphisms of a surface Σ with a fixed area form (or more precisely, the extension of this group given by a choice of Hamiltionian). The essential difficulty is that this group does not have a complexification. However, as explained in Donaldson (1999), Mabuchi (1987), and Semmes (1993), the space $\mathcal{H}$ of Kähler potentials behaves formally like the quotient space $\mathcal{G}^c/\mathcal{G}$ for a fictitious group $\mathcal{G}^c$. Thus the problem we have formulated in Section 5.2 can be viewed as an analogue of the desired kind provided that our potential function V can be seen as an analogue of V_B in the finite-dimensional case.

If we have a path ϕ_t in $\mathcal{H}$ with $\phi_0 = 0$ and a function $\beta : \Sigma \to \mathbb{C}$ we can write down a differential equation for a one-parameter family β_t which corresponds, formally, to the adjoint action of the complexified group $\mathcal{G}^c$, with the initial condition $\beta_0 = \beta$. The equation has the shape

$$\frac{\partial \beta_t}{\partial t} = \nabla \dot{\phi} \bar{\partial} \beta_t.$$

The problem is that this evolution equation will not have solutions, even for a short time, in general. But if we suspend for a moment our assumption that we are working over a compact Riemann surface and suppose that β is a *holomorphic* function then there is a trivial solution $\beta_t = \beta$. So, formally, the functional V_β on $\mathcal{H}$ is given by the L^2 norm of β with respect to the measure $d\mu_\phi$:

$$\int (1 - \Delta_X \phi)|\beta|^2.$$

Even if this integral is divergent, the variation with respect to compactly supported variations in ϕ is well defined, and this is what corresponds to the gradient of V_B appearing in the equations of motion. Moreover, we can integrate by parts to get another formal representation of a functional with the same variation

$$-\int \phi \Delta_X |\beta|^2 = \int \phi |\nabla \beta|^2.$$

Now take the compact Riemann surface Σ to be a two-torus, and identify the space $\mathcal{H}$ with periodic Kähler potentials on the universal cover $\mathbb{C}$. On this cover the identity function β is holomorphic, and we see from the above that the formal expression

$$V_\beta = \int_{\mathbb{C}} \phi$$

is analogous to the function V_B in the finite-dimensional case. Of course the integrand is periodic and so the integral will be divergent but we can return to the compact surface Σ and consider the well-defined functional

$$V_\beta(\phi) = \int_\Sigma \phi$$

which will generate the same equations of motion. So we see that, modulo some blurring of the distinction between Σ and its universal cover, the functional we have been considering is indeed analogous to that in the finite-dimensional case.

Using the transformation from the Φ equation to the θ equation, we obtain a relation between Nahm's equations for the Hamiltonian diffeomorphisms of a surface and harmonic functions on $\mathbb{R}^3$. This can be seen in other ways. Most directly, we consider three one-parameter families of functions $h_i(t)$ on a surface Σ with an area form which satisfy

$$\frac{dh_i}{dt} = \{h_j, h_k\}, \tag{5.22}$$

where $\{\ ,\ \}$ is the Poisson bracket. We think of these as a one-parameter family of maps $\underline{h}_t : \Sigma \to \mathbb{R}^3$, and assume for simplicity that these are embeddings, with disjoint images. Then it is a simple exercise to show that (5.22) imply that the images $\underline{h}_t(\Sigma)$ are the level sets of a harmonic function on a domain in $\mathbb{R}^3$. From another point of view, the geometric structure defined by a solution to the Φ equation is an S^1 invariant Kähler metric $\Omega = \omega_0 + i\overline{\partial}\partial\Phi$ on $\Sigma \times S^1 \times (0,1)$ with volume form

$$\Omega^2 = d\tau d\beta \overline{d\tau d\beta}.$$

Since $d\tau d\beta$ is an S^1-invariant holomorphic two-form, what we have is an S^1-invariant *hyperkähler* structure. Then the relation with harmonic functions appears as the Gibbons–Hawking construction for hyperkähler metrics.

The development above is rather limited, since we have only been able to formulate an analogue of our Nahm's equation problem for a single function β. One can go further, and arrive at other interesting free-boundary problems. Consider for example the case when the surface Σ is the two-sphere with the standard area form, and the orientation-reversing map $\sigma : \Sigma \to \Sigma$ given by reflection in the x_1, x_2 plane. Now consider maps $\beta : \Sigma \to \mathbb{C}$ with $\beta = \beta \circ \sigma$ which are diffeomorphisms on each hemisphere. Then the push-forward of the area form on the upper hemisphere defines a two-form ρ_β on $\mathbb{C}$ with support in a topological disc $\beta(\Sigma) \subset \mathbb{C}$. (The form ρ_β will not usually be smooth, but will behave like $d^{-1/2}$ where d is the distance to the boundary of $\beta(\Sigma)$.) Clearly the form ρ_β determines β up to the action of the σ-equivariant Hamiltionian diffeomorphisms of Σ. Suppose that h is a σ-invariant function on Σ. We can regard this as an element of the Lie algebra of $\mathcal{G}^c$ and consider its action on β. This is given by $\Delta_{\mathbb{C}} h$ where h is thought of as a function on $\mathbb{C}$, vanishing outside $\beta(\Sigma)$. So a reasonable candidate for a model of the quotient of the space of maps β by the action of $\mathcal{G}^c$ is given by the following. We consider two-forms ρ supported on topological discs in $\mathbb{C}$, with singularities at the boundary of the kind arising above, and impose the equivalence relation that $\rho_0 \sim \rho_1$ if there is a compactly supported harmonic function F on $\mathbb{C}$ with $\Delta F = \rho_0 - \rho_1$.

Now let $\theta(x_1, x_2, z)$ be a harmonic function on an open set $\Omega \subset \mathbb{R}^3$, with $\theta(x_1, x_2, z) = \theta(x_1, x_2, -z)$. Suppose that Ω is diffeomorphic to $S^2 \times (0,1)$, and that $\theta = 0$ on the inner boundary component Σ_0 and $\theta = 1$ on the outer boundary component Σ_1. Suppose also that the projections of Σ_0, Σ_1 to the (x_1, x_2) plane are diffeomorphisms on each upper hemisphere, mapping to a pair of topological disc $D_0 \subset D_1$. Then the flux of $\nabla\theta$ on each boundary component pushes forward to define a pair of compactly supported two-forms ρ_0, ρ_1 on $\mathbb{C}$. These are equivalent in the sense above, since $\rho_0 - \rho_1 = \Delta_{\mathbb{C}} F$ for the function

$$F(x_1, x_2) = \int z \frac{\partial\theta}{\partial z} dz,$$

where the integral is taken over the intersection of the vertical line through $(x_1, x_2, 0)$ with Ω. Our hypotheses imply that $F \geq 0$, and F is supported on the larger disc D_1.

The question we are lead to is the following:

Question 5.3 *Suppose that $D_0 \subset D_1$ are topological discs in $\mathbb{C}$, that ρ_i are two-forms supported on D_i, and that there is a non-negative function F on $\mathbb{C}$, supported on D_1, with $\rho_0 - \rho_1 = \Delta_{\mathbb{C}} F$ (where the Laplacian is defined in the distributional sense). Do ρ_0, ρ_1 arise from a unique harmonic function θ on a domain in $\mathbb{R}^3$, by the construction above?*

(For simplicity we have not specified precisely what singularities should be allowed in the forms ρ_i: this specification should be a part of the question.)

Hitchin showed in Hitchin (1983) that Nahm's equations form an integrable system. The root of this is the invariance of the conjugacy class given by (5.21), together with the family of similar statements that arise from the $SO(3)$ action on the set-up. In this vein, we can write down infinitely many conserved quantities for the solutions of (5.11) on the Riemannian manifold $\mathcal{H}$. Let f_λ be an eigenfunction of the Laplacian Δ_X, with eigenvalue $\lambda > 0$. Then we have

Proposition 5.3 *For any $\epsilon > 0$, if ϕ_t satisfies (5.11) then the quantity*

$$\int_X \exp\left(\sqrt{\frac{\lambda}{\epsilon}}\,\dot{\phi}\right) f_\lambda(1 - \Delta\phi)\, d\mu,$$

does not vary with t.

This becomes rather transparent in the θ-formulation, using the fact that the function

$$K_\lambda(x, z) = f_\lambda(x) \exp\left(\sqrt{\frac{\lambda}{\epsilon}}\, z\right),$$

satisfies $\Delta_\epsilon K_\lambda = 0$.

Acknowledgements

The author is grateful to Professors Colin Atkinson, Xiuxiong Chen, Darryl Holm, and John Ockendon for helpful discussions.

References

Baiocchi, C. and Capelo, A. (1984). *Variational and Quasivariational Inequalities.* Wiley, New York.

Chen, X.-X.(2000). The space of Kahler metrics. *Journal of Differential Geometry* **56**, 189–234.

Crank, J. (1984). *Free and Moving Boundary Problems.* Oxford University Press, Oxford.

Donaldson, S. (1984). Nahm's equations and the classification of monopoles. *Communications in Mathematical Physics* **96**, 387–407.

Donaldson, S. (1993). Complex cobordism, Ashtekar's equations and diffeomorphisms. In: *Symplectic Geometry*, D. Salamon (ed.). Cambridge University Press, Cambridge, pp. 45–55.

Donaldson, S. (1999). Symmetric spaces, Kahler geometry and Hamiltonian dynamics. In: *Northern California Symplectic Geometry Seminar*, Y. Eliashberg *et al.* (eds.). American Mathematical Society, Providence, RI, pp.13–33.

Donaldson, S. (2002). Holomorphic discs and the complex Monge-Ampère equation. *Journal of Symplectic Geometry* **1**, 171–96.

Elliot, C. and Ockendon, J. (1982). *Weak and Variational Methods for Moving Boundary Problems.* Pitman, Boston, MA.

Hitchin, N. (1983). On the construction of monopoles. *Communications in Mathematical Physics* **89**, 145–90.

Kinderlehrer, D. and Nirenberg, L. (1979). Regularity in free boundary problems. *Annali della Scuola Normale Superiore di Pisa* **4**(2), 373–91.

Mabuchi, T. (1987). Some symplectic geometry on Kahler manifolds, I. *Osaka Journal of Mathematics* **24**, 227–52.

Semmes, S. (1992). Complex Monge-Ampère and symplectic manifolds. *American Journal of Mathematics* **114**, 495–550.

VI

SOME ASPECTS OF THE THEORY OF HIGGS PAIRS

S. Ramanan

Dedicated to Nigel Hitchin

This is an exposition of certain aspects of the theory of Higgs pairs (also called 'Higgs bundles' by some authors), naturally with some emphasis on topics in which I played a part. This is an area to which fundamental contributions have been made by Nigel Hitchin, and I take pleasure in presenting this summary on the occasion of his sixtieth birthday.

6.1 Moduli of vector bundles

Let C be a projective nonsingular curve over $\mathbb{C}$, or what is the same, a compact Riemann surface. We will denote its genus by g. We will assume that $g \geq 2$ throughout. It is well known that isomorphism classes of line bundles (or divisor classes) of degree 0 on C are in one-to-one correspondence with unitary characters of the fundamental group π. Indeed every character of π gives rise to a line bundle, namely, the one associated to the universal covering which is a principal π-bundle. It is this association which gives rise to the above bijection.

Since the abelianization of π is free of rank $2g$, the space of unitary characters of π can be identified with the $2g$-fold product of S^1. Thus our remark above implies that there is a natural bijection of the moduli of line bundles of degree 0 with the $2g$-dimensional (real) torus.

The moduli space of line bundles of fixed degree d comes with a natural complex structure. Indeed, it is a nonsingular projective variety J^d. When $d = 0$, what we get is a natural structure of a projective variety on the torus, in other words, an *abelian variety*. This is the classical *Jacobian variety* J of C.

If we think of a line bundle as a (fibred) family of one-dimensional spaces, parametrized by points of C, the ambiguity in identifying the fibres with a standard one-dimensional vector space, say, $\mathbb{C}$, is given by the action of $\mathbb{C}^*$ on $\mathbb{C}$. We can then ask for a higher rank, or *non-abelian* analogue of this construction. In other words, one considers *vector bundles* of rank n, instead of line bundles and replaces the abelian group $\mathbb{C}^*$ by the non-abelian group $GL(n, \mathbb{C})$.

Again, as before, any unitary representation ρ of π gives rise to a vector bundle E_ρ. This was considered by A. Weil in 1939, who observed that *vector bundles associated to unitary representations ρ and ρ' are isomorphic if and only if ρ is equivalent to ρ'* (Weil 1938). Moreover, any infinitesimal deformation of a unitary representation ρ clearly gives rise to one of E_ρ. If ρ is irreducible, this gives an *injective* map from the infinitesimal deformation space of ρ into that of E_ρ. This is just the tangent space level analogue of the above statement.

Weil also observed that actually, every infinitesimal deformation of E_ρ arises from an infinitesimal deformation of ρ. In other words, if the set of isomorphism classes of all vector bundles of a given rank and degree 0 had the structure of a variety, the map from (the set of equivalence classes of) irreducible unitary representations to the space of vector bundles would be bijective at the infinitesimal level. Finally he also remarked that although not every vector bundle arises from a unitary representation, those that so arise 'play, without doubt, an important part'.

There things stood for a quarter century, till Mumford, thanks to his newly developed tool of 'Geometric Invariant Theory', based on ideas of Hilbert, came up with a new notion. He showed that the bundles that are classifiable in the sense of allowing a natural structure of a variety in the set of their isomorphism classes can be characterized purely from the algebraic geometric point of view. He called them 'stable' bundles. This is defined in terms of the *slope* of a vector bundle. Indeed, to every vector bundle E of rank n, one associates a line bundle, *its determinant*, namely, the nth exterior power $\Lambda^n(E)$. The degree of the latter is called the degree of E as well. The slope is the rational number $\mu(E) = \text{degree}(E)/\text{rank}(E)$.

Definition 6.1 *A vector bundle E is stable if for every proper subbundle F, we have $\mu(F) < \mu(E)$. A vector bundle E is polystable if E is isomorphic to a direct sum $\oplus E_i$ of stable subbundles E_i all of whose slopes are the same (as that of E).*

The fundamental theorem of Narasimhan and Seshadri (1965) resolved the puzzle completely.

Theorem 6.1 (Narasimhan and Seshadri 1965) *A vector bundle of rank n is isomorphic to one associated to a unitary representation $\pi \to U(n)$ if and only if it is polystable of degree 0. This gives a bijection between equivalence classes of unitary representations of π, and isomorphism classes of polystable vector bundles of degree 0. Under this bijection, irreducible representations correspond to stable bundles.*

If in the definition of stable bundles, we replace the strict inequality $<$ by the non-strict inequality $\leq$, we get the notion of *semi-stability*. Every semistable bundle E admits a filtration by subbundles such that the associated graded bundle is polystable. Although the filtration itself is not unique, the associated graded polystable bundles given by any two such filtrations are isomorphic. Hence one can associate to any semistable bundle E, a polystable bundle $Gr(E)$, well defined up to isomorphism.

Definition 6.2 *A vector bundle E is said to be S-equivalent to E' if $Gr(E)$ is isomorphic to $Gr(E')$.*

It turns out that from the moduli point of view, it is better to think of the moduli space as the set of S-equivalence classes of semi-stable bundles, rather than as isomorphism classes of polystable bundles.

Theorem 6.2 (Seshadri 1967) *The set of S-equivalence classes of semi-stable bundles of a given rank n and degree d has a natural structure of a normal (irreducible) projective variety of dimension $n^2(g-1)+1$.*

We will denote the above variety corresponding to rank n and degree d by $U_C(n,d)$ or simply $U(n,d)$ where C is understood. If the rank is 1 and the degree 0, we get back the Jacobian J. If $n=1$ but the degree d is arbitrary, the corresponding moduli space is denoted J^d. Clearly J is a group variety under tensor product, and it acts simply transitively on J^d for all d.

There is a natural morphism of $U(n,d)$ into J^d namely the association of $\det(E)=\Lambda^n(E)$ to E. Notice that $\xi \mapsto \xi^n$ gives a morphism $J \to J$ which is a Galois covering with Galois group J_n, the group $\{\xi \in J : \xi^n = 1\}$ of n-division points of the Jacobian. Under tensor product operation, J (and in particular, J_n) acts on $U(n,d)$. If $SU(n,d)$ denotes the subspace of $U(n,d)$ consisting of bundles with some fixed determinant, then the action of J_n leaves $SU(n,d)$ invariant. Thus the J_n-Galois covering $J \to J$ gives rise to the associated bundle over J with $SU(n,d)$ as fibre. This can be identified with $U(n,d)$. Thus, for the most part, the study of $U(n,d)$ can be reduced to one of $SU(n,d)$ (together with the action of J_n on it).

In the 1960s and 1970s, these varieties were studied intensely, especially by Narasimhan, Newstead, Seshadri, and myself. I will recall here some of the results obtained, in order to facilitate the rest of this survey.

Firstly, when n and d are relatively prime, the varieties $U(n,d)$ and $SU(n,d)$ are smooth projective. Under this assumption, semi-stability and stability of vector bundles are equivalent, and indeed it is true that points of $U(n,d)$ (respectively, $SU(n,d)$) corresponding to stable bundles are always smooth. The tangent space at a stable point E to $U(n,d)$ (respectively, $SU(n,d)$) can be identified with $H^1(C,\mathrm{End}(E))$ (respectively, $H^1(C,\mathrm{ad}(E))$, where $\mathrm{ad}(E)$ denotes the bundle of trace-free endomorphisms). It is easy to verify that $H^0(C,\mathrm{End}(E))$ is one-dimensional (respectively, $H^0(C,\mathrm{ad}(E))=0$). Hence the dimension of the tangent space at E to $U(n,d)$ (respectively, $SU(n,d)$) is $n^2(g-1)+1$ (respectively, $(n^2-1)(g-1)$). Moreover, $SU(n,d)$ is simply connected and its second Betti number is 1. Let us denote by Θ the unique ample line bundle whose Chern class generates $H^2(SU(n,d),\mathbb{Z})$. The determinant of the tangent bundle is ample and is an even multiple of Θ.

6.2 Hecke correspondence

One of the tools which Narasimhan and I came up with (Narasimhan and Ramanan 1975) in the study of $U(n,d)$ is the *Hecke correspondence.* This sets up a correspondence between $U(n,d)$ (respectively, $SU(n,d)$) and $U(n,d-1)$ (respectively, $SU(n,d-1)$). The idea is very simple. Let E be a vector bundle of rank n and degree d. Fix a point a of C and take a nonzero linear form on the fibre E_a. It can be considered as a homomorphism of the locally free sheaf defined by E onto the structure sheaf of the point a. If E is sufficiently stable, which is generically the case, the kernel is also stable and therefore defines a point of $SU(n, d-1)$. Incidentally, the kernels corresponding to two linear forms are isomorphic if they are scalar multiples of each other. Hence the kernel is determined by an element of $P(E_a^*)$.

Two points E, E' of $SU(n, d)$ are *Hecke modifications* of each other if there exist elements $l \in P(E_a^*)$ and $l' \in P(E_a'^*)$ such that $\ker(l)$ is isomorphic to $\ker(l')$. Thus we have a closed subvariety H_C of $SU(n, d) \times SU(n, d)$ which is invariant under the transposition of coordinates.

For a generic $E \in SU(n, d+1)$ with determinant $\xi \otimes \mathcal{O}(a)$ we can thus see that $P(E_a^*)$ is contained in $SU(n, d)$ with determinant ξ. These projective spaces are called *Hecke cycles*. In particular, if we consider a projective line in $P(E_a^*)$ it gives a projective line in $SU(n, d)$ which we call a *Hecke curve*. Thus $SU(n, d)$ contains a lot of rational curves. This is not surprising, since it is easily seen that $SU(n, d)$ are unirational, and indeed (not so easily) even rational if n and d are coprime (King and Schofield 1999).

6.3 Moduli of Higgs pairs

One may ask whether the set of all (i.e. not necessarily unitary) representations of π can also be described purely algebraically. Donaldson (1993) in reproving the result of Narasimhan and Seshadri, started this process of thinking. The upshot was the following.

Let L be a fixed line bundle. Consider pairs (E, α) where E is a vector bundle and α is an *L-twisted* endomorphism, namely, a linear map $E \to E \otimes L$. A subbundle F of E is *invariant* if α maps F into $F \otimes L$. One requires the condition of stability now only for invariant subbundles and arrives at the notion of *stable*, *semi-stable*, and *polystable* pairs. Also one can make sense of S equivalence and construct the moduli of such pairs. We will call it the *moduli space of L-twisted Higgs pairs.*

We can define the *direct sum* (respectively, *tensor product*) of L-twisted Higgs pairs (E, α) and (E, α') to be $(E \oplus E', \alpha \oplus \alpha')$ (respectively, $(E \otimes E', \alpha \otimes 1_{E'} + 1_E \otimes \alpha')$). The *dual* of (E, α) is defined to be $(E^*, -\alpha^t \otimes 1_L)$.

The case when L is the canonical line bundle K, namely, the cotangent bundle of C, is of particular interest. In this case, we will omit the phrase 'K-twisted'. The component α in (E, α) can then be treated as a one-differential form with values in the bundle of endomorphisms of E and is called the *Higgs field.*

Remark 6.1 A Higgs pair (E, α) may be stable or semi-stable even if E is not stable as a vector bundle. However if (E, α) is stable, E cannot be terribly nonstable. For example, if E is of rank 2 and degree 0, and ξ is a line subbundle with $\deg(\xi) > g - 1$, then $\deg(E/\xi) < 1 - g$ and so any homomorphism of ξ into $E/\xi \otimes K$ is zero. In particular, the restriction of α to ξ followed by the projection $E \otimes K \to E/\xi \otimes K$ is zero. Hence ξ is invariant, contradicting the stability of (E, α).

Example 6.1 Let ξ be a 'theta characteristic', namely, a line bundle such that ξ^2 is isomorphic to K. In particular the degree of ξ is $g-1$. Take for E the vector bundle $\xi \oplus \xi^{-1}$. It is of rank 2 and degree 0. Since it has ξ as a subbundle, it is not semi-stable. Now a homomorphism $\alpha : E \to E \otimes K$ can be written in the form of a (2×2) matrix $\begin{pmatrix} \alpha_1 & \alpha_2 \\ \alpha_3 & \alpha_4 \end{pmatrix}$ where α_1 and α_4 are sections of K, α_2 is a section of K^2, and α_3 is a section of $\mathcal{O}$. If α_3 is nonzero, then ξ is not invariant under α. Any line subbundle of E other than ξ admits a nonzero homomorphism into ξ^{-1} and hence has degree at most $-(g-1)$. Hence for any such α, the pair (E, α) is stable.

Remark 6.2 On the other hand, if E is stable, then any α (including 0) gives rise to a Higgs pair (E, α).

I believe that the following theorem may be attributed to Donaldson (1993) and Hitchin (1987) in the case of rank 2 and to Simpson (1992) and Corlette (1988), who generalized it to higher rank and also to bundles on higher dimensional spaces:

Theorem 6.3 *There is a canonical bijection between the set of equivalence classes of completely reducible representations of π in $GL(n)$ and the set of isomorphism classes of polystable Higgs pairs of rank n and degree 0.*

Remark 6.3 We regain the theorem of Narasimhan and Seshadri by noting that under the above bijection, unitarisable representations correspond to those Higgs pairs in which the Higgs field vanishes.

We observed that every representation ρ of π gives rise to a vector bundle E_ρ, associated to the principal π-bundle, namely the universal covering. This is in general *not* the bundle in the Higgs pair associated to ρ in the above correspondence.

The bundle E_ρ comes with a natural flat (holomorphic) connection ∇_ρ. If the given representation is unitarizable, then E_ρ admits a metric along the fibres which is invariant (or flat) for the connection ∇_ρ. If ρ is not unitarizable, such a metric of course does not exist. Nevertheless one seeks 'the most favourable' unitary metric along the fibres of E_ρ.

Using the flat connection, one can trivialize the bundle *and* the connection locally. Any two such trivializations are mediated by a locally constant map into $GL(n, \mathbb{C})$. The metric we seek can then be viewed locally as a differentiable map of a domain in C into the space $\mathcal{P}$ of positive definite Hermitian metrics on $\mathbb{C}^n$. Since $\mathcal{P}$ is a (negatively curved) Hermitian manifold, it makes sense to say that the map is *harmonic*. Clearly harmonicity does not depend on the particular local trivialization that we use. The favourable metric we referred to is one which is harmonic locally. Using ideas of Eells and Sampson (1964), Donaldson (1987) proved the following theorem:

Theorem 6.4 *Complete reducibility of ρ is a necessary and sufficient condition for the existence of a global harmonic metric along the fibres on E_ρ.*

Remark 6.4 Note that when the representation is unitary, the flat metric gives, on local trivialization on a domain, a constant map, which is of course harmonic.

Any unitary metric m on a holomorphic bundle gives rise to a connection ∇_m, called the *Chern connection.* The difference of the flat connection on E_ρ, and ∇_m, is a one-form β on C with values in the bundle $\mathrm{End}(E_\rho)$. The harmonic equation is essentially equivalent to the following. If the complex structure on E_ρ is modified by the $(0,1)$ part of β, we get a holomorphic bundle with respect to which the $(1,0)$ part of β is holomorphic.

As in the case of vector bundles, one can also construct (Nitsure 1991; Simpson 1992) the moduli space $Hg(n,d)$ of Higgs pairs corresponding to a given rank n and degree d. These are normal, irreducible, and quasi-projective (but *not* projective) varieties. If n and d are coprime, they are even smooth. If E is a stable vector bundle, the entire vector space $\Gamma(C,\mathrm{End}(E)\otimes K)$ is contained in $Hg(n,d)$. Since this space is dual to $H^1(C,\mathrm{End}(E))$, the tangent space to the moduli space $U(n,d)$ at E, we conclude that the cotangent bundle over the stable locus in $U(n,d)$ is contained in $Hg(n,d)$. In any family, stable bundles form a Zariski open set and so, the cotangent bundle is in fact a dense, open subset of Hg.

6.4 Higgs pairs and the fundamental group

There is another kind of structure that Higgs pairs constitute which has profound significance. Polystable Higgs pairs of degree 0 form an additive category $\mathcal{H}g$. (In fact, morphisms form vector spaces, and compositions are bilinear.) Using semi-stability, one checks that it is actually an abelian category.

On the other hand, completely reducible π-modules also form such a category $\mathcal{R}ep$. The above bijection can be elevated to a categorical equivalence of $\mathcal{R}ep$ with $\mathcal{H}g$. In fact, this functor respects the structures of direct sum, tensor product, duals, etc. which make sense in $\mathcal{H}g$ as well as in $\mathcal{R}ep$.

One axiomatizes these structures and calls a category so equipped, a *tensor category.*

Finally, one also notices that if we fix a base point x_0 in C, then the functor that associates to the Higgs pair (E,α) the fibre E_{x_0} is an additive (indeed linear) functor of $\mathcal{H}g$ into the category of vector spaces, which respects all the structures we mentioned above. This functor is actually *faithful* in the sense that a morphism $(E,\alpha)\to(E',\alpha')$ is zero if and only if the induced morphism $E_{x_0}\to E'_{x_0}$ at the fibre level is 0. The faithful functor Φ taking (E,α) to E_{x_0} of the Higgs-tensor category into the tensor category of vector spaces, is called a *fibre* functor. A tensor category provided with a fibre functor is called a *Tannaka category.*

Note that the category $\mathcal{R}ep$ of completely reducible representations of a group G is a tensor category in a natural way. Furthermore, the forgetful functor, which associates to a representation ρ of G the underlying vector space V_ρ, is faithful and makes $\mathcal{R}ep$ a Tannaka category as well. The above correspondence is an equivalence of the two Tannaka categories.

What is its importance? To every Tannaka category $\mathcal{C}$ associate the group of automorphisms of its fibre functor. It is called the *Tannaka dual* $Tn(\mathcal{C})$ of the category. If we start with a Lie group G, and consider $\mathcal{R}ep$, then there is a natural homomorphism of G into $Tn(\mathcal{R}ep)$. We get, by definition, for any $g \in G$, and for each representation ρ, an automorphism of the vector space V_ρ, namely, $\rho(g)$. This can actually be regarded as an automorphism of the fibre functor. The map which takes g to this automorphism gives a homomorphism of G into $Tn(\mathcal{R}ep)$. It has the universal property that any completely reducible representation of G 'extends' to a unique representation of $Tn(\mathcal{R}ep)$.

We refer to it therefore as the *Tannakian completion* $Tn(G)$ of G. Tannaka considered this in the case when G is a compact Lie group. Chevalley interpreted Tannaka's results as showing that Tannakian completion $Tn(G)$ of a compact Lie group G is a reductive algebraic group which complexifies G.

The Tannakian completion $Tn(\Gamma)$ of a finitely generated group Γ is a pro-reductive algebraic group. The homomorphism $\Gamma \to Tn(\Gamma)$ is indeed universal for homomorphisms of Γ into reductive algebraic groups.

The category $\mathcal{H}g$ has been defined purely in terms of the algebraic structure unlike the fundamental group which is a priori a highly transcendental construct. We have therefore accessed the Tannaka completion $Tn(\mathcal{H}g) = Tn(\pi)$ of the fundamental group purely by algebraic geometric methods.

The profinite completion of π is a group $Pf(\pi)$, together with a homomorphism of π into it which is universal for homomorphisms into finite groups. It is called the *Grothendieck* fundamental group.

Since each homomorphism of π into a finite group also gives rise to a unique homomorphism of $Tn(\pi)$ into that group, we get a homomorphism of $Tn(\pi)$ into $Pf(\pi)$. Thus the group $Tn(\pi)$ is much richer than $Pf(\pi)$. Incidentally, Nori (1976) recovered the group $Pf(\pi)$ as the Tannaka dual of the category of what he called 'finite' bundles. These are bundles E that satisfy an isomorphism of the type

$$\oplus m_i(\otimes^i E) \simeq \oplus n_j(\otimes^j E), \quad m_i, n_j \in \mathbb{N}.$$

Here we interpret mE as the m-fold direct sum of E and assume that the i's and j's are disjoint. He showed these form a Tannaka category $\mathcal{N}$ and that its Tannaka dual is $Pf(\pi)$.

6.5 Non-abelian Hodge theory

Note that for any $\lambda \in \mathbb{C}^*$ and polystable Higgs pair (E, α), the pair $(E, \lambda\alpha)$ is also polystable. Since this action is compatible with all the structures of the

Tannaka category $\mathcal{H}g$, one easily concludes that there is a natural action of $\mathbb{C}^*$ on $Tn(\pi)$ as well.

The group of homomorphisms of π into $\mathbb{C}$ is isomorphic to $H^1(C, \mathbb{C})$ and the Hodge decomposition can be described as the action of $\mathbb{C}^*$ on it, acting by the natural (identity) character on $H^{1,0}$ and by its inverse (namely, the character $\lambda \to \lambda^{-1}$ on $H^{0,1}$).

From this point of view, the above action of $\mathbb{C}^*$ on $Tn(\pi)$, which mimics the Hodge action, can be thought of as the *non-abelian version* of the Hodge action. It should be pointed out that this action depends on the complex structure of the curve.

Actually, the considerations above can be generalized in two directions. Firstly, we can take a reductive group G and consider principal G-bundles. *Stability* of a principal G-bundle E will then mean that for any reduction of its structure group G to a parabolic subgroup P (which can be interpreted to mean a section s of the bundle E/P), the pull-back of the tangent bundle along the fibres is of positive degree. As for a Higgs pair, it now consists of a principal G-bundle E and a differential form with values in the associated adjoint bundle $\mathrm{ad}(E)$, namely the vector bundle associated to E via the adjoint representation of G in its Lie algebra. We will discuss this generalization in some detail below.

Secondly, we may take, instead of C, an arbitrary smooth, polarized projective variety X of dimension n. This essentially means that we are given an ample line bundle ξ on X. Then one can define the degree of a line bundle L to be the intersection number $c_1(L).\xi^{n-1}$, and proceed as before. A Higgs field is, in the general case, a principal G-bundle E, and a one-form α with values in $ad(E)$ such that the two-form $[\alpha, \alpha]$ vanishes. Thus we can get an algebraic description of the pro-reductive algebraic completion of the fundamental group, but more importantly, an action of $\mathbb{C}^*$ as well.

This would show that many (non-uniform) discrete subgroups of real semi-simple groups cannot be the fundamental group (nor even split quotients of the fundamental group) of smooth projective varieties. This is because one can check, thanks to rigidity theorems on uniform discrete subgroups of Lie groups, that these do not admit an action of $\mathbb{C}^*$ on the above lines. For example, $SL(n, \mathbb{Z})$, $n \geq 3$ are not.

6.6 Hitchin morphism

We have already mentioned that there is a natural inclusion of the cotangent bundle of the moduli space of stable vector bundles as an open dense set in the Higgs moduli space.

As is well known, there is a natural sympectic structure on the cotangent bundle of any smooth variety. Actually this structure extends to the whole of the locus of stable Higgs pairs.

Let (E, α) be a stable Higgs pair. Consider the map

$$\text{End}(E) \to \text{End}(E) \otimes K,$$

given by bracketing with α. We may regard this as a complex of sheaves having only two terms C^0 and C^1. It is easy to check (Biswas and Ramanan 1994) that the first hypercohomology space of this complex is the tangent space at E to the Higgs moduli space. The dual of this complex tensored with K, namely, the Serre dual of the complex, can be identified with the complex itself. In this way, we get a symplectic form on the first hypercohomology of the complex, that is, the tangent space at (E, α) to Hg. It is this that gives the canonical (Hamiltonian) symplectic form at those points of the Higgs moduli which belong to the cotangent bundle of the moduli of vector bundles, that is, when E is itself stable.

Now Hitchin defined a morphism of the Higgs moduli space into the affine space $Ht = \oplus_{i=1}^{i=r}\Gamma(\mathbb{C}, K^i)$ as follows. To any Higgs field α one can associate the *trace* as an element of $\Gamma(K)$. In a similar manner one can lift α to a map $\Lambda^i(\alpha)$ of $\Lambda^i(E)$ into $\Lambda^i(E) \otimes K^i$ and take its trace, interpreted as a section of K^i. In other words, the coefficients of the characteristic polynomial of α make sense as elements of $\Gamma(K^i)$. This gives rise to the *Hitchin morphism* of Hg mentioned above into the Hitchin space Ht.

From the above description of the tangent space and the symplectic form, one can easily conclude that the form vanishes identically on the fibres of the Hitchin morphism. On the other hand, the dimension of Ht is $g + \sum_{i=2}^{i=r}(2i-1)(g-1) = r^2(g-1)+1$. In other words, the generic fibres are *Lagrangian* submanifolds.

It was shown by Nitsure (1991) that the Hitchin morphism is actually a proper map, so that the fibres are projective varieties. By its very nature one can prove that if there is a completely integrable system, the Lagrangian fibres are, generally speaking, (open sets in) tori. In our case, the generic fibres in Hg should perhaps be abelian varieties. If so, what are these?

Fix a point $s = (s_1, \ldots, s_r) \in Ht$. Then one can construct a curve S and a (finite) morphism $\pi : S \to C$ such that $\pi_*(\mathcal{O}) = V = \mathcal{O} \oplus K^{-1} \oplus \cdots \oplus K^{-(r-1)}$ as vector bundles. But the left side comes with an algebra structure over $\mathcal{O}_C$. We only have to identify this algebra structure on V in order to construct S. In fact, S is then Spec(V).

Now we have natural isomorphisms $K^{-i} \otimes K^{-j} \to K^{-(i+j)} \subset V$, if $i + j \leq r-1$. If $i + j \geq r$, we have a homomorphism $K^{-i-j} \to K^{-k}$ for each $k, 0 \leq k \leq r-1$ given by tensoring with s_{i+j-k}. All these add up to a morphism $V \otimes V \to V$, giving the required structure. If s is sufficiently general, it can be checked that S is actually smooth. It is called the *spectral curve* defined by s. The pull-back $\pi^*(K)$ comes with a canonical section which can be described as follows. We have $\pi_*(\pi^*(K)) = K \otimes \pi_*(\mathcal{O}) = K \oplus \mathcal{O} \oplus K^{-1} \oplus \cdots \oplus K^{-(r-2)}$. The canonical section of the second summand, namely, $\mathcal{O}$, gives a section of $\pi_*(\pi^*(K))$ and hence that of $\pi^*(K)$ as claimed.

If ξ is any line bundle on S, then the direct image $\pi_*(\xi)$ is a vector bundle of rank r on C. Also the canonical section of $\pi^*(K)$ gives rise to a homomorphism $\xi \to \xi \otimes \pi^*(K)$ and therefore of $E = \pi_*\xi$ into $\pi_*(\xi \otimes (\pi^*(K)) = E \otimes K$. It is therefore a Higgs pair. It is straightforward to check that its image under the Hitchin map is s. Equally one can check that the Higgs pair thus obtained is semi-stable and that the Hitchin morphism takes this Higgs pair to (s). Thus the fibre over s of the Hitchin map can be identified with the Jacobian (of suitable degree) of S. The intersection of this fibre with the cotangent bundle over the stable locus of the moduli space of vector bundles is a Zariski open subset of the above Jacobian, and is Lagrangian for the Hamiltonian symplectic form.

The above result can be restated slightly differently. Let us choose a base for Ht. Then the Hitchin map can be considered as a map into $\mathbb{C}^l$ given by l functions on the Higgs moduli space and hence on the cotangent bundle of the vector bundle moduli space as well. The key point is that these functions commute in the Poisson algebra. It is rarely that one finds the maximum number of Poisson commuting functions on the cotangent bundle. These functions are not only algebraic, but actually homogenous polynomials on the cotangent spaces. One talks, in this case, of *algebraic complete integrability.*

Beauville, Narasimhan, and I (1989) considered the map of the Jacobian of S into $U(r, 0)$ as well as the corresponding Prym variety (the kernel of the natural 'norm map' from the Jacobian of S onto the Jacobian of C) into $SU(r, 0)$. Note that it is defined only on a big Zariski open subset but it is dominant, that is, has dense image. By looking into the induced map from $\Gamma(SU(r, 0), \Theta)$ into the space of sections of the pull-back of Θ on the Prym variety, it was shown that $\dim \Gamma(SU(r, 0), \Theta) = r^g$. Thus the theory of Higgs pairs has provided many other interesting results on linear systems on $U(n, d)$ as well.

6.7 Quantization

If D is a (linear) differential operator of order $\leq k$, then its (kth order) *symbol* is a function on the cotangent bundle which is a homogeneous polynomial of degree k on the fibres. In the above context therefore one may ask if one can associate in a canonical fashion a basis of commuting differential operators whose symbols are the Poisson commuting functions defined above.

Some motivation or analogy may be in order here. The Hitchin space is the analogue of (actually the dual of) the space of invariant polynomials for the conjugacy action of $GL(n)$ on the vector space of (n, n) matrices. The Hitchin map itself can be thought of as the analogue of associating to any matrix A, the linear form on invariant polynomials which maps any f to the evaluated complex number $f(A)$. One can associate canonically to any homogeneous invariant polynomial f, a bi-invariant differential operator D_f on the Lie group, with symbol f. Moreover, the map $f \mapsto D_f$ is an isomorphism of the algebra of invariant polynomials onto the algebra of bi-invariant differential operators on $GL(n, \mathbb{C})$.

So one might expect a homomorphism of the dual Ht^* of the Hitchin space into the space of differential operators on the moduli space of vector bundles. This is called the *quantization* of the Hitchin map.

However, it is immediately seen that this is too naive. Indeed, the moduli space has no global differential operators of higher order at all! Now the canonical bundle of $SU(n, d)$ is divisible by 2. In other words, there is a (unique) line bundle η on it whose square is the canonical bundle. It turns out that these symbols can be lifted to a commutative subalgebra of differential operators from η to itself. We will be more precise below, where we deal with the general case of a semi-simple group. This is related to the so-called *geometric Langlands programme.*

6.8 Hecke transformation and Hitchin discriminant

We may ask if one can start with a stable Higgs pair (E, α) and modify it by a Hecke transformation. Let $a \in C$ and $l \in E_a^* \setminus \{0\}$. We then get the kernel E' from the exact sequence

$$0 \to E' \to E \to \mathcal{O}_a \to 0.$$

The question is if α also gives rise to a Higgs field on E'. By composing the map $E' \to E$ with $\alpha : E \to K \otimes E$, we get a map $E' \to K \otimes E$. In order for this to factor through a map $E' \to K \otimes E'$, the induced map $E'_a \to K_a \otimes E_a$ has to have its image contained in the kernel of l. This happens for every l if and only if α vanishes at a. In particular, it does not happen for a generic (E, α). Indeed, notice first that even if we start with a Hecke line, and lift it to Hg, the Hitchin morphism (being a map into an affine space) has to be a constant on it. On the other hand, it cannot be contained in a generic Hitchin fibre, since we know that the fibre is an abelian variety and so does not contain any rational curve. So it is clear that Hecke curves do not in general lift to Hg.

Now we will look into the question if any Hecke curve at all lifts to Hg. Again such a lift has to be contained in a Hitchin fibre. So let us assume that this fibre is not generic, that is to say, the spectral curve S is *not smooth.* We will assume for our analysis that the spectral curve has only one nodal singularity. Let S' be its normalization. Denote by b the node in S over $a \in C$ and a_1, a_2 the two points in S' over b. A line bundle on S is given by a line bundle L on S' together with an isomorphism η of L_{a_1} with L_{a_2}. Thus the space of line bundles on S can be regarded as a $\mathbb{C}^*$-bundle over the Jacobian of S'. Consider the associated $\mathbb{P}^1$ bundle over the Jacobian. It comes with two canonical sections. Identify the first section with the second section after translating by the divisor class $a_2 - a_1$. This is the *compactified Jacobian* of C. This can be identified with the Hitchin fibre.

Let E' be the direct image of L on C. Its fibre at a is of the form $V \oplus L_{a_1} \oplus L_{a_2}$, where V is a vector space of dimension $n-2$. It comes with a one-dimensional subspace, namely, the graph of the isomorphism η. The corresponding Hecke transform is the bundle E, obtained as the direct image on C of the line bundle on S given by (L, η).

In other words, the Hitchin fibre corresponding to the spectral curve S is the compactified Jacobian $\tilde{J}$ described above. The explicit Higgs pair given rise to by a point of $\tilde{J}$ is also given above. If we deal with $SU(n,d)$ instead of $U(n,d)$, one has to replace the Jacobian by the Prym variety. The point is of course that we get on this fibre, projective lines parametrized by the Jacobian, or the Prym variety of S'. These lines map to Hecke curves on $SU(n,d)$.

Hwang and I (2004) considered the tangents to Hecke curves passing through the generic point E of $SU(n,d)$ and studied their lifts to the cotangent bundle. The unions of all these are, as a simple consequence of our remark above, dense in the Hitchin discriminant. The intersection of this discriminant with the cotangent space at E is a cone and defines a projective subvariety of $P(T_E^*)$. We showed from the above considerations that the dual of this variety is the variety of all the tangents to Hecke curves through E.

This implies that from $SU(n,d)$ as a variety one can geometrically extract the Hitchin discriminant and hence the variety of Hecke tangents. The latter has as its Albanese image, the curve C, so that we have a geometric description of the curve – a Torelli kind of result.

Indeed, one can work a little more and show the following (Hwang and Ramanan 2004):

Theorem 6.5 *If C and C' are two curves, then there exists a nonconstant morphism from $SU_C(n,d)$ into $SU'_C(n,d)$ if and only if C is isomorphic to C'. Moreover any isomorphism is actually induced by an isomorphism of C with C', tensoring by line bundles of order n and possibly dualizing.*

6.9 Hitchin component

We will now investigate stable pairs (E, α) in which E itself is highly non-stable. Let us first analyse the case of rank 2 bundles with trivial determinant. We observed in Remark 6.1 that every line subbundle of E is of degree $\leq g-1$.

We also gave an example where E admits a line subbundle of degree $g-1$, namely, $E = \xi \oplus \xi^{-1}$, with ξ a theta characteristic and a suitable α.

The (K-twisted) automorphisms of E can be represented as matrices $\left(\begin{smallmatrix}\lambda & \omega \\ 0 & \lambda^{-1}\end{smallmatrix}\right)$ with $\omega \in \Gamma(K)$. Also α can be expressed as $\left(\begin{smallmatrix}\alpha_1 & \alpha_2 \\ \alpha_3 & -\alpha_1\end{smallmatrix}\right)$ where α_1 is a section of K, α_2 and α_3 are sections of K^2 and $\mathcal{O}$, respectively. Here α_3 has to be nonzero. Taking $\lambda^2 = \alpha_3$ and $\omega = -\alpha_1/\lambda$, we see that (E, α) is equivalent to (E, β), where β is a matrix of the form $\left(\begin{smallmatrix}0 & -\gamma \\ 1 & 0\end{smallmatrix}\right)$. Thus fixing the above bundle E and varying γ, we get an embedding of $\Gamma(C, K^2)$ in $Hg(2,0)$. This is called the *Hitchin component.* One should bear in mind though that it depends on the choice of theta characteristic ξ. Thus there are 2^{2g} Hitchin components and the group J_2 acts transitively on the set of these components.

Note that the Hitchin map in this case is simply the map $(E, \beta) \longmapsto \gamma \in \Gamma(C, K^2)$ and that it maps the Hitchin component isomorphically onto Ht. Also this component stays in the complement of the cotangent bundle of the stable locus in $SU_C(2)$.

We will now explain why we call it the Hitchin component. It is a natural question to ask whenever one sees a geometrically defined subvariety in Hg if it has a good meaning in the space of representations. We will now discuss this in the case of the Hitchin component.

Consider those completely reducible representations of π in $SL(2, \mathbb{C})$ which come from representations into $SL(2, \mathbb{R})$. We then of course get oriented real bundles. Even topologically, these have an invariant, namely, the *Euler* class. Since we are discussing the case of compact, oriented (real) manifolds, this invariant is an integer. It was observed by Milnor (1957) and Wood (1976) that the Euler class does not take arbitrary values.

Indeed, we have the *Milnor–Wood inequality* for the Euler class c, namely, $-(g-1) \leq c \leq g-1$. Once we fix the Euler class, the bundle is topologically determined. There is an involution acting on $SL(2, \mathbb{R})$ given by conjugation by the matrix $\left(\begin{smallmatrix}0 & i \\ i & 0\end{smallmatrix}\right)$. If two representations into $SL(2, \mathbb{R})$ are taken to each other by the above involution, then obviously they are equivalent as representations into $SL(2, \mathbb{C})$. It is easy to see that *two completely reducible representations into $SL(2, \mathbb{R})$ are equivalent as representations into $SL(2, \mathbb{C})$ if and if only they are either conjugate in $SL(2, \mathbb{R})$, or conjugate after composing one of them with the involution above.* The Euler invariant of a representation is the negative of that of its transform by this involution.

The Hitchin components correspond to those representations which come from a representation into $SL(2, \mathbb{R})$ with Euler invariant $g-1$ (or equally $-(g-1)$). Hitchin himself called it the *Teichmüller component* since it is the component consisting of representations for which the image of π in $SL(2, \mathbb{R})$ are such that the quotients are compact. Thus these give again curves of genus g.

The representations with a fixed Euler invariant other than $\pm(g-1)$ form however an irreducible variety.

Hitchin wrote down for each non-negative integer $c \leq g-1$, the subset of Higgs pairs in $Hg(2,0)$ that come from representations into $SL(2, \mathbb{R})$ and has Euler invariant c.

6.10 Reductive groups and principal bundles

As we have already remarked, many of the above considerations can be extended to any principal bundle E with a reductive structure group G. We will quickly run through these analogues.

For simplicity, we will confine ourselves to semi-simple groups here. Generally speaking, the corresponding notions for a reductive group G can be reduced to this case.

Let then G be a complex semi-simple algebraic group and $\mathfrak{g}$ its Lie algebra. We are interested in principal G-bundles over C.

Definition 6.3 *Let E be a principal G-bundle. Then it is said to be stable (respectively, semi-stable) if for every reduction τ of the structure group of E from G to any parabolic subgroup P, namely, a section of the associated bundle with G/P as fibre, the pull-back of the tangent bundle along the fibres has positive (respectively, non-negative) degree.*

Using this definition, A. Ramanathan (1996) showed that the isomorphism classes of stable principal G-bundles form a quasi-projective variety of dimension $(\dim\mathfrak{g})(g-1)$. In the case of reductive groups, the dimension is $(\dim\mathfrak{g})(g-1)+\dim\mathfrak{z}$, where $\mathfrak{z}$ is the centre of $\mathfrak{g}$.

A. Ramanathan also showed that if E is semi-stable, then there exists a reduction to a parabolic subgroup P with the following property. Let $R = P/U$ be the reductive group obtained by passing to the quotient by the unipotent radical U of P. Then the R-bundle associated to the reduced P-bundle is stable. Also any character on P gives a line bundle of degree 0. Moreover, the R-bundle depends up to isomorphism only on E.

Two semi-stable bundles are defined to be *S-equivalent* if the corresponding R-bundles are isomorphic. A *polystable* bundle is one which can be reduced to the reductive part of a parabolic subgroup as a stable bundle of degree 0.

The S-equivalence classes of semi-stable G-bundles form a projective variety M_CG or MG in which stable bundles form a dense open set. However, a point corresponding to a stable point can be singular, unlike the case when $G = GL(n)$ or $SL(n)$. One can only say these are orbifold singularities. This is because, in the general case, a stable bundle may admit non-trivial automorphisms. The automorphism group $\mathrm{Aut}(E)$, which is finite, acts on $H^1(C, \mathrm{ad}(E))$, where $\mathrm{ad}(E)$ is the vector bundle associated to the adjoint representation of G in its Lie algebra. The singularity is like this quotient. Of course generically, we have $\mathrm{Aut}(G)$ is isomorphic to the centre of G.

Let U be a maximal compact subgroup of G. A. Ramanathan (1975) proved that a bundle is polystable if and only if it comes from a homomorphism of the fundamental group into U, and that two such homomophisms are conjugate in U if and only if the corresponding polystable bundles are isomorphic.

6.11 Reductive groups and Higgs pairs

We now turn to the corresponding considerations for Higgs pairs. Firstly, a *Higgs pair* consists of a G-bundle E and a section α of $\mathrm{ad}(E) \otimes K$. Given a reduction F of the structure group of E to a subgroup H of G, we have an inclusion of $\mathrm{ad}(F)$ in $\mathrm{ad}(E)$. If α comes from a section of $\mathrm{ad}(F) \otimes K$, we say *the Higgs pair is reduced to* H. Then we define stability, semi-stability, polystability, and S-equivalence as above, but now in terms of reductions of the *Higgs pair* to parabolic subgroups.

One can show (Nitsure 1991) that S-equivalence classes of polystable Higgs pairs form a quasi-projective variety $Hg(G)$. It contains a dense open set which is isomorphic to the cotangent bundle over the locus of those stable bundles in the moduli $M(G)$ of G-bundles, which do not admit non-trivial automorphisms.

We now come to the Hitchin morphism in the general case. It is a well-known theorem of Chevalley that the algebra $I(G)$ of polynomial functions on $\mathfrak{g}$ invariant under the adjoint action is isomorphic to a polynomial ring generated by l homogeneous polynomials of degrees m_i, $i = 1, \ldots, l$, where l is the rank of G. Indeed, let $\mathfrak{h}$ be a Cartan subalgebra of $\mathfrak{g}$. Then one can show that by restricting polynomial functions on $\mathfrak{g}$ to $\mathfrak{h}$, one gets an isomorphism of the algebra of adjoint invariant polynomial functions on $\mathfrak{g}$ with the algebra of Weyl group invariant polynomial functions on $\mathfrak{h}$. Using the fact that the Weyl group is generated by reflections in hyperplanes, Chevalley proved the above theorem.

To any $x \in \mathfrak{g}$, one associates the graded algebra homomorphism from $P(\mathfrak{g}) \to \mathbb{C}[t]$ which takes any $f \in P$ to the polynomial $t \mapsto P(tx)$. By restriction, we also get a map $\mathfrak{g} \to \mathrm{GrAlgHom}(I(G), \mathbb{C}[t])$. This can be elevated to the level of Higgs pairs.

Let (E, α) be a Higgs pair. At every point $a \in C$, the section $\alpha \in \Gamma(\mathrm{ad}(E) \otimes K)$ gives rise to an element of $\mathrm{ad}(E)_a \otimes K_a$ and hence a graded algebra homomorphism from $I(G)$ to $\mathrm{Sym}(K_a)$. This in turn induces a map of $\Gamma((\mathrm{ad}(E) \otimes K) \to \mathrm{GrAlgHom}(I(G), \mathrm{Sym}(K))$. The right side is defined to be the *Hitchin space* $Ht(G)$ and the above morphism from $Hg(G)$ to $Ht(G)$, the *Hitchin morphism.* More concretely, if $I(G)$ is the polynomial algebra on $\{f_1, \ldots, f_l\}$, with f_i's, homogeneous of respective degrees m_i, then $Ht(G)$ is the affine space $\oplus_{i=1}^{i=l}(\Gamma(C, K^{m_i}))$. Note that the Hitchin space has dimension $\sum(2m_i - 1)(g - 1)$, and one knows that $\sum(2m_i - 1) = \dim\mathfrak{g}$.

As we mentioned above, the Higgs moduli space contains, as a Zariski open set, the cotangent bundle of the smooth stable points of $M(G)$. The standard symplectic structure on the cotangent bundle extends to the whole of the smooth locus of $Hg(G)$.

The tangent space to the Hitchin space at (E, α) is the first hypercohomology of the two-term complex

$$\mathrm{ad}(E) \to \mathrm{ad}(E) \otimes K$$

where the map is given by Lie bracketing with α. Using this, one can identify the symplectic form and show that the fibres of the Hitchin morphism are Lagrangian. This would prove the complete integrability in general (Biswas and Ramanan 1994).

Nitsure showed in Nitsure (1991) that the Hitchin map is proper. One can seek therefore to identify the abelian varieties which are generic fibres of the Hitchin morphism. This has been done in a beautiful paper of Donagi (1993).

The Hitchin map has also been quantized in the general case (Frenkel 2004; Beilinson and Drinfeld, to appear). I should point out here that the moduli space may be gainfully replaced by the moduli 'stack' and it is in this set-up that the subject is dealt with in the above references. Since the moduli stack is smooth, many of the circumlocutions in the statements can thereby be avoided. Moreover, in the context of the Langlands programme, the moduli stack is more natural, but I prefer to leave things as they are, rather than launch into the theory of algebraic stacks here.

The algebra D of differential operators on the moduli space $M(G)$ from the square root of the canonical bundle to itself turns out to be commutative and their symbols give the space of functions on $Ht(G)$. In other words, the algebra of their symbols has as its spectrum the space $Ht(G)$.

One may then ask for the spectrum of D itself. What is it natually isomorphic to? This turns out to be an affine space, called the space of *opers*.

Here there is another subtlety. There is a canonical isomorphism (or rather duality) between the algebra of adjoint invariant polynomials on the Lie algebra of a semi-simple group G and those of its *Langlands dual* ^LG. The Langlands dual is a group associated to the root system, which is dual to that of G. If G is simply connected, ^LG is the adjoint group with the dual root system. Also the centre of G and the fundamental group of ^LG are canonically dual to each other. In general, the Langlands dual of G/A, where A is a central subgroup of G, is defined to be the covering group of ^LG corresponding to the quotient A^* of $\pi_1(^LG)$.

For simplicity, let us assume here that G is simply connected. We have remarked that the algebra of adjoint invariant polynomials on $\mathfrak{g}$ and $^L\mathfrak{g}$ are isomorphic to their respective Weyl group invariant polynomials on their Cartan algebras. The Cartan algebras are canonically dual, compatible with the Weyl group actions.

A *G-oper* consists of a principal G-bundle E provided with a connection ∇, and a reduction F of E to a Borel subgroup B of G. The connection is *not* supposed to be compatible with the reduction. On the contrary, one can associate to the connection its *second fundamental form* ψ which measures the extent to which ∇ does not come from an F-connection. This is defined for example, locally, by taking a connection ∇_{loc} on F and taking the difference $\psi = \nabla - \nabla_{\text{loc}}$. It is a one-form with values in $\text{ad}(E)$. Using the natural map $\mathfrak{g} \to \mathfrak{g}/\mathfrak{b}$, we get a section φ of $K \otimes F_\rho$, where F_ρ is the vector bundle associated to F via the representation of B in $\mathfrak{g}/\mathfrak{b}$. This form is obviously independent of ∇_{loc} and therefore glues into a global form. It is called the second fundamental form of ∇. We will assume actually that it is highly nondegenerate in the following sense.

Consider the set in $\mathfrak{g}/\mathfrak{b}$ which gives an open orbit O for B. Since by definition B acts on this set, the bundle associated to F with O as fibre makes sense and is a subbundle of $\mathrm{ad}(E)$. We will denote this by F_O. We assume that φ takes values in F_O.

It is fairly straightforward to check that if a connection ∇ of the type required exists, then the corresponding B-reduction is unique. Indeed, it is the Harder–Narasimhan reduction (Ramanathan 1996) of the bundle E. Moreover, it is also easy to check that the bundle E itself is uniquely determined under our assumptions (if G is the adjoint group).

We will start with an example. Let ξ be a theta characteristic. There is then a canonical non-trivial extension

$$0 \to \xi \to P \to \xi^{-1} \to 0$$

given by the generator of $H^1(C, K)$. The bundle P can be considered an $SL(2, \mathbb{C})$ bundle with a reduction to a Borel subgroup B_0. It is clear that it is indecomposable of degree 0. Hence by Weil's theorem, it admits a connection. Although P depends on the choice of the theta characteristic, the corresponding $PSL(2)$-bundle is entirely canonical.

In general, that is to say, for any semi-simple Lie group G, Kostant considered all non-trivial homomorphisms of $SL(2)$ into it, up to conjugacy. He isolated one such conjugacy class as particularly interesting, calling it the *principal TDS*. TDS stands for three-dimensional simple Lie group. If we fix one such homomorphism κ, we get the associated principal G-bundle $\kappa(P)$. It comes with a reduction to a Borel subgroup B of G. Thus, we have a principal G-bundle, a B-reduction which admits a G-connection. Its Harder–Narasimhan filtration is essentially the reduction to B.

The set of (equivalence classes of) opers is an affine space with the torsor vector space $Ht(G)$. The conclusion is that the sought for spectrum of D is the set of opers for the Langlands dual group. There does not yet seem to be a simple understanding of this nice relationship. For further details one may consult Beilinson and Drinfeld (to appear).

6.12 Real forms

Again let G be any semi-simple group and we will stick to the above notation. Let B be a Borel subgroup in G and $U = [B, B]$ its unipotent radical. Consider a principal TDS G_0 in G whose Borel subgroup B_0 is contained in B. Restrict the adjoint representation of G in its Lie algebra $\mathfrak{g}$ to G_0, and consider the U_0-fixed subspace L in it. If f generates the nilpotent radical of the opposite Borel subalgebra in $\mathfrak{g}_0$, then the coset $f + L$ containing f maps isomorphically on the space $\mathrm{GrAlgHom}_0(I(G), \mathbb{C}[t])$. Also, the Kostant subspace $\mathbb{C}f \oplus L$ splits into a direct sum of one-dimensional spaces under the diagonal matrices of G_0.

These are acted upon via the character $\lambda \mapsto \lambda^{-2}$ on $\mathbb{C}f$ and as characters $\lambda \to \lambda^{2m_i-2}$ on these one-dimensional subspaces in L.

We considered the $SL(2,\mathbb{C})$-bundle $E = \xi^{-1} \oplus \xi$ and determined all the Higgs pairs with that as the underlying bundle. Take the Kostant homomorphism of $SL(2,\mathbb{C})$ in G and consider the associated bundle E_G. We will now consider Higgs fields on the bundle E_G. Now $\mathrm{ad}(E_G)$ can be identified with the bundle associated to E via the restriction of the adjoint representation of G to $SL(2)$. Since the bundle E comes from the line bundle ξ by the inclusion of $\mathbb{C}^*$ as diagonal matrices in $SL(2)$, the Kostant subbundle also makes sense in $\mathrm{ad}(E_G)$. Clearly this bundle is isomorphic to the direct sum $\xi^{-2} \oplus_{i=1}^{i=l} (\xi)^{2m_i-2}$. Thus, $\mathrm{ad}(E_G) \otimes K$ contains the trivial bundle, and Higgs fields on E_G of the form $\alpha = 1 + \sum \beta_i, \beta_i \in \Gamma(K^{m_i})$ therefore make sense. Then one can show that the set of all Higgs pairs (E_G, α) with α, as above, are mapped by the Hitchin morphism isomorphically on the Hitchin space. This is 'the' generalized *Hitchin component.* It of course depends on the choice of ξ.

Note that there is a canonical conjugacy class of real forms G_r of G, namely, the *Chevalley* or *split* form. Reductive representations of π in G_r give a closed subspace of $Rep(\pi, G)$. As we have seen this space is not connected. In some cases, such as $SL(2,\mathbb{R})$, there is a topological invariant (called the Toledo invariant) which takes different values on different components. The invariants which serve to nail down all components is, to my knowledge, not fully understood yet. One of the components in this case is the Hitchin component we have described above.

We have thus some understanding of the Higgs pairs corresponding to the compact and the split forms. It is natural to consider the other real forms as well. A real form is defined as the fixed point set of a complex anti-holomorphic involution of G, but the fixed points in $Hg(G)$ constitute a complex subvariety. The relation between Higgs pairs and representations can then be understood in the following way. Let σ be a complex conjugation defining a real form G_σ of G. Then there is a suitable compact real form H of G which is invariant under σ. Now extend $\sigma|H$ as a complex analytic involution of the complexification H^c of H in G. Let $\mathfrak{m}$ be the intersection of $\mathfrak{h}$ and $\mathfrak{g}_\sigma$. Then there is a *Cartan* decomposition of the vector space $\mathfrak{g}_\sigma$ as $\mathfrak{m} \oplus \mathfrak{p}$. Thus we have $\mathfrak{g} = \mathfrak{h}^c = \mathfrak{m}^c \oplus \mathfrak{p}^c$. Then we have

Theorem 6.6 *There is a bijection between the set of equivalence classes of polystable G-Higgs pairs (E, α) where E comes from a principal M^c-bundle V and α from a section of $K \otimes V_{\mathfrak{p}^c}$ on the one hand, and the set of conjugacy classes of reductive homomorphisms of the fundamental group into G which factor through G_σ, on the other.*

Remark 6.5 Since the normalizer of G_σ in G may contain G_σ properly, the map $Rep(G_\sigma) \to Rep(G)$ is not injective in general. This happens already for $SL(2,\mathbb{R}) \to SL(2,\mathbb{C})$, since the matrix $\left(\begin{smallmatrix} 0 & 1 \\ 1 & 0 \end{smallmatrix}\right)$ normalizes $SL(2,\mathbb{R})$.

The connected components of the space of representations into classical real forms have been studied in Bradlow, Garcia-Prada, and Gothen (2003). Let us take a look at the example of $SL(2)$-Higgs pairs. Notice that the complex analytic involution $(E, \alpha) \to (E, -\alpha)$ of the moduli of Higgs pairs fixes the moduli of vector bundles embedded in Hg by taking $\alpha = 0$. This variety of course corresponds to representations into $SU(2)$. But the above involution fixes other varieties *including* the Hitchin component! On the representation side, we note that representations into $SU(2)$ *as well as those into* $SL(2, \mathbb{R})$ are fixed under this involution. On the Higgs side, recall that the Hitchin component is defined by $E = \xi \oplus \xi^{-1}$ (with $\xi^2 \simeq K$) with α of the form $\left(\begin{smallmatrix} 0 & 1 \\ -\gamma & 0 \end{smallmatrix}\right)$ with $\gamma \in \Gamma(C, K^2)$. It is easy to see that the diagonal automorphism with i and $-i$ on the diagonal takes α to $-\alpha$, so that (E, α) and $(E, -\alpha)$ are indeed equivalent.

The involution $(E, \alpha) \to (E, -\alpha)$ obviously fixes the Higgs pairs given by representations into $SU(n)$ for all n, since α is 0 for these pairs. On the other hand, it is easily seen that points in the Hitchin component are *not* fixed by this involution if $n \geq 3$. Why is this so?

A little reflection immediately gives us the clue. The two anti-holomorphic involutions of $SL(n, \mathbb{C})$ given by $A \to \overline{A}$ and $A \to \overline{(A')^{-1}}$ give rise to the two real forms $SL(n, \mathbb{R})$ and $SU(n, \mathbb{C})$ as fixed points, respectively. Consider two complex conjugations as equivalent if they differ by an inner automorphism. Then the maps induced on the set of equivalence classes of representations coming from either of these two real forms are clearly the same. In the case of $SL(2)$, the two anti-holomorphic involutions are indeed equivalent, but in the case of $SL(n)$'s for $n \geq 3$, they are not!

One may also consider other involutions and determine their fixed points, and this gives rise to a nice interplay between the geometry of etale coverings of C, the equivalence among real forms, etc. These are being studied in Garcia-Prada and Ramanan (in preparation).

Acknowledgement

I would like finally to thank the referee for reading the typescript carefully, pointing out many typographical errors, and catching at least one misleading statement.

References

Beauville, A., Narasimhan, M. S. and Ramanan, S. (1989). Spectral curves and the generalized theta divisor. *J. Reine Angew. Math.* **398**, 801–28.

Beilinson, A. and Drinfeld, V. (2005). Opers—Quantization of Hitchin integrable systems and Hecke eigensheaves. *math.* AG/0501398.

Biswas, I. and Ramanan, S. (1994). An infinitesimal study of the moduli of Hitchin pairs. *J. Lond. Math. Soc.* **49**, 219–31.

Bradlow, S. B., Garcia-Prada, O. and Gothen, P. B. (2003). Surface group representations and $U(p, q)$-Higgs bundles. *J. Diff. Geom.* **64**, 111–70.

Corlette, K. (1988). Flat G-bundles with canonical metrics. *J. Diff. Geom.* **28**, 361–82.

Donagi, R. (1993). Decompositions of spectral covers. *Astérisque* **218**, 145–75.

Donaldson, S. K. (1987). Twisted harmonic maps and self-duality equations. *Proc. Lond. Math. Soc.* **55**, 127–31.

Donaldson, S. K. (1993). A new proof of a theorem of Narasimhan and Seshadri. *J. Diff. Geom.* **18**, 269–77.

Drezet M. and Narasimhan, M. S. (1989). Groupe de Picard des variétés de modules de fibré semi-stable sur les courbes algèbriques. *Inv. Math.* **97**, 53–94.

Eells, J. and Sampson, J. H. (1964). Harmonic mappings of Riemannian manifolds. *Am. J. Math.* **86**, 109–60.

Frenkel, E. (2004). Recent advances in the Langlands program. *Bull. Am. Math. Soc.* **41**, 151–84.

Garcia-Prada, O. and Ramanan, S. Involutions on Higgs moduli space on curves, in preparation.

Goldman, W. M. (1988). Topological components of spaces of representations. *Inv. Math.* **93**, 557–607.

Hitchin, N. J. (1987). The self-duality equations on a Riemann surface. *Proc. Lond. Math. Soc.* **55**, 59–126.

Hitchin, N. J. (1992). Lie groups and Teichmüller space. *Topolgy* **31**, 449–73.

Hwang, J.-M. and Ramanan, S. (2004). Hecke curves and Hitchin discriminant. *Ann. Sci. Ecole Norm. Sup.* **37**, 801–17.

King, A. and Schofield, A. (1999). Rationality of moduli of vector bundles on curves. *Indagationes Mathematicae* **10**, 519–35.

Kostant, B. (1959). The principal three-dimensional subgroup and the Betti numbers of a complex simple group. *Am. J. Math.* **81**, 973–1032.

Kostant, B. (1963). Lie group representations on polynomial rings. *Am. J. Math.* **82**, 327–404.

Milnor, J. W. (1957). On the existence of a connection with curvature zero. *Commun. Math. Helv.* **32**, 216–23.

Mumford, D. (1962). Projective invariants of projective structures and applications. *Proceedings of the International Congress on Mathematics*, Stockholm, Sweden, pp. 526–30.

Narasimhan, M. S. and Ramanan, S. (1975). Deformations of the moduli space of vector bundles over an algebraic curve. *Ann. Math.* **101**, 391–417.

Narasimhan, M. S. and Seshadri, (1965). C. S. Stable and unitary vector bundles on compact Riemann surfaces. *Ann. Math.* **82**, 540–64.

Nitsure, N. (1991). Moduli spaces of semistable pairs on a curve. *Proc. Lond. Math. Soc.* **62**, 275–300.

Nori, M. V. (1976). On the representations of the fundamental group. *Compositio Mathematica* **33**, 29–41.

Ramanan, S. (1973). The moduli space of vector bundles over an algebraic curve. *Math. Ann.* **200**, 69–84.

Ramanathan, A. (1975). Stable and principal bundles on a compact Riemann surface. *Math. Ann.* **213**, 129–52.

Ramanathan, A. (1996). Moduli for principal bundles over algebraic curves. *Proc. Indian Acad. Sci.* **106**, 301–28.

Seshadri, C. S. (1967). Space of unitary vector bundles on a compact Riemann surface. *Ann. Math.* **85**, 303–36.

Simpson, C. T. (1992). Higgs bundles and local systems. *Inst. Hautes Études Sci. Publ. Mat.* **75**, 5–95.

Toledo, D. (1989). Representations of surface groups in complex hyperbolic space. *J. Diff. Geom.* **29**, 125–33.

Uhlenbeck, K. and Yau, S. T. (1986). On the existence of Hermitian-Yang-Mills connections in stable vector bundles. *Commun. Pure Appl. Math.* **39**, 5257–93.

Weil, A. (1938). Généralisation de fonctions abéliennes. *J. Math. Pure Appl.* **17**, 47–87.

Wood, J. W. (1976). Bundles with totally disconnected structure group. *Commun. Math. Helv.* **51**, 183–99.

VII

MIRROR SYMMETRY, HITCHIN'S EQUATIONS, AND LANGLANDS DUALITY

Edward Witten

Dedicated to Nigel Hitchin on the occasion of his 60th birthday

7.1 A-model and B-model

Let G be a compact Lie group and let $G_{\mathbb{C}}$ be its complexification. And let C be a compact-oriented two-manifold without boundary. We write $\mathcal{Y}(G,C)$ (or simply $\mathcal{Y}(G)$ or $\mathcal{Y}$ if the context is clear) for the moduli space of flat $G_{\mathbb{C}}$ bundles $E \to C$, modulo gauge transformations. Equivalently, $\mathcal{Y}(G,C)$ parametrizes homomorphisms[1] of the fundamental group of C to $G_{\mathbb{C}}$.

$\mathcal{Y}(G,C)$ is in a natural way a complex symplectic manifold, that is, a complex manifold with a nondegenerate holomorphic two-form. The complex structure comes simply from the complex structure of $G_{\mathbb{C}}$, and the symplectic form, which we call Ω, comes from the intersection pairing[2] on $H^1(C, \mathrm{ad}(E))$, where $\mathrm{ad}(E)$ is the adjoint bundle associated to a flat bundle E. Since $\mathcal{Y}(G,C)$ is a complex symplectic manifold, in particular it follows that its canonical line bundle is naturally trivial.

Geometric Langlands duality is concerned with certain topological field theories associated with $\mathcal{Y}(G,C)$. The most basic of these are the B-model that is defined by viewing $\mathcal{Y}(G,C)$ as a complex manifold with trivial canonical bundle, and the A-model that is defined by viewing it as a real symplectic manifold with symplectic form[3] $\omega = \mathrm{Im}\,\Omega$.

[1] Actually, it is best to define $\mathcal{Y}(G,C)$ as a geometric invariant theory quotient that parametrizes stable homomorphisms plus equivalence classes of semi-stable ones. This refinement will not concern us here (see Section 7.6.1).

[2] The definition of this intersection pairing depends on the choice of an invariant quadratic form on the Lie algebra of G. It can be shown using Hitchin's $\mathbb{C}^*$ action on the moduli space of Higgs bundles that the A-model that we define shortly is independent of this choice, up to a natural isomorphism. The geometric Langlands duality that one ultimately defines likewise does not depend on this choice.

[3] The usual definition of Ω is such that $\mathrm{Im}\,\Omega$ is cohomologically trivial, while $\mathrm{Re}\,\Omega$ is not. The fact that $\omega = \mathrm{Im}\,\Omega$ is cohomologically trivial is a partial explanation of the fact, mentioned in the last footnote, that the A-model of $\mathcal{Y}$ is invariant under scaling of ω.

These are the topological field theories that are relevant to the most basic form of geometric Langlands duality. However, there is also a generalization that is relevant to what is sometimes called quantum geometric Langlands. From the A-model side, it is obvious that a generalization is possible, since we could use a more general linear combination of $\mathrm{Re}\,\Omega$ and $\mathrm{Im}\,\Omega$ in defining the A-model. What is less evident is that the B-model can actually be deformed, as a topological field theory, into this family of A-models. This rather surprising fact is natural from the point of view of generalized complex geometry (see Hitchin 2003) and has been explained from that point of view in section 4.6 of Gualtieri (2003), as a general statement about complex symplectic manifolds. In sections 5.2 and 11.3 of Kapustin and Witten (2007), it was shown that quantum geometric Langlands is naturally understood in precisely this setting.

Here, however, to keep things simple, we will focus on the most basic B-model and A-model that were just described.

7.2 Mirror symmetry and Hitchin's equations

The next ingredient we need is Langlands or Goddard–Nuyts–Olive duality. To every compact Lie group G is naturally associated to its dual group ${}^LG$. For example, the dual of $SU(N)$ is $PSU(N) = SU(N)/\mathbb{Z}_N$, the dual of E_8 is E_8, and so on. And we must also recall the concept of mirror symmetry between A-models and B-models (e.g. see Hori *et al.* 2003). This is a quantum symmetry of two-dimensional non-linear sigma models whose most basic role is to transform questions of complex geometry into questions of symplectic geometry. The geometric Langlands correspondence does not appear at first sight to be an example of mirror symmetry, but it turns out that it is.

With a little bit of hindsight (the question was first addressed in Hausel and Thaddeus 2002, following earlier works by Bershadsky *et al.* 1995, and Harvey *et al.* 1995), we may ask whether the B-model of $\mathcal{Y}({}^LG, C)$ may be mirror to the A-model of $\mathcal{Y}(G, C)$. Even once this question is asked, it is difficult to answer it without some additional structure. The additional structure that comes in handy is provided by Hitchin's equations (see Hitchin 1987*a*). Until this point, C has simply been an oriented two-manifold (compact and without boundary). But now we pick a complex structure and view C as a complex Riemann surface. Hitchin's equations with gauge group G are equations for a pair (A, ϕ). Here A is a connection on a G-bundle $E \to C$ (we stress that the structure group of E is now the *compact* group G), and ϕ is a one-form on C with values in $\mathrm{ad}(E)$. Hitchin's equations, which are elliptic modulo the gauge group, are the system

$$
\begin{aligned}
F - \phi \wedge \phi &= 0 \\
D\phi = D \star \phi &= 0.
\end{aligned}
\tag{7.1}
$$

Here $\star$ is the Hodge $\star$ operator determined by the complex structure on C. The role of the complex structure of C is that it enables us to write the last of these equations.

A solution of Hitchin's equations has two interpretations. On the one hand, given such a solution, we can define the complex-valued connection $\mathcal{A} = A + i\phi$. Hitchin's equations imply that the corresponding curvature $\mathcal{F} = d\mathcal{A} + \mathcal{A} \wedge \mathcal{A}$ vanishes, so a solution of Hitchin's equations defines a complex-valued flat connection, and thus a point in $\mathcal{Y}(G, C)$.

On the other hand, the $(0,1)$ part of the connection A determines a $\bar{\partial}$ operator on the bundle E (or rather its complexification, which we also call E). There is no integrability condition on $\bar{\partial}$ operators in complex dimension 1, so this $\bar{\partial}$ operator endows E with a complex structure; it becomes a holomorphic $G_{\mathbb{C}}$ bundle over C. Moreover, let us write $\phi = \varphi + \bar{\varphi}$, where φ and $\bar{\varphi}$ are the $(1,0)$ and $(0,1)$ parts of ϕ, respectively. Then Hitchin's equations imply that φ, regarded as a section of $K \otimes \mathrm{ad}(E)$ (with K the canonical line bundle of C), is holomorphic. The pair (E, φ), where $E \to C$ is a holomorphic $G_{\mathbb{C}}$ bundle and $\varphi \in H^0(C, K \otimes \mathrm{ad}(E))$, is known as a Higgs bundle.

We write $\mathcal{M}_H$ for the moduli space of solutions of Hitchin's equations, modulo a gauge transformation. The fact that a solution of these equations can be interpreted in two different ways means that $\mathcal{M}_H$ is endowed with two different natural complex structures. In one complex structure, which has been called I, $\mathcal{M}_H$ parametrizes isomorphism classes of semi-stable Higgs bundles (E, φ). In another complex structure, J, it parametrizes equivalence classes of flat $G_{\mathbb{C}}$-bundles or in other words homomorphisms $\rho : \pi_1(C) \to G_{\mathbb{C}}$. I, J, and $K = IJ$ fit together to a natural hyper-Kähler structure on $\mathcal{M}_H$, as described in Hitchin (1987*a*). In particular, there are holomophic two-forms $\Omega_I, \Omega_J, \Omega_K$ and Kähler forms $\omega_I, \omega_J, \omega_K$. These are all related by $\Omega_I = \omega_J + i\omega_K$, and cyclic permutations of this statement, as is usual in hyper-Kähler geometry.

In complex structure J, $\mathcal{M}_H$ is the same as the variety $\mathcal{Y}$ that we described earlier. The natural holomorphic symplectic form Ω of $\mathcal{Y}$ is the same as $i\Omega_J$. And the real symplectic form $\omega = \mathrm{Im}\,\Omega$ used in defining the A-model coincides with ω_K. Complex structure J and the holomorphic symplectic form $\Omega_J = \omega_K + i\omega_I$ do not depend on the chosen complex structure on C, in contrast to the rest of the hyper-Kähler structure of $\mathcal{M}_H$.

Remark 7.1 As an aside, one may ask how closely related ϕ, known in the present context as the Higgs field, is to the Higgs fields of particle physics. Thus, to what extent is the terminology that was introduced in Hitchin (1987*a*) actually justified? The main difference is that Higgs fields in particle physics are scalar fields, while ϕ is a one-form on C (valued in each case in some representation of the gauge group). However, although Hitchin's equations were first written down and studied directly, they can be obtained from $\mathcal{N} = 4$ supersymmetric gauge theory via a sort of twisting procedure (similar to the procedure that leads from $\mathcal{N} = 2$ supersymmetric gauge theory to Donaldson theory). In this twisting procedure, some of the Higgs-like scalar fields of $\mathcal{N} = 4$ super Yang–Mills theory are indeed converted into the Higgs field that enters in Hitchin's equations. This gives a reasonable justification for the terminology.

As we will explain next, it is possible, with the aid of Hitchin's equations, to answer the question of whether the B-model of $\mathcal{Y}(^LG, C)$ is mirror to the A-model of $\mathcal{Y}(G, C)$. This in fact was first pointed out in Hausel and Thaddeus (2002), and used in Kapustin and Witten (2007) as a key ingredient in understanding the geometric Langlands correspondence.

7.3 Hitchin fibration

We will have to use the Hitchin fibration. This is the map, holomorphic in complex structure I, that takes a Higgs bundle (E, φ) to the characteristic polynomial of φ. For example, for $G = SU(2)$, (E, φ) is mapped simply to the quadratic differential $\det \varphi$. The target of the Hitchin fibration is thus in this case the space $\boldsymbol{B} = H^0(C, K^2)$ that parametrizes quadratic differentials. This has a natural analog for any G. From the standpoint of complex structure I, the generic fiber of the map $\pi : \mathcal{M}_H \to \boldsymbol{B}$ is a complex abelian variety (or to be slightly more precise, in general a torsor for one). The fibers are Lagrangian from the standpoint of the holomorphic symplectic form Ω_I. Such a fibration by complex Lagrangian tori turns $\mathcal{M}_H$ into a completely integrable Hamiltonian system in the complex sense (Hitchin 1987*b*).

There is, however, another way to look at the Hitchin fibration, as first described in Hausel and Thaddeus (2002). Let us go back to the A-model defined with the real symplectic structure ω. Since the fibers of $\pi : \mathcal{M}_H \to \boldsymbol{B}$ are Lagrangian for $\Omega_I = \omega_J + i\omega_K$, they are in particular Lagrangian for $\omega = \omega_K$. Moreover, being holomorphic in complex structure I, these fibers are actually area-minimizing in their homology class—here areas are computed using the hyper-Kähler metric on $\mathcal{M}_H$. So the Hitchin fibration, from the standpoint of the A-model, is actually a fibration of $\mathcal{M}_H$ by special Lagrangian tori.

Mirror symmetry is believed to arise from T-duality on the fibers of a special Lagrangian fibration (see Strominger *et al.* 1996). Generally, it is very difficult to explicitly exhibit a non-trivial special Lagrangian fibration. The present example is one of the few instances in which this can actually be done, with the aid of the hyper-Kähler structure of $\mathcal{M}_H$ and its integrable nature. Non-trivial special Lagrangian fibrations are hard to understand because it is difficult to elucidate the structure of the singularities. In the hyper-Kähler context, the fact that the fibers are holomorphic in a different complex structure makes everything far more accessible.

Once we actually find a special Lagrangian fibration, what we are supposed to do with it, in order to give an example of mirror symmetry, is to construct the dual special Lagrangian fibration, which should be mirror to the original one. The mirror map exchanges the symplectic structure on one side with the complex structure on the other side.

In the present context, there is a very beautiful description of the dual fibration: it is, as first shown in Hausel and Thaddeus (2002), simply the Hitchin fibration of the dual group. Thus, one considers $\mathcal{M}_H(^LG, C)$, the moduli space of solutions of Hitchin's equation for the dual group ${}^LG$. It turns out that the bases of the Hitchin fibrations for G and ${}^LG$ can be identified in a natural way. The resulting picture is something like this:

$$\begin{array}{ccc} \mathcal{M}_H({}^LG, C) & & \mathcal{M}_H(G, C) \\ & \searrow \quad \swarrow & \\ & \boldsymbol{B} & \end{array}$$

In complex structure I, the fibers over a generic point $b \in \boldsymbol{B}$ are, roughly speaking, dual abelian varieties (more precisely, they are torsors for dual abelian varieties).

Alternatively, the fibers are special Lagrangian submanifolds in the symplectic structure $\omega = \omega_K$. From this second point of view, the same picture leads to a mirror symmetry between the B-model of $\mathcal{M}_H({}^LG, C)$ in complex structure J and the A-model of $\mathcal{M}_H(G, C)$ with symplectic structure ω_K.

As we have just explained, the tools that make this mirror symmetry visible are the hyper-Kähler structure of $\mathcal{M}_H$ and its Hitchin fibration. Those structures depend on the choice of a complex structure on C, but in fact, the resulting mirror symmetry does not really depend on this choice. This was shown in Kapustin and Witten (2007) in the process of deriving this example of mirror symmetry from a four-dimensional topological field theory. The topological field theory in question is obtained by twisting of $\mathcal{N} = 4$ super Yang–Mills theory.

7.3.1 *A few hints*

There are a few obstacles to overcome to go from this instance of mirror symmetry to the usual formulation of geometric Langlands duality. Unfortunately, it will not be practical here to give more than a few hints.

One key point is that in the usual formulation, the dual of a B-brane on $\mathcal{M}_H({}^LG, C)$ is supposed to be not an A-brane on $\mathcal{M}_H(G, C)$—which is what we most naturally get from the above construction—but a sheaf of $\mathcal{D}$-modules on $\mathcal{M}(G, C)$, the moduli space of G-bundles over C (a sheaf of $\mathcal{D}$-modules is by definition a sheaf of modules for the sheaf $\mathcal{D}$ of differential operators on $\mathcal{M}(G, C)$). The link between the two statements is explained in section 11 of Kapustin and Witten (2007), using the existence of a special A-brane on $\mathcal{M}_H(G, C)$ that is intimately related to differential operators on $\mathcal{M}(G, C)$. This relation is possible because, as explained in Hitchin (1987*a*), $\mathcal{M}_H(G, C)$ contains $T^*\mathcal{M}^{\mathrm{st}}(G, C)$ as a Zariski open set; here $\mathcal{M}^{\mathrm{st}}(G, C)$ is the subspace of $\mathcal{M}(G, C)$ parametrizing strictly stable bundles.

Another key point is the following. A central role in the usual formulation is played by the geometric Hecke operators, which act on holomorphic G-bundles over C and therefore also on $\mathcal{D}$-modules on $\mathcal{M}(G, C)$. They have a natural role in the present story, but this is one place that one misses something if one attempts to express this subject just in terms of two-dimensional sigma models and mirror symmetry. This particular instance of mirror symmetry actually originates from a duality in an underlying four-dimensional gauge theory. Once this is understood, basic facts about the Wilson and 't Hooft line operators of gauge theory lead to the usual statements about Hecke eigensheaves, as explained in sections 9 and 10 of Kapustin and Witten (2007). The geometric Hecke operators are naturally reinterpreted in this context in terms of the Bogomolny equations of three-dimensional gauge theory, which are of great geometrical as well as physical interest and have been much studied, for example in Atiyah and Hitchin (1988).

A proper formulation of some of these statements leads to another important role for four dimensions. The usual formulation of geometric Langlands involves $\mathcal{D}$-modules not on the moduli space of semi-stable G-bundles but on the moduli stack of all G-bundles. The main reason for this is that one cannot see the action of the Hecke operators if one considers only semi-stable bundles. As we will explain in Section 7.6, the role of stacks in the standard description can be understood as a strong clue for an alternative approach that starts in four-dimensional gauge theory.

7.4 Ramification

Before getting back to stacks, however, I want to give an idea of what is called "ramification" in the context of geometric Langlands.

A simple generalization of what we have said so far is to consider flat bundles not on a closed oriented two-manifold C but on a punctured two-manifold $C' = C\backslash p$; that is, C' is C with a point p omitted.

We pick a conjugacy class $\mathcal{C} \subset G_{\mathbb{C}}$, and we let $\mathcal{Y}(G, C'; \mathcal{C})$ denote the moduli space of homomorphisms $\rho : \pi_1(C') \to G_{\mathbb{C}}$, up to conjugation, such that the monodromy around p is in the conjugacy class $\mathcal{C}$.

Many statements that we made before have natural analogs in this punctured case. In particular, $\mathcal{Y}(G, C'; \mathcal{C})$ has a natural structure of a complex symplectic manifold. It has a natural complex structure and holomorphic symplectic form Ω. Just as in the unpunctured case, we can define a B-model of $\mathcal{Y}(G, C'; \mathcal{C})$. Also, viewing $\mathcal{Y}(G, C'; \mathcal{C})$ as a real symplectic manifold with symplectic form $\omega = \mathrm{Im}\,\Omega$, we can define an A-model. The B-model and the A-model are both completely independent of the complex structure of C'.

Next, introduce the dual group ${}^LG$ and let ${}^L\mathcal{C}$ denote a conjugacy class in its complexification. Again, the space $\mathcal{Y}({}^LG, C'; {}^L\mathcal{C})$ has a natural B-model and A-model.

Based on what we have said so far, one may wonder if, for some map between $\mathcal{C}$ and ${}^L\mathcal{C}$, there might be a mirror symmetry between $\mathcal{Y}(G, C'; \mathcal{C})$ and $\mathcal{Y}({}^LG, C'; {}^L\mathcal{C})$. The answer to this question is "not quite," for a number of reasons. One problem is that there is no natural correspondence between conjugacy classes in $G_{\mathbb{C}}$ and in ${}^LG_{\mathbb{C}}$. A more fundamental problem is that the B-model of $\mathcal{Y}(G, C'; \mathcal{C})$ varies holomorphically with the conjugacy class $\mathcal{C}$, but the A-model of the same space does not. To find a version of the statement that has a chance of being right, we have to add additional parameters to find a mirror-symmetric set.

In any event, regardless of what parameters one adds, it is very difficult to answer the question about mirror symmetry if C' is viewed simply as an oriented two-manifold with a puncture. To make progress, just as in the unramified case (i.e. the case without punctures), it is very helpful to endow C' with a complex structure and to use Hitchin's equations. This actually also helps us in finding the right parameters, because an improved set of parameters appears just in trying to give a natural formulation of Hitchin's equations on a punctured surface. Let z be a local parameter near the puncture and write $z = re^{i\theta}$. In the punctured case, it is natural (see Simpson 1990) to introduce variables α, β, γ taking values in the Lie algebra $\mathfrak{t}$ of a maximal torus $T \subset G$, and consider solutions of Hitchin's equations on C' whose behavior near $z = 0$ is as follows:

$$A = \alpha \, d\theta + \dots \tag{7.2}$$

$$\phi = \beta \frac{dr}{r} - \gamma \, d\theta + \dots . \tag{7.3}$$

The ellipses refer to terms that are less singular near $z = 0$.

All the usual statements about Hitchin's equations have close analogs in this situation. The moduli space of solutions of Hitchin's equations with this sort of singularity is a hyper-Kähler manifold $\mathcal{M}_H(G, C; \alpha, \beta, \gamma)$. In one complex structure, usually called J, it coincides with $\mathcal{Y}(G, C; \mathcal{C})$, where $\mathcal{C}$ is the conjugacy class that contains[4] $U = \exp(-2\pi(\alpha - i\gamma))$. In another complex structure, often called I, $\mathcal{M}_H(G, C; \alpha, \beta, \gamma)$ parametrizes Higgs bundles (E, φ), where $\varphi \in H^0(C', K \otimes \mathrm{ad}(E))$ has a pole at $z = 0$, with $\varphi \sim \frac{1}{2}(\beta + i\gamma)(dz/z)$. Moreover, there is a Hitchin fibration, and most of the usual statements about the unramified case—those that we have explained and those that we have omitted here—have close analogs. For a much more detailed explanation, and references to the original literature, see Gukov and Witten (2006).

The variables α, β, γ are a natural set of parameters for the classical geometry. However, quantum mechanically, there is one more natural variable η (analogous to the usual θ angles of gauge theory), as described in section 2.3 of Gukov and Witten (2006). With the complete set of parameters $(\alpha, \beta, \gamma, \eta)$ at hand, it is possible to formulate a natural duality statement, according to which $\mathcal{M}_H({}^L G, C; {}^L\alpha, {}^L\beta, {}^L\gamma, {}^L\eta)$ is mirror to $\mathcal{M}_H(G, C; \alpha, \beta, \gamma, \eta)$, with a certain map between the parameters, described in section 2.4 of Gukov and Witten (2006). The main point of this map is that $(\alpha, \eta) = ({}^L\eta, -{}^L\alpha)$. Since the monodromy U depends on ${}^L\alpha$, this shows that the dual of the monodromy involves the quantum parameter η that is invisible in the classical geometry. In the A-model, η becomes the imaginary part of the complexified Kähler class.

[4] For simplicity, we assume that U is regular. The more involved statement that holds in general is explained in Gukov and Witten (2006).

This duality statement leads, after again mapping A-branes to $\mathcal{D}$-modules, to a statement of geometric Langlands duality for this situation similar to what has been obtained via algebraic geometry and two-dimensional conformal field theory in Frenkel and Gaitsgory (2005).

Remark 7.2 We pause here to explain one very elementary fact about the classical geometry that will be helpful as background for Section 7.5. In complex structure J, a solution of Hitchin's equations with the singularity of (7.2) describes a flat $G_{\mathbb{C}}$ bundle $E \to C'$ with monodromy around the puncture p. E can be extended over p as a holomorphic bundle, though of course not as a flat one, and moreover from a holomorphic point of view, E can be trivialized near p. The flat connection on $E \to C'$ is then represented, in this gauge, by a holomorphic $(1,0)$-form on C' (valued in the Lie algebra of $G_{\mathbb{C}}$) with a simple pole at p:

$$\mathcal{A} = dz\left(\frac{\alpha - i\gamma}{iz} + \dots\right), \tag{7.4}$$

where the omitted terms are regular at $z = 0$. The singularity of the connection at $z = 0$ is a simple pole because the ansatz (7.2) for Hitchin's equations only allows a singularity of order $1/|z|$. A holomorphic connection with such a simple pole is said to have a regular singularity.

In geometric Langlands, what is usually called tame ramification is, roughly speaking, the case that we have just arrived at: a holomorphic bundle $E \to C$ that has a holomorphic connection form with a regular singularity. Actually, the phrase "tame ramification" is sometimes taken to refer to the case that the residue of the simple pole is nilpotent, while in (7.4) we seem to be in the opposite case of semi-simple residue. In Gukov and Witten (2006), it is explained that, with some care, mirror symmetry for $\mathcal{M}_H(G, C'; \alpha, \beta, \gamma, \eta)$ is actually a sufficient framework to understand geometric Langlands for a connection with a simple pole of any residue. For example, the case of a nilpotent residue can be understood by setting ${}^L\alpha = {}^L\gamma = 0$ (or $\gamma = \eta = 0$ in the dual description).

7.5 Wild ramification

Based on an analogy with number theory, geometric Langlands is usually formulated not only for the case of tame ramification. One goes on to inquire about an analogous duality statement involving a holomorphic bundle $E \to C$ with a holomorphic connection that has a pole of any order. In other words, after trivializing the holomorphic structure of E near a point $p \in C$, the connection looks like

$$\mathcal{A} = dz\left(\frac{T_n}{z^n} + \frac{T_{n-1}}{z^{n-1}} + \dots + \frac{T_1}{z} + \cdots\right), \tag{7.5}$$

where regular terms are omitted. A meromorphic connection with a pole of degree greater than 1 is said to have an irregular singularity.

Trying to formulate a duality statement for this situation poses, at first sight, a severe challenge for the approach to geometric Langlands described here. Our basic point of view is that the fundamental duality statements depend on C only as an oriented two-manifold. A complex structure on C is introduced only as a tool to answer certain natural questions that can be asked without introducing the complex structure.

From this point of view, tame ramification is natural because a simple pole in this sense has a clear topological meaning. A meromorphic connection with a simple pole at a point $p \in C$ is a natural way to encode the monodromy about p of a flat connection on $C' = C \backslash p$. And this monodromy, of course, is a purely topological notion. But what could possibly be the topological meaning of a connection with a pole of degree greater than 1?

A closely related observation is that T_1 is the residue of the pole in $\mathcal{A}$ at $z = 0$, and so is independent of the choice of local coordinate z. However, the coefficients $T_2, \ldots, T_n$ of the higher order poles most definitely do depend on the choice of a local coordinate. How can we hope to include them in a theory that is supposed to depend on C only as an oriented two-manifold?

Moreover, if the plan is to formulate a duality conjecture of a topological nature and then prove it using Hitchin's equations, we face the question of whether Hitchin's equations are compatible with an irregular singularity. Hitchin's equations for a pair $\Phi = (A, \phi)$ are schematically of the form $d\Phi + \Phi^2 = 0$. If near $z = 0$, we have a singularity with $|\Phi| \sim 1/|z|^n$, then $|d\Phi| \sim 1/|z|^{n+1}$ and $|\Phi|^2 \sim 1/|z|^{2n}$. For $n = 1$, $d\Phi$ and $|\Phi|^2$ are comparable in magnitude, and therefore Hitchin's equations look reasonable. However, for $n > 1$, we have $|\Phi|^2 >> |d\Phi|$, and it looks like the non-linear term in Hitchin's equations will be too strong.

Both questions, however, have natural answers. The answer to the first question is that, despite appearances, one actually can associate to a connection with irregular singularity something that goes beyond the ordinary monodromy and has a purely topological meaning. The appropriate concept is an extended monodromy that includes Stokes matrices as well as the ordinary monodromy. Stokes matrices are part of the classical theory of ordinary differential equations with irregular singularity (e.g. see Wasow 1965).

Assuming for brevity that the leading coefficient T_n of the singular part of the connection is regular semi-simple, one can make a gauge transformation to conjugate $T_1, \ldots, T_n$ to the maximal torus. Then one defines a moduli space $\mathcal{Y}(G, C; T_1, \ldots, T_n)$ that parametrizes, up to a gauge transformation, pairs consisting of a holomorphic $G_{\mathbb{C}}$-bundle over C and a connection with an irregular singularity of the form described in (7.5). As shown in Boalch (2001), it turns out that this space $\mathcal{Y}(G, C; T_1, \ldots, T_n)$ is in a natural way a complex symplectic manifold, with a complex symplectic structure that depends on C only as an oriented two-manifold. This can be proved by adapting to the present setting the gauge theory definition of the symplectic structure, formulated in Atiyah and

Bott (1982). Moreover, the complex symplectic structure of $\mathcal{Y}(G, C; T_1, \ldots, T_n)$ is independent of $T_2, \ldots, T_n$ (as long as T_n remains semi-simple).

At this point, the important concept of isomonodromic deformation, introduced by Jimbo *et al.* (1981), comes into play. There is a natural way to vary the parameters $T_2, \ldots, T_n$, without changing the generalized monodromy data that is parametrized by $\mathcal{Y}(G, C; T_1, \ldots, T_n)$. Moreover, as has been proved quite recently in Boalch (2001), the complex symplectic structure of the space of generalized monodromy data is invariant under isomonodromic deformation. Thus, roughly speaking, one can define a complex symplectic manifold $\mathcal{Y}(G, C; T_1, n)$ that depends only on T_1 and the integer $n \geq 1$.

The fact that the parameters $T_2, \ldots, T_n$ turn out to be, in the sense just described, inessential is certainly welcome, since as we have already observed, they have no evident topological meaning. Now we are in a situation very similar to what we had in the unramified and tamely ramified cases. Given $\mathcal{Y}(G, C; T_1, n)$ as a complex symplectic manifold, with complex symplectic form Ω, we can define its B-model or its A-model using the real symplectic form $\omega = \operatorname{Im} \Omega$. Of course, we can do the same for the dual group, defining another complex symplectic manifold $\mathcal{Y}({}^L G, C; {}^L T_1, n)$, with its own B-model and A-model. And, just as in the unramified case, we can ask if these two models are mirror to each other.

Even before trying to answer this question, we should refine it slightly. Because of the constraint that T_n should be regular semi-simple, it is not quite correct to simply forget about T_n. There can be monodromies as T_n varies. We think of T_n as taking values in $\mathfrak{t}_{\mathbb{C}}^{\mathrm{reg}} \otimes K_p^{n-1}$, with notation as follows: $\mathfrak{t}_{\mathbb{C}}$ is the Lie algebra of a maximal torus in $G_{\mathbb{C}}$, $\mathfrak{t}_{\mathbb{C}}^{\mathrm{reg}}$ is its subspace consisting of regular elements, and K_p is the fiber at p of the cotangent bundle to C. The fundamental group of $\mathfrak{t}_{\mathbb{C}}^{\mathrm{reg}}$ is known as the braid group of G; we call it $B(G)$. Because of the monodromies, one really needs to choose a basepoint $* \in \mathfrak{t}_{\mathbb{C}}^{\mathrm{reg}}$ to define $\mathcal{Y}({}^L G, C; {}^L T_1, n)$; to be more precise, we can denote this space as $\mathcal{Y}({}^L G, C; {}^L T_1, n, *)$. The group $B(G)$ acts via monodromies on both the B-model and the A-model of $\mathcal{Y}(G, C; T_1, n, *)$. Dually, the corresponding braid group $B({}^L G)$ acts on the B-model and the A-model of $\mathcal{Y}({}^L G, C; {}^L T_1, n, *)$. However, the two groups $B(G)$ and $B({}^L G)$ are naturally isomorphic; indeed, modulo a choice of an invariant quadratic form, there is a natural map from $\mathfrak{t}_{\mathbb{C}}^{\mathrm{reg}}$ to ${}^L\mathfrak{t}_{\mathbb{C}}^{\mathrm{reg}}$, so the two spaces have the same fundamental group and a choice of basepoint in one determines a basepoint in the other, up to homotopy. A better (but still not yet precise) question is whether there is a mirror symmetry between $\mathcal{Y}(G, C; T_1, n, *)$ and $\mathcal{Y}({}^L G, C; {}^L T_1, n, *)$ that commutes with the braid group.

We expect as well that this mirror symmetry depends on C only as an oriented two-manifold, and so commutes with the mapping class group. We can think of the mapping class group of C and the braid group as playing quite parallel roles. In fact, because of the appearance of the fiber K_p of the canonical bundle in the last paragraph, these two groups do not simply commute with each other; the group that acts is an extension of the mapping class group by $B(G)$.

Just as in the tamely ramified case, to get the right mirror symmetry conjecture, we need to extend the parameters slightly to get a mirror-symmetric set. But we also face the fundamental question of whether Hitchin's equations are compatible with wild ramification. As explained above, the non-linearity of Hitchin's equations makes this appear doubtful at first sight. But happily, it turns out that all is well, as shown in Biquard and Boalch (2004). The key point is that, again with T_n assumed to be regular semi-simple, we can assume that the singular part of the connection is abelian. Though Hitchin's equations are non-linear, they become linear in the abelian case, and once abelianized, they are compatible with a singularity of any order. Using this as a starting point, it turns out that, for any n, one can develop a theory of Hitchin's equations with irregular singularity that is quite parallel to the more familiar story in the unramified case. For example, the moduli space $\mathcal{M}_H$ of solutions of the equations is hyper-Kähler. In one complex structure, $\mathcal{M}_H$ parametrizes flat connections with a singularity similar to that in (7.5); in another complex structure, it parametrizes Higgs bundles (E, φ) in which φ has an analogous pole of order n. There is a Hitchin fibration, and all the usual properties have close analogs.

All this gives precisely the right ingredients to use Hitchin's equations to establish the desired mirror symmetry between the two moduli spaces. See Witten (2007) for a detailed explanation in which this classical geometry is embedded in four-dimensional gauge theory. Many of the arguments are quite similar to those given in the tame case in Gukov and Witten (2006). The construction makes it apparent that the duality commutes with isomonodromic deformation.

Finally, one might worry that the assumption that T_n is regular semi-simple may have simplified things in some unrealistic way. This is actually not the case. For one thing, the analysis in Biquard and Boalch (2004) requires only that $T_2, \ldots, T_n$ should be simultaneously diagonalizable (in some gauge), and in particular semi-simple, but not that T_n is regular. But even if these coefficients are not semi-simple, there is no essential problem. In the classical theory of ordinary differential equations, it is shown that given any such equation with an irregular singularity at $z = 0$, after possibly passing to a finite cover of the punctured z-plane and changing the extension of a holomorphic bundle over the puncture at $z = 0$, one can reduce to the case that the irregular part of the singularity has the properties assumed in Biquard and Boalch (2004). Given this, one can adapt all the relevant arguments concerning geometric Langlands duality to the more general case, as is explained in section 6 of Witten (2007).

7.6 Four-dimensional gauge theory and stacks

To a physicist, it is natural, in studying dualities involving gauge theory, to begin in four dimensions, which is often found to be the natural setting for gauge theory duality. There is a simple reason for this. The curvature, which is one of the most fundamental concepts in gauge theory, is a two-form. In d dimensions, the dual of a two-form is a $(d-2)$-form, so it is only a two-form if $d = 4$. This suggests that $d = 4$ is the most natural dimension in which the dual of a gauge theory might be another gauge theory.

Moreover, $\mathcal{N} = 4$ supersymmetric Yang–Mills theory, originally constructed in Brink *et al.* (1977), is a natural place to start, as it has the maximal possible supersymmetry, and has the celebrated duality whose origins go back to the early work of Montonen and Olive (1977). It indeed turns out that geometric Langlands has a natural origin in a twisted version of $\mathcal{N} = 4$ super Yang–Mills theory in four dimensions. The twisting is quite analogous to the twisting of $\mathcal{N} = 2$ super Yang–Mills theory that leads to Donaldson theory.

That particular motivation may seem opaque to some, and instead I will adopt here a different approach in explaining why it is natural to begin in four dimensions for understanding geometric Langlands, instead of relying only on the B-model and the A-model of $\mathcal{M}_H(G,C)$.

First of all, the B-model and the A-model of any space X are both twisted versions of a quantum sigma model that governs maps $\Phi : \Sigma \to X$, where Σ is a two-manifold (or better, a supermanifold of bosonic dimension 2). Since the A-model involves in its most elementary form a counting of holomorphic maps $\Phi : \Sigma \to X$ that obey appropriate conditions, the roles of Σ and Φ are clear in the A-model. Mirror symmetry indicates that it must be correct to also formulate the B-model in terms of maps $\Phi : \Sigma \to X$, and this is done in the usual formulation by physicists.

In the present case, we are interested, roughly speaking, in the B-model and A-model of $\mathcal{M}_H(G,C)$, for some compact Lie group G and two-manifold C. Therefore, roughly speaking, we want to study a sigma model of maps $\Phi : \Sigma \to \mathcal{M}_H(G,C)$, where as before Σ is an auxiliary two-manifold.

The reason that this description is rough is that $\mathcal{M}_H(G,C)$ has singularities,[5] and the sigma model of target $\mathcal{M}_H(G,C)$ is therefore not really well defined. Therefore, a complete description cannot be made purely in terms of a sigma model in which the target space is $\mathcal{M}_H(G,C)$, viewed as an abstract manifold. We need a more complete description that will tell us how to treat the singularities. What might this be?

By definition, a point in $\mathcal{M}_H(G,C)$ determines up to gauge-equivalence a pair (A,ϕ) obeying Hitchin's equations. A and ϕ are fields defined on C, so let us write them more explicitly as $(A(y),\phi(y))$, where y is a coordinate on C.

Now suppose that we have a map $\Phi : \Sigma \to \mathcal{M}_H(G,C)$, where Σ is a Riemann surface with a local coordinate x. Such a map is described by a pair $(A(y),\phi(y))$ that also depends on x. So we can describe the map Φ via fields $(A(x,y),\phi(x,y))$ that depend on both x and y. We would like to interpret these fields as fields on the four-manifold $M = \Sigma \times C$. The pair $(A(x,y),\phi(x,y))$ is not quite a natural

[5] Moreover, these singularities are worse than orbifold singularities. Orbifold singularities would cause no difficulty. See Frenkel and Witten (2007) for a discussion of orbifold singularities in geometric Langlands.

set of fields on M but can be naturally completed to one. For example, $A(x, y)$ is locally a one-form tangent to the second factor in $M = \Sigma \times C$; to get a four-dimensional gauge field, we should relax the condition that A is tangent to the second factor. Similarly, we can extend ϕ to an adjoint-valued one-form on $\Sigma \times C$. $\mathcal{N} = 4$ super Yang–Mills theory, or rather its twisted version that is related to geometric Langlands, is obtained by completing this set of fields to a supersymmetric combination in a minimal fashion.

In $\mathcal{N} = 4$ super Yang–Mills theory, there are no singularities analogous to the singularities of $\mathcal{M}_H(G, C)$. The space of gauge fields, for example, is an affine space, and the other fields (such as ϕ) take values in linear spaces. The problems with singularities that make it difficult to define a sigma model of maps $\Phi : \Sigma \to \mathcal{M}_H(G, C)$ have no analog in defining gauge theory on $M = \Sigma \times C$ (or any other four-manifold). The relation between the two is that the two-dimensional sigma model is an approximation to the four-dimensional gauge theory. The approximation breaks down when one runs into the singularities of $\mathcal{M}_H(G, C)$. Any question that involves those singularities should be addressed in the underlying four-dimensional gauge theory.

But away from singularities, it suffices to consider only the smaller set of fields that describe a map $\Phi : \Sigma \to \mathcal{M}_H(G, C)$. Many questions do not depend on the singularities and for these questions the description via two-dimensional sigma models and mirror symmetry is adequate.

7.6.1 *Stacks*

To conclude, we will make contact with the counterpart of this discussion in the usual mathematical theory. We start with bundles rather than Higgs bundles because this case will be easier to explain.

In the usual mathematical theory, the right-hand side of the geometric Langlands correspondence is described in terms of $\mathcal{D}$-modules on, roughly speaking, the moduli space of all holomorphic $G_{\mathbb{C}}$ bundles on the Riemann surface C.

However, instead of the moduli space $\mathcal{M}(G, C)$ of semi-stable holomorphic $G_{\mathbb{C}}$ bundles $E \to C$, one considers $\mathcal{D}$-modules on the "stack" $\mathrm{Bun}_G(C)$ of all such bundles. The main reason for this is that to define the action of Hecke operators, it is necessary to allow unstable bundles. Unstable bundles are related to the nonorbifold singularities of $\mathcal{M}(G, C)$.

What is a stack? Roughly, it is a space that can everywhere be locally described as a quotient. The trivial case is a stack that can actually be described globally as a quotient. Interpreting $\mathrm{Bun}_G(C)$ as a global quotient would mean finding a pair $(Y, W_{\mathbb{C}})$, consisting of a smooth algebraic variety Y and a complex Lie group $W_{\mathbb{C}}$ acting on Y, with the following properties. Isomorphism classes of holomorphic $G_{\mathbb{C}}$ bundles $E \to C$ should be in one-to-one correspondence with $W_{\mathbb{C}}$ orbits on Y, and for every $E \to C$, its automorphism group should be isomorphic to the subgroup of $W_{\mathbb{C}}$ leaving fixed the corresponding point in Y.

A pair $(Y, W_{\mathbb{C}})$ representing in this way the stack $\mathrm{Bun}_G(C)$ does not exist if Y and $W_{\mathbb{C}}$ are supposed to be finite-dimensional. Indeed, the $G_{\mathbb{C}}$-bundle $E \to C$ can be arbitrarily unstable, so there is no upper bound on the dimension of its automorphism group. So no finite-dimensional $W_{\mathbb{C}}$ can contain all such automorphism groups as subgroups.

However as shown in Atiyah and Bott (1982), taking G to be of adjoint type for simplicity, there is a natural infinite-dimensional pair $(Y, W_{\mathbb{C}})$. One simply takes Y to be the space of all connections on a given G-bundle $E \to C$ which initially is defined only topologically. One defines W to be the group of all gauge transformations of the bundle E; thus, if E is topologically trivial, we can identify W as the group $\mathrm{Maps}(C, G)$. Then we take $W_{\mathbb{C}}$ to be the complexification of W, or in other words $\mathrm{Maps}(C, G_{\mathbb{C}})$. (This complexification acts on Y as follows. We associate to a connection A the corresponding $\bar{\partial}$ operator $\bar{\partial}_A$. Then a complex-valued gauge transformation acts by $\bar{\partial}_A \to g\bar{\partial}_A g^{-1}$.)

Suppose then that we were presented with the problem of making sense of the supersymmetric sigma model of maps $\Phi : \Sigma \to \mathcal{M}(G, C)$, given the singularities of $\mathcal{M}(G, C)$. (This is a practice case for our actual problem, which involves $\mathcal{M}_H(G, C)$ rather than $\mathcal{M}(G, C)$.) Our friends in algebraic geometry would tell us to replace $\mathcal{M}(G, C)$ by the stack $\mathrm{Bun}_G(C)$. We interpret this stack as the pair $(Y, W_{\mathbb{C}})$, where Y is the space of all connections on a G-bundle $E \to C$ and $W_{\mathbb{C}}$ is the complexified group of gauge transformations. The connected components of the stack correspond to the topological choices for E.

By a supersymmetric sigma model with target a pair $(Y, W_{\mathbb{C}})$, with $W_{\mathbb{C}}$ a complex Lie group acting on a complex manifold Y, we mean[6] in the finite-dimensional case a gauge-invariant supersymmetric sigma model in which the gauge group is W (a maximal compact subgroup of $W_{\mathbb{C}}$) and the target is Y. Actually, to define this sigma model, we want Y to be a Kähler manifold with a W-invariant (but of course not $W_{\mathbb{C}}$-invariant) Kähler structure. The sigma model action contains a term which is the square of the moment map for the action of W. This term is minimized precisely when the moment map vanishes. The combined operation of setting the moment map to zero and dividing by W is equivalent classically to dividing by $W_{\mathbb{C}}$.

To write down the term in the action that involves the square of the moment map (and in fact, to write down the kinetic energy of the gauge fields) one needs an invariant and positive definite quadratic form on the Lie algebra of W. If W is finite-dimensional, existence of such a form is equivalent to W being compact. However, the appropriate quadratic form also exists in the infinite-dimensional case that $W = \mathrm{Maps}(C, G)$ for some space C. (An element of the Lie algebra of W is a $\mathfrak{g}$-valued function ϵ on C, and the quadratic form is defined by $\int_C d\mu\,(\epsilon, \epsilon)$, where $(,)$ is an invariant positive-definite quadratic form on $\mathfrak{g}$, and μ is a suitable measure on C.)

[6] For a discussion of this construction in relation to stacks, see Pantev and Sharpe (2006).

Now, suppose we construct the two-dimensional sigma model of maps from a Riemann surface Σ to the stack $\mathrm{Bun}_G(C)$, understood as above. What is the group of gauge transformations of the sigma model? In general, in a gauge theory on any space Σ with gauge group W, the group of gauge transformations (of a topologically trivial W-bundle, for simplicity) is the group of maps from Σ to W, or $\mathrm{Maps}(\Sigma, W)$. In our case, W is in turn $\mathrm{Maps}(C, G)$. So $\mathrm{Maps}(\Sigma, W)$ is the same as $\mathrm{Maps}(M, G)$, where $M = \Sigma \times C$. But this is simply the group of gauge transformations in gauge theory on M, with gauge group G. In the present case, Σ and C are two-manifolds and M is a four-manifold. We have arrived at four-dimensional gauge theory. If we chase through the definitions a little more, we learn that the supersymmetric sigma model of maps $\Phi : \Sigma \to \mathrm{Bun}_G(C)$ should be understood as four-dimensional $\mathcal{N} = 2$ supersymmetric gauge theory, with gauge group G, on the four-manifold $M = \Sigma \times C$. (This is the theory that after twisting is related to Donaldson theory.)

Now let us return to the original problem. Geometric Langlands duality is a statement about the B-model and A-model not of $\mathcal{M}(G, C)$ but of $\mathcal{M}_H(G, C)$, the corresponding moduli space of Higgs bundles, and its analog for the dual group ${}^LG$. To deal with the singularities, we want to "stackify" this situation. We are now in a hyper-Kähler context and the appropriate concept of a stack should incorporate this. (What algebraic geometers would call the stack of Higgs bundles does not quite do justice to the situation, since it emphasizes one complex structure too much.) Since quaternionic Lie groups do not exist, we cannot ask to construct $\mathcal{M}_H(G, C)$ as the quotient of a smooth space by a quaternionic Lie group. However, the notion of a symplectic quotient does have a good analog in the hyper-Kähler world, namely, the hyper-Kähler quotient, described in Hitchin *et al.* (1987). The analog of what we explained for $\mathcal{M}(G, C)$ is to realize $\mathcal{M}_H(G, C)$ as the hyper-Kähler quotient of a smooth space Y by a group W. It may be impossible to do this with finite-dimensional Y and W, but in the infinite-dimensional world, this problem has a natural solution described in Hitchin's original paper (1987*a*) on the Hitchin equations. (Y is the space of pairs (A, ϕ) on C, and $W = \mathrm{Maps}(C, G)$.) Taking this as input and interpreting what it should mean to have a sigma model whose target is the hyper-Kähler stack corresponding to $\mathcal{M}_H(G, C)$, one arrives at the twisted version of $\mathcal{N} = 4$ super Yang–Mills theory that was the starting point in Kapustin and Witten (2007).

Acknowledgement

This chapter was supported in part by NSF Grant Phy-0503584.

References

Atiyah, M. F. and Bott, R. (1982). The Yang-Mills equations over Riemann surfaces. *Phil. Trans. R. Soc. London* **A308**, 166–86.

Atiyah, M. F. and Hitchin, N. J. (1988). *The Physics and Geometry of Monopoles.* Princeton University Press, Princeton, NJ.

Bershadsky, M., Johansen, A., Sadov, V. and Vafa C. (1995). Topological reduction of $4-D$ SYM to $2-D$ sigma models. *Nucl. Phys.* **B448**, 166–86.

Biquard, O. and Boalch, P. (2004). Wild non-abelian Hodge theory on curves. *Compositio Mathematica* **140**, 179–204.

Boalch, P. (2001). Symplectic manifolds and isomonodromic deformations. *Adv. Math.* **163**, 137–205.

Brink, L., Schwarz, J. H. and Scherk, J. (1977). Supersymmetric Yang-Mills theories. *Nucl. Phys.* **B121**, 77.

Frenkel, E. and Gaitsgory, D. (2005). Local geometric Langlands correspondence and affine Kac-Moody algebras. math.RT/0508382.

Frenkel, E. and Witten, E. (2007). Geometric endoscopy and mirror symmetry. arXiv.0710.5939.

Gualtieri, M. (2003). Generalized complex geometry. DPhil thesis, Oxford University, Oxford, math.DG/0401221.

Gukov, S. and Witten, E. (2006). Gauge theory, ramification, and the geometric Langlands program. hep-th/0612073.

Harvey, J. A., Moore, G. W. and Strominger A. (1995). Reducing S duality to T duality. *Phys. Rev.* **D52**, 7161–7.

Hausel, T. and Thaddeus, M. (2002). Mirror symmetry, Langlands duality, and the Hitchin system. math.AG/0205236.

Hitchin, N. J. (1987*a*). The self-duality equations on a Riemann surface. *Proc. London Math. Soc. (3)* **55**, 67.

Hitchin, N. J. (1987*b*). Stable bundles and integrable systems. *Duke Math. J.* **54**, 91–114.

Hitchin, N. J. (2003). Generalized Calabi-Yau manifolds. *Q. J. Math.* **54**, 281–308, math.DG/0209099.

Hitchin, N. J., Karlhede, A., Lindstrom, U. and Rocek M. (1987). Hyperkähler metrics and supersymmetry. *Commun. Math. Phys.* **108**, 535–89.

Hori, K. *et al.*, (2003). editors. *Mirror Symmetry.* American Mathematical Society, Providence, RI.

Jimbo, M., Miwa, T. and Ueno, K. (1981). Monodromy preserving deformations of linear differential equations with rational coefficients. *Physica* **2D**, 407–48.

Kapustin, A. and Witten, E. (2007). Electric-magnetic duality and the geometric Langlands program. *Commun. Number Theory Phys.* **1**, 1–236, hep-th/0604151.

Montonen, C. and Olive, D. (1977). Magnetic monopoles as gauge particles?. *Phys. Lett.* **B72**, 117.

Pantev, T. and Sharpe, E. (2006). GLSM's for gerbes (and other toric stacks). *Adv. Theor. Math. Phys.* **10**, 77–121, hep-th/0502053.

Simpson, C. (1990). Harmonic bundles on noncompact curves. *J. Am. Math. Soc.* **3**, 713–70.

Strominger, A., Yau, S.T. and Zaslow, E. (1996). Mirror symmetry is T duality. *Nucl. Phys.* **B479**, 243–59, hep-th/9606040.

Wasow, W. (1965). *Asymptotic Expansions for Ordinary Differential Equations.* Wiley, New York.

Witten, E. (2007). Gauge theory and wild ramification. arXiv:0710.0631.

VIII

HIGGS BUNDLES AND GEOMETRIC STRUCTURES ON SURFACES

William M. Goldman

Dedicated to Nigel Hitchin on the occasion of his 60th birthday

8.1 Introduction

In the late 1980s Hitchin (1987) and Simpson (1988) discovered deep connections between representations of fundamental groups of surfaces and algebraic geometry. The fundamental group $\pi = \pi_1(\Sigma)$ of a closed orientable surface Σ of genus $g > 1$ is an algebraic object governing the topology of Σ. For a Lie group G, the space of conjugacy classes of representations $\pi \to G$ is a natural algebraic object $\mathsf{Hom}(\pi, G)/G$ whose geometry, topology, and dynamics intimately relates to the topology of Σ and the various geometries associated with G. In particular $\mathsf{Hom}(\pi, G)/G$ arises as a moduli space of locally homogeneous geometric structures as well as flat connections on bundles over Σ.

Giving Σ a conformal structure profoundly affects π and its representations. This additional structure induces further geometric and analytic structure on the deformation space $\mathsf{Hom}(\pi, G)/G$. Furthermore this analytic interpretation allows Morse-theoretic methods to compute the algebraic topology of these non-linear finite-dimensional spaces.

For example, when $G = \mathsf{U}(1)$, the space of representations is a torus of dimension $2g$. Giving Σ a conformal structure denotes the resulting Riemann surface by X. The classical Abel–Jacobi theory identifies representations $\pi_1(X) \longrightarrow \mathsf{U}(1)$ with topologically trivial holomorphic line bundles over X. The resulting *Jacobi variety* is an abelian variety, whose structure strongly depends on the Riemann surface X. However the underlying *symplectic manifold* depends only on the topology of Σ, and indeed just the fundamental group π. (Compare Goldman 1984).

Another important class of representations of π arises from introducing the local structure of *hyperbolic geometry* to Σ. Giving Σ a Riemannian metric of curvature -1 determines a representation ρ in the group $G = \mathsf{Isom}^+(\mathsf{H}^2) \cong \mathsf{PSL}(2, \mathbb{R})$. These representations, which we call *Fuchsian,* are characterized as *embeddings* of π onto *discrete* subgroups of G. Equivalence classes of

Fuchsian representations comprise the Fricke–Teichmüller space $\mathfrak{F}(\Sigma)$ of marked hyperbolic structures on Σ, which embeds in $\mathsf{Hom}(\pi, G)/G$ as a connected component. This component is a cell of dimension $6g-6$ upon which the mapping class group acts properly.

The theory of Higgs bundles, pioneered by Hitchin and Simpson, provides an analytic approach to studying surface group representations and their deformation space. The purpose of this chapter is to describe the basic examples of this theory, emphasizing relations to deformation and rigidity of geometric structures. In particular we report on some very recent developments when G is a real Lie group, either a split real semisimple group or an automorphism group of a Hermitian symmetric space of noncompact type.

In the 20 years since the appearance of Hitchin's and Simpson's work, many other developments directly arose from their works. These relate to variations of Hodge structures, spectral curves, integrable systems, Higgs bundles over noncompact Riemann surfaces and higher dimensional Kähler manifolds, and the finer topology of the deformation spaces. None of these topics are discussed here. It is an indication of the power and the depth of these ideas that so many mathematical subjects have been profoundly influenced by the pioneering work of Hitchin and Simpson.

8.2 Representations of the fundamental group

8.2.1 *Closed surface groups*

Let $\Sigma = \Sigma_g$ be a closed orientable surface of genus $g > 1$. Orient Σ, and choose a smooth structure on Σ. Ignoring basepoints, denote the fundamental group $\pi_1(\Sigma)$ of Σ by π. The familiar decomposition of Σ as a $4g$-gon with $2g$ identifications (depicted in Figures 8.1 and 8.2) of its sides leads to a presentation

$$\pi = \langle A_1, B_1, \ldots, A_g, B_g \mid [A_1, B_1] \cdots [A_g, B_g] = 1 \rangle \tag{8.1}$$

where $[A, B] := ABA^{-1}B^{-1}$.

8.2.2 *Representation variety*

Denote the set of representations $\pi \xrightarrow{\rho} G$ by $\mathsf{Hom}(\pi, G)$. Evaluation on a collection $\gamma_1, \ldots, \gamma_N \in \pi$ defines a map

$$\begin{aligned} \mathsf{Hom}(\pi, G) &\longrightarrow G^N \\ \rho &\longmapsto \begin{bmatrix} \rho(\gamma_1) \\ \vdots \\ \rho(\gamma_N) \end{bmatrix} \end{aligned} \tag{8.2}$$

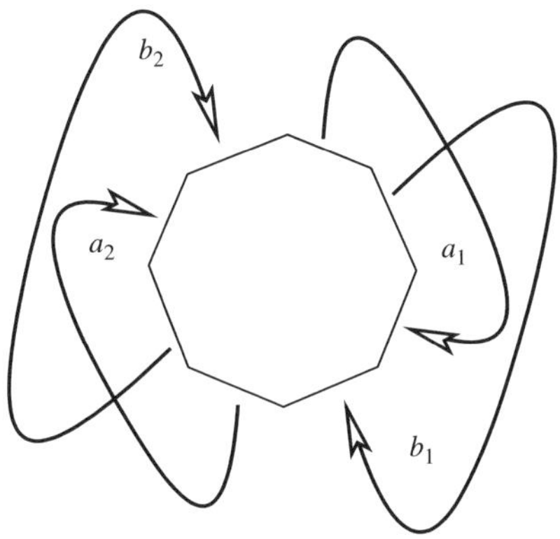

Figure 8.1 The pattern of identifications for a genus 2 surface. The sides of an octagon are pairwise identified to construct a surface of genus 2. The 8 vertices identify to a single 0-cell in the quotient, and the 8 sides identify to four 1-cells, which correspond to the 4 generators in the standard presentation of the fundatmental group.

which is an embedding if $\gamma_1, \ldots, \gamma_N$ generate π. Its image consists of N-tuples

$$(g_1, \ldots, g_N) \in G^N$$

satisfying equations

$$R(g_1, \ldots, g_N) = 1 \in G$$

where $R(\gamma_1, \ldots, \gamma_N)$ are defining relations in π satisfied by $\gamma_1, \ldots, \gamma_N$. If G is a linear algebraic group, these equations are polynomial equations in the matrix entries of g_i. Thus the evaluation map (8.2) identifies $\mathsf{Hom}(\pi, G)$ as an algebraic subset of G^N. The resulting algebraic structure is independent of the generating

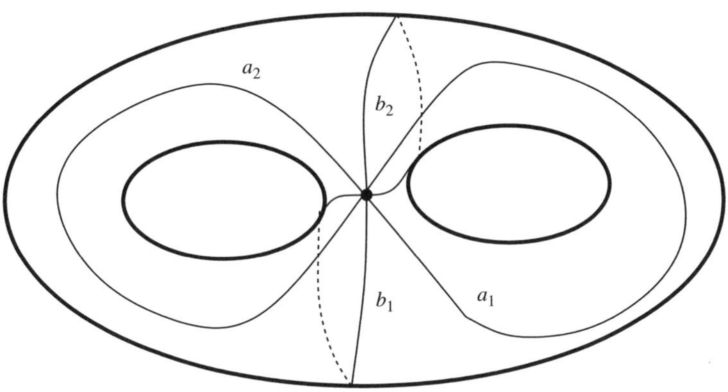

Figure 8.2 The genus 2 surface as an identification space.

set. In particular $\mathsf{Hom}(\pi, G)$ inherits both the Zariski and the classical topology. We consider the classical topology unless otherwise noted.

In terms of the standard presentation (8.1), $\mathsf{Hom}(\pi, G)$ identifies with the subset of G^{2g} consisting of

$$(\alpha_1, \beta_1, \ldots \alpha_g, \beta_g)$$

satisfying the single G-valued equation

$$[\alpha_1, \beta_1] \cdots [\alpha_g, \beta_g] = 1.$$

8.2.3 *Symmetries*

The product $\mathsf{Aut}(\pi) \times \mathsf{Aut}(G)$ acts naturally by left and right compositions, on $\mathsf{Hom}(\pi, G)$: An element

$$(\phi, \alpha) \in \mathsf{Aut}(\pi) \times \mathsf{Aut}(G)$$

transforms $\rho \in \mathsf{Hom}(\pi, G)$ to the composition

$$\pi \xrightarrow{\phi^{-1}} \pi \xrightarrow{\rho} G \xrightarrow{\alpha} G.$$

The resulting action preserves the algebraic structure on $\mathsf{Hom}(\pi, G)$.

8.2.4 *Deformation space*

For any group H, let $\mathsf{Inn}(H)$ denote the normal subgroup of $\mathsf{Aut}(H)$ comprising *inner autormorphisms.* The quotient group $\mathsf{Aut}(H)/\mathsf{Inn}(H)$ is the *outer automorphism group,* denoted $\mathsf{Out}(H)$.

We will mainly be concerned with the quotient

$$\mathsf{Hom}(\pi, G)/G := \mathsf{Hom}(\pi, G)/\big(\{1\} \times \mathsf{Inn}(G)\big),$$

which we call the *deformation space.* For applications to differential geometry, such as moduli spaces of flat connections (gauge theory) or locally homogeneous geometric structures, it plays a more prominent role than the representation variety $\mathsf{Hom}(\pi, G)$. Although $\mathsf{Inn}(G)$ preserves the algebraic structure, $\mathsf{Hom}(\pi, G)/G$ will generally not admit the structure of an algebraic set.

Since the $\mathsf{Inn}(G)$ action on $\mathsf{Hom}(\pi, G)$ absorbs the $\mathsf{Inn}(\pi)$ action on $\mathsf{Hom}(\pi, G)$, the outer automorphism group $\mathsf{Out}(\pi)$ acts on $\mathsf{Hom}(\pi, G)/G$. By a theorem of M. Dehn and J. Nielsen (compare Nielsen 1927 and Stillwell 1987), $\mathsf{Out}(\pi)$ identifies with the *mapping class group*

$$\mathsf{Mod}(\Sigma) := \pi_0\big(\mathsf{Diff}(\Sigma)\big).$$

One motivation for this study is that the deformation spaces provide natural objects upon which mapping class groups act (Goldman 2006).

8.3 Abelian groups and rank 1 Higgs bundles

The simplest groups are commutative. When G is abelian, then the commmutators $[\alpha, \beta] = 1$ and the defining relation in (8.1) is vacuous. Thus

$$\mathsf{Hom}(\pi, G) \longleftrightarrow G^{2g}.$$

Furthermore $\mathsf{Inn}(G)$ is trivial so

$$\mathsf{Hom}(\pi, G)/G \longleftrightarrow G^{2g}$$

as well.

8.3.1 *Symplectic vector spaces*

Homological machinery applies. By the Hurewicz theorem and the universal coefficient theorem,

$$\mathsf{Hom}(\pi, G) \cong \mathsf{Hom}(\pi/[\pi,\pi], G) \cong \mathsf{Hom}(H_1(\Sigma), G) \cong H^1(\Sigma, G)$$

(or $H^1(\pi, G)$ if you prefer group cohomology). In particular when $G = \mathbb{R}$, then $\mathsf{Hom}(\pi, G)/G$ is the real vector space

$$H^1(\Sigma, \mathbb{R}) \cong \mathbb{R}^{2g}$$

which is naturally a *symplectic vector space* under the cup-product pairing

$$H^1(\Sigma, \mathbb{R}) \times H^1(\Sigma, \mathbb{R}) \longrightarrow H^2(\Sigma, \mathbb{R}) \cong \mathbb{R},$$

the last isomorphism corresponding to the orientation of Σ.

Similarly when $G = \mathbb{C}$, the representation variety and the deformation space

$$\mathsf{Hom}(\pi, G)/G = \mathsf{Hom}(\pi, G) \longleftrightarrow H^1(\pi, \mathbb{C}) \cong H^1(\Sigma, \mathbb{C})$$

is a *complex-symplectic vector space*, that is, a complex vector space with a complex-bilinear symplectic form.

The mapping class group action factors through the action on homology of Σ, or equivalently the abelianization of π, which is the homomorphism

$$\mathsf{Mod}(\Sigma) \longrightarrow \mathsf{Sp}(2g, \mathbb{Z}).$$

8.3.2 *Multiplicative characters: $G = \mathbb{C}^*$*

Representations $\pi \longrightarrow \mathbb{C}^*$ correspond to *multiplicative characters*, and are easily understood using the universal covering

$$\begin{aligned} \mathbb{C} &\longrightarrow \mathbb{C}^* \\ z &\longmapsto \exp(2\pi i z) \end{aligned}$$

with kernel $\mathbb{Z} \subset \mathbb{C}$. Such a representation corresponds to a *flat complex line bundle over* Σ. The deformation space $\mathsf{Hom}(\pi, G)$ identifies with the quotient

$$H^1(\Sigma, \mathbb{C})/H^1(\Sigma, \mathbb{Z}).$$

Restricting to unit complex numbers $G = \mathsf{U}(1) \subset \mathbb{C}^*$ identifies $\mathsf{Hom}(\pi, G)$ with the $2g$-dimensional *torus*

$$H^1(\Sigma, \mathbb{R})/H^1(\Sigma, \mathbb{Z}),$$

the quotient of a real symplectic vector space by an integer lattice, $\mathsf{Mod}(\Sigma)$ acts on this torus by *symplectomorphisms.*

8.3.3 *Jacobi variety of a Riemann surface*

The classical Abel–Jacobi theory (compare, e.g., Farkas and Kra 1980) identifies unitary characters $\pi_1(X) \longrightarrow \mathsf{U}(1)$ of the fundamental group of a Riemann surface X with topologically trivial holomorphic line bundles over X. In particular $\mathsf{Hom}(\pi, G)$ identifies with the *Jacobi variety* $\mathsf{Jac}(X)$.

While the basic structure of $\mathsf{Hom}(\pi, G)$ is a $2g$-dimensional compact real torus with a parallel symplectic structure, the conformal structure on X provides much stronger structure. Namely, $\mathsf{Jac}(X)$ is a *principally polarized abelian variety,* a projective variety with the structure of an abelian group. Indeed this extra structure, by Torelli's theorem, is enough to recover the Riemann surface X.

In particular the analytic/algebraic structure on $\mathsf{Jac}(X)$ is definitely *not* invariant under the mapping class group $\mathsf{Mod}(\Sigma)$. However the symplectic structure on $\mathsf{Hom}(\pi, G)$ is independent of the conformal structure X and is invariant under $\mathsf{Mod}(\Sigma)$.

The complex structure on $\mathsf{Jac}(X)$ is the effect of the complex structure on the tangent bundle TX (equivalent to the Hodge $\star$-operator). The Hodge theory of harmonic differential forms finds unique harmonic representatives for cohomology classes, which uniquely extend to *holomorphic differential forms.* Higgs bundle theory is *nonabelian Hodge theory* (Simpson 1992) in that it extends this basic technique from ordinary one-dimensional cohomology classes to flat connections.

When $G = \mathbb{C}^*$, then $\mathsf{Hom}(\pi, G)$ acquires a complex structure J coming from the complex structure on $\mathbb{C}^*$. This depends only on the topology Σ, in fact just its fundamental group π. Cup product provides a holomorphic symplectic structure Ω on this complex manifold, giving the moduli space the structure of a *complex-symplectic manifold.*

As for the $\mathsf{U}(1)$-case above, Hodge theory on the Riemann surface X determines another complex structure by I; then these two complex structures anticommute:

$$IJ + JI = 0,$$

generating a quaternionic action on the tangent bundle with $K := IJ$. The symplectic structure arising from cup product is not holomorphic with respect to I; instead it is *Hermitian* (of Hodge type $(1,1)$) with respect to I, extending the Kähler structure on $\mathsf{Jac}(X)$. Indeed with the structure I, $\mathsf{Hom}(\pi, \mathbb{C}^*)$ identifies with the *cotangent bundle* $T^*\mathsf{Jac}(X)$ with Kähler metric defined by

$$g(X, Y) := \Omega(X, IY).$$

The triple (Ω, I, J) defines a *hyper-Kähler structure refining* the complex-symplectic structure. If one thinks of a complex-symplectic structure as a G-structure where $G = \mathsf{Sp}(2g, \mathbb{C})$, then a hyper-Kähler refinement is a reduction of the structure group to the maximal compact $\mathsf{Sp}(2g, \mathbb{C}) \supset \mathsf{Sp}(2g)$. The more common definition of a hyper-Kähler structure involves the Riemannian metric g which is Kählerian with respect to all three complex structures I, J, K; alternatively it is characterized as a Riemannian manifold of dimension $4g$ with holonomy reduced to $\mathsf{Sp}(2g) \subset \mathsf{SO}(4g)$.

For a detailed exposition of the theory of rank 1 Higgs bundles on Riemann surfaces, compare Goldman and Xia (2008).

8.4 Stable vector bundles and Higgs bundles

Narasimhan and Seshadri (1965) generalized the Abel–Jacobi theory above to identify $\mathsf{Hom}(\pi, G)/G$ with a moduli space of holomorphic objects over a Riemann surface X, when $G = \mathsf{U}(n)$. (This was later extended by Ramanathan 1975 to general compact Lie groups G.)

A notable new feature is that, unlike line bundles, not every topologically trivial holomorphic rank n vector bundle arises from a representation $\pi \longrightarrow \mathsf{U}(n)$. Furthermore equivalence classes of all holomorphic $\mathbb{C}^n$ bundles does not form an algebraic set.

Narasimhan and Seshadri define a degree 0 holomorphic $\mathbb{C}^n$-bundle V over X to be *stable* (respectively, *semistable*) if and only if every holomorphic vector subbundle of V has negative (respectively, nonpositive) degree. Then a holomorphic vector bundle arising from a unitary representation ρ is semistable, and the bundle is stable if and only if the representation is irreducible. Furthermore every such semistable bundle arises from a unitary representation. Narasimhan and Seshadri show the moduli space $\mathcal{M}_{n,0}(X)$ of semistable bundles of degree 0 and rank n over X is naturally a projective variety, thus defining such a structure on $\mathsf{Hom}(\pi, G)/G$. The Kähler structure depends heavily on the Riemann surface X, although the symplectic structure depends only on the topology Σ.

It is useful to extend the notions of stability to bundles which may not have degree 0. In particular we would like stability to be preserved by tensor product with holomorphic line bundles. Define a holomorphic vector bundle V to be *stable* if every holomorphic subbundle $W \subset V$ satisfies the inequality

$$\frac{\mathsf{deg}(W)}{\mathsf{rank}(W)} < \frac{\mathsf{deg}(V)}{\mathsf{rank}(V)}.$$

Semistability is defined similarly by replacing the strict inequality by a weak inequality.

In trying to extend this correspondence to the complexification $G = \mathsf{GL}(n, \mathbb{C})$ of $\mathsf{U}(n)$, one might consider the *cotangent bundle* $T^*\mathcal{M}_{n,0}(X)$ of the Narasimhan–Seshadri moduli space, and relate it to representations $\pi \rightarrow \mathsf{GL}(n, \mathbb{C})$. In particular since cotangent bundles of Kähler manifolds tend to be hyper-Kähler, relating $\mathsf{Hom}(\pi, G)/G$ to $T^{\mathcal{M}}_{n,0}(X)$ might lead to a hyper-Kähler geometry on $\mathsf{Hom}(\pi, G)/G$.

Thus a neighborhood of the $\mathsf{U}(n)$ representations in the space of $\mathsf{GL}(n, \mathbb{C})$ corresponds to a neighborhood of the zero-section of $T^*\mathcal{M}_{n,0}(X)$. In turn, elements in this neighborhood identify with pairs (V, Φ) where V is a semistable holomorphic vector bundle and Φ is a tangent covector to V in the space of holomorphic vector bundles. Such a tangent covector is with a *Higgs field,* by definition, an $\mathsf{End}(V)$-valued holomorphic one-form on X.

Although one can define a hyper-Kähler structure on the moduli space of such pairs, the hyper-Kähler metric is incomplete and not all irreducible linear representations arise. To rectify this problem, one must consider Higgs fields on possibly unstable vector bundles.

Following Hitchin (1987) and Simpson (1988), define a *Higgs pair* to be a pair (V, Φ) where V is a (not ncessarily semistable) holomorphic vector bundle and the Higgs field Φ a $\mathsf{End}(V)$-valued holomorphic one-form. Define (V, Φ) to be *stable* if and only if for all Φ-invariant holomorphic subbundles $W \subset V$,

$$\frac{\mathsf{deg}(W)}{\mathsf{rank}(W)} < \frac{\mathsf{deg}(V)}{\mathsf{rank}(V)}.$$

The Higgs bundle (V, Φ) is *polystable* if and only if $(V, \Phi) = \bigoplus_{i=1}^{l}(V_i, \Phi_i)$ where each summand (V_i, Φ_i) is stable and

$$\frac{\mathsf{deg}(V_i)}{\mathsf{rank}(V_i)} = \frac{\mathsf{deg}(V)}{\mathsf{rank}(V)}$$

for $i = 1, \ldots, l$.

The following basic result follows from Hitchin (1987) and Simpson (1988), with a key ingredient (the *harmonic metric*) supplied by Corlette (1988) and Donaldson (1987):

Theorem 8.1 *The following natural bijections exist between equivalences classes:*

$$\left\{\begin{array}{l}\text{Stable Higgs pairs} \\ (V, \Phi) \text{ over } \Sigma\end{array}\right\} \longleftrightarrow \left\{\begin{array}{l}\text{Irreducible representations} \\ \pi_1(\Sigma) \xrightarrow{\rho} \mathsf{GL}(n, \mathbb{C})\end{array}\right\}$$

$$\left\{\begin{array}{l}\text{Polystable Higgs pairs} \\ (V, \Phi) \text{ over } \Sigma\end{array}\right\} \longleftrightarrow \left\{\begin{array}{l}\text{Reductive representations} \\ \pi_1(\Sigma) \xrightarrow{\rho} \mathsf{GL}(n, \mathbb{C})\end{array}\right\}$$

When the Higgs field $\Phi = 0$, this is just the Narasimhan–Seshadri theorem, identifying stable holomorphic vector bundles with irreducible $\mathsf{U}(n)$ representations. Allowing the Higgs field Φ to be nonzero, even when V is unstable, leads to a rich new class of examples, which can now be treated using the techniques of geometric invariant theory.

8.5 Hyperbolic geometry: G = $\mathsf{PSL}(2,\mathbb{R})$

Another important class of surface group representations are *Fuchsian representations*, which arise by endowing Σ with the local geometry of *hyperbolic space* H^2. Here G is the group of orientation-preserving isometries $\mathsf{Isom}^+(\mathsf{H}^2)$, which, using Poincaré's *upper half-space model*, identifies with $\mathsf{PSL}(2,\mathbb{R})$. Fuchsian representations are characterized in many different equivalent ways; in particular a representation $\pi \xrightarrow{\rho} G = \mathsf{PSL}(2,\mathbb{R})$ is Fuchsian if and only if it is a *discrete embedding*, that is, ρ embeds π isomorphicly onto a discrete subgroup of G.

8.5.1 *Geometric structures*

Let H^2 be the hyperbolic plane with a fixed orientation and $G \cong \mathsf{Isom}^+(\mathsf{H}^2) \cong \mathsf{PSL}(2,\mathbb{R})$ its group of orientation-preserving isometries. A *hyperbolic structure* on a topological surface Σ is defined by a coordinate atlas $\{(U_\alpha, \psi_\alpha)\}_{\alpha\in A}$ where

- The collection $\{U_\alpha\}_{\alpha\in A}$ of coordinate patches *covers* Σ (for some index set A).
- Each coordinate *chart* ψ_α is an orientation-preserving homeomorphism of the coordinate patch U_α onto an open subset $\psi_\alpha(U_\alpha) \subset \mathsf{H}^2$.
- For each connected component $C \subset U_\alpha \cap U_\beta$, there is (necessarily unique) $g_{C,\alpha,\beta} \in G$ such that

$$\psi_\alpha|_C = g_{C,\alpha,\beta} \circ \psi_\beta|_C.$$

The resulting *local* hyperbolic geometry defined on the patches by the coordinate charts is independent of the charts, and extends to a global structure on Σ. The surface Σ with this refined structure of local hyperbolic geometry will be called a *hyperbolic surface* and denoted by M. Such a structure is equivalent to a Riemannian metric of constant curvature -1. The equivalence follows from two basic facts:

- Any two Riemannian manifolds of curvature -1 are locally isometric.
- A local isometry from a connected subdomain of H^2 extends globally to an isometry of H^2.

Suppose M_1, M_2 are two hyperbolic surfaces. Define a *morphism* $M_1 \xrightarrow{\phi} M_2$ as a map ϕ, which, in the preferred local coordinates of M_1 and M_2, is defined by isometries in G. Necessarily a morphism is a local isometry of Riemannian manifolds. Furthermore, if M is a hyperbolic surface and $\Sigma \xrightarrow{f} M$ is a local homeomorphism, there exists a hyperbolic structure on Σ for which f is a morphism. In particular every covering space of a hyperbolic surface is a hyperbolic surface.

In more traditional terms, a morphism of hyperbolic surfaces is just a local isometry.

8.5.2 *Relation to the fundamental group*

While the definitions involving coordinate atlases or Riemannian metrics have certain advantages, another point of view underscores the role of the fundamental group.

Let M be a hyperbolic surface. Choose a universal covering space $\tilde{M} \to M$ and give $\tilde{M}$ the unique hyperbolic structure for which $\tilde{M} \to M$ is a local isometry. Then there exists a *developing map* $\tilde{M} \xrightarrow{\mathsf{dev}_M} \mathsf{H}^2$, a local isometry, which induces the hyperbolic structure on $\tilde{M}$ from that of H^2. The group $\pi_1(M)$ of deck transformations of $\tilde{M} \to M$ acts on H^2 by isometries and dev is equivariant respecting this action: for all $\gamma \in \pi_1(M)$, the figure

$$\begin{array}{ccc} \tilde{M} & \xrightarrow{\mathsf{dev}_M} & \mathsf{H}^2 \\ {\scriptstyle\gamma}\big\downarrow & & \big\downarrow{\scriptstyle\rho(\gamma)} \\ \tilde{M} & \xrightarrow[\mathsf{dev}_M]{} & \mathsf{H}^2. \end{array}$$

commutes. The correspondence $\gamma \longmapsto \rho(\gamma)$ is a homomorphism,

$$\pi_1(M) \xrightarrow{\mathsf{hol}_M} \mathsf{Isom}^+(\mathsf{H}^2),$$

the *holonomy representation* of the hyperbolic surface M. The pair $(\mathsf{dev}_M, \mathsf{hol}_M)$ is unique up to the G-action defined by

$$(\mathsf{dev}_M, \mathsf{hol}_M) \stackrel{g}{\longmapsto} (g \circ \mathsf{dev}_M, \mathsf{Inn}(g) \circ \mathsf{hol}_M)$$

for $g \in \mathsf{Isom}^+(\mathsf{H}^2)$.

If the hyperbolic structure is *complete*, that is, the Riemannian metric is geodesically complete, then the developing map is a *global isometry* $\tilde{M} \approx \mathsf{H}^2$. In that case the π-action on H^2 defined by the holonomy representation ρ is equivalent to the action by deck transformations. Thus ρ defines a proper free action of π on H^2 by isometries. Conversely if ρ defines a proper free isometric π-action, then the quotient

$$M := \mathsf{H}^2/\rho(\pi)$$

is a complete hyperbolic manifold with a preferred isomorphism

$$\pi_1(\Sigma) \xrightarrow{\rho} \rho(\pi) \subset G.$$

This isomorphism (called a *marking*) determines a preferred homotopy class of homotopy equivalences

$$\Sigma \longrightarrow M.$$

8.5.3 *Examples of hyperbolic structures*

We now give three examples of surface group representations in $\mathsf{PSL}(2, \mathbb{R})$. The first example is Fuchsian and corresponds to a hyperbolic structure on a surface of genus 2. The second example is not Fuchsian, but corresponds to a hyperbolic structure with a single *branch point*, that is, a point with local coordinate given by a branched conformal mapping $z \longmapsto z^k$ where $k \geqslant 1$. (The nonsingular case corresponds to $k = 1$.) In our example $k = 2$ and the singular point has a neighborhood isometric to a hyperbolic cone of cone angle 4π.

8.5.3.1 *A Fuchsian example*

Here is a simple example of a hyperbolic surface of genus 2. Figure 8.1 depicts a topological construction for a genus 2 surface Σ. Realizing this topological construction in hyperbolic geometry gives Σ a local hyperbolic geometry as follows. Take a regular octagon P with angles $\pi/4$. Label the sides as

$$A_1^-, B_1^-, A_1^+, B_1^+, A_2^-, B_2^-, A_2^+, B_2^+,$$

a_i pairs B_i^- to B_i^+ and b_i pairs A_i^- to A_i^+, respectively.

Pair the sides by

$$a_1, b_1, a_2, b_2 \in \mathsf{PSL}(2, \mathbb{R})$$

according to the pattern described in Figure 8.1. Given any two oriented geodesic segments in H^2 of equal length, a unique orientation-preserving isometry

Figure 8.3 A regular octagon with vertex angles $\pi/4$ can be realized in the tiling of H^2 by triangles with angles $\pi/2, \pi/4$, and $\pi/8$. The identifications depicted in Figure 8.1 are realized by orientation-preserving isometries. The eight angles of $\pi/4$ fit together to form a cone of angle 2π, forming a coordinate chart for a hyperbolic structure around that point.

maps one to the other. Since the polygon is regular, one can realize all four identifications in $\mathsf{Isom}^+(\mathsf{H}^2)$.

The quotient (compare Figure 8.2) contains three types of points:

- A point in the open 2-cell has a coordinate chart which is the embedding $P \hookrightarrow \mathsf{H}^2$.
- A point on the interior of an edge has a half-disc neighborhood, which together with the half-disc neighborhood of its part, gives a coordinate chart for the corresponding point in the quotient.
- Around the single vertex in the quotient is a cone of angle

$$8(\pi/4) = 2\pi,$$

a disc in the hyperbolic plane.

The resulting identification space is a hyperbolic surface of genus $g = 2$. The above isometries satisfying the defining relation for $\pi_1(\Sigma)$:

$$a_1 b_1 a_1^{-1} b_1^{-1} a_2 b_2 a_2^{-1} b_2^{-1} = 1$$

Figure 8.4 A regular right-angled octagon can also be realized in the tiling of H^2 by triangles with angles $\pi/2, \pi/4$, and $\pi/8$. The identifications depicted in Figure 8.1 are realized by orientation-preserving isometries. The eight angles of $\pi/2$ fit together to form a cone of angle 4π, forming a coordinate chart for a singular hyperbolic structure, branched at one point.

and define a Fuchsian representation

$$\pi_1(\Sigma) \xrightarrow{\rho} \mathsf{PSL}(2, \mathbb{R}).$$

Compare Figure 8.3.

8.5.3.2 *A branched hyperbolic structure*

We can modify the preceding example to include a singular structure, again on a surface of genus 2. Take a regular *right-angled* octagon. Again, labeling the sides as before, side pairings a_1, b_1, a_2, b_2 exist. Now eight right angles compose a neighborhood of the vertex in the quotient space. The quotient space is a hyperbolic structure with one singularity of cone angle $4\pi = 8(\pi/2)$. Since the product of the identification mappings

$$a_1 b_1 a_1^{-1} b_1^{-1} a_2 b_2 a_2^{-1} b_2^{-1}$$

is rotation through 4π (the identity), the holonomy representation $\hat{\rho}$ of the nonsingular hyperbolic surface $\Sigma \setminus \{p\}$ extends:

$$\begin{array}{ccc} \pi_1\big(S \setminus \{p\}\big) & & \\ \downarrow & \searrow^{\hat{\rho}} & \\ \pi_1(\Sigma) & \dashrightarrow_{\rho} & \mathsf{PSL}(2, \mathbb{R}) \end{array}$$

Although $\rho(\pi)$ is discrete, ρ is not injective. Compare Figure 8.4.

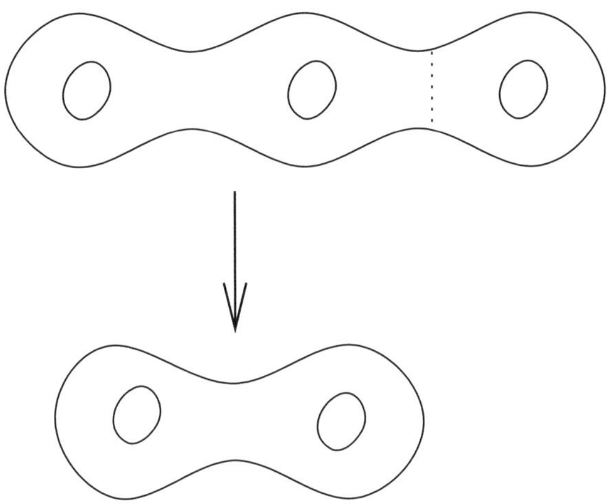

Figure 8.5 A degree 1 map from a genus 3 surface to a genus 2 surface which collapses a handle. Such a map is not homotopic to a smooth map with branch point singularities (such as a holomorphic map).

8.5.3.3 *A representation with no branched structures*

Consider a degree 1 map f from a genus 3 surface Σ to a genus 2 hyperbolic surface M, depicted in Figure 8.5. Let $\pi_1(M) \xrightarrow{\mu} G$ denote the holonomy representation of M and consider the composition

$$\pi = \pi_1(\Sigma) \xrightarrow{f_*} \pi_1(M) \overset{\mu}{\hookrightarrow} G.$$

Then a branched hyperbolic structure with holonomy $\mu \circ f_*$ corresponds to a mapping with branch singularities

$$\Sigma \xrightarrow{F} \mathsf{H}^2/\mathsf{Image}(\mu \circ f_*) = M,$$

inducing the homomorphism

$$\pi = \pi_1(\Sigma) \xrightarrow{f_*} \pi_1(M).$$

In particular $F \simeq f$. However, since $\mathsf{deg}(f) = 1$, any mapping with only branch point singularities of degree 1 must be a homeomorphism, a contradiction.

8.6 Moduli of hyperbolic structures and representations

To understand "different" geometric structures on the "same" surface, one introduces *markings.* Fix a topological type Σ and let the geometry M vary. The fundamental group $\pi = \pi_1(\Sigma)$ is also fixed, and each marked structure determines a well-defined equivalence class in $\mathsf{Hom}(\pi, G)/G$. Changing the marking corressponds to the action of the mapping class group $\mathsf{Mod}(\Sigma) = \mathsf{Out}(\pi)$ on $\mathsf{Hom}(\pi, G)/G$. *Unmarked structures* correspond to the orbits of the $\mathsf{Mod}(\Sigma)$ action.

8.6.1 *Deformation spaces of geometric structures*

A *marked hyperbolic structure on* Σ is defined as a pair (M, f) where M is a hyperbolic surface and f is a homotopy equivalence $\Sigma \to M$. Two marked hyperbolic structures

$$\Sigma \xrightarrow{f} M, \qquad \Sigma \xrightarrow{f'} M'$$

are *equivalent* if and only if there exists an isometry $M \xrightarrow{\phi} M'$ such that

$$\begin{array}{ccc} & & M \\ & \overset{f}{\nearrow} & \downarrow \phi \\ \Sigma & \xrightarrow[f']{} & M' \end{array}$$

homotopy commutes, that is, $\phi \circ f \simeq f'$. The *Fricke space* $\mathfrak{F}(\Sigma)$ of Σ is the space of all such equivalence classes of marked hyperbolic structures on Σ (Bers and Gardiner 1986). The Fricke space is diffeomorphic to $\mathbb{R}^{6g-6}$. The theory of moduli of hyperbolic structures on surfaces goes back at least to Fricke and Klein (1897/1912).

The Teichmüller space $\mathfrak{T}(\Sigma)$ of Σ is defined similarly, as the space of equivalence classes of *marked conformal structures* on Σ, that is, pairs (X, f) where X is a Riemann surface and $\Sigma \xrightarrow{f} X$ is a homotopy equivalence. Teichmüller used quasiconformal mappings to parametrize $\mathfrak{T}(\Sigma)$ by elements of a vector space, define a metric on $\mathfrak{T}(\Sigma)$, and prove analytically that $\mathfrak{T}(\Sigma)$ is a cell. Using these ideas, Ahlfors (1960) proved $\mathfrak{T}(\Sigma)$ is naturally a complex manifold.

Since a hyperbolic structure is a Riemannian metric, every hyperbolic structure has an underlying conformal structure. The *uniformization theorem* asserts that if $\chi(\Sigma) < 0$, then every conformal structure on Σ underlies a unique hyperbolic structure. The resulting identification of conformal and hyperbolic structures identifies $\mathfrak{T}(\Sigma)$ with $\mathfrak{F}(\Sigma)$. As discussed below, $\mathfrak{F}(\Sigma)$ identifies with an open subset of $\mathsf{Hom}(\pi, \mathsf{PSL}(2, \mathbb{R}))/\mathsf{PSL}(2, \mathbb{R})$ which has no apparent complex structure. Thus the complex structure on $\mathfrak{T}(\Sigma)$ is more mysterious when $\mathfrak{T}(\Sigma)$ is viewed as a space of hyperbolic structures. For a readable survey of classical Teichmüller theory, see Bers (1972), Abikoff (1980), Nag (1988), Imayoshi and Taniguchi (1992), Gardiner-Lakic (2000) or Hubbard (2006).

8.6.2 *Fuchsian components of* $\mathsf{Hom}(\pi, G)/G$

To every equivalence class of marked hyperbolic structures is associated a well-defined element

$$[\rho] \in \mathsf{Hom}(\pi, G)/G.$$

A representation $\pi \xrightarrow{\rho} G$ is *Fuchsian* if and only if it arises as the holonomy of a hyperbolic structure on Σ. Equivalently, it satisfies the three conditions:

- ρ is injective.
- Its image $\rho(\pi)$ is a discrete subgroup of G.
- The quotient $G/\rho(\pi)$ is compact.

The first condition asserts that ρ is an *embedding*, and the second two conditions assert that $\rho(\pi)$ is a *cocompact lattice.* Under our assumption $\partial\Sigma = \emptyset$, the third condition (compactness of $G/\rho(\pi)$) follows from the first two. In general, we say that ρ is a *discrete embedding* (or *discrete and faithful*) if ρ is an embedding with discrete image (the first two conditions).

Theorem 8.2 *Let* $G = \mathsf{Isom}(\mathsf{H}^2) = \mathsf{PGL}(2, \mathbb{R})$ *and* Σ *a closed connected surface with* $\chi(\Sigma) < 0$. *Fricke space, the subset of* $\mathsf{Hom}(\pi, G)/G$ *consisting of* G-*equivalence classes of Fuchsian representations, is a connected component of* $\mathsf{Hom}(\pi, G)/G$.

This result follows from three facts:

- Openness of Fricke space (Weil 1962)
- Closedness of Fricke space (Chuckrow 1968)
- Connectedness of Fricke space

Chuckrow's theorem is a special case of a consequence of the Kazhdan–Margulis uniform discreteness (compare Raghunathan 1972 and Goldman and Millson 1987). These ideas go back to Bieberbach and Zassenhaus in connection with the classification of Euclidan crystallographic groups. Uniform discretness applies under very general hypotheses, to show that discrete embeddings form a closed subset of the representation variety. For the proof of connectedness, see, for example, Jost (2006, section 4.3), Buser (2002), Thurston (1979) or Ratcliffe (2006) for elementary proofs using Fenchel–Nielsen coordinates. Connectedness also follows from the uniformization theorem, together with the identification of Teichmüller space $\mathfrak{T}(\Sigma)$ as a cell.

When $G = \mathsf{Isom}^+(\mathsf{H}^2) = \mathsf{PSL}(2,\mathbb{R})$, the situation slightly complicates, due to the choice of orientation. Assume Σ is orientable, and orient it. Orient H^2 as well. Let $\Sigma \xrightarrow{f} M$ be a marked hyperbolic structure on Σ. The orientation of M induces an orientation of $\tilde{M}$ which is invariant under $\pi_1(M)$. However, the developing map dev_M may or may not preserve the (arbitrarily chosen) orientations of $\tilde{M}$ and H^2. Accordingly $\mathsf{Isom}^+(\mathsf{H}^2)$-equivalence classes of Fuchsian representations in G fall into two classes, which we call *orientation-preserving* and *orientation-reversing*, respectively. These two classes are interchanged by inner automorphisms of orientation-reversing isometries of H^2.

Theorem 8.3 *Let $G = \mathsf{Isom}^+(\mathsf{H}^2) = \mathsf{PSL}(2,\mathbb{R})$ and Σ a closed connected oriented surface with $\chi(\Sigma) < 0$. The set of G-equivalence classes of Fuchsian representations forms two connected components of $\mathsf{Hom}(\pi, G)/G$. One component corresponds to orientation-preserving Fuchsian representations and the other to orientation-reversing Fuchsian representations.*

8.6.3 *Characteristic classes and maximal representations*

Characteristic classes of flat bundles determine invariants of representations. In the simplest cases (when G is compact or reductive complex), these determine the connected components of $\mathsf{Hom}(\pi, G)$.

8.6.3.1 *Euler class and components of* $\mathsf{Hom}(\pi, \mathsf{PSL}(2,\mathbb{R}))$

The components of $\mathsf{Hom}(\pi, G)$ were determined in Goldman (1988) using an invariant derived from the Euler class of the oriented H^2-bundle

$$\begin{array}{ccc} \mathsf{H}^2 & \longrightarrow & (\mathsf{H}^2)_\rho \\ & & \downarrow \\ & & \Sigma \end{array}$$

associated to a representation $\pi \xrightarrow{\rho} \mathsf{PSL}(2,\mathbb{R})$ as follows. The total space is the quotient

$$(\mathsf{H}^2)_\rho := (\tilde{\Sigma} \times \mathsf{H}^2)/\pi$$

where π acts diagonally on $\tilde{\Sigma} \times \mathsf{H}^2$ by deck transformations on $\tilde{\Sigma}$ and via ρ on H^2. Isomorphism classes of oriented H^2-bundles over Σ are determined by the *Euler class*, which lives in $H^2(\Sigma, \mathbb{Z})$. The orientation of Σ identifies this cohomology group with $\mathbb{Z}$. The resulting map

$$\mathsf{Hom}(\pi, G) \xrightarrow{\mathsf{Euler}} H^2(S; \mathbb{Z}) \cong \mathbb{Z}$$

satisfies

$$|\mathsf{Euler}(\rho)| \leqslant |\chi(S)| = 2g - 2 \tag{8.3}$$

(Milnor 1958 and Wood 1976). Call a representation *maximal* if equality holds in (8.3), that is, $\mathsf{Euler}(\rho) = \pm\chi(\Sigma)$.

The following converse was proved in Goldman (1980) (compare also Hitchin 1987 and Goldman 1988).

Theorem 8.4 *ρ is maximal if and only if ρ is Fuchsian.*

Suppose M is a branched hyperbolic surface with branch points $p_1, \ldots, p_l$ where p_i is branched of order k_i, where each k_i is a positive integer. In other words, each p_i has a neighborhood which is a hyperbolic cone of cone angle $2\pi k_i$. Consider a marking $\Sigma \to M$, determining a holonomy representation ρ. Then

$$\mathsf{Euler}(\rho) = \chi(\Sigma) + \sum_{i=1}^{l} k_i.$$

Consider the two examples for genus 2 surfaces:

- The first (Fuchsian) example (Section 8.5.3.1) arising from a regular octagon with $\pi/4$ angles has Euler class $-2 = \chi(\sigma)$.
- In the second example (Section 8.5.3.2), the structure is branched at one point, so that $l = k_1 = 1$ and the Euler class equals $-1 = \chi(\Sigma) + 1$.

8.6.4 *Quasi-Fuchsian representations:* $G = \mathsf{PSL}(2, \mathbb{C})$

When the representation

$$\pi \longrightarrow \mathsf{PSL}(2, \mathbb{R}) \hookrightarrow \mathsf{PSL}(2, \mathbb{C})$$

is deformed inside $\mathsf{PSL}(2, \mathbb{C})$, the action on $\mathbb{C}\mathsf{P}^1$ is topologically conjugated to the original Fuchsian action (Figure 8.6). Furthermore there exists a Hölder ρ-equivariant embedding $S^1 \hookrightarrow \mathbb{C}\mathsf{P}^1$, whose image Λ has Hausdorff dimension >1, unless the deformation is still Fuchsian. The space of such representations is the *quasi-Fuchsian space* $\mathcal{QF}(\Sigma)$. By Bers (1960), $\mathcal{QF}(\Sigma)$ naturally identifies with

$$\mathfrak{T}(\Sigma) \times \overline{\mathfrak{T}(\Sigma)} \approx \mathbb{R}^{12g-12}.$$

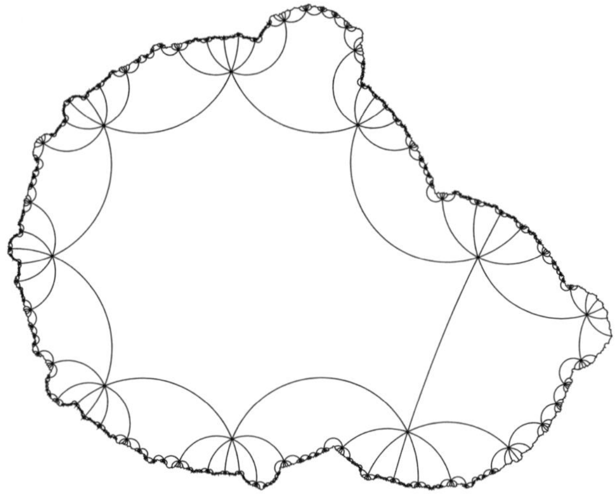

Figure 8.6 A quasi-Fuchsian subgroup of $\mathsf{PSL}(2,\mathbb{C})$ obtained by deforming the genus 2 surface with a fundamental domain, the regular octagon with $\pi/4$ angles in $\mathbb{C}\mathsf{P}^1$. The limit set is a nonrectifiable Jordan curve, but the new action of $\pi_1(\Sigma)$ is topologically conjugated to the original Fuchsian action.

Bers's correspondence is the following. The action of ρ on the complement $\mathbb{C}\mathsf{P}^1 \setminus \Lambda$ is properly discontinuous, and the quotient

$$\big(\mathbb{C}\mathsf{P}^1 \setminus \Lambda\big)/\rho(\pi)$$

consists of two Riemann surfaces, each with a canonical marking determined by ρ. Furthermore these surfaces possess opposite orientations, so the pair of marked conformal structures determine a point in $\mathfrak{T}(\Sigma) \times \overline{\mathfrak{T}(\Sigma)}$. Bonahon (1986) and Thurston (1979) proved that the closure of $\mathcal{QF}(\Sigma)$ in $\mathsf{Hom}(\pi, G)/G$ equals the set of equivalence classes of discrete embeddings. The frontier $\partial\mathcal{QF}(\Sigma) \subset \mathsf{Hom}(\pi, G)/G$ is nonrectifiable, and is near non-discrete representations.

However, the two connected components of $\mathsf{Hom}(\pi, G)/G$ are distinguished by the characteristic class (related to the second Stiefel–Whitney class w_2) which detects whether a representation in $\mathsf{PSL}(2,\mathbb{C})$ lifts to the double covering $\mathsf{SL}(2,\mathbb{C}) \to \mathsf{PSL}(2,\mathbb{C})$ (Goldman 1988). Contrast this situation with $\mathsf{PSL}(2,\mathbb{R})$ where the discrete embeddings form connected components, characterized by maximality.

8.6.4.1 *Higher rank Hermitian spaces: the Toledo invariant*

Vladimir Turaev (1984, 1985) and Domingo Toledo (1989) generalized the Euler class of flat $\mathsf{PSL}(2,\mathbb{R})$-bundles to flat G-bundles, where G is the autormorphism group of a Hermitian symmetric space X of noncompact type.

Let $\pi \xrightarrow{\rho} G$ be a representation and let

$$\begin{array}{ccc} X & \longrightarrow & (X)_\rho \\ & & \downarrow \\ & & \Sigma \end{array}$$

be the corresponding flat (G, X)-bundle over Σ. Then the G-invariant Kähler form ω on X defines a closed exterior two-form ω_ρ on the total space $(X)_\rho$. Let $\Sigma \xrightarrow{s} (X)_\rho$ be a smooth section. Then the integral

$$\int_\Sigma s^*\omega_\rho$$

is independent of s, depends continuously on ρ, and, after suitable normalization, assumes integer values. The resulting *Turaev-Toledo invariant*

$$\mathsf{Hom}(\pi, G) \xrightarrow{\tau} \mathbb{Z}$$

satisfies

$$|\tau(\rho)| \leqslant (2g-2)\mathsf{rank}_{\mathbb{R}}(G)$$

(Domic and Toledo 1987, and Clerc and Ørsted 2003). Define ρ to be *maximal* if and only if

$$|\tau(\rho)| = (2g-2)\mathsf{rank}_{\mathbb{R}}(G).$$

Theorem 8.5 (Toledo 1989) $\pi \xrightarrow{\rho} \mathsf{U}(n,1)$ *is maximal if and only if ρ is a discrete embedding preserving a complex geodesic, that is, ρ is conjugate to a representation with*

$$\rho(\pi) \subset \mathsf{U}(1,1) \times \mathsf{U}(n-1).$$

This rigidity has a curious consequence for the local geometry of the deformation space. Let $G := \mathsf{U}(n,1)$ and

$$G_0 = \mathsf{U}(1,1) \times \mathsf{U}(n-1) \subset G.$$

Then, in an appropriate sense,

$$\dim \mathsf{Hom}(\pi, G)/G = 2g + (2g-2)\big((n+1)^2 - 1\big) = (2g-2)(n+1)^2 + 2,$$

but Toledo's rigidity result implies that the component of maximal representations has strictly lower dimension:

$$\dim \mathsf{Hom}(\pi, G_0)/G_0 = 4g + (2g-2)3 + (2g-2)((n-1)^2 - 1)$$

with codimension

$$8(n-1)(g-1) - 2.$$

Compare Goldman and Millson (1988).

8.6.5 *Teichmüller space: marked conformal structures*

The *Teichmüller space* $\mathfrak{T}(\Sigma)$ of Σ is the deformation space of marked conformal structures on Σ.

A *marked conformal structure* on Σ is a pair (X, f) where X is a Riemann surface and f is a homotopy equivalence $\Sigma \to X$. Marked conformal structures

$$\Sigma \xrightarrow{f} X, \qquad \Sigma \xrightarrow{f'} X'$$

are *equivalent* if and only if there exists a biholomorphism $X \xrightarrow{\phi} X'$ such that

$$\begin{array}{ccc} & & X \\ & \overset{f}{\nearrow} & \downarrow{\scriptstyle \phi} \\ \Sigma & \xrightarrow[f']{} & X' \end{array}$$

homotopy commutes.

Theorem 8.6 (Uniformization) *Let X be a Riemann surface with $\chi(X) < 0$. Then there exists a unique hyperbolic metric whose underlying conformal structure agrees with X.*

Since every hyperbolic structure possesses an underlying conformal structure, Fricke space $\mathfrak{F}(\Sigma)$ maps to Teichmüller space $\mathfrak{T}(\Sigma)$. By the uniformization theorem, $\mathfrak{F}(\Sigma) \to \mathfrak{T}(\Sigma)$ is an isomorphism. It is both common and tempting to confuse these two deformation spaces. In the present context, however, it seems best to distinguish between the representation/hyperbolic structure and the conformal structure.

For example, each Fuchsian representation determines a marked hyperbolic structure, and hence an underlying marked conformal structure. An equivalence class of Fuchsian representations thus determines a special point in Teichmüller space. This contrasts sharply with other representations which do *not* generally pick out a preferred point in $\mathfrak{T}(\Sigma)$. This preferred point can be characterized as the unique minimum of an energy function on Teichmüller space.

The construction, due to Tromba (1992), is as follows. Given a hyperbolic surface M and a homotopy equivalence $X \xrightarrow{f} M$, then by Eels and Sampson (1964) a unique *harmonic map* $X \xrightarrow{F} M$ exists homotopic to f. The harmonic map is conformal if and only if M is the uniformization of X. In general the nonconformality is detected by the *Hopf differential* $\mathsf{Hopf}(F) \in H^0(X, K_X^2)$, defined as the $(2,0)$ part of the pull-back by F of the complexified Riemannian metric on M. The resulting mapping

$$\begin{aligned} \mathfrak{F}(X) &\longrightarrow H^0(X, K_X^2) \\ (f, M) &\longmapsto \mathsf{Hopf}(F) \end{aligned}$$

is a diffeomorphism.

Fixing M and letting the marked complex structure (f, X) vary over $\mathfrak{T}(\Sigma)$ yields an interesting invariant discussed in Tromba (1992), and extended in Goldman and Wentworth (2005) and Labourie (to appear). The *energy of the harmonic map* $F = F(f, X, M)$ is a real-valued function on $\mathfrak{T}(\Sigma)$. In the present context it is the square of the L^2-norm of $\mathsf{Hopf}(F)$.

Theorem 8.7 (Tromba 1992) *The resulting function $\mathfrak{T}(\Sigma) \to \mathbb{R}$ is proper, convex, and possesses a unique minimum at the uniformaization structure X.*

For more applications of this energy function to surface group representations, compare Goldman and Wentworth (2007), where properness is proved for convex cocompact discrete embeddings, and Labourie (2008), where the above result is extended to quasi-isometric embeddings $\pi \hookrightarrow G$.

8.6.6 *Holomorphic vector bundles and uniformization*

Let $\pi \xrightarrow{\rho} \mathsf{PSL}(2, \mathbb{R})$ be a Fuchsian representation corresponding to a marked hyperbolic structure $\Sigma \xrightarrow{f} M$. A *spin structure* on Σ determines a lifting of ρ to

$$\pi \xrightarrow{\tilde{\rho}} \mathsf{SL}(2, \mathbb{R}) \subset \mathsf{SL}(2, \mathbb{C})$$

and hence a flat $\mathbb{C}^2$-bundle $(\mathbb{C}^2)_\rho$ over Σ.

Choose a marked Riemann surface X corresponding to a point in Teichmüller space $\mathfrak{T}(\Sigma)$. Since locally constant maps are holomorphic for *any* complex structure on Σ, the flat bundle $(\mathbb{C}^2)_\rho$ has a natural holomorphic structure; denote the corresponding holomorphic rank 2 vector bundle over X by $E_\rho \to X$.

In trying to fit such a structure into a moduli problem over X, the first problem is that this holomorphic vector bundle is *unstable* and does not seem susceptible to geometric invariant theory techniques. Indeed, its instability intimately relates to its role in uniformization. Namely, the developing map

$$\tilde{M} \xrightarrow{\mathsf{dev}} \mathbb{C}\mathsf{P}^1$$

determines a holomorphic line bundle $L \subset E_{\tilde{\rho}}$. Since $\mathsf{deg}(E_{\tilde{\rho}}) = 0$, and dev is nonsingular, the well-known isomorphism

$$T(\mathbb{C}\mathsf{P}^1) \cong \mathsf{Hom}(\gamma, \gamma^{-1})$$

where $\gamma \to \mathbb{C}\mathsf{P}^1$ is the tautological line bundle implies that

$$L^2 \cong K_X$$

and $\mathsf{deg}(L) = g - 1 > 0$. Therefore $E_{\tilde{\rho}}$ is unstable. In fact, $E_{\tilde{\rho}}$ is a non-trivial extension

$$L \longrightarrow E_{\tilde{\rho}} \longrightarrow E_{\tilde{\rho}}/L \cong L^{-1}$$

determined by the fundamental cohomology class ε in

$$H^1(X, \mathsf{Hom}(L^{-1}, L) \cong H^1(X, K) \cong \mathbb{C}$$

defining Serre duality (compare Gunning (1967)).

One resolves this difficulty by changing the question. Replace the extension class ε by an auxiliary holomorphic object, a Higgs field

$$\Phi \in H^0(X; K_X \otimes \mathsf{End}(E)),$$

for the vector bundle $E := L \oplus L^{-1}$ so that the *Higgs pair* (E, Φ) is *stable* in the appropriate sense. In our setting the Higgs field corresponds to the everywhere nonzero holomorphic section of the trivial holomorphic line bundle

$$\mathbb{C} \cong K_X \otimes \mathsf{Hom}(L, L^{-1}) \subset K_X \otimes \mathsf{End}(E).$$

Now the only Φ-invariant holomorphic subbundle of E is L^{-1} which is negative, and the pair (E, Φ) is stable.

8.7 Rank 2 Higgs bundles

Now we impose a conformal structure on the surface to obtain extra structure on the deformation space $\mathsf{Hom}(\pi, G)/G$. As before Σ denotes a fixed oriented smooth surface, and X a Riemann surface with a fixed marking $\Sigma \to X$.

8.7.1 *Harmonic metrics*

Going from ρ to (V, Φ) involves finding a *harmonic metric*, which may be regarded as a ρ-equivariant harmonic map

$$\widetilde{M} \xrightarrow{\tilde{h}} \mathsf{GL}(n, \mathbb{C})/\mathsf{U}(n)$$

into the symmetric space $\mathsf{GL}(n, \mathbb{C})/\mathsf{U}(n)$. The metric h determines a reduction of structure group of $E_{\tilde{\rho}}$ from $\mathsf{GL}(n, \mathbb{C})$ to $\mathsf{U}(n)$, giving $E_{\tilde{\rho}}$ a Hermitian structure. Let A denote the unique connection on $E_{\tilde{\rho}}$ which is unitary with respect to h. The harmonic metric determines the Higgs pair $(V, \bar{\partial}_V, \Phi)$ as follows:

- The Higgs field Φ is the holomorphic $(1,0)$-form $\partial h \in \Omega^1\big(\mathsf{End}(V)\big)$, where the tangent space to $\mathsf{GL}(n, \mathbb{C})/\mathsf{U}(n)$ is identified with a subspace of $h^*\mathsf{End}(V)$.
- The holomorphic structure d''_A on V arises from conformal structure Σ and the Hermitian connection A.

The Higgs pair satisfies the *self-duality equations* with respect to the Hermitian metric h:

$$\begin{aligned} d''_A(\Phi) &= 0 \\ F(A) + [\Phi, \Phi^*] &= 0 \end{aligned} \qquad (8.4)$$

Here $F(A)$ denotes the curvature of A, and Φ^* denotes the adjoint of Φ with respect to h. Conversely, Hitchin and Simpson show that every stable Higgs pair determines a Hermitian metric satisfying (8.4).

8.7.2 *Higgs pairs and branched hyperbolic structures*

Choose an integer d satisfying

$$0 \leqslant d < 2g - 2.$$

Hitchin identifies the component $\mathsf{Euler}^{-1}(2 - 2g + d)$ with Higgs pairs (V, Φ) where

$$V = L_1 \oplus L_2$$

is a direct sum of line bundles L_1 and L_2 defined as follows. Choose a square root $K_X^{1/2}$ of the canonical bundle K_X and let $K_X^{-1/2}$ be its inverse. Let $D \geqslant 0$ be an effective divisor of degree d. Define line bundles

$$L_1 := K_X^{-1/2} \otimes D$$
$$L_2 := K_X^{1/2}.$$

Define a Higgs field

$$\Phi = \begin{bmatrix} 0 & s_D \\ Q & 0 \end{bmatrix}$$

where

- s_D is a holomorphic section of the line bundle corresponding to D, which determines the component of Φ in
$$K_X \otimes \mathsf{Hom}(L_2, L_1) \cong D \subset \Omega^1\big(\Sigma, \mathsf{End}(V)\big).$$
- $Q \in H^0(\Sigma, K_X^2)$ is a holomorphic quadratic differential with $\mathsf{div}(Q) \geqslant D$, which determines the component of Φ in
$$K_X \otimes \mathsf{Hom}(L_1, L_2) \cong K_X^2 \subset \Omega^1\big(\Sigma, \mathsf{End}(V)\big).$$

Then (V, Φ) is a stable Higgs pair.

When $Q = 0$, this Higgs bundle corresponds to the uniformization representation. In general, when $d = 0$, the harmonic metric is a diffeomorphism (Schoen and Yau 1978) Q is its *Hopf differential.*

The Euler class of the corresponding representation equals

$$\mathsf{deg}(L_2) - \mathsf{deg}(L_1) = 2 - 2g + d.$$

Theorem 8.8 (Hitchin 1987) *The component* $\mathsf{Euler}^{-1}(2 - 2g + d)$ *identifies with a holomorphic vector bundle over the symmetric power* $\mathsf{Sym}^d(X)$. *The fiber* *over* $D \in \mathsf{Sym}^d(X)$ *is the vector space*

$$\{Q \in H^0(X, K_X^2) \mid \mathsf{div}(Q) \geqslant D\} \cong \mathbb{C}^{3(g-1)-d}.$$

The quadratic differential Q corresponds to the *Hopf differential* of the harmonic metric h. When $Q = 0$, the harmonic metric is *holomorphic*, and defines a developing map for a branched conformal structure, with branching defined by D.

When $e = 2 - 2g$, then $d = 0$ and the space $\mathfrak{F}(X)$ of *Fuchsian representations* identifies with the vector space $H^0\left(X, K_X^2\right) \cong \mathbb{C}^{3(g-1)}$.

8.7.3 *Uniformization with singularities*

McOwen 1993 and Troyanov 1991 proved a general uniformization theorem for hyperbolic structures with conical singularities. Specifically, let $D = (p_1) + \cdots + (p_k)$ be an effective divisor, with $p_i \in X$. Choose real numbers $\theta_i > 0$ and introduce singularities in the conformal structure on X by replacing a coordinate chart at p_i with a chart mapping to a cone with cone angle θ_i. The following uniformization theorem describes when there is a singular hyperbolic metric in this singular conformal structure.

Theorem 8.9 (McOwen 1993, Troyanov 1991) *If*

$$2 - 2g + \sum_{i=1}^{k} (\theta_i - 2\pi) > 0,$$

there exists a unique singular hyperbolic surface conformal to X *with cone angle* θ_i *at* p_i.

When the θ_i are multiples of 2π, then this structure is a branched structure (and the above theorem follows from Hitchin (1987)). The moduli space of such branched conformal structures forms a bundle $\mathfrak{S}^d$ over $\mathfrak{T}(\Sigma)$ where the fiber over a marked Riemann surface $\Sigma \to X$ is the symmetric power $\mathsf{Sym}^d(X)$ where

$$d = \frac{1}{2\pi} \sum_{i=1}^{k} (\theta_i - 2\pi).$$

The resulting *uniformization map*

$$\mathfrak{S}^d \xrightarrow{\mathfrak{U}} \mathsf{Euler}^{-1}(2 - 2g + d) \subset \mathsf{Hom}(\pi, G)/G$$

is *homotopy equivalence*, which is not surjective, by the example in Section 8.5.3.3.

Conjecture 8.1 *Every representation with non-discrete image lines in the image of* $\mathfrak{U}$

8.8 Split $\mathbb{R}$-forms and Hitchin's Teichmüller component

When G is a split real form of a semisimple Lie group, Hitchin (1992) used Higgs bundle techniques to determine an interesting connected component of $\mathsf{Hom}(\pi, G)/G$, which is *not detected* by characteristic classes. A *Hitchin component* of $\mathsf{Hom}(\pi, G)$ is the connected component containing a composition

$$\pi \overset{\rho_0}{\hookrightarrow} \mathsf{SL}(2,\mathbb{R}) \overset{\mathsf{K}}{\longrightarrow} G$$

where ρ_0 is Fuchsian and K is the representation corresponding to the *three-dimensional principal subgroup* discovered by Kostant (1959). When $G = \mathsf{SL}(n,\mathbb{R})$, then Kostant's representation K is the irreducible n-dimensional representation corresponding to the symmetric power $\mathsf{Sym}^{n-1}(\mathbb{R}^2)$.

The compositions $\mathsf{K} \circ \rho_0$ above determine a subset of $\mathsf{Hom}(\pi, G)/G$ which identifies with the Fricke–Teichmüller space, and Hitchin's main result is that each Hitchin component is a cell of (the expected) dimension $(2g-2)\dim G$.

For example, if $G = \mathsf{SL}(n,\mathbb{R})$, then Hitchin identifies this component with the $2(g-1)(n^2-1)$-cell:

$$\begin{aligned} H^0\left(X; K_X^2\right) \oplus & H^0\left(X; K_X^3\right) \oplus \cdots \oplus H^0\left(X; K_X^n\right) \\ & \cong \mathbb{C}^{3(g-1)} \oplus \mathbb{C}^{5(g-1)} \oplus \cdots \oplus \mathbb{C}^{(2n-1)(g-1)}. \end{aligned}$$

When n is odd, Hitchin proves there are exactly three components. The second Stiefel–Whitney characteristic class is nonzero on exactly one component; it is zero on two components, one of which is the Hitchin–Teichmüller component.

8.8.1 *Convex $\mathbb{R}\mathsf{P}^2$-structures: $G = \mathsf{SL}(3,\mathbb{R})$*

When $G \cong \mathsf{PGL}(3,\mathbb{R}) \cong \mathsf{SL}(3,\mathbb{R})$, Hitchin (1992) conjectured that his component corresponded to the deformation space $\mathfrak{C}(\Sigma)$ of *marked convex $\mathbb{R}\mathsf{P}^2$-structures,* proved in Goldman (1990) to be a cell of dimension $16(g-1)$. In Choi and Goldman (1993) proves this conjecture. A *convex $\mathbb{R}\mathsf{P}^2$-manifold* is a quotient Ω/Γ where $\Omega \subset \mathbb{R}\mathsf{P}^2$ is a convex domain and Γ a discrete group of collineations acting properly and freely on Ω (Figures 8.7 and 8.8). If $\chi(M) < 0$, then necessarily Ω is *properly convex* (contains no complete affine line), and its boundary $\partial\Omega$ is a $C^{1+\alpha}$ strictly convex curve, for some $0 < \alpha \leqslant 1$. Furthermore $\alpha = 1$ if and only if $\partial\Omega$ is a conic and the $\mathbb{R}\mathsf{P}^2$-structure arises from a hyperbolic structure. These facts are due to Kuiper (1954) and Benzécri (1960) and have recently been extended and amplified to compact quotients of convex domains in $\mathbb{R}\mathsf{P}^{n-1}$ by Benoist (2000, 2004).

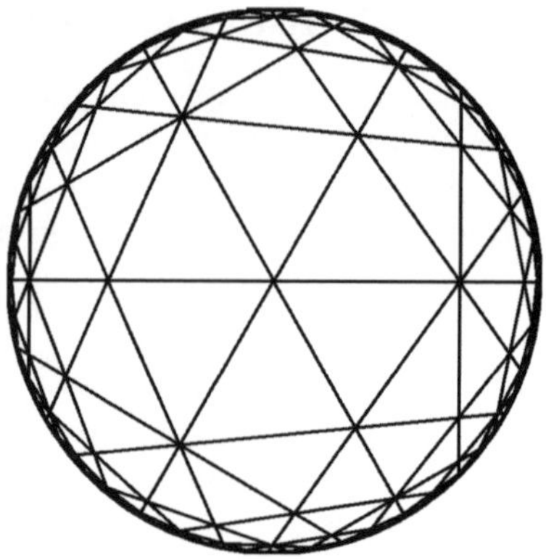

Figure 8.7 A triangle tesselation in the hyperbolic plane, drawn in the Beltrami–Klein projective model. Its holonomy representation is obtained by composing a Fuchsian representation in $\mathsf{SL}(2,\mathbb{R})$ with the irreducible representation $\mathsf{SL}(2,\mathbb{R}) \longrightarrow \mathsf{SL}(3,\mathbb{R})$.

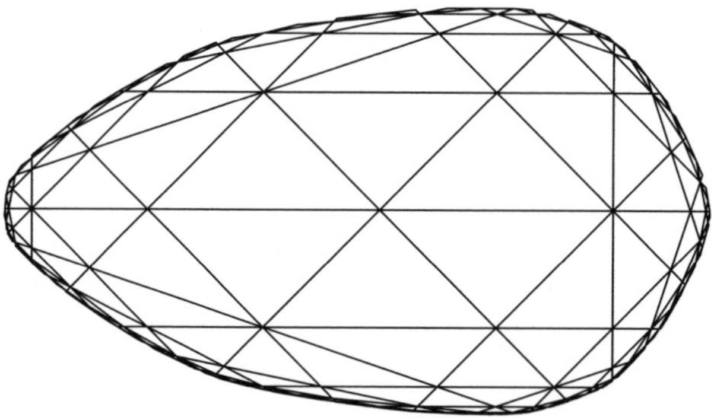

Figure 8.8 A deformation of a Fuchsian representation preserving an exotic convex domain. The boundary is a strictly convex C^1 curve which is not C^2.

8.8.2 *Higgs bundles and affine spheres*

The Higgs bundle theory of Hitchin (1992) identifies, for an arbitrary Riemann surface X, the Hitchin component $\mathfrak{C}(\Sigma)$ with the complex vector space

$$H^0\left(X, K_X^2\right) \oplus H^0\left(X, K_X^3\right) \cong \mathbb{C}^{8g-8}$$

and the component in $H^0\left(X, K_X^2\right)$ of the Higgs field corresponds to the Hopf differential of the harmonic metric. Using the theory of *hyperbolic affine spheres* developed by Calabi, Loewner–Nirenberg, Cheng–Yau, Gigena, Sasaki, Li, Wang, and Labourie (1997, to appear) and Loftin (2001) proved

Theorem 8.10 *The deformation space $\mathfrak{C}(\Sigma)$ naturally identifies with the holomorphic vector bundle over $\mathfrak{T}(\Sigma)$ whose fiber over a marked Riemann surface $\Sigma \to X$ is $H^0\left(X, K_X^3\right)$.*

For every such representation, there exists a unique conformal structure so that

$$\tilde{\Sigma} \xrightarrow{\tilde{h}} \mathsf{SL}(3,\mathbb{R})/\mathsf{SO}(3)$$

is a *conformal map*, that is, the component of the Higgs field in $H^0\left(\Sigma, K_X^2\right)$—the Hopf differential $\mathsf{Hopf}(h)$—vanishes. This defines the projection $\mathfrak{C}(\Sigma) \to \mathfrak{T}(\Sigma)$. The zero section corresponds to the *Fuchsian* $\mathbb{R}\mathsf{P}^2$ structures, that is, the $\mathbb{R}\mathsf{P}^2$ structures arising from hyperbolic structures on Σ.

It is natural to attempt to generalize this as follows. For any split real form G, and Riemann surface X with $\pi_1(X) \cong \pi$, Hitchin (1992) identifies a certain direct sum of holomorphic line bundles $\mathfrak{V}_X$ naturally associated to X so that a Hitchin component of $\mathsf{Hom}(\pi, G)/G$ identifies with the complex vector space

$$H^0\left(X, K_X^2\right) \oplus H^0(X, \mathfrak{V}_X).$$

However, this identification depends crucially on the Riemann surface X and fails to be $\mathsf{Mod}(\Sigma)$-invariant. Generalizing the Labourie–Loftin Theorem 8.10, we conjecture that each Hitchin component of $\mathsf{Hom}(\pi, G)/G$ identifies naturally with the total space of a holomorphic vector bundle $\mathfrak{E}(\Sigma)$ over $\mathfrak{T}(\Sigma)$, whose fiber over a marked Riemann surface X equals $H^0(X, \mathfrak{V}_X)$.

8.8.3 *Hyperconvex curves*

Labourie (2006) discovered an important property of the Hitchin component:

Theorem 8.11 (Labourie) *A representation in the Hitchin component for $G = \mathsf{SL}(n,\mathbb{R})$ is a discrete quasi-isometric embedding*

$$\pi \overset{\rho}{\hookrightarrow} \mathsf{SL}(n,\mathbb{R})$$

with reductive image.

A crucial ingredient in his proof is the following notion. A curve $S^1 \xrightarrow{f} \mathbb{R}\mathsf{P}^{n-1}$ is *hyperconvex* if and only if for all $x_1, \ldots, x_n \in S^1$ distinct,

$$f(x_1) + \cdots + f(x_n) = \mathbb{R}^n.$$

Theorem 8.12 (Guichard 2005, Guichard 2008, Labourie 2006) *ρ is Hitchin if and only if ρ preserves hyperconvex curve.*

Recently Fock and Goncharov (2006, 2007) have studied this component of representations, using global coordinates generalizing Thurston and Penner's *shearing coordinates.* In these coordinates the Poisson structure admits a particularly simple expression, leading to a quantization. Furthermore they find a *positive structure* which leads to an intrinsic characterization of these semi-algebraic subsets of $\mathsf{Hom}(\pi, G)/G$. Their work has close and suggestive connections with cluster algebras and K-theory.

8.9 Hermitian symmetric spaces: maximal representations

We return now to the maximal representations into groups of Hermitian type, concentrating on the unitary groups $\mathsf{U}(p,q)$ and the symplectic groups $\mathsf{Sp}(n,\mathbb{R})$.

8.9.1 *Unitary groups* $\mathsf{U}(p,q)$

The Milnor–Wood inequality (8.3) may be the first example of the *boundedness* of a cohomology class. In a series of papers (Burger and Monod 2002; Burger and Iozzi 2002, 2007; Iozzi 2002; Burger *et al.* 2003, preprint), Burger, Monod, Iozzi, and Wienhard place the local and global rigidities in the context of the Turaev-Toledo invariant being a *bounded cohomology class.* A consequence of these powerful methods for surface groups is the following, announced in Burger *et al.* (2003):

Theorem 8.13 (Burger *et al.* 2003) *Let X be a Hermitian symmetric space, and maximal representation*

$$\pi \xrightarrow{\rho} G.$$

- *The Zariski closure L of $\rho(\pi)$ is reductive.*
- *The symmetric space associatied to L is a Hermitian symmetric tube domain, totally geodesicly embedded in the symmetric space of G.*
- *ρ is a discrete embedding.*

Conversely, if X is a tube domain, then there exists a maximal ρ with $\rho(\pi)$ Zariski-dense.

For example, if $G = \mathsf{U}(p,q)$, where $p \leqslant q$, then ρ is conjugate to the normalizer $\mathsf{U}(p,p) \times \mathsf{U}(q-p)$ of $\mathsf{U}(p,p)$ in $\mathsf{U}(p,q)$. As in the rank 1 case (compare Section 8.6.4.1), the components of maximal representations have strictly smaller dimension. (In earlier work Hernández Lamoneda 1991 considered the case of $\mathsf{U}(2,q)$.)

Furthermore every maximal representation deforms into the composition of a Fuchsian representation $\pi \xrightarrow{\rho} \mathsf{SU}(1,1)$ with the diagonal embedding

$$\mathsf{SU}(1,1) \subset \mathsf{U}(1,1) \xrightarrow{\Delta} \overbrace{\mathsf{U}(1,1) \times \cdots \times \mathsf{U}(1,1)}^{p} \subset \mathsf{U}(p,p) \subset \mathsf{U}(p,q).$$

At roughly the same time, Bradlow, García-Prada, and Gothen (2003) investigated the space of Higgs bundles using infinite-dimensional Morse theory, in a similar way to Hitchin (1987). Their critical point analysis also showed that maximal representations formed components of strictly smaller dimension. They found that the number of connected components of $\mathsf{Hom}(\pi, \mathsf{U}(p,q))$ equals

$$2(p+q)\,\mathsf{min}(p,q)\,(g-1) \;+\; \mathsf{gcd}(p,q).$$

(For a survey of these techniques and other results, compare Bradlow *et al.* 2006 as well as their recent column Bradlow *et al.* 2007.)

8.9.2 *Symplectic groups* $\mathsf{Sp}(2n,\mathbb{R})$

The case $G = \mathsf{Sp}(2n,\mathbb{R})$ is particularly interesting, since G is both $\mathbb{R}$-split and of Hermitian type. Gothen (2001) showed there are $3 \cdot 2^{2g} + 2g - 4$ components of maximal representations when $n = 2$. For $n > 2$, there are $3 \cdot 2^{2g}$ components of maximal representations (García-Prada *et al.*, 2008). For $n = 2$, the components of the nonmaximal representations are just the preimages of the Turaev–Toledo invariant, comprising $1 + 2(2g - 3) = 4g - 5$ components. Thus the total number of connected components of $\mathsf{Hom}\big(\pi, \mathsf{Sp}(4,\mathbb{R})\big)$ equals

$$2\big(3 \cdot 2^{2g} + 2g - 4\big) + 4g - 5 \;=\; 6 \cdot 4^{g} + 10g - 13.$$

The Hitchin representations are maximal and comprise 2^{2g+1} of these components. They correspond to deformations of compositions of Fuchsian representations $\pi \xrightarrow{\rho_0} \mathsf{SL}(2,\mathbb{R})$ with the *irreducible* representation

$$\mathsf{SL}(2,\mathbb{R}) \longrightarrow \mathsf{Aut}\big(\mathsf{Sym}^{2n-1}(\mathbb{R}^2)\big) \hookrightarrow \mathsf{Sp}(2n,\mathbb{R})$$

where $\mathbb{R}^{2n} \cong \mathsf{Sym}^{2n-1}(\mathbb{R}^2)$ with the symplectic structure induced from $\mathbb{R}^2$.

Another class of maximal representations arises from deformations of compositions of a Fuchsian representation $\pi \xrightarrow{\rho_0} \mathsf{SL}(2,\mathbb{R})$ with the *diagonal embedding*

$$\mathsf{SL}(2,\mathbb{R}) \overset{\Delta}{\hookrightarrow} \overbrace{\mathsf{SL}(2,\mathbb{R}) \times \cdots \times \mathsf{SL}(2,\mathbb{R})}^{n} \hookrightarrow \mathsf{Sp}(2n,\mathbb{R}).$$

More generally, the diagonal embedding extends to a representation

$$\mathsf{SL}(2,\mathbb{R}) \times \mathsf{O}(n) \overset{\tilde{\Delta}}{\hookrightarrow} \mathsf{Sp}(2n,\mathbb{R})$$

corresponding to the $\mathsf{SL}(2,\mathbb{R}) \times \mathsf{O}(n)$-equivariant decomposition of the symplectic vector space

$$\mathbb{R}^{2n} = \mathbb{R}^2 \otimes \mathbb{R}^n$$

as a tensor product of the symplectic vector space $\mathbb{R}^2$ and the Euclidean inner product space $\mathbb{R}^n$. Deformations of compositions of Fuchsian representations into $\mathsf{SL}(2,\mathbb{R}) \times \mathsf{O}(2)$ with $\tilde{\Delta}$ provide 2^{2g} more components of maximal representations.

For $n > 2$, these account for all the maximal components. This situation is more complicated when $n = 2$. In that case, $4g - 5$ components of maximal representations into $\mathsf{Sp}(4,\mathbb{R})$ do not contain representations into smaller compact extensions of embedded subgroups isomorphic to $\mathsf{SL}(2,\mathbb{R})$. In particular the image of every representation in such a maximal component is *Zariski dense* in $\mathsf{Sp}(4,\mathbb{R})$, in contrast to the situation for $\mathsf{U}(p,q)$ and $\mathsf{Sp}(2n,\mathbb{R})$ for $n > 2$ (see Guichard and Wienhard in preparation, and also Bradlow, García-Prada, and Gothen 2009, for more details).

8.9.3 *Geometric structures associated to Hitchin representations*

Fuchsian representations into $\mathsf{SL}(2,\mathbb{R})$ correspond to hyperbolic structures on Σ, and Hitchin representations into $\mathsf{SL}(3,\mathbb{R})$ correspond to convex $\mathbb{R}\mathsf{P}^2$ structures on Σ. What geometric structures correspond to other classes of surface group representations?

Guichard and Wienhard (to appear) associate to a Hitchin representation in $\mathsf{SL}(4,\mathbb{R})$, an $\mathbb{R}\mathsf{P}^3$ structure on the *unit tangent bundle* $T_1(\Sigma)$ of a rather special type. The trajectories of the geodesic flow on $T_1(\Sigma)$ (for any hyperbolic metric on Σ) develop into projective lines. The leaves of the weak-stable foliations of this structure develop into convex subdomains of projective planes in $\mathbb{R}\mathsf{P}^3$. The construction of this structure uses the hyperconvex curve in $\mathbb{R}\mathsf{P}^3$. This *convex-foliated structure* is a geometric structure corresponding to Hitchin representations in $\mathsf{SL}(4,\mathbb{R})$.

For the special case of Hitchin representations into $\mathsf{Sp}(4,\mathbb{R})$ (which are readily Hitchin representations into $\mathsf{SL}(4,\mathbb{R})$), the convex-foliated structures are characterized by a duality. Furthermore the symplectic structure on $\mathbb{R}^4$ induces a contact structure on $T^1(\Sigma)$ which is compatible with the convex-foliated $\mathbb{R}\mathsf{P}^3$ structure. In addition, another geometric structure on another circle bundle over Σ arises naturally, related to the local isomorphism $\mathsf{Sp}(4,\mathbb{R}) \longrightarrow \mathsf{O}(3,2)$ and the identfication of the Grassmannian of Lagrangian subspaces of the symplectic vector space$\mathbb{R}^4$ with the conformal compactification of Minkowski $(2+1)$-space, the $(2+1)$-*Einstein universe* (compare Borbot *et al.*, to appear for an exposition of this geometry). The interplay between the contact $\mathbb{R}\mathsf{P}^3$ geometry, flat conformal Lorentzian structures, the dynamics of geodesics on hyperbolic surfaces, and the resuting deformation theory promises to be a fascinating extension of ideas rooted in the work of Nigel Hitchin.

Acknowledgments

Goldman gratefully acknowledges partial support from National Science Foundation grants DMS-070781, DMS-0103889, and DMS-0405605. This chapter was presented at the Geometry Conference honoring Nigel Hitchin, *Consejo Superior de Investigaciones Científicas*, Madrid, Spain, 5 September 2006. We gratefully acknowledge support from the Oswald Veblen Fund at the Institute for Advanced Study where this work was completed.

I am grateful to Nigel Hitchin for the inspiration of these ideas, and to him and Simon Donaldson for the opportunity to study with them at the Maths Institute in Oxford in 1989. Over the years my knowledge of the subject has benefitted enormously from conversations with Yves Benoist, Steve Bradlow, Marc Burger,

Suhyoung Choi, Kevin Corlette, Oscar García-Prada, Peter Gothen, Olivier Guichard, Alessandra Iozzi, Misha Kapovich, François Labourie, John Loftin, John Millson, Ignasi Mundet i Riera, Walter Neumann, Carlos Simpson, Ser-Peow Tan, Richard Wentworth, Anna Wienhard, Graeme Wilkin, Mike Wolf, Scott Wolpert, and Eugene Xia. I also thank the students in the Experimental Geometry Lab at the University of Maryland, in particular Ryan Hoban, Rachel Kirsch, and Anton Lukyanenko for help producing the illustrations for this chapter. I am grateful to the anonymous referee for several helpful suggestions. Finally I wish to thank the Mathematical Sciences Research Institute, the Institute for Advanced Study, and the Mathematics Department at Princeton University for their hospitality in 2007–2008.

References

Abikoff, W. (1980). The real-analytic theory of Teichmüller spaces. *Lecture Notes in Mathematics* Vol. **820**, Springer-Verlag, Berlin, Germany.

Ahlfors, L. (1960). The complex analytic struture of the space of closed Riemann surfaces. in *Analytic Functions.* Princeton University Press Princeton, NJ, pp. 45–66.

Barbot, T., Charette, V., Drumm, T., Goldman, W. and Melnick, K. (2008). A primer on the $(2+1)$ Einstein universe in *Recent developments in pseudo-Riemannian geometry.* ESI Lect. Math. Phys., Math. Soc., Zurich, 179–229.

Benoist, Y. (2000). Automorphismes des cônes convexes. *Inv. Math.* **141**(1), 149–93.

Benoist, Y. Convexes divisibles. I, Algebraic groups and arithmetic, *Tata Inst. Fund. Res.*, Mumbai (2004) 339–74. II, *Duke Math. J.* **120**(1) (2003), 97–120. III, *Ann. Sci. École Norm. Sup. (4)* **38**(5) (2005), 793–832. IV, Structure du bord en dimension 3, *Inv. Math.* **164**(2) (2006), 249–78.

Benzécri, J. P. (1960). Sur les variétés localement affines et projectives. *Bull. Soc. Math. France* **88**, 229–332.

Bers, L. (1960). Simultaneous uniformization. *Bull. Am. Math. Soc.* **66**, 94–7.

Bers, L. (1972). Uniformization, moduli, and Kleinian groups. *Bull. Lond. Math. Soc.* **4**, 257–300.

Bers, L. and Gardiner, F. (1986). Fricke spaces. *Adv. Math.* **62**, 249–84.

Bonahon, F. (1986). Bouts des varités hyperboliques de dimension 3. *Ann. Math. (2)* **124**(1), 71–158.

Bradlow, S., García-Prada, O. and Gothen, P. (2003). Surface group representations and $U(p,q)$-Higgs bundles. *J. Diff. Geom.* **64**(1), 111–70.

Bradlow, S., García-Prada, O. and Gothen, P. (2006). Maximal surface group representations in isometry groups of classical Hermitian symmetric spaces. *Geom. Ded.* **122**, 185–213.

Bradlow, S., García-Prada, O. and Gothen, P. (2007). What is a Higgs bundle?. *Notices Am. Math. Soc.* **54**(8), 980–1.

Bradlow, S., García-Prada, O. and Gothen, P. (2009). Deformations of maximal representations in $\mathbf{Sp}(4, \mathbb{R})$. arXiv:0903.5496.

Burger, M. and Iozzi, A. (2002). Boundary maps in bounded cohomology. *Geom. Funct. Anal.* **12**, 281–92.

Burger, M., Iozzi, A., Labourie, F. and Wienhard, A. (2005). Maximal representations of surface groups: symplectic Anosov structures, Pure Appl. Math. Q., Special Issue: In Memory of Armand Borel Part 2 of 3, **1**(3), 555–601.

Burger, M. and Iozzi, A. (2007). Bounded differential forms, generalized Milnor–Wood inequality and applications to rigidity theory. *Geom. Ded.* **125**, 1–23.

Burger, M., Iozzi, A. and Wienhard, A. (2003). Surface group representations with maximal Toledo invariant. *C. R. Acad. Sci. Paris, Sér. I.* **336**, 387–90.

Burger, M., Iozzi, A. and Wienhard, A. Representations of surface groups with maximal Toledo invariant. Preprint. math.DG/0605656.

Burger, M. and Monod, N. (2002). Continuous bounded cohomology and applications to rigidity theory. *Geom. Funct. Anal.* **12**, 219–80.

Buser, P. (1992). Geometry and spectra of compact Riemann surfaces. *Progress in Mathematics.* Vol. **106**, Birkhuser Boston, Inc., Boston, MA.

Choi, S. and Goldman, W. (1993). Convex real projective structures on closed surfaces are closed. *Proc. Am. Math. Soc.* **118**(2), 657–61.

Chuckrow, V. (1968). On Schottky groups with applications to Kleinian groups. *Ann. Math. (2)* **88**, 47–61.

Clerc, J. L. and Ørsted, B. (2003). The Gromov norm of the Kähler class and the Maslov index. *Asian J. Math.* **7**, 269–96.

Corlette, K. (1988). Flat G-bundles with canonical metrics. *J. Diff. Geom.* **28**, 361–82.

Domic, A. and Toledo, D. (1987). The Gromov norm of the Kaehler class of symmetric domains. *Math. Ann.* **276**(3), 425–32.

Donaldson, S. (1987). Twisted harmonic maps and the self-duality equations. *Proc. Lond. Math. Soc. (3)* **55**(1), 127–31. MR MR 887285 (88g:58040).

Eells, J. and Sampson, J. (1964). Harmonic mappings of Riemannian manifolds. *Am. J. Math.* **85**, 109–60.

Farkas, H. and Kra, I. (1980). Riemann surfaces. *Graduate Texts in Mathematics.* Vol. **71**, Springer-Verlag, Berlin/Heidelberg/New York.

Fock, V. and Goncharov, A. (2006). Moduli spaces of local systems and higher Teichmüller theory. *Publ. Math. Inst. Hautes Études Sci.* **103**, 1–211. math.AG/0311149, 2003.

Fock, V. and Goncharov, A. (2007). Moduli spaces of convex projective structures on surfaces. *Adv. Math.* **208**(1), 249–273. math.DG/0405348, 2004.

Fricke, R. and Klein, F. *Vorlesungen der Automorphen Funktionen*, Teubner, Leipzig, Germany, Vol. I (1897), Vol. II (1912).

Gardiner, F. P. and Lakic, N. (2000). Quasiconformal Teichmüller theory. *Mathematical Surveys and Monographs.* Vol. **76**, American Mathematical Society Providence, RI.

García-Prada, O. and Mundet i Riera, I. (2003). Representations of the fundamental group of a closed orientable surface in $\mathsf{Sp}(4, \mathbb{R})$. *Topology* **43**, 831–55.

García-Prada, O., Gothen, P. and Mundet i Riera, I. (2008). Higgs bundles and surface group representations in the real symplectic group. Preprint arXiv:0809.0576v3.

Goldman, W. (1980). Discontinuous groups and the Euler class, Doctoral thesis. University of California, Berkeley, CA.

Goldman, W. (1984). The symplectic nature of fundamental groups of surfaces. *Adv. Math.* **54**, 200–25.

Goldman, W. (1988). Topological components of spaces of representations. *Inv. Math.* **93**(3), 557–607.

Goldman, W. (1990). Convex real projective structures on compact surfaces. *J. Diff. Geom.* **31**(3), 791–845.

Goldman, W. (2006). Mapping class group dynamics on surface group representations, in *Problems on Mapping Class Groups and Related Topics*. B. Farb (ed.), *Proceedings of the Symposium on Pure Mathematics*, Vol. **74**, American Mathematical Society, Providence, RI, pp. 189–214. math.GT/0509114.

Goldman, W. and Millson, J. J. (1987). Local rigidity of discrete groups acting on complex hyperbolic space. *Inv. Math.* **88**, 495–520.

Goldman, W. and Millson, J. J. (1988). The deformation theory of representations of fundamental groups of Kähler manifolds. *Publ. Math. IHES.* **67**, 43–96.

Goldman, W. and Wentworth, R. (2007). Energy of twisted harmonic maps of Riemann surfaces. *In the tradition of Ahlfors-Bers, IV* H. Masur and R. Canary (eds.), Contemp. Math., 432, Amer. Math. Soc., Providence, RI, 45–61. math.DG/0506212.

Goldman, W. and Xia, E. (2008). Rank one Higgs bundles and fundamental groups of Riemann surfaces. *Memoirs of the Am. Math. Soc.* **904**, 1–69. math.DG/0402429.

Gothen, P. (2001). Components of representations and stable triples. *Topology* **40**, 823–50.

Guichard, O. (2005). Une dualité pour les courbes hyperconvexes. *Geom. Ded.* **112**, 141–64. MR MR2163895.

Guichard, O. (2008). Composantes de Hitchin et representations hyperconvexes de groupes de surface. *J. Diff. Geom.* **80**(3), 391–431.

Guichard, O. and Wienhard, A. (2008). Convex foliated projective structures and the Hitchin component for $\mathsf{SL}(4,\mathbb{R})$. *Duke Math. J.* **144**(3), 381–445. math/0702184.

Guichard, O. and Wienhard, A. Topological invariants for Anosov representations, (Submitted).

Gunning, R. (1967). Special coordinate coverings of Riemann surfaces. *Math. Ann.* 170, 67–86.

Hernández Lamoneda, L. (1991). Maximal representations of surface groups in bounded symmetric domains. *Trans. Am. Math. Soc.* **324**, 405–20.

Hitchin, N. (1987). The self-duality equations on Riemann surfaces. *Proc. Lond. Math. Soc.* **55**, 59–126.

Hitchin, N. (1992). Lie groups and Teichmüller space. *Topology* **31**(3), 449–73.

Hubbard, J. H. (2006). *Teichmüller Theory, and Applications to Geometry, Topology and Dynamics, Volume 1*, Matrix Editions, Ithaca, NY.

Imayoshi, Y. and Taniguchi, M. (1992). *An Introduction to Teichmüller Spaces*, Springer-Verlag, Tokyo, Japan.

Iozzi, A. (2002). Bounded cohomology, boundary maps, and representations into $\mathrm{Homeo}_+(S^1)$ and $\mathsf{SU}(1,n)$. in *Rigidity in Dynamics and Geometry, Cambridge, UK, 2000*, Springer Verlag, Berlin, Germany, pp. 237–60.

Jost, J. (2006). Compact Riemann surfaces. in *An Introduction to Contemporary Mathematics*, 3rd edition, Universitext, Springer-Verlag, Berlin, Germany.

Kostant, B. (1959). The principal three-dimensional subgroup and the Betti numbers of a complex simple Lie group. *Am. J. Math.* **81**, 973–1032.

Kuiper, N. (1954). On convex locally projective spaces. *Convegno Internazionale di Geometria Differenziale*, Italy, pp. 200–213.

Labourie, F. (1991). Existence d'applications harmoniques tordues valeurs dans les variétés à courbure négative. *Proc. Am. Math. Soc.* **111**(3), 877–82.

Labourie, F. (1997). $\mathbb{R}\mathsf{P}^2$-structures et differentielles cubiques holomorphes, in *Proceedings of the GARC Conference in Differential Geometry*, Seoul National University.

Labourie, F. (2006). Anosov flows, surface group representations and curves in projective space. *Inv. Math.* **165**(1), 51–114. math.DG/0401230.

Labourie, F. (2007). Flat projective structures on surfaces and cubic holomorphic differentials, Special Issue: Volume in honor of the 60th birthday of Gregory Margulis, Part 1 Pure Appl. Math. Q. **3**(4), 1057–1099. math.DG/0611250.

Labourie, F. (2008). Cross ratios, Anosov representations and the energy functional on Teichmüller space. *Ann. Sci. École Norm. Sup.* (4), **41**(3), 437–469. math.DG/0512070.

Loftin, J. (2001). Affine spheres and convex $\mathbb{R}\mathsf{P}^{n-1}$-manifolds. *Am. J. Math.* **123**, 255–74.

McOwen, R. (1993). Prescribed curvature and singularities of conformal metrics on Riemann surfaces. *J. Math. Anal. Appl.* **177**(1), 287–98.

Milnor, J. (1958). On the existence of a connection with curvature zero. *Commun. Math. Helv.* **32**, 215–23.

Nag, S. (1988). *The Complex Analytic Theory of Teichmüller Spaces.* Wiley-Interscience, New York.

Narasimhan, M. S. and Seshadri, C. S. (1965). Stable and unitary vector bundles on a compact Riemann surface. *Ann. Math.* **82**, 540–67.

Nielsen, J. (1927). Untersuchungen zur Topologie der geschlossenen zweiseitigen Flächen I. *Acta Math.* **50**, 189–358.

Ramanathan, A. (1975). Stable principal bundles on a compact Riemann surface. *Math. Ann.* **213**, 129–52.

Raghunathan, M. (1972). *Discrete Subgroups of Lie Groups*, Springer Verlag, New York.

Ratcliffe, J. (2006). Foundations of hyperbolic manifolds. *Graduate Texts in Mathematics*, Vol. **149**, Springer, New York.

Schoen, R. and Yau, S. T. (1978). Univalent harmonic maps between surfaces. *Invent. Math.* **44**, 265–78.

Simpson, C. (1988). Constructing variations of Hodge structure using Yang-Mills theory and applications to uniformization. *J. Am. Math. Soc.* **1**(4), 867–918.

Simpson, C. (1992). Higgs bundles and local systems. *Inst. Hautes Études Sci. Publ. Math.* **75**, 5–95.

Stillwell, J. (1987). The Dehn-Nielsen theorem, Appendix to *Papers on Group Theory and Topology, by Max Dehn*, Springer Verlag, Berlin/Heidelberg/New York.

Thurston, W. (1979). The geometry and topology of 3-manifolds, Princeton University Lecture Notes (unpublished). Available at http://www.msri.org/communications/books/gt3m.

Toledo, D. (1989). Representations of surface groups in complex hyperbolic space. *J. Diff. Geom.* **29**(1), 125–33.

Tromba, A. (1992). Teichmüller theory in Riemannian geometry. *ETHZ Lectures in Mathematics*, Birkhäuser, Basel, Switzerland.

Troyanov, M. (1991). Prescribing curvature on compact surfaces with conical singularities. *Trans. Am. Math. Soc.* **324**(2), 793–821.

Turaev, V. G. (1985). The first symplectic Chern class and Maslov indices. (Russian) Studies in topology, V. Zap. Nauchn. Sem. Leningrad. *Otdel. Mat. Inst. Steklov.* (LOMI) **143**, 110–129, 178.

Turaev, V. G. (1984). A cocycle of the symplectic first Chern class and Maslov indices. (Russian) *Funktsional. Anal. i Prilozhen.* **18** (1), 43–48.

Weil, A. (1962). On discrete subgroups of Lie groups II. *Ann. Math.* **75**, 578–602.

Wood, J. (1976). Bundles with totally disconnected structure group. *Commun. Math. Helv.* **51**, 183–99.

IX

LOCALITY OF HOLOMORPHIC BUNDLES, AND LOCALITY IN QUANTUM FIELD THEORY

Graeme Segal

Dedicated to Nigel Hitchin, on his 60th birthday

Nigel Hitchin has been a close colleague of mine for most of my mathematical life, and I have profited enormously from my contact with him. His wonderful skill in asking the right questions and in obtaining deep results without sacrificing concreteness or simplicity has both inspired me and filled me with envy. Having seen, however, how many others at this meeting are better qualified than I am to talk about Nigel's work, I decided it would be best to keep to my own terrain, and talk about locality in quantum field theory. All the same, many of the ideas involved are well exemplified in the study of bundles on Riemann surfaces which Nigel is famous for, and I shall begin there, especially as the question of locality relates to an aspect of his work that has not been talked about so far at this meeting, namely, its role in so-called 'geometric Langlands theory'.

We can approach the subject by contrasting two opposite ways of looking at holomorphic vector bundles E on a compact Riemann surface Σ. At one extreme, if we remove any finite set σ of points from Σ then E is trivial on the remaining surface $\Sigma - \sigma$, so we can think of all the 'twisting' of the bundle as being concentrated into tiny neighbourhoods of the points of σ. At the other extreme, we can try to spread the twist as evenly as possible over all of Σ.

The classical case of line bundles is very simple – perhaps misleadingly so. On the one hand, any line bundle L can be constructed from a *divisor* $\mathbf{z} = n_1 z_1 + \cdots + n_k z_k$ – an element of the free abelian group on the set of points of Σ – by attaching the trivial bundle L_0 on $\Sigma - \{z_1, \ldots, z_k\}$ to trivial bundles L_i in the neighbourhood of each z_i by means of the clutching function $\zeta_i^{n_i}$, where ζ_i is a local parameter at z_i. (The resulting bundle $L(\mathbf{z})$ depends up to *canonical* isomorphism only on the divisor $\mathbf{z}$.) On the other hand, for any choice of Riemannian metric on Σ, any line bundle can be given a unique unitary connection with constant curvature, so that it looks exactly the same in the neighbourhood of any point of Σ. The isomorphism classes of holomorphic line bundles on Σ naturally form a commutative complex Lie group $\mathrm{Pic}(\Sigma)$, and the correspondence between the two ways of looking at a bundle amounts to the classical theorem that $\mathrm{Pic}(\Sigma)$ is – in the holomorphic category – the free abelian group on Σ, traditionally called its 'Albanese variety', that is, that the map

$\Sigma \to \text{Pic}(\Sigma)$ given by $z \mapsto L(z)$ is universal[1] among holomorphic maps from Σ to a commutative group.

We encounter two main obstacles when we try to formulate an analogue of this attractive picture for higher dimensional bundles. The first is that the isomorphism classes of n-dimensional holomorphic bundles on a Riemann surface Σ do not form a nice space; and the second is that the bundles trivialized in the complement of a given point do not form an abelian group. For brevity I shall refer to these as the problems of 'noncommutative geometry' and of 'algebraic structures up to homotopy', respectively, and shall say a little about each in turn.

9.1 Noncommutative geometry

If Σ is the Riemann sphere then every n-dimensional holomorphic bundle E on Σ is a sum of line bundles, of degrees $k_1 \geq k_2 \geq \cdots \geq k_n$ say, and E is determined up to isomorphism by the n-tuple of degrees. Equivalently, E can be obtained by attaching trivial bundles on $\Sigma - \{\infty\}$ and $\Sigma - \{0\}$ by a holomorphic attaching function which is a homomorphism $\lambda : \mathbb{C}^\times \to \text{GL}_n(\mathbb{C})$, unique up to conjugation in $\text{GL}_n(\mathbb{C})$. Now the conjugacy classes of homomorphisms from $\mathbb{C}^\times$ to a complex reductive group G – and hence the isomorphism classes of G-bundles on the sphere – are in one-to-one correspondence with the finite-dimensional holomorphic irreducible representations of the Langlands dual group ${}^LG$. (If G is $\text{GL}_n(\mathbb{C})$ then ${}^LG = G$.) This is the starting point of geometric Langlands theory, but it is not my subject here. Let us notice, however, that whereas the isomorphism classes of representations form a countable discrete set the behaviour of the bundles is quite different. When a holomorphic bundle $E^{(t)}$ on the sphere depends holomorphically on a parameter $t \in T$ we find that – if T is connected – the isomorphism class of $E^{(t)}$ is constant on a dense open subset of the parameter space T, but 'jumps' when t belongs to certain submanifolds: the space T is *stratified* by the isomorphism class of $E^{(t)}$. Thus the countable set $\mathcal{B}_\Sigma$ of isomorphism classes of bundles consists of a sequence of connected components

[1] This may seem strange. The set-theoretical free abelian group F_Σ generated by the points of Σ fits into an exact sequence

$$K_\Sigma^\times \to F_\Sigma \to \text{Pic}_\Sigma \to 1,$$

where K_Σ is the field of rational functions on Σ. The group F_Σ is the disjoint union of a sequence of finite-dimensional algebraic varieties $F_\Sigma^{(n)}$, where $F_\Sigma^{(n)}$ consists of all $\Sigma n_k x_k$ such that $\Sigma|n_k| = n$. Furthermore, F_Σ has a natural topology in which the closure of $F_\Sigma^{(n)}$ is compact, and is the union of the $F_\Sigma^{(m)}$ for $m \leq n$. But we cannot say that F_Σ is any kind of algebraic variety: in fact it is easy to see that if U is a neighbourhood of the identity element of F_Σ then any continuous $f : U \to \mathbb{C}$ for which $f|U \cap F_\Sigma^{(n)}$ is holomorphic for each n has to be constant along the orbits of $K_\Sigma^\times$.

indexed by the first Chern class of E, but each connected component, though infinite, is the closure of a single point.

To make better sense of unpromising 'spaces' such as $\mathcal{B}_\Sigma$ a number of different approaches are commonly used. In algebraic geometry the main candidates are

1. To regard $\mathcal{B}_\Sigma$ as a *stack*, that is, to work with the *category* – actually a *groupoid* – of bundles and their isomorphisms rather than just with the set $\pi_0(\mathcal{B}_\Sigma)$ of isomorphism classes of objects
2. the approach of *geometric invariant theory*, which picks out a class of 'stable' bundles whose isomorphism classes do form a nice space, in fact an algebraic variety

Here I am going to talk about the first approach, which is more obviously related to quantum field theory, especially in the treatment of the Langlands theory by Kapustin and Witten (2007). Nigel Hitchin's own main tool, however, was geometric invariant theory.

Stepping back a little from algebraic geometry, one can say that the category of bundles on a space resembles the category of representations of a group. For example, the Hitchin moduli space associated to a surface Σ and a compact group G is – among other things – the space of conjugacy classes of homomorphisms from the fundamental group $\pi_1(\Sigma)$ to the complexified group $G_{\mathbb{C}}$. The 'space' of irreducible unitary representations of a group Γ is an archetypal example in Connes's theory of noncommutative geometry (Connes 1994). He observes that the set of irreducible representations is the set $\mathrm{Spec}(A_\Gamma)$ of irreducible $*$-representations of a C*-algebra A_Γ associated to Γ. If A is a commutative algebra its irreducible representations are the algebra-homomorphisms $A \to \mathbb{C}$, which form a topological space $\mathrm{Spec}(A)$ on which A is an algebra of continuous complex-valued functions. Connes's idea is to think of the geometry of the set of irreducible representations of the group Γ as *defined* by the noncommutative algebra A_Γ rather than by the – often too small – commutative algebra of continuous functions on $\mathrm{Spec}(A_\Gamma)$, which in fact is the *centre* of A_Γ.

To relate this picture to the stacks or groupoids of algebraic geometry we think of a groupoid as halfway between a group and a space. More precisely, a groupoid $\mathcal{B}$ consists of two sets together with some maps between them: a set $\mathcal{B}_0$ of objects, and a set $\mathcal{B}_1$ of morphisms which is the disjoint union of the sets $\mathcal{B}(x, y)$ of morphisms from x to y, where x and y run through all the objects in $\mathcal{B}_0$. We are concerned here, however, with *topological* groupoids, for which the sets $\mathcal{B}_0$ and $\mathcal{B}_1$ have topologies and the structural maps between them are continuous. At one extreme, if $\mathcal{B}_0$ is a point then we have a topological group; at the other, if $\mathcal{B}_0 = \mathcal{B}_1$, we have no morphisms except identities, and the groupoid is simply a space. As Connes has emphasized, a groupoid $\mathcal{B}$ has a groupoid-algebra $A_\mathcal{B}$, which interpolates between the group-algebra of a group and the commutative algebra of functions on a space. (The algebra $A_\mathcal{B}$ is devised so that an $A_\mathcal{B}$-module is the same thing as a functor from $\mathcal{B}$ to vector spaces. For a discrete groupoid, $A_\mathcal{B}$ has a vector space basis e_f indexed by the set

$\mathcal{B}_1$ of morphisms, and composition is given by $e_f e_g = e_{f \circ g}$ when f and g are composable, and $e_f e_g = 0$ otherwise. For the actual groupoids at hand, we must make a choice of what we mean by the algebra, just as for a Lie group we can define different group algebras depending on the geometric context in which we are working.)

The groupoids $\mathcal{B}$ we are concerned with all arise from the action of a group G on a space $\mathcal{B}_0$, so that $\mathcal{B}_1 = G \times \mathcal{B}_0$, and the set of morphisms from x to y is $\{g \in G : gx = y\}$. I shall denote this groupoid by $\mathcal{B}_0//G$. The algebra $A_{\mathcal{B}}$ in this case is the twisted group algebra $G \ltimes A_0$, where A_0 is the commutative algebra of functions on $\mathcal{B}_0$.

The essential features of a groupoid $\mathcal{B}$ are the (often not very nice) space $\pi_0(\mathcal{B})$ of isomorphism classes of its objects, and the groups $\mathrm{Aut}(x)$ of automorphisms of the objects, which are the obstructions to $\mathcal{B}$'s being a space. These features depend only on the equivalence class of the groupoid (in the sense of category theory), and we are interested in groupoids only up to equivalence. In fact we are content (cf. Segal 1973, 2.8) even with what I shall call 'weak equivalences': functors $T : \mathcal{B} \to \mathcal{B}'$ which induce bijections $\pi_0(\mathcal{B}) \to \pi_0(\mathcal{B}')$ and $\mathrm{Aut}(x) \to \mathrm{Aut}(T(x))$ for each object x, and which are 'covering maps' (in the sense that they have local cross-sections) on the space of objects and morphisms. Thus if G acts freely on $\mathcal{B}_0$, making it a locally trivial principal G-bundle on the space $\mathcal{B}_0/G$, then we are not interested in the difference between the groupoid $\mathcal{B}_0//G$ and the space $\mathcal{B}_0/G$. Similarly, whenever we decompose a closed surface Σ into two pieces Σ_1 and Σ_2 intersecting in a curve S we can identify the set of isomorphism classes of n-dimensional bundles on Σ with the (non-Hausdorff) double-coset space

$$G_{\Sigma_1} \backslash G_S / G_{\Sigma_2},$$

where G_{Σ_i} is the group of holomorphic maps from Σ_i to $G = \mathrm{GL}_n(\mathbb{C})$, and G_S is the group of smooth maps $S \to G$, for any bundle can be trivialized on a non-closed surface. We then have three groupoids:

1. $\mathcal{B}^{(1)}$ formed by the action of G_{Σ_1} on the homogeneous space G_S/G_{Σ_2}
2. $\mathcal{B}^{(2)}$ formed by the action of G_{Σ_2} on $G_{\Sigma_1} \backslash G_S$
3. $\mathcal{B}^{(12)}$ formed by the action of $G_{\Sigma_1} \times G_{\Sigma_2}$ on G_S

All three are weakly equivalent, and are weakly equivalent to the groupoids obtained from any other way of decomposing Σ into pieces. According to the 'stack' point of view, any of them can be taken as the 'space' of bundles on Σ.

The relation of equivalence of groupoids translates into *Morita equivalence* of algebras. Two algebras are Morita equivalent if their categories of left modules are equivalent. Recall that if A and B are algebras then an (A, B)-bimodule F defines an additive functor $F_* : \mathcal{M}_B \to \mathcal{M}_A$, where $\mathcal{M}_A$ and $\mathcal{M}_B$ are the categories of left A and B modules, by $F_*(M) = F \otimes_B M$. In particular, for the

(A, A)-bimodule A we have $A_* = \mathrm{id}_{\mathcal{M}_A}$, and if G is a (B, C)-bimodule we have

$$F_* \circ G_* = (F \otimes_B G)_* : \mathcal{M}_C \to \mathcal{M}_A.$$

On the level of discrete groupoids, at least, there is a dictionary

$$\begin{array}{rcl} \text{groupoid } \mathcal{B} & \longmapsto & \text{algebra } A_{\mathcal{B}} \\ \text{functor } T : \mathcal{B} \to \mathcal{B}' & \longmapsto & (A_{\mathcal{B}}, A_{\mathcal{B}'})\text{-bimodule } F_T \\ \text{natural transformation } \Phi : T \to T' & \longmapsto & \text{isomorphism } \Phi_* : F_T \to F_{T'}. \end{array}$$

In the light of the previous discussion we see that a 'noncommutative space' is defined not by an algebra but by a Morita equivalence class of algebras, and so it is more naturally described by a *linear category* – the category of left modules for the algebra. In holomorphic geometry, however, the algebras that arise, even when they are commutative, are not semisimple, and their module categories do not have very convenient properties. It is therefore better to go further, and replace the module categories by the categories of cochain complexes of modules. That is why the geometric Langlands correspondence is stated in terms of the linear categories of complexes of coherent sheaves[2] – or of $\mathcal{D}$-modules – on the moduli spaces of bundles.

Yet another version – slightly more general still – of the notion of noncommutative space arises from quantum field theory, and it is perhaps the most natural one in the geometric Langlands theory. I shall sketch it below. Before leaving the present discussion, however, it may be worth making another remark.

A generalized space defined by a topological groupoid has a *homotopy type*, just like an ordinary space. For – like any topological category (cf. Segal 1968) – the groupoid $\mathcal{B}_0//G$ has a 'realization' $|\mathcal{B}_0//G|$ as a space, and equivalent groupoids have homotopy-equivalent realizations. A generalized space defined by a noncommutative ring – or, better, by a linear category $\mathcal{C}$ – has no such homotopy type. The best one can do is consider the *stable* homotopy type (or *spectrum*) $\mathbb{K}_{\mathcal{C}}$ defined by applying the usual K-theory construction to $\mathcal{C}$: this is (cf. Segal 1977) the 'space' whose homotopy groups are the K-groups of the category.

In the case of the stack $\mathcal{B}_\Sigma$ of holomorphic G-bundles on a Riemann surface Σ the space $|\mathcal{B}_\Sigma|$ has the same homotopy type as the realization of the corresponding topological groupoid $\mathcal{B}_\Sigma^{sm}$ of smooth bundles, which in turn has the homotopy type of the space $\mathrm{Map}(\Sigma; BG)$ of continuous maps from Σ to the classifying space of G. (This follows at once from the fact that for a non-closed Riemann surface Σ_i the space of holomorphic maps from Σ_i to G has the homotopy type of the space of continuous maps.) On the other hand, we know from the beautiful work of Atiyah and Bott 1982 that the moduli space of holomorphic bundles in the sense of geometric invariant theory can be identified with the minimum level of the Yang–Mills functional on the space of smooth bundles.

[2] *Sheaves* rather than modules, because the algebraic varieties are not affine. But that is a standard technicality.

9.2 Algebraic structures up to homotopy

Let us consider the space $\hat{\mathcal{B}}_\Sigma$ of pairs (E, e) consisting of a holomorphic vector bundle E on Σ equipped with a meromorphic trivialization e – that is, E is trivialized, in the complement of a finite subset σ of Σ, by n holomorphic sections $e_1, \ldots, e_n$ which extend to meromorphic sections on Σ. A pair (E, e) is completely determined by giving σ and, for each $z \in \sigma$, the local information consisting of a bundle in a neighbourhood of z trivialized away from z. The items of local information can be prescribed independently, so $\hat{\mathcal{B}}_\Sigma$ can be regarded as a *labelled configuration space*. The group $\mathrm{GL}_n(K_\Sigma)$ acts on $\hat{\mathcal{B}}_\Sigma$ by changing the meromorphic trivialization, and the orbit space is the space of isomorphism classes of holomorphic bundles. (The isotropy group of any pair (E, e) is just the group of holomorphic automorphisms of E.) Let us now recall a few aspects of the theory of labelled configuation spaces.

For an d-dimensional manifold M let $\check{C}(M)$ denote the manifold of all finite unordered subsets of M, with its natural topology in which it is the disconnected union $\coprod_{k \geq 0} C_k(M)$ of the spaces $C_k(M)$ of subsets with exactly k elements. If P is an arbitrary auxiliary space, we can also form $\check{C}(M; P) = \coprod C_k(M; P)$, whose points are the finite subsets σ of M with each point $z \in \sigma$ 'labelled' by a point p_z of P. The space $C_k(M; P)$ is fibred over $C_k(M)$ with fibre P^k.

The configuration spaces we are interested in, however, have a topology which allows the points of a configuration σ to move continuously into coincidence and the labels at the same time to amalgamate – or 'add' – in some sense. For example, the free abelian group F_Σ is the configuration space of Σ labelled by the group $\mathbb{Z}$, with the usual addition when points merge. We can define such a 'configuration space with amalgamation' whenever the labelling space P has a composition-law which is sufficiently associative and commutative to make it what is called in homotopy theory an *d-fold loop space* – in fact the existence of the amalgamated configuration space $C(M; P)$ for all d-manifolds M is a good way of *defining* an d-fold loop space.[3] To be precise, I shall say, in the style of Beilinson and Drinfeld, that an *amalgamated configuration space* is any space $C(M; P)$ to each of whose points is associated a finite subset σ of M called its *support*, and which is equipped with a map $i_{\mathbf{z}} : P^k \to C(M; P)$, for each sequence $\mathbf{z} = \{z_1, \ldots, z_k\}$ of distinct points of M, with the properties:

(i) the image of $i_{\mathbf{z}}$ consists of configurations with support $\mathbf{z}$, and
(ii) if U is the disjoint union of k small open balls $U_1, \ldots, U_k$ such that U_i contains z_i, and $C_U(M; P)$ denotes the part of $C(M; P)$ with supports in U, then $i_{\mathbf{z}} : P^k \to C_U(M; P)$ is a homotopy equivalence.

[3] The conventional way to define the structure of an d-fold loop space on a space P is to give the amalgamation maps

$$C_k(\mathbb{R}^d; P) \to P$$

for each $k > 0$; but these must of course satisfy various compatibilities.

Although it is not strictly necessary, we may as well assume that $C_U(M;P)$ is an open subset of $C(M;P)$ when U is an open subset of M, and that it can be identified with $C(U;P)$. That is certainly true in our examples, where $C(M;P)$ is simply $\check{C}(M;P)$ with a coarser topology.

In fact we need a slight generalization to include *bundles* of d-fold loop spaces. We shall allow the label of a point $z \in M$ to lie not in a fixed space P but rather in the fibre at z of a bundle P on the manifold M. It is clear how the definition of $C(M;P)$ should be adapted to this case.

When we have a single d-fold loop space P we define its *d-fold classifying space* – or dth 'delooping' – as the space

$$B^d P = C(U;P)/C_V(U;P),$$

where U is an open ball – say $U = \{z \in \mathbb{R}^d : \|z\| < 1\}$ – and V is the annular region $\{z \in U : 1/2 \leq \|z\| < 1\}$, and the notation means that the subspace $C_V(U;P)$ of $C(U;P)$ is collapsed to a single point, which is a natural base point in $B^d P$.

Similarly, when P is a bundle of d-fold loop spaces on M we can define a bundle $B^d P$ whose fibre at $z \in M$ is constructed using a neighbourhood U of z in M.

The main theorem about labelled configuration spaces is the following.

Theorem 9.1 *There is a natural map*

$$C(M;P) \longrightarrow \Gamma_{\text{cpt}}(M;B^d P),$$

where Γ_{cpt} denotes the space of cross-sections which are equal to the base point outside of a compact subset of M. If the composition law of P makes the set of components $\pi_0(P)$ into a group then the map is a homotopy equivalence.

Notice that applying the theorem when M is an open ball tells us that P is homotopy-equivalent to the d-fold based loop space of $B^d P$.

The equivalences are defined by the *scanning map* (see McDuff 1975, 1977 and Segal 1979) which associates to an element $c \in C(M;P)$, the section of $B^d P$ whose value at $z \in M$ is the image of c in $C(M;P)/C_{M-W}(M;P) = C(U;P)/C_V(U;P)$, where $W \subset U$ are two concentric open balls around z, and $V = U - W$. The theorem is very easy to prove, and not at all deep, in the form I have stated it here: in applications the main difficulty may be to verify the hypotheses.

In the application to holomorphic vector bundles on a surface, at a point z on the surface, with neighbourhood U, the labelling space P is the quotient $\mathcal{G}_{\check{U}}/\mathcal{G}_U$, where $\mathcal{G}_{\check{U}}$ is the group of holomorphic maps $U - \{z\} \to \mathrm{GL}_n(\mathbb{C})$ which extend meromorphically over U, and $\mathcal{G}_U$ is the group of holomorphic maps $U \to \mathrm{GL}_n(\mathbb{C})$. This can be identified with $\mathrm{GL}_n(\mathbb{C}(t))/\mathrm{GL}_n(\mathbb{C}[t])$. It is (see Pressley and Segal 1986) the union of a sequence of compact finite-dimensional algebraic varieties, and it has the homotopy type of the based loop space of $\mathrm{GL}_n(\mathbb{C})$. To define the topology of $\hat{\mathcal{B}}_\Sigma$ we first define a topology on the part $\hat{\mathcal{B}}_{\Sigma,U}$ with support in a disjoint union U of open discs by identifying it with $\mathcal{G}_{\partial U}/\mathcal{G}_U$, where $\mathcal{G}_{\partial U}$ is the group of smooth maps $\partial U \to \mathrm{GL}_n(\mathbb{C})$ which are boundary values of meromorphic maps $U \to \mathrm{GL}_n(\mathbb{C})$, and then we give $\hat{\mathcal{B}}_\Sigma$ the finest topology compatible with these. Having checked the hypotheses of the theorem, it tells us that $\hat{\mathcal{B}}_\Sigma$ has the homotopy type of the space of continuous maps from Σ to $B^2P \simeq \mathrm{BGL}_n(\mathbb{C})$, that is, the homotopy type of the space of smooth bundles, as we expect from the stack picture.

Before leaving this topic, let us mention the converse question to the one answered by the theorem. If we are given a space Q, can we model the mapping spaces $\mathrm{Map}(M;Q)$ for varying d-manifolds M – at least up to homotopy – by labelled configuration spaces $C(M;P)$? The answer, clearly, is: if and only if the space Q is d-connected; for the delooping B^dP of any d-fold loop space P is d-connected. If we want to model spaces of maps into less highly connected spaces Q we would have to allow not just 'particles' but also configurations of higher dimensional submanifolds – presumably, up to dimension m if Q is $(d-m)$-connected. (One way to see this is to realize Q as an open manifold with a Morse function with critical points of indices $\geq d-m$, and to make maps $M \to Q$ flow downwards along the gradient flow.)

9.3 Quantum field theory

All the mathematical phenomena I have been discussing play an important role in quantum field theory. In particular, noncommutative geometry enters in two somewhat opposite ways, first because the moduli spaces of theories are noncommutative spaces, and also because field theories can be used to give a new formulation of noncommutative geometry. It is only the second aspect that I am going to talk about. But let us begin at the beginning. . . .

In quantum mechanics a system is described at any time by giving an algebra $\mathcal{A}$ of 'observables' and a linear map $\theta : \mathcal{A} \to \mathbb{C}$ called the 'state'.[4] In quantum field theory, in space-time of dimension d, this picture is enriched by supposing that the observables are spread out over a given space-time manifold M. More precisely, for each $x \in M$ there is given a sub-vector space $\mathcal{O}_x$ of $\mathcal{A}$ formed by the observables which can be measured in the neighbourhood of x. Quantum field theory assumes that – up to some global topological effects[5] to which I shall return at the end of this talk – the complete information about the system is contained in the maps

$$\Theta_k : \mathcal{O}_{x_1} \otimes \cdots \otimes \mathcal{O}_{x_k} \longrightarrow \mathbb{C},$$

[4] It is perhaps more usual to say that $\mathcal{A}$ is an algebra of operators in a Hilbert space $\mathcal{H}$, and that the state is a unit vector ψ in $\mathcal{H}$ related to θ by $\theta(a) = \langle \psi, a\psi \rangle$. But the equivalent description by $(\mathcal{A}, \theta)$ seems more satisfactory to me.

[5] The essential example is the Bohm–Aharonov phenomenon, when an electromagnetic field in a non-simply connected region is described by a flat connection which is undetectable in any simply connected subregion.

for each finite set $\{x_1, \ldots, x_k\}$ of distinct points of M, which are got by multiplying in $\mathcal{A}$ and composing with θ. The Θ_k are traditionally called *vacuum expectation values.*

It is not easy to say what properties the vector spaces $\mathcal{O}_x$ and the functions Θ_k must have for them to constitute a quantum field theory. A first attempt at an answer can be given by defining a d-dimensional theory as a rule which

1. associates a complex topological vector space $\mathcal{H}_Y$ to each compact-oriented Riemannian manifold Y of dimension $d-1$, functorially with respect to diffeomorphisms $Y \to Y'$, and
2. associates a trace-class operator $U_X : \mathcal{H}_{Y_0} \to \mathcal{H}_{Y_1}$ to each oriented Riemannian cobordism X from Y_0 to Y_1.

These data are constrained to satisfy two axioms:

(*a*) *Concatenation*:

$$U_{X' \circ X} = U_{X'} \circ U_X$$

when $X' \circ X$ is the cobordism from Y_0 to Y_2 obtained by concatenating X from Y_0 to Y_1 with X' from Y_1 to Y_2.

(*b*) *Tensoring*: We are given associative natural isomorphisms

$$\mathcal{H}_Y \otimes \mathcal{H}_{Y'} \stackrel{\cong}{\to} \mathcal{H}_{Y \sqcup Y'}$$

$$U_X \otimes U_{X'} = U_{X \sqcup X'}$$

when we have disjoint unions $Y \sqcup Y'$ or $X \sqcup X'$ of $(d-1)$-manifolds or cobordisms.

Notice that it follows from property (*a*) that $\mathcal{H}_Y = \mathbb{C}$ if Y is the empty $(d-1)$-manifold.

When we have a theory in this sense we can reconstruct the local observables and their expectation values. We define the vector space $\mathcal{O}_x$ of observables for each point x in a closed d-manifold M by

$$\mathcal{O}_x = \lim_{\leftarrow} \mathcal{H}_{\partial D},$$

where the inverse limit is over the ordered set of all closed balls D in M which are neighbourhoods of x, ordered by

$$D' > D \Longleftrightarrow D' \subset \overset{\circ}{D},$$

in which case we have a canonical map $U_{D - \overset{\circ}{D'}} : \mathcal{H}_{\partial D'} \to \mathcal{H}_{\partial D}$ defined by the annular cobordism.

If $x_1, \ldots, x_k$ are distinct points of M, and $D_1, \ldots, D_k$ are disjoint discs with x_i in the interior of D_i, let M_0 denote the manifold with boundary obtained by deleting from M the interiors of the discs D_i. We regard M_0 as a cobordism from $\coprod \partial D_i$ to the empty manifold, and define

$$\Theta_k : \mathcal{O}_{x_1} \otimes \cdots \otimes \mathcal{O}_{x_k} \longrightarrow \mathbb{C}$$

as the inverse limit of the maps

$$U_{M_0} : \mathcal{H}_{\partial D_1} \otimes \cdots \otimes \mathcal{H}_{\partial D_k} \to \mathbb{C}$$

as the discs D_i shrink to points around the points x_i.

If the points x_i are all contained in the interior of a disc D then the map Θ_k clearly factorizes through $\mathcal{H}_{\partial D}$. We can interpret this as saying that the local observables have the structure of an *algebra* which is associative and commutative up to homotopy, for if D is a small neighbourhood of a point $x \in M$ then we can regard $\mathcal{H}_{\partial D}$ as a *completion* $\hat{\mathcal{O}}_x$ of $\mathcal{O}_x$ (for the maps $\mathcal{H}_{\partial D'} \to \mathcal{H}_{\partial D}$ in the system defining $\mathcal{O}_x$ are always injective with dense image), and we have a map

$$\mathcal{O}_{x_1} \otimes \cdots \otimes \mathcal{O}_{x_k} \to \hat{\mathcal{O}}_x$$

for any family $x_1, \ldots, x_k$ of distinct points sufficiently close to x. This is an exact linear analogue of the d-fold loop space structures discussed above, and it is what in traditional quantum field theory is called the *operator product expansion*.

It is still not clear, however, that we have made our definition of a quantum field theory sufficiently rigid. One feels that the vector space $\mathcal{H}_Y$ associated to a $(d-1)$-manifold Y should be constructed *locally* from Y, and perhaps more axioms are needed to ensure this. A natural second approximation to the definition is the notion of a *three-tier theory*, which gives an additional layer of structure to allow $(d-1)$-manifolds to be cut into pieces. In a three-tier theory

1. to each compact oriented Riemannian $(d-2)$-manifold Z there is associated a linear category $\mathcal{C}_Z$,
2. to each $(d-1)$-dimensional Riemannian cobordism Y from Z_0 to Z_1 there is associated an additive functor $\mathcal{H}_Y : \mathcal{C}_{Z_0} \to \mathcal{C}_{Z_1}$, and
3. to each d-dimensional Riemannian cobordism X from Y to Y', where Y and Y' are cobordisms from Z_0 to Z_1, there is associated a transformation of functors $U_X : \mathcal{H}_Y \to \mathcal{H}_{Y'}$.

As with the earlier definition, the data are required to satisfy two axioms of concatenation and tensoring. A theory in the earlier two-tier sense is obtained from the three-tier structure by restricting to closed $(d-1)$-manifolds, which can be regarded as cobordisms from the empty $(d-2)$-manifold $\emptyset$ to itself. The tensoring axiom implies that $\mathcal{C}_\emptyset$ is the category of vector spaces, and since any additive functor $\mathcal{C}_\emptyset \to \mathcal{C}_\emptyset$ is given by tensoring with a vector space, we can identify $\mathcal{H}_Y$ with a vector space when Y is closed.

In the form I have just stated, the three-tier definition is too vague to be of much use. In this talk I shall not try to elaborate it, as my purpose is to make just two points. The first is that a three-tier two-dimensional field theory seems to have a good claim as a candidate definition of a 'noncommutative manifold'. For, schematically at least, a natural way to give the data of a three-tier theory is to associate an algebra $\mathcal{A}_Z$ to each $(d-2)$-manifold, and to take $\mathcal{C}_Z$ to be the category of left $\mathcal{A}_Z$-modules; then to a cobordism Y is associated an $(\mathcal{A}_{Z_1}, \mathcal{A}_{Z_0})$-bimodule $\mathcal{H}_Y$, which defines a functor $\mathcal{C}_{Z_0} \to \mathcal{C}_{Z_1}$ by

$$M \mapsto \mathcal{H} \otimes_{\mathcal{A}_{Z_0}} M;$$

and to a cobordism between cobordisms is associated a homomorphism of bimodules. When $d = 2$ this simply means that we have a dual pair of linear categories – the left and right modules for an algebra – associated to a point with its two orientations, while the one-dimensional data expresses the categorical duality, and the two-dimensional data gives us 'trace' or 'integration' maps. The field theory even leads one naturally from categories of modules to categories of cochain complexes of modules, if we assume that the field theory is supersymmetric.

The idea that two-dimensional theories should replace manifolds is, of course, the central proposal of string theory, which models space-time by a two-dimensional conformal field theory, with the category of D-branes in space-time as the category which the field theory associates to a point. It is also what arises in the Kapustin–Witten treatment of geometric Langlands duality. There, one begins from the maximally supersymmetric four-dimensional Yang–Mills theory associated to a compact group G, and observes that, for any compact surface Σ, a four-dimensional theory gives a two-dimensional theory by dimensional reduction along Σ – that is, by composing with the functor $M \mapsto \Sigma \times M$ from i-manifolds to $(i+2)$-manifolds. The two-dimensional theory obtained from Yang–Mills theory for G by reducing along Σ is supposed to associate to a point the category of $\mathcal{D}$-modules on the moduli space of holomorphic $G_{\mathbb{C}}$-bundles on Σ.

My second point is even vaguer. The obvious fear, if one starts to study three-tier theories, is that one will be impelled to believe that a d-dimensional theory should really mean a $(d+1)$-tier d-dimensional theory, which associates a two-category to a manifold of dimension $d-3$, and even worse things to lower dimensional manifolds, until one gets to a $(d-1)$-category associated to a point.

This is not completely mad, as it works well in the one famous example afforded by three-dimensional Chern–Simons theory for a compact group G at a given 'level'. There, the category associated to a circle S is the category of positive energy representations of the loop group of maps $S \to G$ at the specified level, and the two-category associated to a point is the same thing, but remembering its tensor structure coming from the *fusion* of loop group representations. (We think of this as a two-category with just one object, whose linear category of endomorphisms is the category of loop group representations, with fusion as its composition law.)

I cannot believe, however, that genuine – non-topological – quantum field theory will be advanced by higher categories. The algebras-up-to-homotopy formed by the field operators look much more promising. It is interesting that, in the context of homotopy theory, a d-fold loop space P does indeed give rise to a precise analogue of the structure of a $(d+1)$-tier field theory, as follows.

1. To a closed d-manifold X we associate the space $Q_X = C(X;P)$.
2. To a closed $(d-1)$-manifold Y we associate the 'group' – that is, one-fold loop space – $\mathcal{G}_Y = C(I \times Y; P)$, where I is an open interval.
3. To a closed d-manifold X with boundary we associate the $\mathcal{G}_{\partial X}$-space $Q_X = C(X^\circ; P)$, where X° is the interior of X, noticing that if $X = X_1 \cup_Y X_2$ then $Q_X \simeq Q_{X_1} \times_{\mathcal{G}_Y} Q_{X_2}$. And, more generally, to a cobordism from Y_0 to Y_1 we can associate a $(\mathcal{G}_{Y_0} \times \mathcal{G}_{Y_1})$-space, and hence a functor from $\mathcal{G}_{Y_0}$ spaces to $\mathcal{G}_{Y_1}$ spaces.
4. To a closed $(d-2)$-manifold Z we associate the two-fold loop space $\mathcal{G}_Z = C(I \times I \times Z; P)$, noticing that if Y is a $(d-1)$-manifold with boundary then $\mathcal{G}_Y = C(I \times Y^\circ; P)$ is a one-fold loop space on which the two-fold loop space $\mathcal{G}_{\partial Y}$ acts by maps of one-fold loop spaces.

And so on – I shall not spell out all the details.

I hope that a linear and analytical version of this picture is the model for quantum field theory. One point where the analogy may help is in considering whether one should expect that the d-fold algebra of field operators defined in a small d-dimensional ball should determine the theory on an arbitrary space-time, in the way that the d-fold loop space P determines all the spaces $C(X;P)$. We saw that a mapping space $\mathrm{Map}(X;Q)$ is of the form $C(X;P)$ only when the target space Q is d-connected, and I would guess that an analogous distinction applies to field theories: just as the mapping space cannot be modelled by particles, but requires m-dimensional 'objects', if Q is only $(d-m)$-connected, so we know that a field theory in general has non-local observables that can be seen only in topologically non-trivial regions of space-time. It seems, however, that the non-local observables are 'topological' in the sense that they contribute only a finite number of degrees of freedom to the infinite-dimensional physical system that the field theory describes.

References

Atiyah, M. F. and Bott, R. (1982). The Yang-Mills equations over Riemann surfaces. *Phil. Trans. Roy. Soc. London* **308A**, 523–615.

Connes, A. (1994). *Noncommutative Geometry.* Academic Press, San Diego, CA.

Kapustin, A. and Witten, E. (2007). Electro-magnetic duality and the geometric Langlands program. *Commun. Number Theory Phys.* **1**, 1–236. hep-th/0610210.

McDuff, D. (1975). Configuration spaces of positive and negative particles. *Topology* **14**, 91–107.

McDuff, D. (1977). Configuration spaces. In *K-Theory and Operator Algebras.* Proceedings Georgia 1975, Springer Lecture Notes **575**, pp. 88–95.

Pressley, A. and Segal, G. (1986). *Loop Groups.* Oxford University Press, Oxford.

Segal, G. (1968). Classifying spaces and spectral sequences. *Proc. Math. IHES Paris* **34**, 105–12.

Segal, G. (1973). Configuration spaces and iterated loop-spaces. *Invent. Math.*, **21**, 213–21.

Segal, G. (1977). K-homology theory and algebraic K-theory. In *K-Theory and Operator Algebras.* Proceedings Georgia 1975, Springer Lecture Notes **575**, pp. 113–27.

Segal, G. (1979). The topology of spaces of rational functions. *Acta Math* **143**, 39–72.

X

TOEPLITZ OPERATORS AND HITCHIN'S PROJECTIVELY FLAT CONNECTION

Jørgen Ellegaard Andersen

Dedicated to Nigel Hitchin on the occasion of his 60th birthday

10.1 Introduction

Let Z be a $(2+1)$-dimensional topological quantum field theory (TQFT). That is, the two dimensional part of Z is a modular functor with finite label set[1] Λ:

$$Z : \left\{\begin{array}{l}\text{Category of}\\ \text{(extended) closed}\\ \text{oriented surfaces}\\ \text{with } \Lambda\text{-labeled}\\ \text{marked points with}\\ \text{projective tangent}\\ \text{vectors}\end{array}\right\} \rightarrow \left\{\begin{array}{l}\text{Category of finite}\\ \text{dimensional vector}\\ \text{spaces over } \mathbb{C}\end{array}\right\}$$

The three-dimensional part of the TQFT is an association of a vector

$$Z(M, L, \lambda) \in Z(\partial M, \partial L, \partial\lambda)$$

to any compact oriented framed three-manifold M together with an oriented framed link $(L, \partial L) \subseteq (M, \partial M)$ and a Λ-labeling $\lambda : \pi_0(L) \rightarrow \Lambda$:

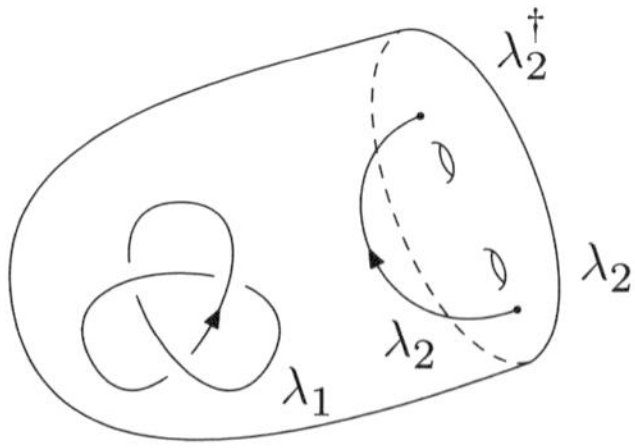

This association has to satisfy the Atiyah–Segal–Witten TQFT axioms (see, e.g., Witten 1989, Atiyah 1990, Segal 1992 and for a more comprehensive presentation of the axioms see Turaev 1994).

[1] The set is also equipped with an involution † and a trivial element which is preserved by the involution.

Witten constructed via path integral techniques a quantization of Chern–Simons theory in $(2+1)$ dimensions and argued in Witten (1989) that this produced a TQFT index by a compact simple Lie group and an integer level k. For the group $SU(n)$ and level k let us denote this TQFT by $Z_k^{(n)}$. Witten (1989) shows that the theory $Z_k^{(2)}$ determines the Jones polynomial of a knot in S^3. Combinatorially this theory was first constructed by Reshetikin and Turaev (1990, 1991) using representation theory of $U_q(\mathrm{sl}(n,\mathbb{C}))$ at $q = e^{(2\pi i)/(k+n)}$. Subsequently the TQFTs $Z_k^{(n)}$ were constructed using skein theory by Blanchet, Habegger, Masbaum, and Vogel in Blanchet *et al.* (1992), (1995) and Blanchet (2000).

For this theory the label set $\Lambda_k^{(n)}$ is a finite subset (depending on k) of the set of finite-dimensional irreducible representations of $SU(n)$. We use the usual labeling of irreducible representations by Young diagrams, so in particular $\square \in \Lambda_k^{(n)}$ is the defining representation for $SU(n)$. Let further $\lambda_0^{(d)} \in \Lambda_k^{(n)}$ be the Young diagram consisting of d columns of length k.

Part of this TQFT consists of the quantum $SU(n)$ representations of the mapping class group. Namely, if Σ is a closed oriented surface of genus g and Γ is the mapping class group of Σ. Let p be a point on Σ. Then the modular functor induces a representation

$$Z_k^{(n,d)} : \Gamma \to \mathbb{P}\,\mathrm{Aut}\big(Z_k^{(n)}(\Sigma, p, \lambda_0^{(d)})\big). \tag{10.1}$$

For a general label of p we would need to choose a projective tangent vector $v_p \in T_p\Sigma/\mathbb{R}_+$ and we would get a representation of the mapping class group of (Σ, p, v_p), but for the special labels $\lambda_0^{(d)}$, the dependence on v_p is trivial and in fact we get a representation of Γ. Furthermore, the curve operators are also part of any TQFT: For $\gamma \subseteq \Sigma - \{p\}$ an oriented simple closed curve and any $\lambda \in \Lambda_k^{(n)}$ we have the operators

$$Z_k^{(n,d)}(\gamma, \lambda) : Z_k^{(n)}(\Sigma, p, \lambda_0^{(d)}) \to Z_k^{(n)}(\Sigma, p, \lambda_0^{(d)}). \tag{10.2}$$

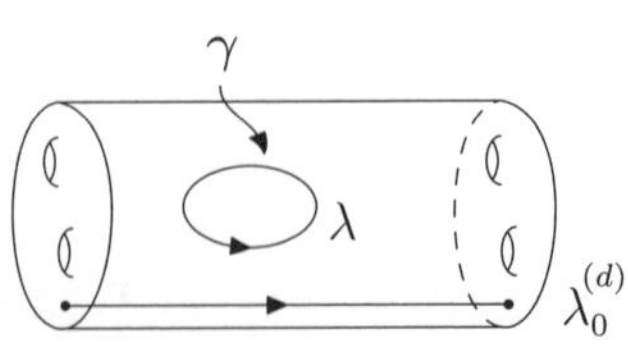

For the curve operators, we can derive an explicit formula using factorization: Let Σ' be the surface obtained from cutting Σ along γ and identifying the two boundary components to two points, say $\{p_+, p_-\}$. Here p_+ is the point corresponding to the "left" side of γ. For any label $\mu \in \Lambda_k^{(n)}$ we get a labeling of the ordered points (p_+, p_-) by the ordered pair of labels $(\mu, \mu^\dagger)$.

Since $Z_k^{(n)}$ is also a modular functor one can factor the space $Z_k^{(n)}(\Sigma, p, \lambda_0^{(d)})$ into a direct sum "along" γ as a direct sum over $\Lambda_k^{(n)}$. That is, we get an isomorphism

$$Z_k^{(n)}(\Sigma, p, \lambda_0^{(d)}) \cong \bigoplus_{\mu \in \Lambda_k^{(n)}} Z^{(k)}(\Sigma', p_+, p_-, p, \mu, \mu^\dagger, \lambda_0^{(d)}). \tag{10.3}$$

Strictly speaking we need a base point on γ to induce tangent directions at $p_\pm$. However, the corresponding subspaces of $Z^{(k)}(\Sigma, p, \lambda_0^{(d)})$ do not depend on the choice of base point. The isomorphism (10.3) induces an isomorphism

$$\mathrm{End}(Z^{(k)}(\Sigma, p, \lambda_0^{(d)})) \cong \bigoplus_{\mu \in \Lambda_k^{(n)}} \mathrm{End}(Z^{(k)}(\Sigma', p_+, p_-, p, \mu, \mu^\dagger, \lambda_0^{(d)})).$$

which also induces a direct sum decomposition of $\mathrm{End}(Z^{(k)}(\Sigma, p, \lambda_0^{(d)}))$ which is independent of the choice of the base point.

The TQFT axioms imply that the curve operator $Z^{(k)}(\gamma, \lambda)$ is diagonal with respect to this direct sum decomposition along γ. One has the formula

$$Z^{(k)}(\gamma, \lambda) = \bigoplus_{\mu \in \Lambda_k^{(n)}} S_{\lambda,\mu}(S_{0,\mu})^{-1} \,\mathrm{Id}_{Z^{(k)}(\Sigma', p_+, p_-, p, \mu, \mu^\dagger, \lambda_0^{(d)})}.$$

Here $S_{\lambda,\mu}$ is the S-matrix[2] of the theory $Z_k^{(n)}$ (see, e.g., Blanchet 2000 for a derivation of this formula).

In this chapter we review some of the results we have established regarding these representations of the mapping class group and the curve operators. For that we need the geometric construction of these theories as discussed below. We will here discuss the applications to TQFT of the general program of combining the theory of Toeplitz operators with the Hitchin connection. In this introduction we will cover the applications. The general setting will be discussed in the rest of this chapter.

10.1.1 *Geometric construction of* $Z_k^{(n)}$

The geometric construction of these TQFTs was proposed by Witten (1989), where he derived via the Hamiltonian approach to quantum Chern–Simons theory, that the geometric quantization of the moduli spaces of flat connections should give the two-dimensional part of the theory. Further, he proposed an alternative construction of the two-dimensional part of the theory

[2] The S-matrix is determined by the isomorphism a modular functor induces from two different way of glueing an annulus to obtain a torus. For its definition see for example Moore and Seiberg (1989), Walker (1991), Segal (1992), or Bakalov and Kirillov (2000) and references therein. It is also discussed in Andersen and Ueno (2006).

via WZW-conformal field theory. This theory has been studied intensively. In particular the work of Tsuchiya, Ueno, and Yamada (1989) provided the major geometric constructions and results needed. In Bakalov and Kirillov (2000) their results was used to show that the category of integrable highest modules of level k for the affine Lie algebra associated to any simple Lie algebra is a modular tensor category. Further in Bakalov and Kirillov (2000) this result is combined with the works of Kazhdan and Lusztig (1993, 1994*a*, 1994*b*) and of Finkelberg (1996) to argue that this category is isomorphic to the modular tensor category associated to the corresponding quantum group, from which Reshetikhin and Turaev constructed their TQFT. Unfortunately, these results do not allow one to conclude the validity of the geometric constructions of the two-dimensional part of the TQFT proposed by Witten. However, in joint work with Ueno (Andersen and Ueno, 2006, 2007*a*, 2007*b*, 2008), we have given a proof, based mainly on the results of Tsuchiya, Ueno and Yamada (1989), that the TUY construction of the WZW-conformal field theory after twist by a fractional power of an abelian theory satisfies all the axioms of a modular functor. Furthermore, we have proved that the full $(2+1)$-dimensional TQFT that results from this is isomorphic to the one constructed by BHMV via skein theory mentioned above. Combining this with the theorem of Laszlo (Laszlo 1998), which identifies (projectively) the representations of the mapping class groups one obtains from the geometric quantization of the moduli space of flat connections with the ones obtained from the TUY constructions, one gets a proof of the validity of the construction proposed by Witten (1989).

Let us now briefly recall the geometric construction of the representations $Z_k^{(n,d)}$ of the mapping class group, as proposed by Witten, using geometric quantization of moduli spaces.

We assume from now on that the genus of the closed oriented surface Σ is at least 2. Let M be the moduli space of flat $SU(n)$ connections on $\Sigma - p$ with holonomy around p equal to $\exp(2\pi i d/n)\,\mathrm{Id} \in SU(n)$. In the case (n,d) are coprime, the moduli space is smooth. In all cases the smooth part of the moduli space has a natural symplectic structure ω. There is a natural smooth symplectic action of the mapping class group Γ of Σ on M. Moreover there is a unique prequantum line bundle $(\mathcal{L}, \nabla, (\cdot,\cdot))$ over (M,ω). The Teichmüller space $\mathcal{T}$ of complex structures on Σ parametrizes naturally and Γ-equivariantly Kähler structures on (M,ω). For $\sigma \in \mathcal{T}$, we denote (M,ω) with its corresponding Kähler structure by M_σ.

By applying geometric quantization to the moduli space M, one gets a certain finite rank bundle over Teichmüller space $\mathcal{T}$, which we will call the *Verlinde* bundle $\mathcal{V}_k$ at level k, where k is any positive integer. The fiber of this bundle over a point $\sigma \in \mathcal{T}$ is $\mathcal{V}_{k,\sigma} = H^0(M_\sigma, \mathcal{L}^k)$. We observe that there is a natural Hermitian structure $\langle\cdot,\cdot\rangle$ on $H^0(M_\sigma, \mathcal{L}^k)$ by restricting the L_2 inner product on global L_2 sections of $\mathcal{L}^k$ to $H^0(M_\sigma, \mathcal{L}^k)$.

The main result pertaining to this bundle is

Theorem 10.1 (Axelrod, Della Pietra, and Witten 1991; Hitchin 1990) *The projectivization of the bundle $\mathcal{V}_k$ supports a natural flat Γ-invariant connection $\hat{\nabla}$.*

This is a result proved independently by Axelrod, Della Pietra, and Witten (1991) and by Hitchin (1990). In Section 10.2 we review our differential geometric construction of the connection $\hat{\nabla}$ in the general setting discussed in Andersen (2006*b*). We obtain as a corollary that the connection constructed by Axelrod, Della Pietra, and Witten projectively agrees with Hitchin's.

Definition 10.1 *We denote by $Z_k^{(n,d)}$ the representation*

$$Z_k^{(n,d)} : \Gamma \to \mathbb{P}\,\mathrm{Aut}\big(Z_k^{(n)}(\Sigma, p, \lambda_0^{(d)})\big)$$

obtained from the action of the mapping class group on the covariant constant sections of $\mathbb{P}(\mathcal{V}_k)$ over $\mathcal{T}$.

The projectively flat connection $\hat{\nabla}$ induces a *flat* connection $\hat{\nabla}^e$ in $\mathrm{End}(\mathcal{V}_k)$. Let $\mathrm{End}_0(\mathcal{V}_k)$ be the subbundle consisting of traceless endomorphisms. The connection $\hat{\nabla}^e$ also induces a connection in $\mathrm{End}_0(\mathcal{V}_k)$, which is invariant under the action of Γ.

10.1.2 *Asymptotic faithfulness*

In Andersen (2006*a*) we proved the following. (See also Freedman, Walker, and Wang for an alternative proof in the $(n, d) = (2, 0)$ case.)

Theorem 10.2 *Assume that n and d are coprime or that $(n, d) = (2, 0)$ when $g = 2$. Then we have that*

$$\bigcap_{k=1}^{\infty} \ker\left(Z_k^{(n,d)}\right) = \begin{cases} \{1, H\} & g = 2,\ n = 2 \text{ and } d = 0 \\ \{1\} & \text{otherwise.} \end{cases}$$

where H is the hyperelliptic involution.

Large parts of the proof of this theorem, which we gave in Andersen (2006*a*), apply to the general setting discussed in the following sections. In particular, we get the general asymptotic faithfulness Theorem 10.19 discussed in the last section of this chapter. Theorem 10.2 follows directly from Theorem 10.19 as argued in the end of section 6 of Andersen (2006*a*). See also Andersen (2008*a*) and Andersen and Christ (2008) for the singular case.

The main ingredient behind this is the theory of *Toeplitz operators* associated to smooth functions on M. For each $f \in C^\infty(M^{(d)})$ and each point $\sigma \in \mathcal{T}$, we have the Toeplitz operator

$$T_{f,\sigma}^{(k)} : H^0\left(M_\sigma, \mathcal{L}_\sigma^k\right) \to H^0\left(M_\sigma, \mathcal{L}_\sigma^k\right)$$

which is given by

$$T^{(k)}_{f,\sigma} = \pi^{(k)}_\sigma(fs)$$

for all $s \in H^0(M_\sigma, \mathcal{L}^k_\sigma)$, where $\pi^{(k)}_\sigma$ is the orthogonal projection onto $H^0(M_\sigma, \mathcal{L}^k_\sigma)$ induced from the L_2 inner product on $C^\infty(M, \mathcal{L}^k)$. We get a smooth section of $\mathrm{End}(\mathcal{V}^{(k)})$ over $\mathcal{T}$

$$T^{(k)}_f \in C^\infty(\mathcal{T}, \mathrm{End}(\mathcal{V}^{(k)}))$$

by letting $T^{(k)}_f(\sigma) = T^{(k)}_{f,\sigma}$ (see Andersen 2006*a*). See Section 10.3 for further discussion of the Toeplitz operators and their connection to deformation quantization.

The sections $T^{(k)}_f$ of $\mathrm{End}(\mathcal{V}^{(k)})$ over $\mathcal{T}$ are not covariant constant with respect to Hitchin's connection $\hat{\nabla}^e$. However, they are asymptotic as k goes to infinity. The precise meaning of this is stated in Theorem 10.18 of Section 10.5, where it is proved in the general setting. Theorem 10.19 then follows in a rather straightforward way from this asymptotic covariant constantness of the Toeplitz operators and Theorem 10.3 due to Bordeman, Meinrenken, and Schlichenmaier (1994).

Theorem 10.3 (Bordeman, Meinrenken, and Schlichenmaier 1994) *For any $f \in C^\infty(M)$ we have that*

$$\lim_{k\to\infty} \left\| T^{(k)}_{f,\sigma} \right\| = \sup_{x\in M} |f(x)|.$$

Since the association of the sequence of Toeplitz operators $T^{(k)}_{f,\sigma}$, $k \in \mathbb{Z}_+$ is linear in f, we see from this theorem that this association is faithful.

10.1.3 *Kashdan's property (T) and the mapping class group*

As an application of TQFT to a problem in geometric group theory, which has been around for some time (see e.g. problem (7.2) in chapter 7, "A short list of open questions," of Bekka, de la Harpe, and Valette 2007), we proved in (Andersen, 2007) that

Theorem 10.4 *The mapping class group of a closed oriented surface of genus at least 2 does not have Kazhdan's property (T).*

In Kazhdan (1967) the author introduced his property (T) for topological groups. A topological group has Kazhdan's property (T) if the trivial representation is isolated in the Fell topology on the space of unitary representations of the group. Alternatively, we can formula Kazhdan's property (T) as follows.

A discrete countable group G has Kazhdan's property (T) if and only if for all unitary Hilbert space representations of G, the existence of an *almost fixed vector* implies the existence of a non-trivial fixed vector.

Definition 10.2 *Let $\rho : G \to U(\mathcal{R})$ be a unitary Hilbert space representation of a discrete countable group G on the Hilbert space $\mathcal{R}$. By an almost fixed vector for ρ we mean a sequence of unit vectors $(v_k) \subset \mathcal{R}$ with the property that*

$$\lim_{k \to \infty} | \rho(g)v_k - v_k | = 0$$

for all $g \in G$.

The Hilbert space representation of the mapping class group we consider is the following.

Let

$$\mathcal{R}_k = \mathrm{End}_0(Z_k^{(2,1)}(\Sigma)),$$

where End_0 means the traceless endomorphisms. Since the TQFTs are unitary, there is a natural unitary structure which is mapping class group invariant on $\mathcal{R}_k$. Now define

$$\tilde{\mathcal{R}} = \bigoplus_{k+2 \text{ prime}}^{\infty} \mathcal{R}_k$$

and let $\mathcal{R}$ be the Hilbert space completion of $\tilde{\mathcal{R}}$.

Theorem 10.5 (Roberts 2001) *The only Γ invariant vector in $\mathcal{R}$ is 0.*

This theorem follows from the fact that the representations $Z_k^{(2,1)}(\Sigma)$ are irreducible as Γ-representation for k, such that $k+2$ is prime. This result was established in the un-twisted case by Roberts (2001) and his proof can be applied word for word also to this case.

The basic idea behind building the required almost fixed vector for $\mathcal{R}$ is to consider coherent states on M_σ, $\sigma \in \mathcal{T}$.

Fix a point $x \in M$. Evaluation of sections at x gives a section of $\mathcal{V}_k^*$ up to scale. Using the Hermitian structure $\langle \cdot, \cdot \rangle$ we get induced a section $e_x^{(k)}$ of $\mathcal{V}_k$ up to scale. For each $\sigma \in \mathcal{T}$, $e_x^{(k)}(\sigma)$ is the coherent state associated to x on M_σ. Let $E_x^{(k)}$ be the section of $\mathrm{End}(\mathcal{V}_k)$ obtained as the orthogonal projection (with respect to $\langle \cdot, \cdot \rangle$) onto the one-dimensional subspace spanned by $e_x^{(k)}$. We observe that $E_x^{(k)}$ only depends on x.

Theorem 10.6 *The sections $E_x^{(k)}$ of $\mathrm{End}(\mathcal{V}_k)$ over $\mathcal{T}$ are asymptotically covariant constant. That is, for any pair of points $\sigma_0, \sigma_1 \in \mathcal{T}$ there exists a constant C such that*

$$\left| P^e_{\sigma_0,\sigma_1}\left(E_x^{(k)}(\sigma_0)\right) - E_x^{(k)}(\sigma_1)\right| \leq \frac{C}{k},$$

where $P^e_{\sigma_0,\sigma_1}$ is the parallel transport from σ_0 to σ_1 in $\mathrm{End}(\mathcal{V}_k)$ and the norm $| \cdot |$ is the one associated to the Hermitian structure on $\mathcal{V}_k^* \otimes \mathcal{V}_k$ induced from $\langle \cdot, \cdot \rangle$ on $\mathcal{V}_k$.

In fact the proof of this theorem is valid in the general setting discussed in the following sections. However, the techniques used in Andersen (2007) go beyond the techniques reviewed in this chapter and we therefore refer to Andersen (2007) for the proof of this theorem.

To produce the almost fixed vectors, we now pick a finite subgroup Λ of $SU(2)$ which contains $-1 \in SU(2)$ and we consider the finite subset X of M, consisting of connections which reduce to Λ. For the dihedral group Λ, we get a non-empty finite subset X of M this way which is invariant under the action of the mapping class group. Let now $E_X^{(k)}$ be the section of $\mathrm{End}(\mathcal{V}_k)$ given by

$$E_X^{(k)} = \sum_{x \in X} E_x^{(k)}.$$

Let $E_{X,0}^{(k)}$ be the traceless part of $E_X^{(k)}$. As we have argued in section 7 of Andersen (2007), for large enough k, $E_X^{(k)} \neq 0$. Hence for large enough k we have a unique (up to unit scale) unit vector in $\mathcal{E}_{X,0}^{(k)} \in \mathcal{R}_k$, which at σ_0 agrees projectively with $E_{X,0}^{(k)}(\sigma_0)$.

Theorem 10.7 *The sequence* $\left\{\mathcal{E}_{X,0}^{(k)}\right\}$ *is an almost fixed vector for the action of* Γ *on* $\mathcal{R}$.

This theorem is proved in Andersen (2007). Theorem 10.4 is of course a consequence of Theorems 10.5 and 10.7.

10.1.4 *Geometric construction of the curve operators*

In the gauge theory picture, the decomposition (10.3) is obtained as follows (see Andersen 2008*b* for details).

One considers a one-parameter family of complex structures $\sigma_t \in \mathcal{T}$, $t \in \mathbb{R}_+$, such that the corresponding family in the moduli space of curves converges in the Mumford–Deligne boundary to a nodal curve, which topologically corresponds to shrinking γ to a point. By the results of Andersen (1998) the corresponding sequence of complex structures on the moduli space M converges to a non-negative polarization on M, whose isotropic foliation is spanned by the Hamiltonian vector fields associated to the holonomy functions of γ. The main result of Andersen (2008*b*) is that the covariant constant sections of $\mathcal{V}^{(k)}$ along the family σ_t converge to distributions supported on the Bohr–Sommerfeld leaves of the limiting non-negative polarization as t goes to infinity. The direct sum of the geometric quantization of the level k Bohr–Sommerfeld levels of this non-negative polarization is precisely the left-hand side of (10.3). A sewing-construction inspired from conformal field theory (see Tsuchiya, Ueno, and Yamada 1989) is then applied to show that the resulting linear map from the right-hand side of (10.3) to the left-hand side is an isomorphism. This is described in detail in Andersen (2008*b*).

In Andersen (2008*b*) we further prove the following important asymptotic result. Let $h_{\gamma,\lambda} \in C^\infty(M)$ be the holonomy function obtained by taking the trace in the representation λ of the holonomy around γ.

Theorem 10.8 *For any one-dimensional oriented submanifold γ and any labeling λ of the components of γ, we have that*

$$\lim_{k \to \infty} \|Z_k^{(n,d)}(\gamma, \lambda) - T_{h_{\gamma,\lambda}}^{(k)}\| = 0.$$

Let us here give the main idea behind the proof of Theorem 10.8 and refer to Andersen (2008*b*) for details. One considers the explicit expression for the S-matrix, as given in formula (13.8.9) in Kac (1995):

$$S_{\lambda,\mu}/S_{0,\mu} = \lambda(e^{-2\pi i \frac{\check{\mu}+\check{\rho}}{k+n}}), \tag{10.4}$$

where ρ is half of the sum of the positive roots and $\check{\nu}$ (ν any element of Λ) is the unique element of the Cartan subalgebra of the Lie algebra of $SU(n)$ which is dual to ν with respect to the Cartan–Killing form $(\cdot, \cdot)$.

From (10.4) one sees that under the isomorphism $\check{\mu} \mapsto \mu$, $S_{\lambda,\mu}/S_{0,\mu}$ makes sense for any $\check{\mu}$ in the Cartan subalgebra of the Lie algebra of $SU(n)$. Furthermore one finds that the values of this sequence of functions (depending on k) are asymptotic to the values of the holonomy function $h_{\gamma,\lambda}$ at the level k Bohr–Sommerfeld sets of the limiting non-negative polarizations discussed above (see Andersen 1998). From this one can deduce Theorem 10.8 (see again Andersen 2008*b* for details). Please see Andersen (2005) for the corresponding result in the abelian case.

10.1.5 *The Nielsen–Thurston classification of mapping classes is determined by TQFT*

Let us now review the result from Andersen (2006*c*), where via the use of curve operators we show that the Nielsen–Thurston classification of mapping classes is determined by TQFT.

The Nielsen–Thurston classification of mapping classes of compact oriented surfaces splits mapping classes into three disjoint types (Thurston 1988) (see also Fathi, Laudenbach, and Poénaru 1979, and Bleiler and Casson 1988). We shall only be interested in the closed surface case here, so we state the Nielsen–Thurston theorem in this case.

Suppose Σ is a closed oriented surface of genus $g \geq 2$ and let Γ be the mapping class group of Σ.

Theorem 10.9 (Nielsen–Thurston) *A mapping class $\phi \in \Gamma$ has exactly one of the following three properties:*

1. *The mapping class ϕ is,* finite-order, *that is, ϕ is a finite-order element in Γ. This is equivalent to ϕ having an automorphism of a Riemann surface as representative.*

2. *The mapping class ϕ is not finite order, but it is* reducible, *meaning there exists a simple closed curve on the surface, whose non-trivial homotopy class is preserved by some power of ϕ.*
3. *The mapping class ϕ is pseudo-Anosov, meaning that there exists $\zeta > 1$, two transverse measured foliations F^s and F^u on Σ and a diffeomorphism f of Σ, which represents ϕ, such that*

$$f_*(F^s) = \zeta^{-1} F^s \text{ and } f_*(F^u) = \zeta F^u.$$

In the pseudo-Anosov case, ζ is uniquely determined by ϕ and it is called the streching factor for ϕ.

In the reducible case one continues the analysis of ϕ, by cutting Σ along the preserved simple closed curve, to get a mapping class of a surface with boundary. This mapping class is then classified in terms of the Nielsen–Thurston classification of mapping classes of surfaces with boundary. The upshot of this is that there is a diffeomorphism f of Σ, which represents ϕ and which preserves a system of simple closed curves, and when one cuts the surface along these curves, f induces a diffeomorphism of the resulting cut surface. For each component of the cut surface, there is a smallest power of f, which preserves the component, and this power of f is a diffeomorphism of the component, which is either finite order or pseudo-Anosov (see Fathi, Laudenbach, and Poénaru 1979 for further details regarding this).

The asymptotic faithfulness property gives us immediately the following theorem:

Theorem 10.10 *For any mapping class $\phi \in \Gamma$ we have that there exists an integer M such that*

$$\left(Z_k^{(n,d)}(\phi)\right)^M \in \mathbb{C}\,\mathrm{Id}$$

for all k if and only if $\phi^M = 1$ (or $\phi^{2M} = 1$, in case $(n, d) = (2, 0)$).

This separates the finite-order ones from the rest. In order to separate the reducibles from the pseudo-Anosov ones, we consider the curve operators. Suppose that γ is a simple closed curve on the surface Σ. When we choose an orientation on γ, we have the holonomy function $h_{\gamma,\Box}$. Note that $h_{\gamma,\Box}$ only depends on the free homotopy class of γ. Further $h_{\gamma,\Box}$ is constant if γ is nul-homotopic.

Theorem 10.11 *For any mapping class $\phi \in \Gamma$ and any homotopy class γ of a simple closed curve on Σ we have that ϕ is reducible along γ, that is,*

$$\phi(\gamma) = \gamma$$

if and only if

$$[Z_k^{(n,d)}(\phi), Z_k^{(n,d)}(\gamma, \Box)] = 0$$

for infinitely many k and any fixed pair (n,d) (and any choice of orientation of γ if $n > 2$).

Using these two conditions, we see immediately how to determine the Nielsen–Thurston classification of a mapping class using TQFT:

1. Given a mapping class ϕ, we first determine if it is finite order or not by checking if there exists an N such that $\rho_k(\phi)^N = 1$ for all k.
2. If not, we check through the non-trivial homotopy classes γ of simple closed curves to see if there exists an integer M such that $Z_k^{(n,d)}(\phi^M)$ commutes with $Z_k^{(n,d)}(\gamma, \Box)$ for infinitely many k for some pair (n,d). If so, ϕ is reducible.
3. If not, then ϕ is pseudo-Anosov.

Proof. It is clear that $\phi(\gamma) = \gamma$ implies that $Z_k^{(n,d)}(\phi)$ commutes with $Z_k^{(n,d)}(\gamma, \Box)$, since

$$Z_k^{(n,d)}(\phi) Z_k^{(n,d)}(\gamma, \Box) Z_k^{(n,d)}(\phi^{-1}) = Z_k^{(n,d)}(\phi(\gamma), \Box). \tag{10.5}$$

Conversely, suppose that

$$\left[Z_k^{(n,d)}(\phi), Z_k^{(n,d)}(\gamma, \Box) \right] = 0.$$

Then

$$Z_k^{(n,d)}(\gamma, \Box) = Z_k^{(n,d)}(\phi(\gamma), \Box)$$

by (10.5). From Theorem 10.8 we conclude that

$$\lim_{k \to \infty} \left\| T_{h_{\gamma,\Box}}^{(k)} - Z_k^{(n,d)}(\phi) T_{h_{\gamma,\Box}}^{(k)} Z_k^{(n,d)}(\phi^{-1}) \right\| = 0.$$

But then from Proposition 10.5 below and Theorem 10.3, we get that $h_{\gamma,\Box} = h_{\phi(\gamma),\Box}$. We now get that $\phi(\gamma) = \gamma$ by proposition 1 in Andersen (2006*c*). The main idea behind proposition 1 from Andersen (2006*c*) is that the holonomy functions extend to holomorphic functions on the $SL(n, \mathbb{C})$-moduli space $\mathcal{M}$. The restriction of the holonomy functions from $\mathcal{M}$ to the real slice $M \subset \mathcal{M}$ is injective. Moreover if two holonomy functions for two simple closed curves agree on $\mathcal{M}$ for some d, then they also agree on $\mathcal{M}$ for $d = 0$. But this space contains a copy of the Teichmüller space of Σ where it is clear that the holonomy function of a simple closed curve determines the curve up to homotopy on Σ. We refer to Andersen (2006*c*) for the full details of this argument. □

We consider it an interesting problem to provide a TQFT formula for the streching factor ζ of any pseudo-Anosov mapping class (see Andersen, Masbaum, and Ueno 2006 for the first steps in this direction). We expect that the asymptotic expansions of the operators $Z_k^{(n,d)}(\gamma, \Box)$ should be related to work of Andersen, Mattes and Reshetikhin (1996, 1998)

10.1.6 *General setting*

Throughout the rest of this chapter, we consider the general setting treated in Andersen (2006*b*). That is, we consider a general prequantizable symplectic manifold (M, ω) with a prequantum line bundle $(L, (\cdot,\cdot), \nabla)$ (as opposed to only considering the moduli spaces). We assume that $\mathcal{T}$ is a complex manifold which holomorphically and rigidly (see Definition 10.4) parameterizes Kähler structures on (M, ω). Under a mild cohomological condition we then establish the existence of the Hitchin connection in this setting as stated in Theorem 10.12 in Section 10.2. In Section 10.3 we recall well-known results about the relation between Toeplitz operators and deformation quantization. In Section 10.4 we define and construct the formal Hitchin connection as stated in Theorem 10.16. We further discuss how this allows us under certain cohomological conditions to construct symmetry invariant star-products (Theorem 10.17). In Section 10.5, we apply these results to prove Theorem 10.18, which gives the asymptotic flatness of the Toeplitz operators with respect to the Hitchin connection. We have here given a somewhat simpler proof of this theorem, than the one given in Andersen (2006*a*). The proof makes use of our notion of *formal trivializations* of the formal Hitchin connection. In Section 10.6 we establish a generalized asymptotic faithfulness, Theorem 10.19, and a generalized condition for reducibility, Theorem 10.20.

These general results form the analytic basis for the TQFT applications discussed above.

Finally, we would like to add that we have several other applications of the theory of Toeplitz operators to the study of the Hitchin connection and TQFT. In particular the Toeplitz operator theory gives the complete asymptotic expansion of the quantum invariant of closed three-manifolds. As a corollary of this we get that the colored Jones polynomials of a knot is an unknot detector. Moreover we also get the corollary that if a closed three-manifold X satisfies that

$$Z_k^{(n)}(X) = Z_k^{(n)}(S^3)$$

then the only group homomorphism from $\pi_1(X)$ to $SU(n)$ is the trivial one. By Perelman's work we know that $\pi_1(X)$ is residually finite, thus it follows that X is simply connected and therefore by Perelman's proof of the Poincaré conjecture, that $X = S^3$. The writing up of these results is in progress.

10.2 The Hitchin connection

In this section we review our construction of the Hitchin connection using the global differential geometric setting of Andersen (2006*b*). This approach is close in spirit to Axelrod, Della Pietra, and Witten (1991), however we do not use any infinite-dimensional gauge theory. In fact the setting is more general than the gauge theory setting in which Hitchin (1990) constructed his original connection. But when applied to the gauge theory situation we get the corollary that Hitchin's connection agrees with Axelrod, Della Pietra, and Witten's.

Hence, we start in the general setting and let (M, ω) be any compact symplectic manifold.

Definition 10.3 *A prequantum line bundle $(\mathcal{L}, (\cdot,\cdot), \nabla)$ over the symplectic manifold (M, ω) consists of a complex line bundle $\mathcal{L}$ with a Hermitian structure $(\cdot,\cdot)$ and a compatible connection ∇ whose curvature is*

$$F_\nabla = \frac{i}{2\pi}\omega.$$

We say that the symplectic manifold (M, ω) is prequantizable if there exists a prequantum line bundle over it.

The condition on the curvature is to be interpreted in terms of the induced principal $U(1)$-connection in the circle bundle of L in the above definition. In terms of covariant derivatives, this means

$$\omega(X, Y) = [\nabla_X, \nabla_Y] - \nabla_{[X,Y]}.$$

Recall that the condition for the existence of a prequantum line bundle is that $[\omega] \in \mathrm{Im}(H^2(M, \mathbb{Z}) \to H^2(M, \mathbb{R}))$ and that the inequivalent choices of prequantum line bundles (if they exist) are parametriced by $H^1(M, \mathbb{R})$ (see e.g. Woodhouse 1992).

We shall assume that (M, ω) is prequantizable and fix a prequantum line bundle $(\mathcal{L}, (\cdot,\cdot), \nabla)$ over (M, ω).

Assume that $\mathcal{T}$ is a smooth manifold which smoothly parametrizes Kähler structures on (M, ω). This means we have a smooth[3] map $I : \mathcal{T} \to C^\infty(M, \mathrm{End}(TM))$ such that (M, ω, I_σ) is a Kähler manifold for each $\sigma \in \mathcal{T}$.

We will use the notation M_σ for the complex manifold (M, I_σ). For each $\sigma \in \mathcal{T}$ we use I_σ to split the complexified tangent bundle $TM_\mathbb{C}$ into the holomorphic and the anti-holomorphic parts, which we denote

$$T_\sigma = E(I_\sigma, i) = \mathrm{Im}(\mathrm{Id} - iI_\sigma)$$

and

$$\bar{T}_\sigma = E(I_\sigma, -i) = \mathrm{Im}(\mathrm{Id} + iI_\sigma),$$

respectively.

[3] Here a smooth map from $\mathcal{T}$ to $C^\infty(M, W)$ for any smooth vector bundle W over M means a smooth section of $\pi_M^*(W)$ over $\mathcal{T} \times M$, where π_M is the projection onto M. Likewise a smooth p-form on $\mathcal{T}$ with values in $C^\infty(M, W)$ is by definition a smooth section of $\pi_\mathcal{T}^*\Lambda^p(\mathcal{T}) \otimes \pi_M^*(W)$ over $\mathcal{T} \times M$. We will also encounter the situation where we have a bundle $\tilde{W}$ over $\mathcal{T} \times M$ and then we will talk about a smooth p-form on $\mathcal{T}$ with values in $C^\infty(M, \tilde{W}_\sigma)$ and mean a smooth section of $\pi_\mathcal{T}^*\Lambda^p(\mathcal{T}) \otimes \tilde{W}$ over $\mathcal{T} \times M$.

The real Kähler metric g_σ on (M_σ, ω) extended complex linear to $TM_{\mathbb{C}}$ is by definition

$$g_\sigma(X,Y) = \omega(X, I_\sigma Y),$$

where $X, Y \in C^\infty(M, TM \otimes \mathbb{C})$.

Suppose V is a vector field on $\mathcal{T}$. Then we can differentiate I along V and we denote this derivative $V[I] : \mathcal{T} \to C^\infty(M, \mathrm{End}(TM_{\mathbb{C}}))$. Differentiating the equation $I^2 = -\,\mathrm{Id}$, we see that $V[I]$ anti-commutes with I. Hence we get that

$$V[I]_\sigma \in C^\infty\left(M, (T_\sigma^* \otimes \bar{T}_\sigma) \oplus (\bar{T}_\sigma^* \otimes T_\sigma)\right)$$

for each $\sigma \in \mathcal{T}$. Let

$$V[I]_\sigma = V[I]''_\sigma + V[I]'_\sigma$$

be the corresponding decomposition such that $V[I]''_\sigma \in C^\infty\left(M, T_\sigma^* \otimes \bar{T}_\sigma\right)$ and $V[I]'_\sigma \in C^\infty\left(M, \bar{T}_\sigma^* \otimes T_\sigma\right)$.

Now we will further assume that $\mathcal{T}$ is a complex manifold and that I is a holomorphic map from $\mathcal{T}$ to the space of all complex structures on M. Concretely, this means that

$$V'[I]_\sigma = V[I]'_\sigma$$

and

$$V''[I]_\sigma = V[I]''_\sigma$$

for all $\sigma \in \mathcal{T}$, where V' means the $(1,0)$ part of V and V'' means the $(0,1)$ part of V over $\mathcal{T}$.

Let us now define $\tilde{G}(V) \in C^\infty(M, TM_{\mathbb{C}} \otimes TM_{\mathbb{C}})$ by

$$V[I] = \tilde{G}(V)\omega,$$

and define $G(V) \in C^\infty(M, T_\sigma \otimes T_\sigma)$ such that

$$\tilde{G}(V) = G(V) + \overline{G}(V)$$

for all real vector fields V on $\mathcal{T}$. We see that $\tilde{G}$ and G are one-forms on $\mathcal{T}$ with values in $C^\infty(M, TM_{\mathbb{C}} \otimes TM_{\mathbb{C}})$ and $C^\infty(M, T_\sigma \otimes T_\sigma)$, respectively. We observe that

$$V'[I] = G(V)\omega,$$

and $G(V) = G(V')$.

It is easy to check that $\tilde{G}$ takes values in $C^\infty(M, S^2(TM_{\mathbb{C}}))$ and therefore that G takes values in $C^\infty(M, S^2(T_\sigma))$.

On $\mathcal{L}^k$ we have the smooth family of $\bar{\partial}$ operators $\nabla^{0,1}$ defined at a $\sigma \in \mathcal{T}$ by

$$\nabla_\sigma^{0,1} = \frac{1}{2}(1 + iI_\sigma)\nabla.$$

For every $\sigma \in \mathcal{T}$ we consider the finite-dimensional subspace of $C^\infty(M, \mathcal{L}^k)$ given by

$$H_\sigma^{(k)} = H^0(M_\sigma, \mathcal{L}^k) = \left\{ s \in C^\infty(M, \mathcal{L}^k) | \nabla_\sigma^{0,1} s = 0 \right\}.$$

Let $\hat{\nabla}^t$ denote the trivial connection in the trivial bundle $\mathcal{T} \times C^\infty(M, \mathcal{L}^k)$. Let $\mathcal{D}(M, \mathcal{L}^k)$ denote the vector space of differential operators on $C^\infty(M, \mathcal{L}^k)$. For any smooth one-form u on $\mathcal{T}$ with values in $\mathcal{D}(M, \mathcal{L}^k)$ we have a connection $\hat{\nabla}$ in $\mathcal{H}^{(k)}$ given by

$$\hat{\nabla}_V = \hat{\nabla}_V^t - u(V)$$

for any vector field V on $\mathcal{T}$.

Lemma 10.1 *The connection $\hat{\nabla}$ in $\mathcal{H}^{(k)}$ preserves the subspaces $H_\sigma^{(k)} \subset C^\infty(M, \mathcal{L}^k)$, for all $\sigma \in \mathcal{T}$ if and only if*

$$\frac{i}{2} V[I] \nabla^{1,0} s + \nabla^{0,1} u(V) s = 0 \tag{10.6}$$

for all vector fields V on $\mathcal{T}$ and all smooth sections s of $H^{(k)}$.

This result is not surprising (see Andersen 2006*b* for a proof this lemma). Observe that if this condition holds, we can conclude that the subspaces $H_\sigma^{(k)} \subset C^\infty(M, \mathcal{L}^k)$, for all $\sigma \in \mathcal{T}$, form a subbundle $H^{(k)}$ of $\mathcal{H}^{(k)}$.

We observe that

$$V''[I] \nabla^{1,0} s = 0,$$

so $u(V'') = 0$ solves (10.6) along the anti-holomorphic directions on $\mathcal{T}$. In other words the $(0,1)$ part of the trivial connection $\hat{\nabla}^t$ induces a $\bar{\partial}$ operator on $H^{(k)}$ and hence makes it a holomorphic vector bundle over $\mathcal{T}$.

This is of course not in general the situation in the $(1,0)$ direction. Let us now consider a particular u and prove that it solves (10.6) under certain conditions.

On the Kähler manifold (M_σ, ω) we have the Kähler metric and we have the Levi-Civita connection ∇ in T_σ. We also have the Ricci potential $F_\sigma \in C_0^\infty(M, \mathbb{R})$. Here

$$C_0^\infty(M, \mathbb{R}) = \left\{ f \in C^\infty(M, \mathbb{R}) \mid \int_M f \omega^m = 0 \right\}$$

and the Ricci potential is the element of $F_\sigma \in C_0^\infty(M, \mathbb{R})$ which satisfies

$$\mathrm{Ric}_\sigma = \mathrm{Ric}_\sigma^H + \frac{1}{2} i \partial_\sigma \bar{\partial}_\sigma F_\sigma,$$

where $\mathrm{Ric}_\sigma \in \Omega^{1,1}(M_\sigma)$ is the Ricci form and Ric_σ^H is its harmonic part. We see that we get this way a smooth function $F : \mathcal{T} \to C_0^\infty(M, \mathbb{R})$.

For any $G \in C^\infty(M, S^2(T_\sigma))$ we get a linear bundle map

$$G : T_\sigma^* \to T_\sigma$$

and we have the operator

$$\Delta_G : C^\infty(M,\mathcal{L}^k) \overset{\nabla^{1,0}_\sigma}{\to} C^\infty\left(M, T^*_\sigma \otimes \mathcal{L}^k\right) \overset{G\otimes \mathrm{Id}}{\to} C^\infty(M, T_\sigma \otimes \mathcal{L}^k)$$
$$\overset{\nabla^{1,0}_\sigma \otimes \mathrm{Id} + \mathrm{Id}\otimes \nabla^{1,0}_\sigma}{\to} C^\infty\left(M, T^*_\sigma \otimes T_\sigma \otimes \mathcal{L}^k\right) \overset{\mathrm{Tr}}{\to} C^\infty(M,\mathcal{L}^k).$$

For any smooth function f on M, we get a vector field

$$Gdf \in C^\infty(M, T_\sigma).$$

Implicit in this definition is the projection from $TM \cong T_\sigma \oplus \bar{T}_\sigma$ to T_σ, which takes df to $\partial_\sigma f$.

Putting these constructions together we consider the following operator for some $n \in \mathbb{Z}$ such that $2k + n \neq 0$:

$$u(V) = \frac{1}{2k+n} o(V) + V'[F] \tag{10.7}$$

where

$$o(V) = \frac{1}{2}\Delta_{G(V)} - \nabla_{G(V)dF} - nV'[F]. \tag{10.8}$$

The connection associated to this u is denoted $\hat{\nabla}$ and we call it the *Hitchin connection* in $\mathcal{H}^{(k)}$.

Definition 10.4 *We say that the complex family I of Kähler structures on (M,ω) is* rigid *if*

$$\bar{\partial}_\sigma(G(V)_\sigma) = 0$$

for all vector fields V on $\mathcal{T}$ and all points $\sigma \in \mathcal{T}$.

We will assume our holomorphic family I is rigid.

Theorem 10.12 *Suppose that I is a rigid family of Kähler structures on the compact symplectic prequantizable manifold (M,ω), which satisfies that there exists $n \in \mathbb{Z}$ such that the first Chern class of (M,ω) is $n[\omega] \in H^2(M,\mathbb{Z})$ and $H^1(M,\mathbb{R}) = 0$. Then u given by (10.7) and (10.8) satisfies (10.6) for all k such that $2k+n \neq 0$*

Hence the Hitchin connection $\hat{\nabla}$ preserves the subbundle $H^{(k)}$ under the stated conditions.

Theorem 10.12 is established in Andersen (2006*b*) through the following three lemmas.

Lemma 10.2 *Assume that the first Chern class of (M,ω) is $n[\omega] \in H^2(M,\mathbb{Z})$. For any $\sigma \in \mathcal{T}$ and for any $G \in H^0(M_\sigma, S^2(T_\sigma))$ we have the following formula:*

$$\nabla^{0,1}_\sigma(\Delta_G(s) - 2\nabla_{GdF_\sigma}(s)) = -i(2k+n)G\omega\nabla^{1,0}_\sigma(s)$$
$$-ik\,\mathrm{Tr}\left(-2G\partial_\sigma F\omega + \nabla^{1,0}_\sigma(G)\omega\right)s,$$

for all $s \in H^0(M_\sigma, \mathcal{L}^k)$.

Lemma 10.3 *We have the following relation*

$$4i\bar{\partial}_\sigma(V'[F]_\sigma) = \mathrm{Tr}(2G(V)\partial(F)\omega - \nabla^{1,0}(G(V))\omega)_\sigma$$

provided $H^1(M,\mathbb{R}) = 0$.

Lemma 10.4 *For any smooth vector field V on $\mathcal{T}$ we have that*

$$(V'[\mathrm{Ric}])^{1,1} = -\frac{1}{2}\partial\,\mathrm{Tr}(\nabla^{1,0}(G(V))\omega). \tag{10.9}$$

Let us here recall how Lemma 10.3 is derived from Lemmas 10.2 and 10.4. By the definition of the Ricci potential,

$$\mathrm{Ric} = \mathrm{Ric}^H + 2id\bar{\partial}F$$

where $\mathrm{Ric}^H = n\omega$ by the assumption. Hence

$$V'[\mathrm{Ric}] = -dV'[I]dF + \frac{1}{2}id\bar{\partial}V'[F]$$

and therefore

$$i\partial\bar{\partial}V'[F] = 2(V'[\mathrm{Ric}])^{1,1} + 2\partial V'[I]\partial F.$$

From the above we conclude that

$$\mathrm{Tr}(2G(V)\partial F\omega - \nabla^{1,0}(G(V))\omega)_\sigma - i\bar{\partial}_\sigma V'[F]_\sigma \in \Omega^{0,1}_\sigma(M)$$

is ∂_σ-closed. By Lemma 10.2 it is also $\bar{\partial}_\sigma$-closed, hence it is a closed one-form on M. But since we assume that $H^1(M,\mathbb{R}) = 0$, we see it is exact, but then it in fact vanishes since it is also of type $(0,1)$ on M_σ.

From the above we conclude that

$$u(V) = \frac{1}{2k+n}\left\{\frac{1}{2}\Delta_{G(V)} - \nabla_{G(V)dF} + 2kV'[F]\right\}$$

solves (10.6) and hence we have established Theorem 10.12.

We arrive at the following theorem:

Theorem 10.13 *Suppose that I is a rigid family of Kähler structures on the symplectic prequantizable compact manifold (M,ω), which satisfies that there exists $n \in \mathbb{Z}$ such that the first Chern class of (M,ω) is $n[\omega] \in H^2(M,\mathbb{Z})$ and $H^1(M,\mathbb{R}) = 0$. Then the Hitchin connection $\hat{\nabla}$ in the bundle $\mathcal{H}^{(k)}$ preserves the subbundle $H^{(k)}$. It is given by*

$$\hat{\nabla}_V = \hat{\nabla}^t_V - u(V)$$

where $\hat{\nabla}^t$ is the trivial connection in $\mathcal{H}^{(k)}$, V is any smooth vector field on $\mathcal{T}$, and $u(V)$ is the second-order differential operator given by (10.7) *and* (10.8).

In Andersen, Gammelgaard, and Lauridsen (2008) we use half-forms and the metaplectic correction to prove the existence of a Hitchin connection in the context of half-form quantization. The assumption that the first Chern class of

(M, ω) is $n[\omega] \in H^2(M, \mathbb{Z})$ is then just replaced by the vanishing of the second Stiefel–Whitney class of M (see Andersen, Gammelgaard, and Lauridsen 2008 for more details).

In Andersen (2006*b*) we prove the following lemma:

Lemma 10.5 *There exist smooth one-forms X_r, Z and functions Y_r, $r = 1, \ldots, R$, on $\mathcal{T}$ with values in $C^\infty(M, T_\sigma)$ such that*

$$\frac{1}{2}\Delta_{G(V)} - \nabla_{G(V)dF} = \sum_{r=1}^{R} \nabla_{X_r(V)}\nabla_{Y_r} + \nabla_{Z(V)} \tag{10.10}$$

for all vector fields V on $\mathcal{T}$.

This gives us the expression

$$u(V) = \frac{1}{2k+n}\left(\sum_{r=1}^{R} \nabla_{X_r(V)}\nabla_{Y_r} + \nabla_{Z(V)} - nV'[F]\right) + V'[F]. \tag{10.11}$$

Suppose Γ is a group which acts by bundle automorphisms of $\mathcal{L}$ over M preserving both the Hermitian structure and the connection in $\mathcal{L}$. Then there is an induced action of Γ on (M, ω). We will further assume that Γ acts on $\mathcal{T}$ and that I is Γ-equivariant. In this case we immediately get the following invariance:

Lemma 10.6 *The natural induced action of Γ on $\mathcal{H}^{(k)}$ preserves the subbundle $H^{(k)}$ and the Hitchin connection.*

We are actually interested in the induced connection $\hat{\nabla}^e$ in the endomorphism bundle $\operatorname{End}(H^{(k)})$. Suppose Φ is a section of $\operatorname{End}(H^{(k)})$. Then for all sections s of $H^{(k)}$ and all vector fields V on $\mathcal{T}$, we have that

$$(\hat{\nabla}^e_V \Phi)(s) = \hat{\nabla}_V \Phi(s) - \Phi(\hat{\nabla}_V(s)).$$

Assume now that we have extended Φ to a section of $\operatorname{Hom}(\mathcal{H}^{(k)}, H^{(k)})$ over $\mathcal{T}$. Then

$$\hat{\nabla}^e_V \Phi = \hat{\nabla}^{e,t}_V \Phi + [\Phi, u(V)] \tag{10.12}$$

where $\hat{\nabla}^{e,t}$ is the trivial connection in the trivial bundle $\operatorname{End}(\mathcal{H}^{(k)})$ over $\mathcal{T}$.

10.3 Toeplitz operators and Berezin–Toeplitz deformation quantization

We shall in this section discuss the Toeplitz operators and their asymptotics as the level k goes to infinity. The properties we need can all be derived from the fundamental work of Boutet de Monvel and Sjöstrand. In Boutet de Monvel and Sjöstrand (1976) they did a microlocal analysis of the Szegö projection, which can be applied to the asymptotic analysis in the situation at hand, as was done by Boutet de Monvel and Guillemin (1981) (in fact in a much more general situation than the one we consider here) and others following them. In particular the applications developed by Schlichenmaier and further by Karabegov and Schlichenmaier to the study of Toeplitz operators in the geometric quantization setting is what will interest us here. Let us first describe the basic setting.

For each $f \in C^\infty(M)$ we consider the prequantum operator, namely, the differential operator $M_f^{(k)} : C^\infty(M, L^k) \to C^\infty(M, L^k)$ given by

$$M_f^{(k)}(s) = fs$$

for all $s \in H^0(M, L^k)$.

These operators act on $C^\infty(M, \mathcal{L}^k)$ and therefore also on the bundle $\mathcal{H}^{(k)}$, however, they do not preserve the subbundle $H^{(k)}$. In order to turn these operators into operators which act on $H^{(k)}$ we need to consider the Hilbert space structure.

Integrating the inner product of two sections against the volume form associated to the symplectic form gives the pre-Hilbert space structure on $C^\infty(M, \mathcal{L}^k)$:

$$\langle s_1, s_2 \rangle = \frac{1}{m!} \int_M (s_1, s_2)\omega^m.$$

We think of this as a pre-Hilbert space structure on the trivial bundle $\mathcal{H}^{(k)}$ which of course is compatible with the trivial connection in this bundle. This pre-Hilbert space structure induces a Hermitian structure $\langle \cdot, \cdot \rangle$ on the finite rank subbundle $H^{(k)}$ of $\mathcal{H}^{(k)}$. The Hermitian structure $\langle \cdot, \cdot \rangle$ on $H^{(k)}$ also induces the operator norm $\| \cdot \|$ on $\mathrm{End}(H^{(k)})$.

Since $H_\sigma^{(k)}$ is a finite-dimensional subspace of $C^\infty(M, \mathcal{L}^k) = \mathcal{H}_\sigma^{(k)}$ and therefore closed, we have the orthogonal projection $\pi_\sigma^{(k)} : \mathcal{H}_\sigma^{(k)} \to H_\sigma^{(k)}$. Since $H^{(k)}$ is a smooth subbundle of $\mathcal{H}^{(k)}$ the projections $\pi_\sigma^{(k)}$ form a smooth map $\pi^{(k)}$ from $\mathcal{T}$ to the space of bounded operators on the L_2 completion of $C^\infty(M, \mathcal{L}^k)$. The easiest way to see this is to consider a local frame for $(s_1, \ldots, s_{\mathrm{Rank}\, H^{(k)}})$ of $H^{(k)}$. Let $h_{ij} = \langle s_i, s_j \rangle$. Let h_{ij}^{-1} be the inverse matrix of h_{ij}. Then

$$\pi_\sigma^{(k)}(s) = \sum_{i,j} \langle s, (s_i)_\sigma \rangle \left(h_{ij}^{-1}\right)_\sigma (s_j)_\sigma. \tag{10.13}$$

From these projections we can construct the Toeplitz operators associated to any smooth function $f \in C^\infty(M)$, $T_{f,\sigma}^{(k)} : \mathcal{H}_\sigma^{(k)} \to H_\sigma^{(k)}$, defined by

$$T_{f,\sigma}^{(k)}(s) = \pi_\sigma^{(k)}(fs)$$

for any element s in $\mathcal{H}_\sigma^{(k)}$ and any point $\sigma \in \mathcal{T}$. We observe that the Toeplitz operators are smooth sections $T_f^{(k)}$ of the bundle $\mathrm{Hom}(\mathcal{H}^{(k)}, H^{(k)})$ and restrict to smooth sections of $\mathrm{End}(H^{(k)})$.

Remark 10.1 Similarly for any pseudo-differential operator A on M with coefficients in $\mathcal{L}^k$ (which may even depend on $\sigma \in \mathcal{T}$), we can consider the associated Toeplitz operator $\pi^{(k)}A$ and think of it as a section of $\mathrm{Hom}(\mathcal{H}^{(k)}, H^{(k)})$. However, whenever we consider asymptotic expansions of such or consider their operator norms, we implicitly restrict them to $H^{(k)}$ and consider them as sections of $\mathrm{End}(H^{(k)})$ or equivalently assume that they have been precomposed with $\pi^{(k)}$.

Suppose we have a smooth section $X \in C^\infty(M, T_\sigma)$ of the holomorphic tangent bundle of M_σ. We then claim that the operator $\pi^{(k)}\nabla_X$ is a zero-order Toeplitz operator. Suppose $s_1 \in C^\infty(M, \mathcal{L}^k)$ and $s_2 \in H^0(M_\sigma, \mathcal{L}^k)$, then we have that

$$X(s_1, s_2) = (\nabla_X s_1, s_2).$$

Now, calculating the Lie derivative along X of $(s_1, s_2)\omega^m$ and using the above, one obtains after integration that

$$\langle \nabla_X s_1, s_2 \rangle = -\langle \Lambda d(i_X \omega) s_1, s_2 \rangle,$$

where Λ denotes contraction with ω. Thus

$$\pi^{(k)}\nabla_X = T^{(k)}_{f_X}, \tag{10.14}$$

as operators from $C^\infty(N, L^k)$ to $H^0(N, L^k)$, where $f_X = -\Lambda d(i_X \omega)$.
Iterating (10.14), we find for all $X_1, X_2 \in C^\infty(M, T_\sigma)$ that

$$\pi^{(k)}\nabla_{X_1}\nabla_{X_2} = T^{(k)}_{f_{X_2} f_{X_1} - X_2(f_{X_1})} \tag{10.15}$$

again as operators from $C^\infty(M, \mathcal{L}^k)$ to $H^0(M_\sigma, \mathcal{L}^k)$.
For $X \in C^\infty(M, T_\sigma)$, the complex conjugate vector field $\bar{X} \in C^\infty(M, \bar{T}_\sigma)$ is a section of the anti-holomorphic tangent bundle, and for $s_1, s_2 \in C^\infty(M, \mathcal{L}^k)$, we have that

$$\bar{X}(s_1, s_2) = (\nabla_{\bar{X}} s_1, s_2) + (s_1, \nabla_X s_2).$$

Computing the Lie derivative along $\bar{X}$ of $(s_1, s_2)\omega^m$ and integrating, we get that

$$\langle \nabla_{\bar{X}} s_1, s_2 \rangle + \langle (\nabla_X)^* s_1, s_2 \rangle = \langle \Lambda d(i_{\bar{X}} \omega) s_1, s_2 \rangle.$$

Hence we see that

$$(\nabla_X)^* = -(\nabla_{\bar{X}} + f_{\bar{X}})$$

as operators on $C^\infty(M, \mathcal{L}^k)$. In particular, we see that

$$\pi^{(k)}(\nabla_X)^*\pi^{(k)} = -T^{(k)}_{f_{\bar{X}}}|_{H^0(M_\sigma, \mathcal{L}^k)} : H^0(M_\sigma, \mathcal{L}^k) \to H^0(M_\sigma, \mathcal{L}^k). \tag{10.16}$$

For two smooth sections X_1, X_2 of the holomorphic tangent bundle T_σ and a smooth function $h \in C^\infty(M)$, we deduce from the formula for $(\nabla_X)^*$ that

$$\pi^{(k)}(\nabla_{X_1})^*(\nabla_{X_2})^* h\pi^{(k)} = \pi^{(k)}\bar{X}_1\bar{X}_2(h)\pi^{(k)} \qquad (10.17)$$
$$+\pi^{(k)} f_{\bar{X}_1}\bar{X}_2(h)\pi + \pi f_{\bar{X}_2}\bar{X}_1(h)\pi^{(k)}$$
$$+\pi^{(k)}\bar{X}_1(f_{\bar{X}_2})h\pi + \pi f_{\bar{X}_1} f_{\bar{X}_2} h\pi^{(k)}$$

as operators on $H^0(M_\sigma, \mathcal{L}^k)$.

The product of two Toeplitz operators associated to two smooth functions will in general not be the Toeplitz operator associated to a smooth function again, but there is an asymptotic expansion of the product in terms of such Toeplitz operators on a compact Kähler manifold by the results of Schlichenmaier (1998, 2000 and 2001).

Theorem 10.14 (Schlichenmaier) *For any pair of smooth functions $f_1, f_2 \in C^\infty(M)$, we have an asymptotic expansion*

$$T^{(k)}_{f_1,\sigma} T^{(k)}_{f_2,\sigma} \sim \sum_{l=0}^{\infty} T^{(k)}_{c^{(l)}_\sigma(f_1,f_2),\sigma} k^{-l},$$

where $c^{(l)}_\sigma(f_1, f_2) \in C^\infty(M)$ are uniquely determined since $\sim$ means the following: For all $L \in \mathbb{Z}_+$ we have that

$$\|T^{(k)}_{f_1,\sigma} T^{(k)}_{f_2,\sigma} - \sum_{l=0}^{L} T^{(k)}_{c^{(l)}_\sigma(f_1,f_2),\sigma} k^{-l}\| = O(k^{-(L+1)})$$

uniformly over compact subsets of $\mathcal{T}$. Moreover, $c^{(0)}_\sigma(f_1, f_2) = f_1 f_2$.

Remark 10.2 It will be useful for us to define new coefficients $\tilde{c}^{(l)}_\sigma(f, g) \in C^\infty(M)$ which correspond to the expansion of the product in $1/(2k+n)$ (where n is some fixed integer):

$$T^{(k)}_{f_1,\sigma} T^{(k)}_{f_2,\sigma} \sim \sum_{l=0}^{\infty} T^{(k)}_{\tilde{c}^{(l)}_\sigma(f_1,f_2),\sigma} (2k+n)^{-l}.$$

Theorem 10.14 is proved in Schlichenmaier (1998, 2000), where it is also proved that the formal generating series for the $c^{(l)}_\sigma(f_1, f_2)$'s gives a formal deformation quantization[4] of the symplectic manifold (M, ω).

We recall the definition of a formal deformation quantization. Introduce the space of formal functions $C^\infty_h(M) = C^\infty(M)[[h]]$ as the space of formal power series in the variable h with coefficients in $C^\infty(M)$. Let $\mathbb{C}_h = \mathbb{C}[[h]]$.

Definition 10.5 *A deformation quantization of (M, ω) is an associative product $*$ on $C^\infty_h(M)$ which respects the $\mathbb{C}_h$-module structure. It is determined by a*

[4] We have the opposite sign convention on the curvature, which means our c_l are $(-1)^l c_l$ in Schlichenmaier (1998, 2001).

sequence of bilinear operators

$$c^{(l)} : C^\infty(M) \otimes C^\infty(M) \to C^\infty(M)$$

defined by

$$f \star g = \sum_{l=0}^{\infty} c^{(l)}(f,g)h^l,$$

where $f, g \in C^\infty(M)$. The deformation quantization is said to be differential if the operators $c^{(l)}$ are bidifferential operators. Considering the symplectic action of Γ on (M, ω), we say that a $*$-product $*$ is Γ-invariant if

$$\gamma^*(f * g) = \gamma^*(f) * \gamma^*(g)$$

for all $f, g \in C^\infty(M)$ and all $\gamma \in \Gamma$.

Theorem 10.15 (Karabegov and Schlichenmaier 2001) *The product $\star_\sigma^{BT}$ given by*

$$f \star_\sigma^{BT} g = \sum_{l=0}^{\infty} (-1)^l c_\sigma^{(l)}(f,g)h^l,$$

where $f, g \in C^\infty(M)$ and $c_\sigma^{(l)}(f,g)$ are determined by Theorem 10.14, is a differentiable deformation quantization of (M, ω).

Definition 10.6 *The Berezin–Toeplitz deformation quantization of the compact Kähler manifold (M_σ, ω) is the product $\star_\sigma^{BT}$.*

Remark 10.3 Let Γ_σ be the σ-stabilizer subgroup of Γ. For any element $\gamma \in \Gamma_\sigma$, we have that

$$\gamma^*\left(T_{f,\sigma}^{(k)}\right) = T_{\gamma^* f,\sigma}^{(k)}.$$

This implies the invariance of $\star_\sigma^{\mathrm{BT}}$ under the σ-stabilizer Γ_σ.

Remark 10.4 We define a new $*$-product by

$$f \tilde{\star}_\sigma^{\mathrm{BT}} g = \sum_{l=0}^{\infty} (-1)^l \tilde{c}_\sigma^{(l)}(f,g)h^l.$$

Then

$$f \tilde{\star}_\sigma^{\mathrm{BT}} g = \left((f \circ \phi^{-1}) \star_\sigma^{\mathrm{BT}} (g \circ \phi^{-1})\right) \circ \phi$$

for all $f, g \in C_h^\infty(M)$, where $\phi(h) = \frac{h}{2+nh}$.

10.4 The formal Hitchin connection

In this section we review the construction of the formal Hitchin connection as defined and constructed in Andersen (2006*b*).

We assume the conditions on (M,ω) and I of Theorem 10.12, thus providing us with a Hitchin connection $\hat{\nabla}$ in $H^{(k)}$ over $\mathcal{T}$ and the associated connection $\hat{\nabla}^e$ in $\mathrm{End}(H^{(k)})$. Let $\mathcal{D}(M)$ be the space of smooth differential operators on M acting on smooth functions on M. Let C_h be the trivial $C_h^\infty(M)$-bundle over $\mathcal{T}$.

Definition 10.7 *A formal connection D is a connection in C_h over $\mathcal{T}$ of the form*

$$D_V f = V[f] + \tilde{D}(V)(f),$$

where $\tilde{D}$ is a smooth one-form on $\mathcal{T}$ with values in $\mathcal{D}_h(M) = \mathcal{D}(M)[[h]]$, f is any smooth section of C_h, V is any smooth vector field on $\mathcal{T}$, and $V[f]$ is the derivative of f in the direction of V.

For a formal connection we get the series of differential operators $\tilde{D}^{(l)}$ given by

$$\tilde{D}(V) = \sum_{l=0}^{\infty} \tilde{D}^{(l)}(V)h^l.$$

From Hitchin's connection in $H^{(k)}$ we get an induced connection $\hat{\nabla}^e$ in the endomorphism bundle $\mathrm{End}(H^{(k)})$. The Teoplitz operators are not covariant constant sections with respect to $\hat{\nabla}^e$. They are so asymptotically in k in the following very precise sense.

Theorem 10.16 *There is a unique formal Hitchin connection D which satisfies that*

$$\hat{\nabla}^e_V T_f^{(k)} \sim T^{(k)}_{(D_V f)(1/(2k+n))} \tag{10.18}$$

for all smooth section f of C_h and all smooth vector fields on $\mathcal{T}$. Moreover

$$\tilde{D} = 0 \mod h.$$

Here $\sim$ means the following: For all $L \in \mathbb{Z}_+$ we have that

$$\left\| \hat{\nabla}^e_V T_f^{(k)} - \left(T^{(k)}_{V[f]} + \sum_{l=1}^{L} T^{(k)}_{\tilde{D}_V^{(l)} f} \frac{1}{(2k+n)^l} \right) \right\| = O(k^{-(L+1)})$$

uniformly over compact subsets of $\mathcal{T}$ for all smooth maps $f : \mathcal{T} \to C^\infty(M)$.

We call this formal connection D the *formal Hitchin connection.* Since $\tilde{D} = 0 \mod h$ we see that if we now fix $f \in C^\infty(M)$ then we get that

$$\left\| \hat{\nabla}^e_V T_f^{(k)} \right\| = O(k^{-1}).$$

The proof of this theorem is given in Andersen (2006*b*). Let us recall formula (19) from Andersen (2006*b*):

$$\tilde{D}(V)(f) = V'[F]f - V'[F]\tilde{\star}^{\mathrm{BT}} f + h(E(V)(f) - H(V)\tilde{\star}^{\mathrm{BT}} f) \tag{10.19}$$

where E is the one-form on $\mathcal{T}$ with values in $\mathcal{D}(M)$ such that

$$T^{(k)}_{E(V)f} = \pi^{(k)} o(V)^* f \pi^{(k)} + \pi^{(k)} f o(V) \pi^{(k)}$$

and H is the one-form on $\mathcal{T}$ with values in $C^\infty(M)$ such that $H(V) = E(V)(1)$.

Lemma 10.7 *The formal operator D_V is a derivation for $\star^{BT}_\sigma$ for each $\sigma \in \mathcal{T}$, that is,*

$$D_V(f \tilde{\star}^{BT} g) = D_V(f) \tilde{\star}^{BT} g + f \tilde{\star}^{BT} D_V(g)$$

for all $f, g \in C^\infty(M)$.

Again this lemma is proved in Andersen (2006*b*), and in fact it follows basically from

$$\hat{\nabla}^e_V \left(T^{(k)}_f T^{(k)}_g \right) = \hat{\nabla}^e_V \left(T^{(k)}_f \right) T^{(k)}_g + T^{(k)}_f \hat{\nabla}^e_V \left(T^{(k)}_g \right).$$

If the Hitchin connection is projectively flat, then the induced connection in the endomorphism bundle is flat and hence so is the formal Hitchin connection by proposition 3 of Andersen (2006*b*).

Definition 10.8 *A formal trivialization of a formal connection D is a smooth map $P : \mathcal{T} \to \mathcal{D}_h(M)$ which modulo h is an isomorphism for all $\sigma \in \mathcal{T}$ and such that*

$$D_V(P(f)) = 0$$

for all vector fields V on $\mathcal{T}$ and all $f \in C^\infty_h(M)$.

Clearly if D is not flat, such a formal trivialization will not exist even locally on $\mathcal{T}$. However, if D is flat, then we have the following result.

Proposition 10.1 *Assume that D is flat and that $\tilde{D} = 0$ mod h. Then locally around any point in $\mathcal{T}$ there exists a formal trivialization. If $H^1(\mathcal{T}, \mathbb{R}) = 0$ then there exist a formal trivialization defined globally on $\mathcal{T}$. If further $H^1_\Gamma(\mathcal{T}, D(M)) = 0$ then we can construct P such that it is Γ-equivariant.*

In this proposition $H^1_\Gamma(\mathcal{T}, D(M))$ simply refers to the Γ-equivariant first de Rham cohomology of $\mathcal{T}$ with coefficients in the real Γ-vector space $D(M)$.

Now suppose we have a formal trivialization P of the formal Hitchin connection D. Then P is constant mod h and we may and will assume that

$$P = \mathrm{Id} \mod h.$$

We can then define a new smooth family of $\star$-products parametrized by $\mathcal{T}$ as follows

$$f \star_\sigma g = P_\sigma^{-1}(P_\sigma(f) \tilde{\star}^{\mathrm{BT}}_\sigma P_\sigma(g))$$

for all $f, g \in C^\infty(M)$ and all $\sigma \in \mathcal{T}$.

Proposition 10.2 *The $\star$-products $\star_\sigma$ are independent of $\sigma \in \mathcal{T}$.*

Theorem 10.17 *Assume that the formal Hitchin connection D is flat and*

$$H^1_\Gamma(\mathcal{T}, D(M)) = 0,$$

then there is a Γ-invariant trivialization P of D and the $*$-product

$$f \star g = P_\sigma^{-1}(P_\sigma(f) \star_\sigma^{BT} P_\sigma(g))$$

is independent of $\sigma \in \mathcal{T}$ and Γ-invariant. If $H^1_\Gamma(\mathcal{T}, C^\infty(M)) = 0$ and the commutant of Γ in $D(M)$ is trivial, then a Γ-invariant differential $*$-product on M is unique.

This theorem is proved in Andersen (2006*a*).

Let $P^{(r)} : \mathcal{T} \to \mathcal{D}(M)[r]$ be defined by

$$P \equiv P^{(r)} \mod h^{r+1}.$$

Proposition 10.3 *For any $f \in C^\infty(M)$ and any vector field V on $\mathcal{T}$ with compact support we have that*

$$\| \nabla_V T^{(k)}_{P^{(r)}(f)(\frac{1}{k})} \| = O(k^{-r-1}).$$

Proof. This follows directly from the definition of a formal trivialization combined with Theorem 10.16. □

10.5 Asymptotic flatness of Toeplitz operators

In this section we will establish that the Toeplitz operators associated to functions on M are asymptotically (in k) covariant constant sections of $(\mathrm{End}(\mathcal{V}^{(k)}), \hat{\nabla}^e)$.

The setting is as in Section 10.2, that means we are under the assumptions of Theorem 10.13. Then we have the following theorem:

Theorem 10.18 *Let σ_0 and σ_1 be two points in $\mathcal{T}$ and P_{σ_0,σ_1} be the parallel transport in the bundle $(\mathrm{End}(\mathcal{V}^{(k)}), \hat{\nabla}^e)$ from σ_0 to σ_1 with respect to $\hat{\nabla}^e$. Then*

$$\|P_{\sigma_0,\sigma_1} T^{(k)}_{f,\sigma_0} - T^{(k)}_{f,\sigma_1}\| = O(k^{-1}),$$

where $\|\cdot\|$ is the operator norm on $H^0\left(M_{\sigma_1}, \mathcal{L}^k_{\sigma_1}\right)$.

As preparation for the proof of this theorem, we need to discuss a certain Hermitian structure on $H^{(k)}$ over $\mathcal{T}$, which is asymptotically flat with respect to Hitchin's connection. Namely, we let

$$\langle s_1, s_2 \rangle_F = \frac{1}{m!} \int_M (s_1, s_2) e^{-F} \omega^m. \tag{10.20}$$

Along any smooth one-parameter curve σ_t in $\mathcal{T}$, we have that

$$\frac{d}{dt}\langle s_1, s_2\rangle_F = \langle \hat{\nabla}^t_{\sigma'_t} s_1, s_2\rangle_F + \langle s_1, \hat{\nabla}^t_{\sigma'_t} s_2\rangle_F - \left\langle \frac{\partial F}{\partial t} s_1, s_2 \right\rangle_F.$$

Let

$$E(s) = \frac{d}{dt}|s|^2_F - \left\langle \hat{\nabla}_{\sigma'_t} s, s\right\rangle_F - \left\langle s, \hat{\nabla}_{\sigma'_t} s\right\rangle_F.$$

Recall that $\hat{\nabla}_v = \hat{\nabla}^t_v - u(v)$ and $u(v) = \frac{1}{2k+n} o(v) + v'[F]$, hence we have for all sections s of $\mathcal{V}_k$ that

$$E(s) = \frac{1}{2k+n}\left(\langle \pi e^{-F} o\left(\sigma'_t\right) s, s\rangle + \langle s, \pi e^{-F} o\left(\sigma'_t\right) s\rangle\right).$$

Hence by combining Theorem 10.3, (10.10), (10.14), and (10.15) we have proved that

Proposition 10.4 *The Hermitian structure (10.20) is asymptotically flat with respect to the connections $\hat{\nabla}$, that is, for any compact subset K of $\mathcal{T}$, there exists a constant C such that for all sections s of $\mathcal{V}_k$ over K, we have that*

$$|E(s)| \leq \frac{C}{k+n}|s|^2_F$$

over K.

We note that this proposition implies the same proposition for sections of $\mathrm{End}(\mathcal{V}_k)$ with respect to the induced Hermitian structure on $\mathrm{End}(\mathcal{V}_k) = \mathcal{V}_k^* \otimes \mathcal{V}_k$, which we also denote $\langle \cdot, \cdot \rangle_F$. We denote the analogous quantity to E for the endomorphism bundle by E_e.

Lemma 10.8 *The Hermitian structure on $\mathcal{H}_k$*

$$\langle s_1, s_2\rangle_F = \frac{1}{m!}\int_M (s_1, s_2) e^{-F}\omega^m$$

and the constant L_2-Hermitian structure on $\mathcal{H}_k$

$$\langle s_1, s_2\rangle = \frac{1}{m!}\int_M (s_1, s_2)\omega^m$$

are equivalent uniformly in k when restricted to $\mathcal{V}_k$ over any compact subset K of $\mathcal{T}$.

The proof of Lemma 10.8 can be found in section 5 of Andersen (2006*a*).

Proof. (Of Theorem 10.12). Let σ_t, $t \in J$, be a smooth one-parameter family of complex structures such that σ_t is a curve in $\mathcal{T}$ between the two points in question. By Lemma 10.8 the Hermitian structures $\langle \cdot, \cdot \rangle$ and $\langle \cdot, \cdot \rangle_F$ on $\mathcal{V}_k$ are equivalent uniformly in k over compact subsets of $\mathcal{T}$. It follows that there exists constants C_1 and C_2, such that we have the following inequalities over the image of σ_t in $\mathcal{T}$ for the operator norm $\|\cdot\|$ and the norm $|\cdot|_F$ on $\mathrm{End}(\mathcal{V}_k)$:

$$\|\cdot\| \le C_1|\cdot|_F \le C_2\sqrt{P_{g,n}(k)}\|\cdot\|, \tag{10.21}$$

where $P_{g,n}(k)$ is the rank of $\mathcal{V}_k$ given by the Verlinde formula. By the Riemann–Roch theorem this is a polynomial in k of degree m.

Because of these inequalities, we choose an integer r bigger than $m/2$ and let $f_r = P^{(r)}(f)\left(\frac{1}{k}\right)$ where P is a trivialization of the formal Hitchin connection along σ_t, $t\in J$, which clearly exists and is unique under the condition that $P_\sigma = \mathrm{Id}$ by Proposition 10.1. Then let $n_k : J \to [0,\infty)$ be given by

$$n_k(t) = |\Theta_k(t)|_F^2$$

where

$$\Theta_k(t) : \mathcal{V}_{k,\sigma_t} \to \mathcal{V}_{k,\sigma_t}$$

is given by

$$\Theta_k(t) = P_{\sigma_0,\sigma_t} T^{(k)}_{(f_r)_0,\sigma_0} - T^{(k)}_{(f_r)_t,\sigma_t}.$$

The functions n_k are differentiable in t and we compute that

$$\begin{aligned}\frac{dn_k}{dt} &= \langle \hat\nabla^e_{\sigma'_t}(\Theta_k(t)), \Theta_k(t)\rangle_F + \langle \Theta_k(t), \hat\nabla^e_{\sigma'_t}(\Theta_k(t))\rangle_F + E_e(\Theta_k(t))\\ &= -\langle \hat\nabla^e_{\sigma'_t} T^{(k)}_{(f_r)_t,\sigma_t}, \Theta_k(t)\rangle_F - \langle \Theta_k(t), \hat\nabla^e_{\sigma'_t} T^{(k)}_{(f_r)_t,\sigma_t}\rangle_F + E_e(\Theta_k(t)).\end{aligned}$$

Using the above, we get the following estimate:

$$\begin{aligned}\left|\frac{dn_k}{dt}\right| &\le 2|\hat\nabla^e_{\sigma'_t} T^{(k)}_{(f_r)_t,\sigma_t}|_F |\Theta_k(t)|_F + |E_e(\Theta_k(t))|\\ &\le 2C_2\sqrt{P_{g,n}(k)}\|\hat\nabla^e_{\sigma'_t} T^{(k)}_{(f_r)_t,\sigma_t}\| n_k^{1/2} + |E_e(\Theta_k(t))|.\end{aligned}$$

Consequently we can apply Propositions 10.3 and 10.4 to obtain that there exists a constant C such that

$$\left|\frac{dn_k}{dt}\right| \le \frac{C}{k}(n_k^{1/2} + n_k).$$

This estimate implies that

$$n_k(t) \le \left(\exp\left(\frac{Ct}{2k}\right) - 1\right)^2.$$

But by (10.21) we get that

$$\left\|P_{\sigma_0,\sigma_1} T^{(k)}_{(f_r)_0,\sigma_0} - T^{(k)}_{(f_r)_1,\sigma_1}\right\| = \|\Theta_k(1)\| \le C_1 n_k(1)^{1/2}.$$

The theorem then follows from these two estimates, since $(f_r)_0 = f$ and

$$\left\| T^{(k)}_{(f_r)_1,\sigma_1} - T^{(k)}_{f,\sigma_1} \right\| = O(k^{-1}).$$

□

10.6 General asymptotic faithfulness

In this section we will review how the asymptotics of Toeplitz operators can be used to prove asymptotic faithfulness of certain representations, as discussed in the gauge theory setting in Andersen (2006*a*). However, we shall discuss the most general setting in which the argument from Andersen (2006*a*) applies.

In addition to the assumptions of Theorem 10.12, we will further assume, throughout this section, that the Hitchin connection $\hat{\nabla}$ is projectively flat and that we have a symmetry group acting on our set-up as discussed in Section 10.2.

We thus get an induced flat connection in the bundle $\mathbb{P}(\mathcal{V}_k)$. This then gives us a projective representation of the mapping class group

$$\rho_k : \Gamma \to \mathrm{Aut}(\mathbb{P}(V_k)),$$

where $\mathbb{P}(V_k)$ denotes the covariant constant sections of $\mathbb{P}(\mathcal{V}_k)$ over $\mathcal{T}$.

First we establish the following proposition:

Proposition 10.5 *For any $\phi \in \Gamma$, $f \in C^\infty(M)$ and $\sigma \in \mathcal{T}$, we have that*

$$\lim_{k\to\infty} \left\| T^{(k)}_{f,\sigma} - \rho_k(\phi) T^{(k)}_{f,\sigma} \rho_k(\phi^{-1}) \right\| = \lim_{k\to\infty} \left\| T^{(k)}_{(f - f\circ\phi),\sigma} \right\|.$$

Proof. Suppose we have a $\phi \in \Gamma$. Then ϕ induces a symplectomorphism of M which we also just denote ϕ and we get the following commutative diagram for any $f \in C^\infty(M)$:

$$\begin{array}{ccccc}
H^0\left(M_\sigma, \mathcal{L}^k_\sigma\right) & \xrightarrow{\phi^*} & H^0\left(M_{\phi(\sigma)}, \mathcal{L}^k_{\phi(\sigma)}\right) & \xrightarrow{P_{\phi(\sigma),\sigma}} & H^0\left(M_\sigma, \mathcal{L}^k_\sigma\right) \\
\downarrow T^{(k)}_{f,\sigma} & & \downarrow T^{(k)}_{f\circ\phi,\phi(\sigma)} & & \downarrow P_{\phi(\sigma),\sigma} T^{(k)}_{f\circ\phi,\phi(\sigma)} \\
H^0\left(M_\sigma, \mathcal{L}^k_\sigma\right) & \xrightarrow{\phi^*} & H^0\left(M_{\phi(\sigma)}, \mathcal{L}^k_{\phi(\sigma)}\right) & \xrightarrow{P_{\phi(\sigma),\sigma}} & H^0\left(M_\sigma, \mathcal{L}^k_\sigma\right),
\end{array}$$

where $P_{\phi(\sigma),\sigma} : H^0\left(M_{\phi(\sigma)}, \mathcal{L}^k_{\phi(\sigma)}\right) \to H^0\left(M_\sigma, \mathcal{L}^k_\sigma\right)$ on the horizontal arrows refers to parallel transport in the bundle $\mathcal{V}^{(k)}$, whereas $P_{\phi(\sigma),\sigma}$ refers to the parallel transport in the endomorphism bundle $\mathrm{End}(\mathcal{V}^{(k)})$ in the last vertical arrow. By the definition of ρ_k, we see that

$$\rho_k(\phi) T^{(k)}_{f,\sigma} \rho_k(\phi^{-1}) = P_{\phi(\sigma),\sigma} T^{(k)}_{f\circ\phi,\phi(\sigma)}.$$

By Theorem 10.18 we get that

$$\begin{aligned}\lim_{k\to\infty}\left\|T^{(k)}_{(f-f\circ\phi),\sigma}\right\| &= \lim_{k\to\infty}\left\|T^{(k)}_{f,\sigma} - T^{(k)}_{f\circ\phi,\sigma}\right\| \\ &= \lim_{k\to\infty}\left\|T^{(k)}_{f,\sigma} - P_{\phi(\sigma),\sigma}T^{(k)}_{f\circ\phi,\phi(\sigma)}\right\| \\ &= \lim_{k\to\infty}\left\|T^{(k)}_{f,\sigma} - \rho_k(\phi)T^{(k)}_{f,\sigma}\rho_k(\phi^{-1})\right\|.\end{aligned}$$

□

From this proposition we get the following generalization of asymptotic faithfulness:

Theorem 10.19 *For any $\phi \in \Gamma$, we have that*

$$\phi \in \bigcap_{k=1}^{\infty} \ker \rho_k$$

if and only if ϕ induces the identity on M.

Proof. Suppose $\phi \in \bigcap_{k=1}^{\infty} \ker \rho_k$, then by Proposition 10.5 and Bordemann, Meinrenken, and Schlichenmaier's Theorem 10.3 we see that

$$f \circ \phi = f$$

for all $f \in C^\infty(M)$, hence we must have that ϕ acts by the identity on M. □

We also get the following generalized reducibility property:

Theorem 10.20 *For any $\phi \in \Gamma$, $f \in C^\infty(M)$ and $\sigma \in \mathcal{T}$, we have that*

$$\lim_{k\to\infty}\left\|[\rho_k(\phi), T^{(k)}_{f,\sigma}]\right\| = 0$$

if and only if

$$f \circ \phi = f.$$

Proof. Let $\phi \in \Gamma$ and assume that

$$\lim_{k\to\infty}\left\|[\rho_k(\phi), T^{(k)}_{h_\gamma,\sigma}]\right\| = 0$$

for some $f \in C^\infty(M)$. We have that

$$\begin{aligned}\left\|T^{(k)}_{f,\sigma} - \rho_k(\phi^{-1})T^{(k)}_{f,\sigma}\rho_k(\phi)\right\| &= \left\|\rho_k(\phi^{-1})\rho_k(\phi)\left(T^{(k)}_{f,\sigma} - \rho_k(\phi^{-1})T^{(k)}_{f,\sigma}\rho_k(\phi)\right)\right\| \\ &\leq \left\|\rho_k(\phi^{-1})\right\|\left\|\left[\rho_k(\phi), T^{(k)}_{f,\sigma}\right]\right\|.\end{aligned}$$

Lemma 10.8 and Proposition 10.4 give a uniform bound on $\|\rho_k(\phi^{-1})\|$. Hence Proposition 10.5 implies that

$$\lim_{k\to\infty} \left\| T^{(k)}_{(f-f\circ\phi^{-1}),\sigma} \right\| = 0.$$

Then by Bordemann, Meinrenken, and Schlichenmaier's Theorem 10.3, we must have that $f = f \circ \phi^{-1}$. The converse follows since we have that

$$\left\| \left[\rho_k(\phi), T^{(k)}_{h_\gamma,\sigma}\right] \right\| \leq \|\rho_k(\phi)\| \left\| T^{(k)}_{f,\sigma} - \rho_k(\phi^{-1}) T^{(k)}_{f,\sigma} \rho_k(\phi) \right\|.$$

□

Acknowledgement

I would like to take this opportunity to thank Professor Nigel Hitchin for introducing me to the subject at hand and further for the many very helpful discussions we have had over the years regarding this interesting and facinating research field of mathematics bordering on Mathematical Physics.

References

Andersen, J. E. (1998). New polarizations on the moduli space and the Thurston compactification of Teichmuller space. *Int. J. Math.* **9**(1), 1–45.

Andersen, J. E. (2005). Geometric quantization and deformation quantization of abelian moduli spaces. *Commun. Math. Phys.* **2005**, 727–45.

Andersen, J. E. (2006*a*). Asymptotic faithfulness of the quantum $SU(n)$ representations of the mapping class groups. *Ann. Math.* **163**, 347–68.

Andersen, J. E. (2006*b*). Hitchin's connection, Toeplitz operators and symmetry invariant deformation quantization. math.DG/0611126.

Andersen, J. E. (2006*c*). The Nielsen-Thurston classification of mapping classes is determined by TQFT. math.QA/0605036.

Andersen, J. E. (2007). Mapping class groups do not have Kazhdan's property (T), math.QA/0706.2184.

Andersen, J. E. (2008*a*). Asymptotic faithfulness of the quantum $SU(n)$ representations of the mapping class groups in the singular case. In preparation.

Andersen, J. E. (2008*b*). Asymptotic in teichmuller space of the Hitchin connection. In preparation.

Andersen, J. E. and Christ, M. (2008). Asymptotic expansion of the Szegö kernel on singular algebraic varieties. In preparation.

Andersen, J. E., Gammelgaard, N. L., and Lauritsen, M. (2008). Hitchin's connection in half-form quantization. math.DG/0711.3995.

Andersen, J. E., Masbaum, G., and Ueno, K. (2006). Topological quantum field theory and the Nielsen-Thurston classification of $M(0,4)$. *Math. Proc. Camb. Philos. Soc.* **141**(3), 477–88.

Andersen, J. E., Mattes, J., and Reshetikhin, N. (1996). The Poisson Structure on the moduli space of flat connections and chord diagrams. *Topology* **35**, 1069–83.

Andersen, J. E., Mattes, J., and Reshetikhin, N. (1998). Quantization of the algebra of chord diagrams. *Math. Proc. Camb. Phil. Soc.* **124**, 451–67.

Andersen, J. E. and Ueno, K. (2006). Modular functors are determined by their genus zero data. math.QA/0611087.

Andersen, J. E. and Ueno, K. (2007*a*). Abelian conformal field theories and determinant bundles. *Int. J. Math.* **18**, 919–93.

Andersen, J. E. and Ueno, K. (2007*b*). Constructing modular functors from conformal field theories. *J. Knot Theory Ramifications* **16**(2), 127–202.

Andersen, J. E. and Ueno, K. (2008). Construction of the Reshetikhin-Turaev TQFT from conformal field theory. In preparation.

Atiyah, M. (1990). The Jones-Witten invariants of knots. *Astérisque* (189–190), Exp. No. 715, 7–16. Séminaire Bourbaki, Vol. 1989/90.

Axelrod, S., Della Pietra, S., and Witten, E. (1991). Geometric quantization of Chern-Simons gauge theory. *J. Diff. Geom.* **33**(3), 787–902.

Bakalov, B. and Kirillov, A. (2000). *Lectures on Tensor Categories and Modular Functors*, Vol. 21. AMS University Lecture Series, Providence, RI.

Bekka, B., de la Harpe, P., and Valette, A. (2007). *Kazhdan's Property (T)*. Cambridge University Press. In press.

Blanchet, C. (2000). Hecke algebras, modular categories and 3-manifolds quantum invariants. *Topology* **39**(1), 193–223.

Blanchet, C., Habegger, N., Masbaum, G., and Vogel, P. (1992). Three-manifold invariants derived from the Kauffman bracket. *Topology* **31**, 685–99.

Blanchet, C., Habegger, N., Masbaum, G., and Vogel, P. (1995). Topological quantum field theories derived from the Kauffman bracket. *Topology* **34**, 883–927.

Bleiler, S. and Casson, A. (1988). *Automorphisms of Sufaces after Nielsen and Thurston*. Cambridge University Press, Cambridge, UK.

Bordeman, M., Meinrenken, E., and Schlichenmaier, M. (1994). Toeplitz quantization of Kähler manifolds and $gl(N), N \to \infty$ limit. *Commun. Math. Phys.* **165**, 281–96.

Boutet de Monvel, L. and Guillemin, V. (1981). *The Spectral Theory of Toeplitz Operators*, Vol. 99 of *Ann. Math. Stud.* Princeton University Press, Princeton, NJ.

Boutet de Monvel, L. and Sjöstrand, J. (1976). Sur la singularité des noyaux de Bergmann et de Szegö. *Asterique* **34–35**, 123–64.

Faltings, G. (1993). Stable G-bundles and projective connections. *J. Alg. Geom.* **2**, 507–68.

Fathi, A., Laudenbach, F., and Poénaru, V. (1991/1979). Travaux de Thurston sur les surfaces, *Astérisque* **66–67**.

Finkelberg, M. (1996). An equivalence of fusion categories. *Geom. Funct. Anal.* **6**, 249–67.

Freedman, M. H., Walker, K., and Wang, Z. (2002). Quantum SU(2) faithfully detects mapping class groups modulo center. *Geom. Topol.* **6**, 523–39.

Hitchin, N. (1990). Flat connections and geometric quantization. *Commun. Math. Phys.* **131**, 347–80.

Kac, V. G. (1995). *Infinite Dimensional Lie Algebras*, 3 edn. Cambridge University Press, Cambridge, UK.

Karabegov, A. V. and Schlichenmaier, M. (2001). Identification of Berezin–Toeplitz deformation quantization. *J. Reine Angew. Math.* **540**, 49–76.

Kazhdan, D. (1967). Connection of the dual space of a group with the structure of its closed subgroups. *Funct. Anal. Appl.* **1**, 64–65.

Kazhdan, D. and Lusztig, G. (1993). Tensor structures arising from affine Lie algebras. I, II. *J. Am. Math. Soc.* **6**(4), 905–47, 949–1011.

Kazhdan, D. and Lusztig, G. (1994*a*). Tensor structures arising from affine Lie algebras. III. *J. Am. Math. Soc.* **7**(2), 335–81.

Kazhdan, D. and Lusztig, G. (1994*b*). Tensor structures arising from affine Lie algebras. IV. *J. Am. Math. Soc.* **7**(2), 383–453.

Laszlo, Y. (1998). Hitchin's and WZW connections are the same. *J. Diff. Geom.* **49**(3), 547–76.

Moore, G. and Seiberg, N. (1989). Classical and quantum conformal field theory. *Commun. Math. Phys.* **123**, 177–254.

Reshetikhin, N. and Turaev, V. (1990). Ribbon graphs and their invariants derived fron quantum groups. *Commun. Math. Phys.* **127**, 1–26.

Reshetikhin, N. and Turaev, V. (1991). Invariants of 3-manifolds via link polynomials and quantum groups. *Invent. Math.* **103**, 547–97.

Roberts, J. (2001). Irreducibility of some quantum representations of mapping class groups. *J. Knot Theory Ramifications* **10**, 763–7.

Schlichenmaier, M. (1998). Berezin-Toeplitz quantization and conformal field theory. PhD thesis.

Schlichenmaier, M. (2000). Deformation quantization of compact Kähler manifolds by Berezin-Toeplitz quantization. In *Conférence Moshé Flato 1999, Vol. II (Dijon)*, Vol. 22 of *Math. Phys. Stud.* pp. 289–306. Kluwer Academic Publication, Dordrecht, the Netherlands.

Schlichenmaier, M. (2001). Berezin-Toeplitz quantization and Berezin transform. In *Long Time Behaviour of Classical and Quantum Systems (Bologna, 1999)*, Vol. 1 of *Ser. Concr. Appl. Math.*, pp. 271–87. World Scientific Publication, River Edge, NJ.

Segal, G. (1992). *The Definition of Conformal Field Theory.* Oxford University Preprint.

Thurston, W. (1988). On the geometry and dynamics of diffeomorphisms of surfaces. *Bull. Am. Math. Soc.* **19**, 417–31.

Tsuchiya, A., Ueno, K., and Yamada, Y. (1989). Conformal field theory on universal family of stable curves with gauge symmetries. *Adv. Stud. Pure Math.* **19**, 459–566.

Turaev, V. G. (1994). *Quantum Invariants of Knots and 3-Manifolds*, Vol. 18 of *de Gruyter Stud. Math.* Walter de Gruyter & Co., Berlin, Germany.

Walker, K. (1991). On Witten's 3-manifold invariants. Preliminary version #2, preprint.

Witten, E. (1989). Quantum field theory and the Jones polynomial. *Commun. Math. Phys.* **121**, 351–98.

Woodhouse, N. J. (1992). *Geometric Quantization.* Oxford University Press, Oxford.

XI

TOWARDS A NON-LINEAR SCHWARZ'S LIST

Philip Boalch

Dedicated to Nigel Hitchin for his 60th birthday

11.1 Introduction

The main theme of this chapter is 'icosahedral' solutions of (ordinary) differential equations, a topic that seems suitable for a 60th birthday conference. We will however try to go beyond the icosahedron, to see what comes next, and consider various symmetry groups each of which could be thought of as the next in a sequence, following the icosahedral group.

To fix ideas let us give a classical example. Recall the icosahedral rotation group of order 60:

$$A_5 \cong \mathrm{PSL}_2(\mathbb{F}_5) \cong \Delta_{235} \cong \langle\ a, b, c \ \big|\ a^2 = b^3 = c^5 = abc = 1\ \rangle.$$

This is described via three generators a, b, and c whose product is the identity, and so it is natural to look for ordinary differential equations (ODEs) on the three-punctured sphere $\mathbb{P}^1(\mathbb{C}) \setminus \{0, 1, \infty\}$ with monodromy group A_5. Now A_5 is a three-dimensional rotation group so naturally lives in $\mathrm{SO}_3(\mathbb{R})$ which is a subgroup of $\mathrm{SO}_3(\mathbb{C})$ which is isomorphic to $\mathrm{PSL}_2(\mathbb{C})$. Thus we are led to search for connections

$$\nabla = d - \left(\frac{A_1}{z} + \frac{A_2}{z-1}\right) dz, \quad A_i \in \mathfrak{sl}_2(\mathbb{C}) \tag{11.1}$$

on rank 2 holomorphic vector bundles over the three-punctured sphere with *projective* monodromy group equal to A_5.

Such connections are essentially the same as Gauss hypergeometric equations, and H. Schwarz (1873) classified all such equations having finite monodromy groups. The list he produced has 15 rows, 1 for the family of dihedral groups, 2 rows for each of the tetrahedral and octahedral groups, and 10 rows for the icosahedral group (see Table 11.1).

Note added in proof: Lisovyy and Tykhyy have recently announced (arXiv:0809.4873) that the 'Non-linear Schwarz's list' constructed here is in fact complete.

Table 11.1. Schwarz's list:

	No.	λ''	μ''	ν''	$\frac{\text{Inhalt}}{\pi}$	Polyeder
aa∗	I	$\frac{1}{2}$	$\frac{1}{2}$	ν	ν	Regelmässige Doppelpyramide
abb	II	$\frac{1}{2}$	$\frac{1}{3}$	$\frac{1}{3}$	$\frac{1}{6}=A$	Tetraeder
bbb	III	$\frac{2}{3}$	$\frac{1}{3}$	$\frac{1}{3}$	$\frac{1}{3}=2A$	
abg	IV	$\frac{1}{2}$	$\frac{1}{3}$	$\frac{1}{4}$	$\frac{1}{12}=B$	Würfel und Oktaeder
bgg	V	$\frac{2}{3}$	$\frac{1}{4}$	$\frac{1}{4}$	$\frac{1}{6}=2B$	
abc	VI	$\frac{1}{2}$	$\frac{1}{3}$	$\frac{1}{5}$	$\frac{1}{30}=C$	Dodekaeder und Ikosaeder
bbd	VII	$\frac{2}{5}$	$\frac{1}{3}$	$\frac{1}{3}$	$\frac{1}{15}=2C$	
bcc	VIII	$\frac{2}{3}$	$\frac{1}{5}$	$\frac{1}{5}$	$\frac{1}{15}=2C$	
acd	IX	$\frac{1}{2}$	$\frac{2}{5}$	$\frac{1}{5}$	$\frac{1}{10}=3C$	
bcd	X	$\frac{3}{5}$	$\frac{1}{3}$	$\frac{1}{5}$	$\frac{2}{15}=4C$	
ddd	XI	$\frac{2}{5}$	$\frac{2}{5}$	$\frac{2}{5}$	$\frac{1}{5}=6C$	
bbc	XII	$\frac{2}{3}$	$\frac{1}{3}$	$\frac{1}{5}$	$\frac{1}{5}=6C$	
ccc	XIII	$\frac{4}{5}$	$\frac{1}{5}$	$\frac{1}{5}$	$\frac{1}{5}=6C$	
abd	XIV	$\frac{1}{2}$	$\frac{2}{5}$	$\frac{1}{3}$	$\frac{7}{30}=7C$	
bdd	XV	$\frac{3}{5}$	$\frac{2}{5}$	$\frac{1}{3}$	$\frac{1}{3}=10C$	

Source: From Schwarz (1873).

A key point here is that the Gauss hypergeometric equation is *rigid* so the full monodromy representation (of the fundamental group of the three-punctured sphere into $\mathrm{PSL}_2(\mathbb{C})$) is determined by the conjugacy classes of the monodromy around each of the punctures. Thus in Schwarz's list it is sufficient to list these local monodromy conjugacy classes in order to specify the possible monodromy representations (and from this it is easy to find a hypergeometric equation with given monodromy). To ease recognition, to the left of the table we have listed the triples of conjugacy classes which occur, labelling the four non-trivial conjugacy classes of A_5 by a, b, c, and d, representing rotations by $\frac{1}{2}, \frac{1}{3}, \frac{1}{5}$, and $\frac{2}{5}$ of a turn, respectively. (In the octahedral case one may also have rotations by a quarter of a turn, which we label by g.)

11.1.1 *Naive generalizations*

Our basic aim is to discuss three naive generalizations of Schwarz's list, as follows. The first two arise simply by looking for *non-rigid* connections that are natural generalizations of the hypergeometric connections considered above, obtained by adding an extra singularity – the two cases are generalizations of two ways one may view the hypergeometric equation as a connection. First of all we can simply add another pole at some point t:

$$(\mathbf{A}) \qquad \nabla = d - \left(\frac{A_1}{z} + \frac{A_2}{z-t} + \frac{A_3}{z-1}\right) dz, \quad A_i \in \mathfrak{sl}_2(\mathbb{C})$$

and keep the coefficients in $\mathfrak{sl}_2(\mathbb{C})$.

Secondly, we recall that the connection one obtains immediately upon choosing a cyclic vector for the hypergeometric equation is as in (11.1) but with A_1, A_2 both rank 1 matrices (in $\mathfrak{gl}_2(\mathbb{C})$). Then the monodromy group will be a complex reflection group (generated by two two-dimensional complex reflections[1]) and the natural generalization is then to consider connections of the form

$$(\mathbf{B}) \qquad \nabla = d - \left(\frac{B_1}{z} + \frac{B_2}{z-t} + \frac{B_3}{z-1}\right) dz, \quad B_i \in \mathfrak{gl}_3(\mathbb{C})$$

with each B_i having rank 1, so the monodromy group will be generated by three three-dimensional complex reflections. This is a very natural condition as we will see.

Questions A, B: Find the analogue of Schwarz's list for connections (**A**) or (**B**).

These questions can now be answered and lead to two 'non-rigid Schwarz's lists', that is, to classifications of possible monodromy representations with finite image (up to equivalence) and the construction of connections realizing such representations. We should emphasize that the main focus has been the construction of such connections with given monodromy representation for any value of t (which is a tricky business in this non-rigid case), rather than just the classification.

Example (of type (B)). The full symmetry group of the icosahedron is the icosahedral reflection group of order 120:

$$H = H_3 \cong \langle r_1, r_2, r_3 \mid r_i^2 = 1, (r_1r_2)^2 = (r_2r_3)^3 = (r_3r_1)^5 = 1 \rangle$$
$$\subset O_3(\mathbb{R}) \subset \mathrm{GL}_3(\mathbb{C}).$$

This is generated by three reflections (whose product is *not* the identity) and so it is natural to look for connections on rank 3 bundles over a four-punctured sphere with monodromy H (generated by three reflections about three of the punctures – that is, connections of the form (**B**) with each of the three residues B_i having trace $\frac{1}{2}$ so the corresponding reflections are of order 2). There turn out to be three inequivalent triples of generating reflections of H, two of which are related by an outer automorphism. The problem is to write down connections

[1] That is, arbitrary automorphisms of the form 'one plus rank 1', not necessarily of order 2 or orthogonal.

of the desired form for any value t of the final pole position. One triple of generating reflections is intimately related to K. Saito's flat structure (1993) for H (or icosahedral Frobenius manifold) and appears in Dubrovin's article (1995, appendix E). The other two triples were dealt with around 1997: see Dubrovin and Mazzocco (2000); one is similar to the first case (since related to it by an outer automorphism) but the final triple turned out to be much trickier, and writing out the family of connections in this case involved a specific elliptic curve which took about 10 pages of 40 digits integers to write down (see the preprint version of *op. cit.* on the mathematics arxiv). We will eventually see below that this elliptic solution is in fact equivalent to a solution with a simple parametrization, agreeing with Hitchin's philosophy that 'nice problems should have nice solutions'.

Remark. Before moving on to the third generalization let us add some other historical comments. The 'non-naive' generalizations of the Gauss hypergeometric equation are the equations satisfied by the ${}_nF_{n-1}$ hypergeometric functions (the Gauss case being that of $n = 2$). The corresponding Schwarz's list appears in Beukers and Heckman (1989). In terms of connections this amounts to considering connections (11.1) on rank n vector bundles, still with three singularities on $\mathbb{P}^1$, but with A_1 of rank $n - 1$ and A_2 of rank 1; *these connections are still rigid.*

Some work in the non-rigid case has been done (besides that we will recall below) by considering generalizations of the hypergeometric equation as an equation (rather than as a connection); for example, the algebraic solutions of the Lamé equation were studied in Beukers and van der Waall (2004) (Lamé equations are basically the second order Fuchsian equations with four singular points on $\mathbb{P}^1$ such that three of the local monodromies are of order 2). In general connections of type (**A**) with such monodromy representations will not come from a Lamé equation (since upon choosing a cyclic vector the corresponding equations will in general have additional apparent singularities; this can also be seen by counting dimensions). Indeed it turns out (*op. cit.*) that Lamé equations only have finite monodromy for special configurations of the four poles.

11.1.2 *Non-linear analogue: the Painlevé VI equation*

One reason hypergeometric equations are interesting is that they provide the simplest explicit examples of *Gauss–Manin connections*. Indeed this is one reason Gauss was interested in them: he observed that the periods of a family of elliptic curves satisfy a (Gauss) hypergeometric equation. (The modern interpretation of this is as the explicit form of the natural flat connection on the vector bundle of first cohomologies over the base of the family of elliptic curves, written with respect to the basis given by the holomorphic one-forms – and their derivatives – on the fibres.) Nowadays there is still much interest in such linear differential equations 'coming from geometry'.

Thus the non-linear analogue of the Gauss hypergeometric equation should be the explicit form of the simplest *nonabelian* Gauss–Manin connection (i.e. the explicit form of the natural connection on the bundle of first nonabelian cohomologies of some family of varieties). The simplest interesting case corresponds to taking the universal family of four-punctured spheres and taking cohomology with coefficients in $\mathrm{SL}_2(\mathbb{C})$ (one needs a non-trivial family of varieties with nonabelian fundamental groups). This leads to the Painlevé VI equation ($\mathrm{P_{VI}}$), which is a second-order non-linear differential equation whose solutions, like those of the hypergeometric equation, branch only at $0, 1, \infty \in \mathbb{P}^1$. In particular we may study the (non-linear) monodromy of solutions of $\mathrm{P_{VI}}$, by examining how solutions vary upon analytic continuation along paths in the three-punctured sphere.

Thus, since Schwarz lists fundamental solutions of hypergeometric equations having finite monodromy, our main question is to construct the analogue of Schwarz's list for $\mathrm{P_{VI}}$:

Question C: What are the solutions of Painlevé VI having finite monodromy?

This question is still open; there is as yet no full classification – the main effort (at least of the present author) has been towards finding and constructing interesting solutions. So far all known finite-branching solutions are actually algebraic (cf. Iwasaki 2008). Currently we are at the reasonably happy state of affairs that all such solutions known to exist have actually been constructed. In what follows I will explain various aspects of the problem, and in particular show how the non-rigid lists (**A**) and (**B**) map to the list of (**C**). Some key points, demonstrating the richness and variety of solutions, are

- There are algebraic solutions of $\mathrm{P_{VI}}$ not related to finite subgroups of the coefficient group $\mathrm{SL}_2(\mathbb{C})$.
- There are 'generic' solutions of $\mathrm{P_{VI}}$ with finite monodromy; that is, not lying on any of the reflection hyperplanes of the affine F_4 Weyl group of symmetries of $\mathrm{P_{VI}}$.
- There are entries on the list of (**C**) which do not come from either (**A**) or (**B**).

In particular we will see $\mathrm{P_{VI}}$ solutions related to the groups A_6, $\mathrm{PSL}_2(\mathbb{F}_7)$ and Δ_{237}.

11.2 What is Painlevé VI?

There are various viewpoints, and simply giving the explicit equation is perhaps the least helpful introduction to it. In brief, Painlevé VI is

- The explicit form of the simplest nonabelian Gauss–Manin connection
- The equation controlling the 'isomonodromic deformations' of certain logarithmic connections/Fuchsian systems on $\mathbb{P}^1$
- The most general second-order ODE with the so-called 'Painlevé property'

- A certain dimensional reduction of the anti-self-dual Yang–Mills equations (see e.g. Mason and Woodhouse (1996)),
- An equation related to certain elliptic integrals with moving endpoints (cf. Fuchs (1905) and Manin (1998))
- The second-order ODE for a complex function $y(t)$

$$\frac{d^2y}{dt^2} = \frac{1}{2}\left(\frac{1}{y}+\frac{1}{y-1}+\frac{1}{y-t}\right)\left(\frac{dy}{dt}\right)^2 - \left(\frac{1}{t}+\frac{1}{t-1}+\frac{1}{y-t}\right)\frac{dy}{dt}$$
$$+\frac{y(y-1)(y-t)}{t^2(t-1)^2}\left(\alpha+\beta\frac{t}{y^2}+\gamma\frac{(t-1)}{(y-1)^2}+\delta\frac{t(t-1)}{(y-t)^2}\right)$$

where $\alpha, \beta, \gamma, \delta \in \mathbb{C}$ are constants

The Painlevé property means that any local solution $y(t)$ defined in a disc in the three-punctured sphere $\mathbb{P}^1 \setminus \{0, 1, \infty\}$ extends to a meromorphic function on the universal cover of $\mathbb{P}^1 \setminus \{0, 1, \infty\}$. It is this property that enables us to speak of the monodromy of P_{VI} solutions. Concerning solutions there is a basic trichotomy (see Watanabe (1998)):

$$\text{A solution of } P_{VI} \text{ is either } \begin{cases} \text{a `new' transcendental function, or} \\ \text{a solution of a first-order Riccati equation, or} \\ \text{an algebraic function.} \end{cases}$$

In particular if one is interested in constructing new explicit solutions of P_{VI} then, since the Riccati solutions are all well understood, the algebraic solutions are the first place to look.

The standard approach to P_{VI} is as isomonodromic deformations of rank 2 logarithmic connections with four poles on $\mathbb{P}^1$, as the poles move (generic such connections are of the form (**A**), and then t parametrizes the possible pole configurations). In particular one can see the four constants in P_{VI} directly in terms of the eigenvalues of the residues of the connection: if we set θ_i to be the difference of the eigenvalues (in some order) of the residue A_i ($i = 1, 2, 3, 4$, where $A_4 = -\sum_1^3 A_i$ is the residue at infinity) then the relation to the constants is

$$\alpha = (\theta_4 - 1)^2/2, \quad \beta = -\theta_1^2/2, \quad \gamma = \theta_3^2/2, \quad \text{and } \delta = (1 - \theta_2^2)/2.$$

Before going into more detail let us also mention one further property of P_{VI}: it admits a group of symmetries isomorphic to the affine Weyl group of type F_4 (see Okamoto (1987) or the exposition in Boalch (2006)). Indeed treating $\theta = (\theta_1, \ldots, \theta_4) \in \mathbb{C}^4$ as the set of parameters for P_{VI} is useful since the affine F_4 Weyl group of symmetries acts in the standard way on this $\mathbb{C}^4$. (We will see below that these four parameters may also be interpreted as coordinates on the moduli space of cubic surfaces.)

11.2.1 *Conceptual approach to Painlevé VI*

Consider the universal family of smooth four-punctured rational curves with labelled punctures. Write $B := \mathcal{M}_{0,4} \cong \mathbb{P}^1 \setminus \{0, 1, \infty\}$ for the base, $\mathcal{F}$ for the standard fibre, and $\mathcal{C}$ for the total space:

$$\begin{array}{ccc} \mathcal{C} & \longleftarrow & \mathcal{F} \cong \mathbb{P}^1 \setminus 4 \text{ points} \\ \downarrow & & \\ B & & \end{array}$$

Now replace each fibre $\mathcal{F}$ by $\mathrm{H}^1(\mathcal{F}, G)$ where $G = \mathrm{SL}_2(\mathbb{C})$. Here we will use two viewpoints/realizations of this nonabelian cohomology set H^1:

1. Betti: Moduli of fundamental group representations

$$\mathrm{H}^1(\mathcal{F}, G) \cong \mathrm{Hom}(\pi_1(\mathcal{F}), G)/G$$

2. De Rham: Moduli of connections on holomorphic vector bundles over $\mathcal{F}$

These two viewpoints are related by the Riemann–Hilbert correspondence (the nonabelian De Rham functor), taking connections to their monodromy representations. The point is that algebraically these realizations of H^1 are very different and the Riemann–Hilbert map is transcendental (things written in algebraic coordinates on one side will look a lot more complicated from the other side).

Thus we get two non-linear fibrations over the base B, with fibres the De Rham or Betti realizations of $\mathrm{H}^1(\mathcal{F}, G)$, respectively:

$$\begin{array}{ccc} \mathcal{M}_{\mathrm{De\,Rham}} & \xrightarrow{\text{Riemann–Hilbert}} & \mathcal{M}_{\mathrm{Betti}} \\ \downarrow & & \downarrow \\ B & = & B \end{array}$$

As in the case with abelian coefficients one still gets a natural connection on these cohomology bundles. The surprising fact is that it is algebraic on *both* sides (approximating the De Rham side in terms of logarithmic connections to give it an algebraic structure Nitsure 1993). Thus when written explicitly we will get *non-linear algebraic* differential equations 'coming from geometry'. (See Simpson 1994, section 8 for more on these connections in the case of families of projective varieties.)

The two standard descriptions of the abelian Gauss–Manin connection generalize to descriptions of this non-linear connection. In the Betti picture we may identify two nearby fibres of $\mathcal{M}_{\mathrm{Betti}}$ simply by keeping the monodromy representations (points of the fibres) constant: moving around in B amounts to deforming the configuration of four points in $\mathbb{P}^1$ and it is easy to see how to identify the fundamental groups of the four-punctured spheres as the punctures are deformed – use the same generating loops. This 'isomonodromic' description, preserving the monodromy representation, is the nonabelian analogue of keeping the periods of one-forms constant.

On the De Rham side the non-linear connection can be described in terms of extending a connection on a vector bundle over a fibre $\mathcal{F}$, to a flat connection on a vector bundle over a family of fibres and then restricting to another fibre, much as the abelian case is described in terms of closed one-forms (linear connections replacing one-forms and flatness replacing the notion of closedness).

Each of these descriptions has a use: the De Rham viewpoint lends itself to giving an explicit form of the non-linear connection (essentially amounting to the condition for the flatness of the connection over the family of fibres). The Betti viewpoint is more global and allows us to study the monodromy of the non-linear connection, as an explicit action on fibres of $\mathcal{M}_{\text{Betti}}$.

11.2.2 *Explicit non-linear equations*

The De Rham bundle $\mathcal{M}_{\text{De Rham}}$ is well approximated by the space of logarithmic connections with four poles on the trivial rank 2 holomorphic bundle (with trivial determinant) over $\mathbb{P}^1$. Call the space of such connections $\mathcal{M}^*$ and observe it parametrizes connections of the form ($\mathbf{A}$), and that these are determined by the value of $t \in B$ and the residues:

$$\mathcal{M}^* \cong B \times \left\{ (A_1, \ldots, A_4) \mid A_i \in \mathfrak{g}, \sum A_i = 0 \right\} / G.$$

Here $G = \mathrm{SL}_2(\mathbb{C})$ does not act on B and acts by diagonal conjugation on the residues A_i. In general this quotient will not be well behaved, but it has a natural Poisson structure and the generic symplectic leaves will be smooth complex symplectic surfaces. Clearly $\mathcal{M}^*$ is trivial as a bundle over B (projecting onto the configuration of poles), but the nonabelian Gauss–Manin connection is different to this trivial connection and was computed about 100 years ago by Schlesinger (essentially in the way stated above it seems). The non-linear connection is given by *Schlesinger's equations*, which in the case at hand are

$$\frac{dA_1}{dt} = \frac{[A_2, A_1]}{t}, \qquad \frac{dA_3}{dt} = \frac{[A_2, A_3]}{t-1}$$

together with a third equation for dA_2/dt easily deduced from the fact that A_4 remains constant. If the residues of the connection satisfy these equations then the corresponding monodromy representation remains constant as t varies. (They are easily derived from the vanishing of the curvature of the 'full' connection $d - \left(A_1 \frac{dz}{z} + A_2 \frac{dz-dt}{z-t} + A_3 \frac{dz}{z-1}\right)$.)

To get from here to P_{VI} one chooses specific functions x, y on $\mathcal{M}^*$ which restrict to coordinates on each generic symplectic leaf and writes down the connection in these (carefully chosen) coordinates (see Boalch 2005, pp.199–200 for a discussion of the formulae, which are from Jimbo and Miwa 1981). This leads to two coupled non-linear first-order equations, and eliminating x leads to the second-order Painlevé VI equation for $y(t)$. It was first written down in full generality by R. Fuchs (1905) (whose father L. Fuchs was also the father of 'Fuchsian equations').

11.2.3 *Monodromy of Painlevé VI*

Since the Betti and De Rham realizations are analytically isomorphic, we see the monodromy of solutions to P_{VI} thus corresponds to the monodromy of the connection on $\mathcal{M}_{\text{Betti}}$. This amounts to an *action* of the fundamental group of the base B on a fibre, and this action can be described explicitly.

Let $\mathcal{M}_t = \text{Hom}(\pi_1(\mathbb{P}^1 \setminus \{0,t,1,\infty\}), G)/G$ be the fibre of $\mathcal{M}_{\text{Betti}}$ at some fixed point $t \in B$. The key point is that $\pi_1(B) \cong \mathcal{F}_2$ (the free nonabelian group on two generators) may be identified with the pure mapping class group of the four punctured sphere $\mathbb{P}^1 \setminus \{0,t,1,\infty\}$. As such it has a natural action on $\mathcal{M}_t$ (by pushing forward loops generating the fundamental group), and this action is the desired monodromy action.

Explicitly, upon choosing appropriate generating loops of $\pi_1(\mathbb{P}^1 \setminus \{0,t,1,\infty\})$ we see $\mathcal{M}_t$ may be described directly in terms of monodromy matrices:

$$\mathcal{M}_t \cong \left\{ (M_1, \ldots, M_4) \;\middle|\; M_i \in G,\ M_4 \cdots M_1 = 1 \right\} /G$$

which in turn is simply the quotient G^3/G of three copies of G by diagonal conjugation by $G = \text{SL}_2(\mathbb{C})$. In fact this quotient has been studied classically: the ring of G invariant functions on G^3 has seven generators and one relation, embedding the affine quotient variety as a hypersurface in $\mathbb{C}^7$. The particular equation for this hypersurface appears on p. 366 of Fricke and Klein (1897). The Painlevé VI parameters essentially specify the conjugacy classes of the four monodromies M_i, and serve here to fibre the six-dimensional hypersurface G^3/G into a four-parameter family of surfaces. Looking at the explicit equation shows they are affine *cubic* surfaces. In turn Iwasaki (2002) has recently pointed out that this family of cubics may be quite simply related to the explicit family of Cayley (1849) and so contains the generic cubic surface.

The desired action of the free group $\mathcal{F}_2$ on the Betti spaces is given by the squares of the following 'Hurwitz' action:

$$\omega_1(M_1, M_2, M_3) = \left(M_2, M_2 M_1 M_2^{-1}, M_3\right)$$

$$\omega_2(M_1, M_2, M_3) = \left(M_1, M_3, M_3 M_2 M_3^{-1}\right).$$

More explicitly if we consider simple positive loops l_1, l_2 in B based at $\frac{1}{2}$ encircling $0, 1$, respectively, then the monodromy of the connection on $\mathcal{M}_{\text{Betti}}$ around l_i is given by ω_i^2 (with respect to certain generators of $\pi_1\left(\mathbb{P}^1 \setminus \left\{0, \frac{1}{2}, 1, \infty\right\}\right)$. In turn it is possible to write this action directly as an action on the ring of invariant function on G^3.

11.3 Algebraic solutions from finite subgroups of $\mathbf{SL_2}(\mathbf{C})$

11.3.1 *What exactly is an algebraic solution?*

The obvious definition is simply an algebraic function $y(t)$ which solves $\mathrm{P_{VI}}$ for some value of the four parameters. Thus it will be specified by some polynomial equation

$$F(y,t)=0$$

and a four-tuple θ of parameters. In practice however such polynomials F can be quite unwieldy and are difficult to transform under the affine Weyl symmetry group, making it difficult to see if in fact two solutions are equivalent. This leads to our preferred definition.

Definition. *An algebraic solution of* $\mathrm{P_{VI}}$ *is a compact, possibly singular, algebraic curve* Π *together with two rational functions* $y, t : \Pi \to \mathbb{P}^1$:

$$\begin{array}{ccc} \Pi & \xrightarrow{\ y\ } & \mathbb{P}^1 \\ {\scriptstyle t}\big\downarrow & & \\ \mathbb{P}^1 & & \end{array}$$

such that

- t *is a Belyi map (i.e. its branch locus is a subset of* $\{0, 1, \infty\}$*).*
- y*, when viewed as a function of* t *away from the ramification points of* t*, solves* $\mathrm{P_{VI}}$ *for some value of the four parameters.*

In principle it is straightforward to go between the two definitions, but in practice it is useful to look for a good model of Π (and the model given by the closure of the zero locus of the polynomial F is usually a bad choice).

11.3.2 $(\mathbf{A}) \mapsto (\mathbf{C})$

Suppose we have a linear connection $(\mathbf{A})$ with finite monodromy. Its monodromy representation will be specified by a triple $(M_1, M_2, M_3) \in G^3$ generating a finite subgroup $\Gamma \subset G$ (where $G = \mathrm{SL_2}(\mathbb{C})$ as above). This linear connection specifies the initial value (and first derivative) of a solution to $\mathrm{P_{VI}}$. This $\mathrm{P_{VI}}$ solution will have finite monodromy, since we know the branching of $\mathrm{P_{VI}}$ solutions corresponds to the $\mathcal{F}_2$ action on conjugacy classes of triples in G^3, and the orbit through (M_1, M_2, M_3) will be finite, since the action is within triples of generators of Γ.

Thus we see that finite $\mathcal{F}_2$ orbits (in G^3/G) correspond to $\mathrm{P_{VI}}$ solutions with a finite number of branches, and the points of such $\mathcal{F}_2$ orbits correspond to the individual branches of the $\mathrm{P_{VI}}$ solution. In particular the size of the orbit, the number of branches, is the degree of the map $t : \Pi \to \mathbb{P}^1$. (Indeed the $\mathcal{F}_2$ action on such a finite orbit itself gives the full permutation representation of the Belyi map $t : \Pi \to \mathbb{P}^1$, and in particular, by the Riemann–Hurwitz formula, determines the genus of the 'Painlevé curve' Π.)

Said differently it is useful to define a *topological algebraic* P_{VI} *solution* (or henceforth for brevity a *topological solution*) to be a finite $\mathcal{F}_2$ orbit in G^3/G. (The classification of such orbits is still open and is the main step in classifying all finite branching P_{VI} solutions.) In these terms the first paragraph above points out that one obtains 'obvious' topological solutions upon taking any triple of generators of any finite subgroup of G.

For example (omitting discussion of how they were actually constructed), here are some solutions corresponding to certain triples of generators of the binary tetrahedral and octahedral subgroups, due to Dubrovin (1995) and Hitchin (2003) (in different but equivalent forms):

Tetrahedral solution of degree 3

$$y = \frac{(s-1)(s+2)}{s(s+1)}, \quad t = \frac{(s-1)^2(s+2)}{(s+1)^2(s-2)}, \quad \text{and} \quad \theta = (2,1,1,2)/3,$$

Octahedral solution of degree 4

$$y = \frac{(s-1)^2}{s(s-2)}, \quad t = \frac{(s+1)(s-1)^3}{s^3(s-2)}, \quad \text{and} \quad \theta = (1,1,1,1)/4.$$

In both cases Π is a rational curve (with parameter s). Although written in this compact form, one should bear in mind these formulae represent a whole (isomonodromic) family of connections (**A**) as t varies. An explicit elliptic solution appears in Hitchin (1995*a*) and may be written as

Elliptic dihedral solution

$$y = \frac{(3\,s-1)\left(s^2-4\,s-1\right)\left(s^2+u\right)\left(s\,(s+2)-u\right)}{\left(3\,s^3+7\,s^2+s+1\right)\left(s^2-u\right)\left(s\,(s-2)+u\right)},$$

$$t = \frac{\left(s^2+u\right)^2\left(s\,(s+2)-u\right)\left(s\,(s-2)-u\right)}{\left(s^2-u\right)^2\left(s\,(s+2)+u\right)\left(s\,(s-2)+u\right)},$$

where the pair (s,u) lives on the elliptic curve $u^2 = s\left(s^2+s-1\right)$ and $\theta = (1,1,1,1)/2$. This solution has degree 12 and corresponds to a triple of generators of the binary dihedral group of order 20.

It turns out (see Boalch 2006*a*, remark 16) that the icosahedral solutions of Dubrovin and Mazzocco (2000) fit into this framework as well and correspond to (certain) triples of generators of the binary icosahedral group, although in the first instance they arose from the icosahedral reflection group as described earlier. Note that remark 0.1 of Dubrovin and Mazzocco (2000) describes a relation between their solutions of P_{VI} and a certain *folding* of Schwarz's list; this is different to the relation just mentioned – in particular problem (**A**) demands an *extension* of Schwarz's list.

11.4 Beyond Platonic Painlevé VI solutions

My starting point in this project was simply the observation that there should be more algebraic solutions to P_{VI} than those coming from finite subgroups of $SL_2(\mathbb{C})$. Dubrovin (1995) had shown how to relate three-dimensional real orthogonal reflection groups to a certain one-parameter family of the full four-dimensional family of P_{VI} equations (namely, the family having parameters $\theta = (0, 0, 0, *)$) and this was used in Dubrovin and Mazzocco 2000 to classify algebraic solutions having parameters in this one-parameter family. (Some aspects of *op. cit.* were subsequently extended by Mazzocco 2001 to classify rational solutions – i.e. those with only one branch, cf. also Yuan and Li 2002.) The further observation was that if one is able to get away from the orthogonality condition here then one will relate *any* P_{VI} equation to a three-dimensional *complex* reflection group.

Theorem 11.1 (Boalch 2003, 2005) *The isomonodromic deformations of type* (**B**) *connections (on rank 3 vector bundles) are also controlled by the Painlevé VI equation, and all* P_{VI} *equations arise in this way.*

Thus a solution to P_{VI} can also be viewed as specifying an isomonodromic family of rank 3 Fuchsian connections. It turns out that the formulae to go from a P_{VI} solution $y(t)$ to such an isomonodromic family are more symmetrical than in the previous case (type (**A**)) so we will recall them here. (For the analogous formulae for (**A**) see Jimbo and Miwa 1981 and in Harnad's dual picture – the formula for which should be compared to that below – see Harnad 1994 and also Mazzocco 2002, which was kindly pointed out by the referee.) First the parameters: let $\lambda_i = \mathrm{Tr}(B_i)$ for $i = 1, 2, 3$ and let μ_i be the eigenvalues, in some order, of $B_1 + B_2 + B_3$ (which is minus the residue at infinity), so that $\sum \lambda_i = \sum \mu_i$.

Theorem 11.2 (Boalch 2006*b*) *If* $y(t)$ *solves Painlevé VI with parameters* θ *where*

$$\theta_1 = \lambda_1 - \mu_1, \quad \theta_2 = \lambda_2 - \mu_1, \quad \theta_3 = \lambda_3 - \mu_1, \quad \textit{and} \quad \theta_4 = \mu_3 - \mu_2,$$

and we define $x(t)$ *via*

$$x = \frac{1}{2}\left(\frac{(t-1)y' - \theta_1}{y} + \frac{y' - 1 - \theta_2}{y - t} - \frac{t\,y' + \theta_3}{y - 1}\right)$$

then the family of logarithmic connections (**B**) *will be isomonodromic as* t *varies, where*

$$B_1 = \begin{pmatrix} \lambda_1 & b_{12} & b_{13} \\ 0 & 0 & 0 \\ 0 & 0 & 0 \end{pmatrix}, \quad B_2 = \begin{pmatrix} 0 & 0 & 0 \\ b_{21} & \lambda_2 & b_{23} \\ 0 & 0 & 0 \end{pmatrix}, \quad \textit{and} \quad B_3 = \begin{pmatrix} 0 & 0 & 0 \\ 0 & 0 & 0 \\ b_{31} & b_{32} & \lambda_3 \end{pmatrix}$$

$$b_{12} = \lambda_1 \ - \mu_3 y + (\mu_1 - xy)(y-1), \qquad b_{32} = (\mu_2 - \lambda_2 - b_{12})/t,$$
$$b_{13} = \lambda_1 t - \mu_3 y + (\mu_1 - xy)(y-t), \qquad b_{23} = (\mu_2 - \lambda_3)t - b_{13},$$
$$b_{21} = \lambda_2 + \frac{\mu_3(y-t) - \mu_1(y-1) + x(y-t)(y-1)}{t-1}, \qquad b_{31} = (\mu_2 - \lambda_1 - b_{21})/t.$$

The implication of this for algebraic solutions should now be clear: the monodromy of a P_{VI} solution is also described by an action of the free group $\mathcal{F}_2$ on (conjugacy classes of) triples of three-dimensional complex reflections (r_1, r_2, r_3) (with the same formula as before, just replace M_i by r_i). Thus in this context the 'obvious' topological solutions (i.e. finite $\mathcal{F}_2$ orbits) come from taking a triple of generating reflections of a finite complex reflection group in $GL_3(\mathbb{C})$. Such finite complex reflection groups were classified by Shephard and Todd (1954) and apart from the familiar real orthogonal reflection groups there is an infinite family plus four exceptional complex groups, the Klein reflection group (of order 336, a two-fold cover of Klein's simple group isomorphic to $PSL_2(\mathbb{F}_7) \hookrightarrow PGL_3(\mathbb{C})$), two Hessian groups, and the Valentiner group (of order 2,160, a six-fold cover of $A_6 \hookrightarrow PGL_3(\mathbb{C})$).

The infinite family of groups and the two Hessian groups do not seem to lead to interesting new solutions, but by computing the $\mathcal{F}_2$ orbits (determining the topology of Π) it is easy to see that the Klein group yields a genus 0 degree 7 solution and the Valentiner group has three inequivalent triples of generating reflections, each leading to genus 1 solutions with degrees 15,15, and 24, respectively. These are new solutions, previously undetected. (The 24 appearing here led to a certain amount of trepidation, given that the 10 page elliptic solution of Dubrovin and Mazzocco 2000 had degree 18.)

11.4.1 *Construction*

Of course finding the topological solution is not the same as finding an explicit isomonodromic family of connections; one needs to solve a family of Riemann–Hilbert problems inverting the transcendental Riemann–Hilbert map for each value of t. (Indeed my original plan was to just prove the existence of new interesting solutions, in Boalch (2003), but a certain stubbornness, and some inspiration from reading about Klein's work finding explicit 3×3 matrices generating his simple group, convinced us to go further.)

The two main steps in the method we finally got to work are as follows. (This is a generalization of the method used by Dubrovin and Mazzocco 2000.)

1. Jimbo's asymptotic formulae. Jimbo (1982) found an exact formula for the leading asymptotics at $t = 0$ of the branch of the P_{VI} solution $y(t)$ corresponding to any sufficiently generic linear monodromy representation (M_1, M_2, M_3). (This formula was obtained by considering the degeneration of the isomonodromic family of connections (**A**) as $t \to 0$; in the limit the four-punctured sphere degenerates into a stable curve with two components, each

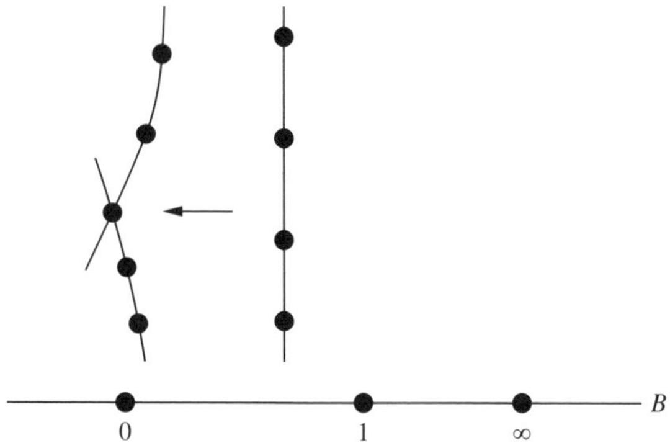

Figure 11.1 Degeneration to two hypergeometric systems.

with three marked points. The connections (**A**) degenerate into hypergeometric systems on each component, with known monodromy (Figure 11.1). Since these are rigid it is easy to solve their Riemann–Hilbert problems explicitly and this gives the leading asymptotics of the isomonodromic family and thus of the P_{VI} solution.)

This is useful for us because, as Jimbo mentions, one may substitute the leading asymptotics back into the P_{VI} equation to get arbitrarily many terms of the precise asymptotic expansion of the solution at 0. If the solution is algebraic, then this is its Puiseux expansion, a sufficient number of terms of which will determine the entire solution.

It turns out there was a typo in Jimbo (1982), which meant the entire method did not work (indeed the fact it did not work led to the questioning of Jimbo's formula and hence the correction in Boalch 2005). (Note the special parameters of Dubrovin and Mazzocco 2000 are not covered by Jimbo's result; rather they adapted the argument of Jimbo 1982 to their case.)

2. Relating (A) and (B). Since Jimbo's formula requires a monodromy representation of a connection of type (**A**), and we are starting with a triple of 3×3 complex reflections (the monodromy representation of a connection of type (**B**)), the second step is that we need to see how to go between these two pictures (on both the De Rham and Betti sides of the Riemann–Hilbert correspondence). This will be described in the following subsection.

11.4.2 *Relating connections* (**A**) *and* (**B**)

We wish to sketch how to convert a connection (**B**) on a rank 3 vector bundle into a connection of the form (**A**) on a rank 2 bundle. On the other side of the Riemann–Hilbert correspondence this amounts to an $\mathcal{F}_2$-equivariant map from triples of complex reflections to triples of elements of $G = \mathrm{SL}_2(\mathbb{C})$ (as in Boalch 2005, section 2).

Of course the monodromy groups change in a highly non-trivial way under this procedure. For example, the Klein reflection group becomes the triangle group $\Delta_{237} \subset G$, which is an infinite group, and the Valentiner group becomes the binary icosahedral group (leading to an unexpected relation between A_6 and A_5).

After this procedure was put on the arxiv (Boalch 2005) we learnt (Dettweiler and Reiter 2007) that it is essentially a case of the middle convolution functor used by Katz (1996), although our construction using the complex analytic Fourier–Laplace transform is different from that of Katz (using l-adic methods) and from the work of Dettweiler and Reiter (2000).

The basic picture which emerges is as follows (see the figure below), and ought to be better known. It was obtained essentially by a careful reading of Balser, Jurkat and Lutz (1981), although the basic idea of relating irregular and Fuchsian systems by the Laplace transform dates back to Birkhoff and Poincaré. (Dubrovin 1995, 1999 used an orthogonal analogue in relation to Frobenius manifolds, also using Balser, Jurkat, and Lutz 1981. Moreover the top triangle is essentially a case of 'Harnad duality' (Harnad 1994) so for $n = 3$ we knew we would obtain all $\mathrm{P_{VI}}$ equations.)

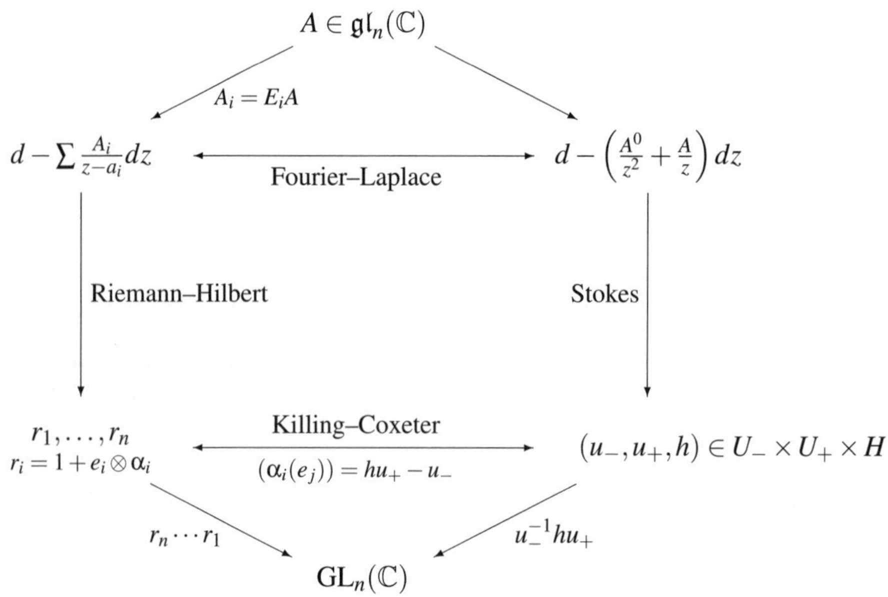

The idea is to describe a transcendental map from $\mathfrak{gl}_n(\mathbb{C})$ to $\mathrm{GL}_n(\mathbb{C})$ in two different ways (the two paths down the left and the right from the top to the bottom of the figure).

Choose n distinct complex numbers $a_1, \ldots, a_n$ and define $A^0 = \mathrm{diag}(a_1, \ldots, a_n)$. Roughly speaking (on a dense open patch) the left-hand column arises by defining $A_i = E_i A$ (setting to zero all but the ith row of A) and constructing the logarithmic connection $d - \sum \frac{A_i}{z-a_i} dz$ having rank 1 residues at each a_i. Then taking the monodromy of this yields n complex reflections r_i (and if bases of solutions are chosen carefully one can naturally define vectors e_i and one-forms α_i such that $r_i = 1 + e_i \otimes \alpha_i$ and that the e_i form a basis). Then the map to $\mathrm{GL}_n(\mathbb{C})$ is given by taking the product of $r_n \cdots r_1$ of these reflections, written in the e_i basis.

Now the key algebraic fact, which dates back at least to Killing (1889) (see Coleman 1989), is that any such product of complex reflection lies in the big cell of $\mathrm{GL}_n(\mathbb{C})$ and so may be factored as the product of a lower triangular and an upper triangular matrix. We write this product as $u_-^{-1} h u_+$ with $u_\pm \in U_\pm$ the unipotent triangular subgroups, and $h \in H$ diagonal:

$$r_n \cdots r_2 r_1 = u_-^{-1} h u_+. \tag{11.2}$$

Further, although this relation between the reflections and $u_\pm$ looks to be highly non-linear, one can relate them in an almost linear fashion: the matrix $hu_+ - u_-$ is the matrix with entries $\alpha_i(e_j)$.

On the other hand it turns out that the same map can be defined by taking the *Stokes data* of the irregular connection $d - \left(\frac{A^0}{z^2} + \frac{A}{z}\right) dz$. Indeed the map on the right-hand side generalizes (Boalch 2002) to any complex reductive group G in place of $\mathrm{GL}_n(\mathbb{C})$, but only for $\mathrm{GL}_n(\mathbb{C})$ is the alternative 'logarithmic' viewpoint available. Thus $u_\pm$ are also the two Stokes matrices of this irregular connection (the natural analogue of monodromy data for such connections); the exact definition is not important here. (The element h is the so-called formal monodromy, explicitly it is simply $\exp(2\pi i \Lambda)$ where Λ is the diagonal part of A.) The two connections are related (see Balser, Jurkat, and Lutz 1981) by the Fourier–Laplace transform: this is more than just formal, and by relating bases of solutions on both sides the stated relation between the Stokes and monodromy data is obtained. (In both cases the resulting element of $\mathrm{GL}_n(\mathbb{C})$ is the monodromy around $z = \infty$ in a suitable basis.) In summary we see that the 'Betti' incarnation of the Fourier–Laplace transform is the relation of Killing–Coxeter.

Now to apply this in the current context we consider the effect of adding a scalar λ to $A \in \mathfrak{gl}_n(\mathbb{C})$. On the right-hand side this corresponds to tensoring the irregular connection by the logarithmic connection $d - \lambda dz/z$ on the trivial line bundle, and Balser, Jurkat, and Lutz (1981) showed that the Stokes data is changed only by scaling h by $s := \exp(2\pi i \lambda)$, fixing $u_\pm$. On the logarithmic side this corresponds to a non-trivial convolution operation, changing the monodromy representation in a non-trivial way. Of course using the Killing–Coxeter identity we now see precisely how the complex reflections vary. (It is perhaps worth noting that this scalar shift is essentially the inverse of the spectral parameter introduced by Killing 1889, p. 20, appearing in the characteristic polynomial of the Killing–Coxeter matrix (11.2): $\det(u_-^{-1} s h u_+ - 1) = \det(s h u_+ - u_-)$.)

If we set $n = 3$ then the logarithmic connections appearing are of the form (**B**), upon taking $a_1, a_2, a_3 = 0, t, 1$. Then we may choose the scalar shift such that the resulting element of $GL_3(\mathbb{C})$ has 1 as an eigenvalue. This implies that the connections are reducible and we can take the irreducible rank 2 sub- or quotient connection. Projecting to $\mathfrak{sl}_2$ gives the desired connection of type (**A**) (see Boalch 2005). (Note that there is a choice involved here, of which eigenvalue to shift to 1.)

11.4.3 *New solutions*

Thus in summary the procedure now is as follows: take a triple of generating reflections of a finite complex reflection group in $GL_3(\mathbb{C})$. Push it down to the 2×2 framework using the scalar shift to obtain a triple (M_1, M_2, M_3) of elements of $SL_2(\mathbb{C})$ in an isomorphic $\mathcal{F}_2$ orbit. Apply Jimbo's formula to get the leading asymptotics of the corresponding P_{VI} solutions at $t = 0$ on each branch (i.e. for each triple in the $\mathcal{F}_2$ orbit). (Converting the values which arise into exact algebraic numbers.) Substitute these leading terms back into P_{VI} to obtain arbitrarily many terms of the Puiseux expansion at 0 of each solution branch. Use these expansions to determine the polynomial $F(y,t)$ defining the solution (assuming it is algebraic). Find a parametrization of the resulting algebraic curve (e.g. using M. van Hoeij's wonderful Maple algebraic curves package).

For example, for the Klein complex reflection group of order 336 this works perfectly (Boalch 2005) and the resulting solution is

Klein solution

$$y = -\frac{\left(5\,s^2 - 8\,s + 5\right)\left(7\,s^2 - 7\,s + 4\right)}{s\,(s-2)\,(s+1)\,(2\,s-1)\,(4\,s^2 - 7\,s + 7)},$$

$$t = \frac{\left(7\,s^2 - 7\,s + 4\right)^2}{s^3\left(4\,s^2 - 7\,s + 7\right)^2}, \qquad \text{and } \theta = (2,2,2,4)/7$$

which has 7 branches. One may of course now substitute this back into the formula of Theorem 11.2 (with $\lambda = (1,1,1)/2$ and $\mu = (3,5,13)/14$) to obtain an explicit family of logarithmic connections having monodromy equal to the Klein reflection group generated by reflections (see Boalch 2006*b*, section 3).

When converted to connections of type (**A**) these 'Klein connections' have infinite (projective) monodromy group equal to the triangle group Δ_{237} (cf. Boalch 2006*c*, appendix B). On the other hand it turns out (Boalch 2006*a*) that for the Valentiner connections, even though they are much trickier to construct directly, we can still compute immediately that they become connections of type (**A**) with binary icosahedral monodromy. They are also inequivalent to those appearing in the work of Dubrovin and Mazzocco related to the real orthogonal icosahedral reflection group (which lead to unipotently generated monodromy with one choice of the scalar shift, but finite binary icosahedral monodromy with a different choice, cf. Boalch 2006*a*, remark 16).

Thus it seemed like a good idea to examine precisely what P_{VI} solutions arise upon taking arbitrary triples of generators (M_1, M_2, M_3) of the binary

Table 11.2. Icosahedral solutions 11–52

	Degree	Genus	Walls	Type		Degree	Genus	Walls	Type
11	2	0	2	$b^2\,c^2$	32	10	0	3	d^4
12	2	0	2	$b^2\,d^2$	33	12	0	0	$a\,b\,c\,d$
13	2	0	2	$c^2\,d^2$	34	12	1	1	$a\,b\,c^2$
14	3	0	1	$b\,c^2\,d$	35	12	1	1	$a\,b\,d^2$
15	3	0	1	$b\,c\,d^2$	36	12	1	1	$b^2\,c\,d$
16	4	0	2	$a\,c^3$	37	15	1	2	$b^3\,c$
17	4	0	2	$a\,d^3$	38	15	1	2	$b^3\,d$
18	4	0	2	$c^3\,d$	39	15	1	2	$b^2\,c^2$
19	4	0	2	$c\,d^3$	40	15	1	2	$b^2\,d^2$
20	5	0	1	$b^2\,c\,d$	41	18	1	3	b^4
21	5	0	2	$c^2\,d^2$	42	20	1	1	$a\,b^2\,c$
22	6	0	1	$b\,c^2\,d$	43	20	1	1	$a\,b^2\,d$
23	6	0	1	$b\,c\,d^2$	44	20	1	3	$a^2\,c^2$
24	8	0	1	$a\,c^2\,d$	45	20	1	3	$a^2\,d^2$
25	8	0	1	$a\,c\,d^2$	46	24	1	2	$a\,b^3$
26	9	1	2	$b\,c^3$	47	30	2	2	$a^2\,b\,c$
27	9	1	2	$b\,d^3$	48	30	2	2	$a^2\,b\,d$
28	10	0	2	$a^2\,c\,d$	49	36	3	3	$a^2\,b^2$
29	10	0	2	$b^3\,c$	50	40	3	3	$a^3\,c$
30	10	0	2	$b^3\,d$	51	40	3	3	$a^3\,d$
31	10	0	3	c^4	52	72	7	3	$a^3\,b$

icosahedral group. Thus we looked at all triples of generators and quotiented by the relation coming from the affine F_4 symmetries of P_{VI}. The resulting table has 52 rows (which is quite small considering there are 26,688 conjugacy classes of generating triples). The first 10 rows correspond to the 10 icosahedral rows of Schwarz's list and thus the projective monodromy around one of the four punctures is the identity (these correspond to the P_{VI} solution $y = t$). The remaining rows are as in Table 11.2 (this is abridged from Boalch 2006*a*). (Note that the right notion of equivalence in the linear non-rigid problem (**A**) seems to be the 'geometric equivalence' of Boalch 2006*a*, section 4 – however this coincides with equivalence under the affine F_4 Weyl group, in this case.)

Thus there are lots of other icosahedral solutions the largest having genus 7 and 72 branches. (The column 'Type' indicates the set of conjugacy classes of local monodromy of the corresponding connections of type (**A**), as we marked on Schwarz's list. The column 'Walls' indicates the number of reflection hyperplanes for the affine F_4 Weyl group that the solution's parameters θ lie on.) A few of these solutions had appeared before: those with degree <5 are simple deformations of previous solutions, solutions 21 and 26 are in Kitaev (2005) and the Dubrovin–Mazzocco icosahedral solutions are equivalent to those on rows 31,32, and 41. On the other hand the Valentiner solutions are quite far down the list on rows 37, 38, and 46.

The above method of constructing solutions using Jimbo's asymptotic formula applies only to sufficiently generic monodromy representations but it turns out that most of the rows of this table have some representative (in their affine F_4 orbit) to which Jimbo's formula maybe applied (on every branch). Thus we could start working down the list constructing new solutions. An initial goal was to get to solution 33: this solution purports to be on *none* of the reflection hyperplanes and the folklore was that all explicit solutions to Painlevé equations must lie on some reflection hyperplane. The folklore was wrong:

'Generic' solution/Icosahedral solution 33

$$y = -\frac{9s(s^2+1)(3s-4)(15s^4-5s^3+3s^2-3s+2)}{(2s-1)^2(9s^2+4)(9s^2+3s+10)},$$

$$t = \frac{27s^5(s^2+1)^2(3s-4)^3}{4(2s-1)^3(9s^2+4)^2}, \quad \text{and}$$

$$\theta = (2/5, 1/2, 1/3, 4/5).$$

So far this looks to be the only example of a 'classical' solution of any of the Painlevé equations that does not lie on a reflection hyperplane (of the full symmetry group). Apart from being in the interior of a Weyl alcove this solution is generic in another sense: a randomly chosen triple of generators of the binary icosahedral group is most likely to lead to it (more of the 26,688 triples of generators correspond to this row than to any other). Notice also that this solution has type *abcd*; there is one local monodromy in each of the four non-trivial conjugacy classes of A_5.

At this stage we were approaching solution 41 which we knew took 10 pages to write down. So we stopped and looked around to see if there were other interesting (even just topological) solutions. (The tetrahedral and octahedral cases could all now be fully dealt with Boalch 2006*c*.)

11.5 Pullbacks

In his 1884 book on the icosahedron (see Klein 1956), Klein showed that all second-order Fuchsian differential equations with finite monodromy are (essentially) pullbacks of a hypergeometric equation along a rational map f:

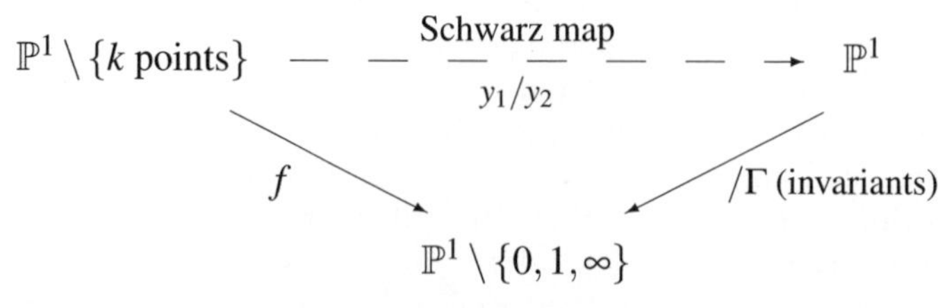

In particular ($k = 3$) all the icosahedral entries on Schwarz's list may be obtained by pulling back the '235' hypergeometric equation (on row VI of Schwarz's list).

In our context, an isomonodromic family of connections of type (**A**) amounts to a family of Fuchsian equations with five singularities (at $0, t, 1, \infty$, plus an apparent singularity at another point y).[2] Klein's theorem says each element of this family arises as the pullback of the 235 hypergeometric equation along a rational map, so the family corresponds to a family of rational maps.

Thus finding a P_{VI} solution corresponding to a family of connections (**A**) with finite monodromy amounts to giving a certain family of rational maps $f : \mathbb{P}^1 \to \mathbb{P}^1$. To construct such P_{VI} solutions one may try to find such families of rational maps, such that each map pulls back a hypergeometric equation to an equation with the right number of singular points – or to one that can be put in this form after using elementary transformations to remove extraneous apparent singularities. (This is not straightforward; e.g. given a finite monodromy representation of a connection (**A**) it is not immediate even what degree such a map f will have.)

An important further observation (due to C. Doran 2001 and A. Kitaev 2002) is that any such family of rational maps will lead to algebraic solutions of Painlevé VI regardless of whether or not the hypergeometric equation being pulled back has finite monodromy (provided the equation upstairs has the right number of poles); the algebraicity follows from that of the family of rational maps.

Andreev and Kitaev (2002); Kitaev (2002, 2005) have used this to construct some P_{VI} solutions, essentially by starting to enumerate all such rational maps (this leads to a few new solutions, but most in fact turn out to be equivalent to each other or to ones previously constructed – see Section 11.7).

On the other hand, Doran had the idea that interesting P_{VI} solutions should come from hypergeometric equations with interesting monodromy groups. Thus (amongst other things) Doran (2001) studied the possible hypergeometric equations with monodromy a *hyperbolic arithmetic triangle group* which may be pulled back to yield P_{VI} solutions. Indeed in Doran (2001, corollary 4.6), he lists such possible triangle groups and the degrees and ramification indices of the corresponding rational maps f, although no new solutions were actually constructed. We picked up on this thread in Boalch (2006*c*, section 5): it was found that all but one entry on Doran's list corresponded to a known explicit solution (although were perhaps unknown when Doran 2001 was published). The remaining entry was for a family of degree 10 rational maps f pulling back the 237 triangle group with ramification indices (partitions of 10):

$$[2, 2, 2, 2, 2], \ [3, 3, 3, 1], \ [7, 1, 1, 1]$$

over $0, 1, \infty$ (where the hypergeometric system has projective monodromy of orders 2, 3, and 7, respectively), as well as minimal ramification $[1^8, 2]$ over another variable point. As explained in Boalch (2006*c*) one can get from here to

[2] This is the same y appearing in P_{VI} – that is, the function y on the space of connections (**A**) is the position of the apparent singularity that appears when the connection is converted into a Fuchsian equation (Fuchs 1905).

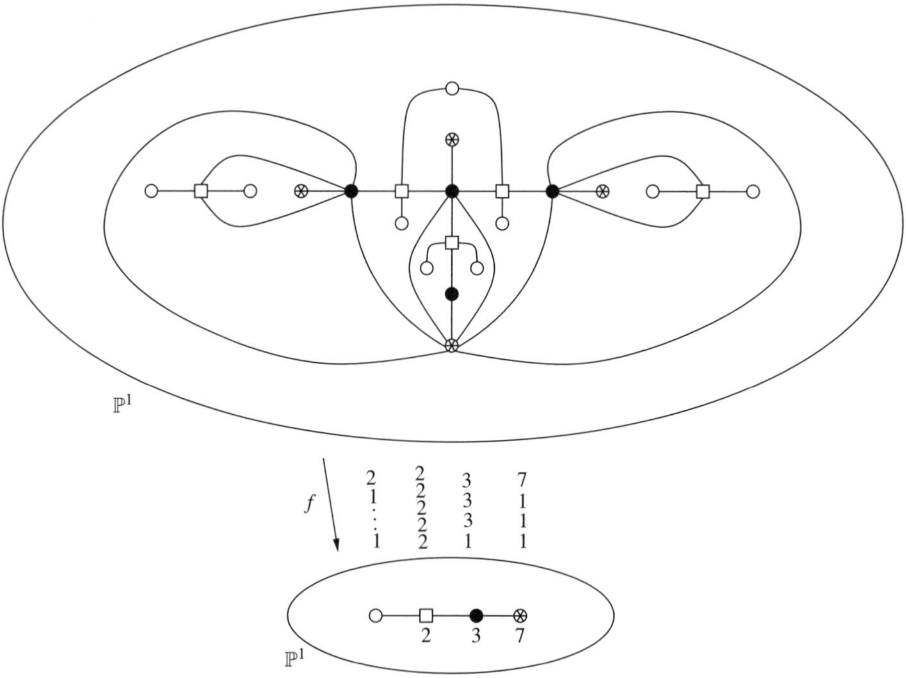

Figure 11.2 237 degree 10 rational map f.

a topological P_{VI} solution by drawing a picture: we wish to find such a rational map f topologically – that is, describe the topology of a branched cover $f : \mathbb{P}^1 \to \mathbb{P}^1$ with this ramification data. This may be done by playing 'join the dots' (completely in the spirit of Grothendieck's Dessins d'Enfants) and yields a covering figure as required. One figure so obtained is shown in Figure 11.2. (Note that, in the context of Painlevé equations, the idea of drawing pictures such as Figure 11.2 first appeared in Kitaev (2005).)

The upper copy of $\mathbb{P}^1$ is thus divided into 10 connected components and f maps each component isomorphically onto the complement of the interval drawn on the lower $\mathbb{P}^1$ (the lines and the vertices upstairs are the preimages of the lines and vertices downstairs). In particular the figure shows how loops upstairs map to words in the generators of the fundamental group $\pi_1(\mathbb{P}^1 \setminus \{0, 1, \infty\})$ downstairs. In this way we can compute by hand the monodromy of the equation upstairs obtained by pulling back a hypergeometric equation with monodromy Δ_{237}. This yields the triple:

$$M_1 = caca^{-1}c^{-1}, \quad M_2 = c, \quad \text{and} \quad M_3 = c^{-1}a^{-1}cac$$

(where a, b, and c are lifts to $\mathrm{SL}_2(\mathbb{C})$ of standard generators of Δ_{237} with $cba = 1$), which we know a priori lives in a finite $\mathcal{F}_2$ orbit. One finds immediately that the orbit through the conjugacy class of this triple has size 18 and constitutes a genus 1, degree 18 topological P_{VI} solution.

Now it turns out that Jimbo's formula may be applied to every branch of this solution, and proceeding as before we obtain (Boalch 2006*c*) the solution explicitly:

Elliptic 237 solution

$$y=\frac{1}{2}-\frac{\left(3\,s^8-2\,s^7-4\,s^6-204\,s^5-536\,s^4-1738\,s^3-5064\,s^2-4808\,s-3199\right)u}{4\,(s^6+196\,s^3+189\,s^2+756\,s+154)\,(s^2+s+7)\,(s+1)},$$

$$t=\frac{1}{2}-\frac{\left(s^9-84\,s^6-378\,s^5-1512\,s^4-5208\,s^3-7236\,s^2-8127\,s-784\right)u}{432\,s\,(s+1)^2\,(s^2+s+7)^2},$$

where $u^2=s\,(s^2+s+7)$ and $\theta=(2/7,2/7,2/7,1/3)$. (This solution, or rather an inequivalent 'Galois conjugate' of it, has also been obtained independently by Kitaev 2006, p. 219 by directly computing such a family of rational maps – apparently also influenced by Doran's list.)

11.6 Final steps

11.6.1 *Up to degree 24*

We now have an example of a degree 18 elliptic solution to Painlevé VI with a quite simple form. This leads immediately to the suspicion that the 10-page Dubrovin–Mazzocco solution is just written at a bad value of the parameters. Indeed using the method we have been 'tweaking' while working down the icosahedral table enables us to guess good a priori choices of the parameters θ within the corresponding affine F_4 equivalence class (row 41 in Table 11.2) i.e. so that the expression for the polynomial F will be 'small'. Choosing such parameters and constructing the solution from scratch at those parameters yields

Theorem 11.3 (Boalch 2006*a*) *The Dubrovin–Mazzocco icosahedral solution is equivalent to the solution*

$$y=\frac{1}{2}-\frac{8\,s^7-28\,s^6+75\,s^5+31\,s^4-269\,s^3+318\,s^2-166\,s+56}{18\,u\,(s-1)\,(3\,s^3-4\,s^2+4\,s+2)}\quad\text{and}$$

$$t=\frac{1}{2}+\frac{(s+1)\left(32\,(s^8+1)-320\,(s^7+s)+1112\,(s^6+s^2)-2420\,(s^5+s^3)+3167\,s^4\right)}{54\,u^3\,s\,(s-1)}$$

on the elliptic curve

$$u^2=s\,(8\,s^2-11\,s+8)$$

with $\theta=(1,1,1,1)/3$. *In particular this elliptic curve is birational to that defined by the 10-page polynomial.*

Substituting this into the formula of Theorem 11.2 with $\lambda = (1,1,1)/2$ and $\mu = (1,3,5)/6$ now gives explicitly the third (and trickiest) family of connections of type (**B**) with monodromy the icosahedral reflection group.

This can be pushed further with more tweaking to get up to degree 24 (row 46 in Table 11.2), that is, to obtain the largest Valentiner solution (Boalch 2006*a*) (the main further tricks used are described in (Boalch 2006*c*, appendix C). In particular this finishes the construction of all *elliptic* icosahedral solutions. Intriguingly, one finds that the resulting elliptic icosahedral Painlevé curves Π become singular only on reduction modulo the primes 2, 3, and 5 (except for rows 44 and 45 – we will see another reason in the following subsection that these are abnormal). Similarly the elliptic Painlevé curve related to the 237 triangle group becomes singular only on reduction modulo 2, 3, and 7.

11.6.2 *Quadratic/Landen/folding transformations*

Now the happy fact is that the remaining icosahedral solutions may be obtained from earlier solutions by a trick, first introduced in the context of P_{VI} by Kitaev (1991) and a simpler equivalent form was found by Ramani, Grammaticos, and Tamizhmani (2000). Manin (1998) refers to some equivalent transformations as Landen transformations. (Landen has clear precedence since the original Landen transformations were rediscovered by Gauss!) Tsuda, Okamoto, and Sakai (2005) call them folding transformations.

In any case the basic idea is simple: if one has a connection (**A**) with two local projective monodromies of order 2 (say at $0, \infty$) then one can pull it back along the map $z \mapsto z^2$ and obtain a connection with only apparent singularities at $0, \infty$ (which can be removed) and four genuine singularities. This can be normalized into the form (**A**), and the key point is that this works in families and maps isomonodromic deformations of the original connections to isomonodromic deformations of the resulting connections – that is, it transforms certain solutions of P_{VI} into different, generally inequivalent, solutions. Of course this is not a genuine symmetry of P_{VI} since special parameters are required, but it is precisely what is needed to construct the remaining solutions.

Indeed observe that each of the rows of the icosahedral table with degree >24 have type $a^2\xi\eta$ for some $\xi, \eta \in \{a, b, c, d\}$ – that is, they have two projective monodromies of order 2. Pulling back along the squaring map will transform the corresponding connections into connections of type $\xi^2\eta^2$. It turns out (in this icosahedral case) the corresponding P_{VI} solutions have half the degree, and we obtain an algebraic relation between the solutions. This program is carried out in Boalch (2007) and the remaining icosahedral solutions are obtained (see also Kitaev and Vidūnas 2007). (Notice also that the elliptic solutions on rows 44 and 45 are related in this way to earlier, genus zero solutions.) For example, in Boalch (2007) we found an explicit equation for the genus 7 algebraic curve naturally attached to the icosahedron, on which the largest (degree 72) icosahedral solution is defined: it may be modelled as the plane octic with affine equation

Genus 7 icosahedral Painlevé curve

$$9\,(p^6\,q^2+p^2\,q^6)+18\,p^4\,q^4+$$
$$4\,(p^6+q^6)+26\,(p^4\,q^2+p^2\,q^4)+8\,(p^4+q^4)+57\,p^2\,q^2+$$
$$20\,(p^2+q^2)+16=0.$$

11.7 Conclusion

Thus in conclusion we have filled in a number of rows of what could be called the *non-linear Schwarz's list*. Whether or not there will be other rows remains to be seen. So far this list of known algebraic solutions to $\mathrm{P_{VI}}$ takes the following shape (we will use the letters d and g to denote the degree and genus of solutions, and consider solutions up to equivalence under Okamoto's affine F_4 symmetry group. Some non-trivial work has been done to establish which of the published solutions are equivalent to each other and which were genuinely new). See also Boalch (2006*d*).

First there are the **rational solutions** ($d = 1$), studied by Mazzocco (2001) and Yuan and Li (2002), which fit into the set of Riccati solutions classified by Watanabe (1998). (Beware that 'rational' here means the solution is a rational function of t, which implies, but is by no means equivalent to, having a rational parameterization.)

Then there are **three continuous families** of solutions $g = 0, d = 2, 3, 4$. The degree 2 family is $y = \sqrt{t}$ which, as one may readily verify, solves $\mathrm{P_{VI}}$ for a family of possible parameter values. Similarly the degree 3 tetrahedral solution, and the degree 4 octahedral and dihedral solutions (of Dubrovin 1995 and Hitchin 1995*a*, 2003) fit into such families, as discussed in Ben Hamed and Gavrilov (2005); Boalch (2006*a*) and Cantat and Loray (2007). In general in such a family $y(t)$ may depend on the parameters of the family. Ben Hamed and Gavrilov (2005) showed that any family with $y(t)$ *not* depending on the parameters is equivalent to one of the above cases and recently Cantat and Loray (2007) showed that any solution with two, three, or four branches is in such family.

Next there is **one discrete family** (d, g unbounded, $\theta = (0, 0, 0, 1) \sim (1, 1, 1, 1)/2$). Indeed this $\mathrm{P_{VI}}$ equation was solved completely by Picard (1889, p. 299), Fuchs (1905), and in a different way by Hitchin (1995*b*). Algebraic (determinantal) formulae for the algebraic solutions amongst these appear in Hitchin (1995*a*), using links with the Poncelet problem – in this framework they are dihedral solutions (controlling connections of type (**A**) with binary dihedral monodromy).

Finally there are **45 exceptional solutions**, which collapse down to **30** if we identify solutions related by quadratic transformations. The possible genera are $0, 1, 2, 3, 7$, and the highest degree is 72. Of these 30 solutions 7 have previously appeared: 1 is due to Dubrovin (1995), 2 to Dubrovin and Mazzocco (2000), and 4 to Kitaev (3 in Kitaev 2005, plus – in Kitaev 2006 – a Galois conjugate of the elliptic 237 solution already mentioned). Two of these exceptional solutions are octahedral, 1 is the Klein solution, 3 are the elliptic 237 solution (and its 2 Galois conjugates), and the remaining 24 are icosahedral.

References

Andreev, F. V. and Kitaev, A. V. (2002). Transformations $\mathrm{RS}_4^2(3)$ of the ranks ≤ 4 and algebraic solutions of the sixth Painlevé equation. *Commun. Math. Phys.* **228**(1), 151–76. nlin.SI/0107074.

Balser, W., Jurkat, W. B. and Lutz, D. A. (1981). On the reduction of connection problems for differential equations with an irregular singularity to ones with only regular singularities, I., *SIAM J. Math. Anal.* **12**(5), 691–721.

Ben Hamed, B. and Gavrilov, L. (2005). Families of Painlevé VI equations having a common solution. *Int. Math. Res. Not.* No. 60, 3727–52. math/0507002.

Beukers, F. and Heckman, G. (1989). Monodromy for the hypergeometric function ${}_nF_{n-1}$. *Invent. Math.* **95**(2), 325–54.

Beukers, F. and van der Waall, A. (2004). Lamé equations with algebraic solutions. *J. Differential Equations* **197**(1), 1–25.

Boalch, P. P. (2002). G-bundles, isomonodromy and quantum Weyl groups. *Int. Math. Res. Not.* No. 22, 1129–66, math.DG/0108152.

Boalch, P. P. (2003). Painlevé equations and complex reflections. *Ann. Inst. Fourier* **53**(4), 1009–22.

Boalch, P. P. (2005). From Klein to Painlevé via Fourier, Laplace and Jimbo. *Proc. London Math. Soc.* **90**(3), no. 3, 167–208, math.AG/0308221.

Boalch, P. P. (2006*a*). The fifty-two icosahedral solutions to Painlevé VI. *J. Reine Angew. Math.* **596**, 183–214, math.AG/0406281.

Boalch, P. P. (2006*b*). Six results on Painlevé VI. *Sémin. Congr.*, Vol. 14, Society of Mathematics France, Paris, math.AG/0503043, pp. 1–20.

Boalch, P. P. (2006*c*). Some explicit solutions to the Riemann–Hilbert problem. *IRMA Lect. Math. Theor. Phys.* **9**, 85–112, math.DG/0501464.

Boalch, P. P. (2006*d*). Survey of the known algebraic solutions of Painlevé VI. Two talks given at the Newton Institute 71pp of transparencies on their webpage www.tinyurl.com/boalchINI1, 2.

Boalch, P. P. (2007). Higher genus icosahedral Painlevé curves. *Funk. Ekvac. (Kobe)* **50**(1), 19–32, math.AG/0506407.

Cantat, S. and Loray, F. (2007). Holomorphic dynamics, Painlevé VI equation and character varieties, arXiv:0711.1579. Preprint Nov.

Cayley, A. (1849). On the triple tangent planes of surfaces of the third order. *Dublin Math. J.* No. 4, 118–38.

Coleman, A. J. (1989). Killing and Coxeter transformations of Kac-Moody algebras. *Invent. Math.* **95**, 447–77.

Dettweiler M. and Reiter, S. (2000). An algorithm of Katz and its application to the inverse Galois problem. *J. Symbolic Comput.* **30**(6), 761–98.

Dettweiler M. and Reiter, S. (2007). Painlevé equations and the middle convolution. *Adv. Geom.* **7**(3), 317–30. (See also preprint dated October 1, 2004).

Doran, C. F. (2001). Algebraic and geometric isomonodromic deformations. *J. Differential Geom.* **59**(1), 33–85.

Dubrovin, B. (1995). Geometry of 2D topological field theories. *Integrable Systems and Quantum Groups*, M. Francaviglia and S. Greco (eds.), Vol. 1620, Springer Lecture Notes in Mathematics, pp. 120–348. hep-th/9407018.

Dubrovin, B. (1999). Painlevé transcendents in two-dimensional topological field theory. *The Painlevé Property*, Springer, New York, pp. 287–412, math/9803107.

Dubrovin B. and Mazzocco, M. (2000). Monodromy of certain Painlevé-VI transcendents and reflection groups. *Invent. Math.* **141**(1), 55–147, math.AG/9806056, SISSA preprint 1997.

Fricke R. and Klein, F. (1897). *Vorlesungen über die Theorie der automorphen Funktionen. I*, Druck und Verlag von B. G. Teubner, Leipzig, Germany, p. 366.

Fuchs, R. (1905). Sur quelques équations différentielles linéaires du seconde ordre. *C. R. Acad. Sci. Paris* **141**, 555–8.

Harnad, J. (1994). Dual isomonodromic deformations and moment maps to loop algebras. *Commun. Math. Phys.* **166**, 337–65. rep-th/9301076.

Hitchin, N. J. (1995*a*). Poncelet polygons and the Painlevé equations. *Geometry and analysis (Bombay, 1992)*, Tata Institute of Fundamental Research, Bombay, pp. 151–85.

Hitchin, N. J. (1995*b*). Twistor spaces, Einstein metrics and isomonodromic deformations. *J. Differential Geom.* **42**(1), 30–112.

Hitchin, N. J. (2003). A lecture on the octahedron. *Bull. London Math. Soc.* **35**, 577–600.

Iwasaki, K. (2008). *Finite branch solutions to Painlevé VI around a fixed singular point. Adv. Math* **217**(5), 1889–1934. arXiv:0704.0679.

Iwasaki, K. (2002). A modular group action on cubic surfaces and the monodromy of the Painlevé VI equation. *Proc. Japan Acad., Ser. A* **78**, 131–5.

Jimbo, M. (1982). Monodromy problem and the boundary condition for some Painlevé equations. *Publ. Res. Inst. Math. Sci.* **18**(3), 1137–61, MR 85c:58050.

Jimbo, M. and Miwa, T. (1981). Monodromy preserving deformations of linear differential equations with rational coefficients II. *Physica* **2D**, 407–48.

Katz, N. M. (1996). Rigid local systems. *Annals of Mathematics Studies*, Vol. 139, Princeton University Press, Princeton, NJ.

Killing, W. (1889). Die Zusammensetzung der stetigen endlichen Transformationsgruppen II. *Math. Ann.* **33**, 1–48. Available via http://dz1.gdz-cms.de.

Kitaev, A. V. (1991). Quadratic transformations for the sixth Painlevé equation. *Lett. Math. Phys.* **21**(2), 105–11.

Kitaev, A. V. (2002). Special functions of isomonodromy type, rational transformations of the spectral parameter, and algebraic solutions of the sixth Painlevé equation. *Algebra i Analiz* **14**(3), 121–39. nlin.SI/0102026.

Kitaev, A. V. (2005). Grothendieck's dessins d'enfants, their deformations, and algebraic solutions of the sixth Painlevé and Gauss hypergeometric equations. *Algebra i Analiz* **17**(1), 224–75, nlin.SI/0309078.

Kitaev, A. V. (2006). Remarks towards a classification of $\mathrm{RS}_4^2(3)$-transformations and algebraic solutions of the sixth Painlevé equation. *Sémin. Congr.*, Vol. 14, Society of Mathematics France, Paris, pp. 199–227. math. CA/0503082.

Kitaev A. V. and Vidūnas, R. (2007). *Quadratic transformations of the sixth Painlevé equation with application to algebraic solutions*, math.CA/0511149. Math. Nachr. 280(16), 1834–1855.

Klein, F. (1956). *Lectures on the Icosahedron and the Solution of Equations of the Fifth Degree*, Dover Publications Inc., New York.

Manin, Yu. I. (1998). Sixth Painlevé equation, universal elliptic curve, and mirror of $\mathbf{P}^2$, Geometry of differential equations. *AMS Transl. Ser. 2*, Vol. 186, AMS, Providence, RI, pp. 131–51. alg-geom/9605010.

Mason L. J. and Woodhouse, N. M. J. (1996). *Integrability, Self-Duality and Twistor Theory*, O.U.P., Oxford.

Mazzocco, M. (2001). Rational solutions of the Painlevé VI equation. *J. Phys. A* **34**(11), 2281–94. See also nlin.SI/0007036 v2, March 2004.

Mazzocco, M. (2002). Painlevé sixth equation as isomonodromic deformations equation of an irregular system. *The Kowalevski Property* (Leeds, 2000), CRM Proc. Lecture Notes, Vol. 32, American Mathematical Society, Providence, RI, pp. 219–38.

Nitsure, N. (1993). Moduli of semistable logarithmic connections. *J. Am. Math. Soc.* **6**(3), 597–609.

Okamoto, K. (1987). Studies on the Painlevé equations. I. Sixth Painlevé equation P_{VI}. *Ann. Mat. Pura Appl. (4)* **146**, 337–81.

Picard, E. (1889). Mémoire sur la théorie des fonctions algébriques de deux variables. *J. de Math. Pures et Appliquées* (dit '*Journal de Liouville*') **4**(5), 135–320. Available via Gallica-Math.

Ramani, A. Grammaticos, B. and Tamizhmani, T. (2000). Quadratic relations in continuous and discrete Painlevé equations. *J. Phys. A* **33**(15), 3033–44.

Saito, K. (1993). On a linear structure of the quotient variety by a finite reflexion group. *Publ. Res. Inst. Math. Sci.* **29**(4), 535–79.

Schwarz, H. A. (1873). Über diejenigen Fälle, in welchen die Gaussische hypergeometrische reihe eine algebraische Funktion ihres vierten Elements darstellt. *J. Reine Angew. Math.* **75**, 292–335.

Shephard G. C. and Todd, J. A. (1954). Finite unitary reflection groups. *Can. J. Math.* **6**, 274–304.

Simpson, C. T. (1994). Moduli of representations of the fundamental group of a smooth projective variety, I, II, *Publ. Math. I.H.E.S.* **79** and **80**, 47–129, and 5–79.

Tsuda, T., Okamoto, K. and Sakai, H. (2005). Folding transformations of the Painlevé equations. *Math. Ann.* **331**, 713–38.

Watanabe, H. (1998). Birational canonical transformations and classical solutions of the sixth Painlevé equation. *Ann. Scuola Norm. Sup. Pisa (4)* **27**(3–4), 379–425.

Yuan, W. and Li, Y. (2002). Rational solutions of Painlevé equations. *Can. J. Math.* **54**(3), 648–70.

XII

AN INTRODUCTION TO BUNDLE GERBES

Michael K. Murray

Dedicated to Nigel Hitchin on the occasion of his 60th birthday

Over the years I have had many interesting mathematical conversations with Nigel and regularly came away with a solution to a problem or a new idea. While preparing this chapter I was trying to recall when he first told me about gerbes. I thought for awhile that age was going to get the better of my memory as many conversations seemed to have blurred together. But then I discovered that the annual departmental research reports really do have their uses. In July 1992 I attended the 'Symposium on gauge theories and topology' at Warwick and reported in the 1992 Departmental Research Report that:

I ... had discussions with Nigel Hitchin about 'gerbes'. These are a generalisation of line bundles ...

Further searching of my electronic files revealed an order for Brylinski's book *Loop Spaces, Characteristic Classes and Geometric Quantization* on 29 April 1993. I recall that the book took some months to makes its away across the sea to Australia during which time I pondered the advertising material I had which said that gerbes were fibrations of groupoids. Trying to interpret this led to a paper on bundle gerbes which I submitted to Nigel in his role as a London Mathematical Society Editor on 25 July 1994. The Departmental Research Report of the same year reports that:

This year I began some work on a geometric construction called a bundle gerbe. These provide a geometric realisation of the three dimensional cohomology of a manifold.

My sincere thanks to Nigel for introducing me to gerbes and for the many other fascinating insights into mathematics that he has given me over the years.

12.1 Introduction

The theory of gerbes began with Giraud (1971) and was popularized in the book by Brylinski (1993). A short introduction by Nigel Hitchin (2003) in the 'What is a gerbe?' series can be found in the *Notices of the AMS.* Gerbes provide a geometric realization of the three-dimensional cohomology of a manifold in a manner analogous to the way a line bundle is a geometric realization of two-dimensional cohomology. Part of the reason for their recent popularity is applications to string theory in particular the notion of the B-field. Strings on a manifold are elements in the loop space of the manifold and we would expect their quantization to involve a Hermitian line bundle on the loop space arising from a two class on the loop space. That two class can arise as the transgression of some three class on the underlying manifold. Gerbes provide a geometrization of this process. String theory however is not the only application of gerbes and we refer the interested reader to the related work of Hitchin (2001, 2006) which applies gerbes to generalized geometry and to reviews such as (Carey *et al.* 2000) and (Mickelsson 2006) which give applications of gerbes to other problems in quantum field theory.

As with everything else in the theory of gerbes, the relationship of bundle gerbes to gerbes is best understood by comparison with the case of Hermitian line bundles or equivalently $U(1)$ (principal) bundles. There are basically three ways of thinking about $U(1)$ bundles over a manifold M:

1. A certain kind of locally free sheaf on M
2. A co-cycle $g_{\alpha\beta}\colon U_\alpha \cap U_\beta \to U(1)$ for some open cover $\mathcal{U} = \{U_\alpha \mid \alpha \in I\}$ of M
3. A principal $U(1)$ bundle $P \to M$

In the case of gerbes over M we can think of these as

1. A certain kind of sheaf of groupoids on M (Giraud and Brylinski)
2. A co-cycle $g_{\alpha\beta\gamma}\colon U_\alpha \cap U_\beta \cap U_\gamma \to U(1)$ for some open cover $\mathcal{U} = \{U_\alpha \mid \alpha \in I\}$ of M or alternatively a choice of $U(1)$ bundle $P_{\alpha\beta} \to U_\alpha \cap U_\beta$ for each double overlap (Hitchin and Chatterjee)
3. A bundle gerbe (Murray)

Note that we are slightly abusing the definition of gerbe here as what we are considering are gerbes with band the sheaf of smooth functions from M into $U(1)$. There are more general kinds of gerbes on M just as there are more general kinds of sheaves on M beyond those arising as the sheave of sections of a Hermitian line bundle.

Recall some of the basic facts about $U(1)$ bundles on a manifold M:

1. If $P \to M$ is a $U(1)$ bundle there is a dual bundle $P^* \to M$ and if $Q \to M$ is another $U(1)$ bundle there is a *product* $P \otimes Q \to M$.

2. If $f\colon N \to M$ is a smooth map there is a *pullback* bundle $f^*(P) \to N$ and this behaves well with respect to dual and product. That is, $f^*(P^*)$ and $(f^*(P))^*$ are isomorphic as also are $f^*(P \otimes Q)$ and $f^*(P) \otimes f^*(Q)$.
3. Associated to a $U(1)$ bundle $P \to M$ is a characteristic class, the chern class, $c(P) \in H^2(M, \mathbb{Z})$, which is natural with respect to pullback, that is, $f^*(c(P)) = c(f^*(P))$ and additive with respect to product and dual, that is, $c(P \otimes Q) = c(P) + c(Q)$ and $c(P^*) = -c(P)$.
4. $P \to M$ is called *trivial* if it is isomorphic to $M \times U(1)$ or equivalently admits a global section. P is trivial if and only if $c(P) = 0$.
5. There is a notion of a *connection* on $P \to M$. Associated to a connection A on P is a closed two-form F_A called the *curvature* of A with the property that $F_A/2\pi i$ is a de Rham representative for the image of $c(P)$ in real cohomology.
6. If $\gamma\colon S^1 \to M$ is a loop in M and $P \to M$ a $U(1)$ bundle with connection A then parallel transport around γ defines the *holonomy*, $\mathrm{hol}(\mathrm{A}, \gamma)$ of A around γ which is an element of $U(1)$. If γ is the boundary of a disc $D \subset M$ then we have

$$\mathrm{hol}(A, \partial D) = \exp\left(\int_D F_A\right).$$

A gerbe is an attempt to generalize all the above facts about $U(1)$ bundles to some new kind of mathematical object in such a way that the characteristic class is in three-dimensional cohomology. Obviously for consistency other dimensions then have to change. In particular the curvature should be a three-form and holonomy should be over two-dimensional submanifolds. It turns out to be useful to consider the general case of any dimension of cohomology which we call a p-gerbe. For historical reasons a p-gerbe has a characteristic class in $H^{p+2}(M, \mathbb{Z})$ so the interesting values of p are $-2, -1, 0, 1, \ldots$ with $U(1)$ bundles corresponding to $p = 0$.

A p-gerbe then is some mathematical object which represents $(p+2)$-dimensional cohomology. To make completely precise what representing $(p+2)$-dimensional cohomology means would take us to far afield from the present topic, but we give a sketch here to motivate the behaviour we are looking for in p-gerbes. To this end we will assume our p-gerbes P live in some category $\mathcal{G}$ and there is a (forgetful) functor $\Pi\colon \mathcal{G} \to$ **Man** the category of manifolds. The functor Π and the category $\mathcal{G}$ have to satisfy

1. If P is a p-gerbe there is a dual p-gerbe P^* and if Q is another p-gerbe there is a *product* p-gerbe $P \otimes Q$. In other words $\mathcal{G}$ is monoidal and has a dual operation.
2. If $f\colon N \to M$ is a smooth map and $\Pi(P) = M$ there is a *pullback* p-gerbe $f^*(P)$ and a morphism $\hat{f}\colon f^*(P) \to P$ such that $\Pi(f^*(P)) = N$ and $\pi(\hat{f}) = f$. Pullback should behave well with respect to dual and product. That is,

$f^*(P^*)$ and $(f^*(P)^*$ should be isomorphic as also should be $f^*(P \otimes Q)$ and $f^*(P) \otimes f^*(Q)$.

3. Associated to a p-gerbe P is a characteristic class $c(P) \in H^{p+2}(\Pi(P), \mathbb{Z})$, which is natural with respect to pullback, that is, $f^*(c(P)) = c(f^*(P))$ and additive with respect to product and dual, that is, $c(P \otimes Q) = c(P) + c(Q)$ and $c(P^*) = -c(P)$.
4. As well as the notion of P and Q being isomorphic there is a possibly weaker notion of *equivalence* where P and Q are equivalent if and only if $c(P) = c(Q)$. We say P is *trivial* if $c(P) = 0$.
5. There is a notion of a *connective structure* A on P. Associated to a connective structure A on P is a closed $(p+2)$-form ω on $\Pi(P)$ called the *$(p+2)$-curvature* of A with the property that $\omega/2\pi i$ is a de Rham representative for the image of $c(P)$ in real cohomology.
6. If $X \subset \Pi(P)$ is an oriented $(p+1)$-dimensional submanifold of $\Pi(P)$ we should be able to define the holonomy of the connective structure $\text{hol}(A, X) \in U(1)$ over X. Moreover if $Y \subset \Pi(M)$ is an oriented $(p+2)$-dimensional submanifold with boundary then we want to have that

$$\text{hol}(\text{A}, \partial Y) = \exp\left(\int_Y F_\omega\right).$$

Clearly by construction the category of $U(1)$ bundles, with the forgetful functor which assigns to a $U(1)$ bundle its base manifold, is an example of a 0-gerbe. Before we consider other examples we need some facts about bundles with structure group an abelian Lie group H. If $P \to M$ is an H bundle on M then by choosing local sections of P for an open cover $\mathcal{U} = \{U_\alpha \mid \alpha \in I\}$ we can construct transition functions $h_{\alpha\beta} \colon U_\alpha \cap U_\beta \to H$ and in the usual way this defines a class $c(P) \in H^1(M, H)$ where here we abuse notation and write H for what is really the sheaf of smooth functions with values in H. It is a standard fact that isomorphism classes of H bundles are in bijective correspondence with $H^1(M, H)$ in this manner. If $P \to M$ is an H bundle we can define its *dual* as follows. Let P^* be isomorphic to P as a manifold with projection to M and for convenience let $p^* \in P^*$ denote $p \in P$ thought of as an element of P^*. Then define a new H action on P^* by $p^*h = (ph^{-1})^*$. It is ovious that if $h_{\alpha\beta}$ are transition functions for P then $h^*_{\alpha\beta} = h^{-1}_{\alpha\beta}$ are transition functions for P^*. In particular we have that $c(P^*) = -c(P)$ if we write the group structure on $H^1(M, H)$ additively. If Q is another H bundle we can form the fibre product $P \times_M Q$ and let H act on it by $(p, q)h = (ph, qh^{-1})$. Denote the orbit of (p, q) under this action by $[p, q]$ and define an H action by $[p, q]h = [p, qh] = [ph, q]$. The resulting H bundle is denoted by $P \otimes Q \to M$. If $h_{\alpha\beta}$ are transition functions for P and $k_{\alpha\beta}$ are transition functions for Q then $h_{\alpha\beta}k_{\alpha\beta}$ are transition functions for $P \otimes Q$ and thus $c(P \otimes Q) = c(P) + c(Q)$. Notice that these constructions will not generally work for non-abelian groups because in such a case the action of H on P^* is not a right action and the action on $P \otimes Q$ is not even well-defined.

Example 12.1 The simplest example is that of functions $f\colon M \to \mathbb{Z}$ where we define the functor Π by $\Pi(f) = M$. The degree of f is the class induced in $H^0(M, \mathbb{Z})$ so functions from M to $\mathbb{Z}$ are -2-gerbes over M. Product and dual are pointwise addition and negation. There is no sensible notion of connective structure.

Example 12.2 Consider next principal $\mathbb{Z}$ bundles $P \to M$. Clearly we want the functor Π to be $\Pi(P) = M$ and pullbacks are well known to exist. As $\mathbb{Z}$ is abelian the constructions above apply and there are duals and products. The isomorphism class of a bundle is determined by a class in $H^1(M, \mathbb{Z})$ so $\mathbb{Z}$ bundles are $p = -1$ gerbes. A $\mathbb{Z}$ bundle is trivial as a -1-gerbe if and only if it is trivial as a $\mathbb{Z}$ bundle.

It is not immediately obvious what a connective structure on a $\mathbb{Z}$ bundle is but it turns out that the correct notion is that of a $\mathbb{Z}$ equivariant map $\hat{\phi}\colon P \to i\mathbb{R}$ where the action of $n \in \mathbb{Z}$ on $i\mathbb{R}$ is addition of $2\pi i n$ so that $\hat{\phi}(pn) = \hat{\phi}(p) + 2\pi i n$. The map $\hat{\phi}$ then descends to a map $\phi\colon M \to S^1$ and the class of the bundle is the degree of this map. The pullback of the standard one-form on $\mathbb{R}$, that is, $d\hat{\phi}$ is a one-form on P which descends to a one-form $\phi^{-1}d\phi$ on M. The de Rham class $(\phi^{-1}d\phi)/2\pi i$ is the image of the class of the bundle in real cohomology.

We expect holonomy to be over a $-1+1 = 0$-dimensional submanifold. If $m_1, \ldots, m_r$ is a collection of points in M with each m_i oriented by some $\epsilon_i \in \{\pm 1\}$ let us denote by $\sum \epsilon_i m_i$ their union as an oriented zero-dimensional submanifold of M. Then we define

$$\operatorname{hol}\left(\hat{\phi}, \sum \epsilon_i m_i\right) = \prod_{i=1}^{\mathrm{r}} \phi(\mathrm{m}_i)^{\epsilon_i}.$$

In the case of an oriented one-dimensional submanifold $X \subset M$ with ends $-X_0$ and $+X_1$ the fundamental theorem of calculus tells us that

$$\operatorname{hol}\left(\hat{\phi}, X_1 - X_0\right) = \exp\left(\int_X d\phi\right).$$

Notice that if we express a $\mathbb{Z}$ bundle locally in terms of transitions functions these are maps of the form $f_{\alpha\beta}\colon U_\alpha \cap U_\beta \to \mathbb{Z}$. That is, over each double overlap we have a -2-gerbe.

Example 12.3 It is clear from the above example that maps $\phi\colon M \to U(1)$ are also -1-gerbes with a connective structure. The dual and product are just pointwise inverse and pointwise product. The class is the degree and the connective structure is included automatically as part of ϕ.

We can also forget that there is a natural connective structure and just regard maps $\phi\colon M \to U(1)$ as -1-gerbes. In that case the natural notion of isomorphism between two maps $\phi, \chi\colon M \to U(1)$ would be equality. However two such maps have the same degree if and only if they are homotopic. So the notion of equivalence of maps $\phi\colon M \to U(1)$, thought of as -1-gerbes (without connective structure), should be homotopy and is different to the notion of isomorphism.

Example 12.4 As we have remarked $U(1)$ bundles are, of course, $p = 0$ gerbes. Notice that locally a $U(1)$ bundle is given by transition functions $g_{\alpha\beta}\colon U_\alpha \cap U_\beta \to U(1)$, that is, on each double overlap we have a -1-gerbe (with connective structure).

We will see below that this pattern of a p-gerbe being defined as a $(p-1)$-gerbe on double overlaps of some open cover is exploited by Hitchin and Chatterjee to give a definition of a 1-gerbe. But first we need some additional background material.

12.2 Background

We will be interested in surjective submersions $\pi\colon Y \to M$ which we regard as generalizations of open covers. In particular if $\mathcal{U} = \{U_\alpha \mid \alpha \in I\}$ is an open cover we have the disjoint union

$$Y_{\mathcal{U}} = \{(x, \alpha) \mid x \in U_\alpha\} \subset M \times I$$

with projection map $\pi(x, \alpha) = x$. The surjective morphism $\pi\colon Y_{\mathcal{U}} \to M$ is called the *nerve* of the open cover $\mathcal{U}$.

A morphism of surjective submersions $\pi\colon Y \to M$ and $p\colon X \to M$ is a map $\rho\colon Y \to X$ covering the identity, that is, $p \circ \rho = \pi$. Any surjective submersion $\pi\colon Y \to M$ admits local sections so there is an open cover $\mathcal{U}$ of M and local sections $s_\alpha\colon U_\alpha \to Y$ of π. These local sections define a morphism $s\colon Y_{\mathcal{U}} \to Y$ by $s(x, \alpha) = s_\alpha(x)$. Indeed any morphism $Y_{\mathcal{U}} \to Y$ will be of this form. If $\mathcal{V} = \{V_\alpha \mid \alpha \in J\}$ is a refinement of $\mathcal{U}$, that is, there is a map $\rho\colon J \to I$ such that for every $\alpha \in J$ we have $V_\alpha \subset U_\rho(\alpha)$, we have a morphism of surjective submersions $Y_{\mathcal{V}} \to Y_{\mathcal{U}}$ defined by $(\alpha, x) \mapsto (\rho(\alpha), x)$.

Given a surjective morphism $\pi\colon Y \to M$ we can form the p-fold fibre product

$$Y^{[p]} = \{(y_1, \ldots, y_p) \mid \pi(y_1) = \cdots = \pi(y_p)\} \subset Y^p.$$

The submersion property of π implies that $Y^{[p]}$ is a submanifold of Y^p. There are smooth maps $\pi_i\colon Y^{[p]} \to Y^{[p-1]}$, for $i = 1, \ldots, p$, defined by omitting the ith element. We will be interested in two particular examples.

Example 12.5 If $\mathcal{U}$ is an open cover of M then the pth fibre product $Y_{\mathcal{U}}^{[p]}$ is the disjoint union of all the *ordered* p-fold intersections. For example, if $\mathcal{U} = \{U_1, U_2\}$ is an open cover of M then $Y_{\mathcal{U}}^{[2]}$ is the disjoint union of $U_1 \cap U_2$, $U_2 \cap U_1$, $U_1 \cap U_1$, and $U_2 \cap U_2$.

Example 12.6 If $P \to M$ is a principal G bundle then $P \to M$ is a surjective submersion. It is easy to show that $P^{[p]} = P \times G^{p-1}$. In particular $P^{[2]} = P \times G$ and we shall need later the related fact that there is a map $g\colon P^{[2]} \to G$ defined by $p_1 g(p_1, p_2) = p_2$.

Let $\Omega^q(Y^{[p]})$ be the space of differential q-forms on $Y^{[p]}$. Define

$$\delta\colon \Omega^q(Y^{[p-1]}) \to \Omega^q(Y^{[p]}) \quad \text{by} \quad \delta = \sum_{i=1}^{p} (-1)^{p-1}\pi_i^*.$$

These maps form the *fundamental complex*

$$0 \to \Omega^q(M) \xrightarrow{\pi^*} \Omega^q(Y) \xrightarrow{\delta} \Omega^q(Y^{[2]}) \xrightarrow{\delta} \Omega^q(Y^{[3]}) \xrightarrow{\delta} \dots$$

and from Murray (1996) we have

Proposition 12.1 *The fundamental complex is exact for all $q \geq 0$.*

Note that if $Y = Y_{\mathcal{U}}$ then this proposition is a well-known result about the Čech de Rham double complex (see e.g. Bott and Tu's book 1982).

Finally we need some notation. Let H be an abelian group. If $g\colon Y^{[p-1]} \to H$ we define $\delta(g)\colon Y^{[p]} \to H$ by

$$\delta(g) = (g \circ \pi_1) - (g \circ \pi_2) + (g \circ \pi_3) \cdots .$$

If $P \to Y^{[p-1]}$ is an H bundle we define an H bundle $\delta(P) \to Y^{[p]}$ by

$$\delta(P) = \pi_1^*(P) \otimes (\pi_2^*(P))^* \otimes \pi_3^*(P) \otimes \cdots .$$

It is easy to check that $\delta(\delta(g)) = 1$ and that $\delta(\delta(P))$ is canonically trivial.

12.3 Bundle gerbes

Definition 12.1 *A* bundle gerbe *(Murray 1996) over M is a pair (P, Y) where $Y \to M$ is a surjective submersion and $P \to Y^{[2]}$ is a $U(1)$ bundle satisfying*

1. *There is a* bundle gerbe multiplication *which is a smooth isomorphism*

$$m\colon \pi_3^*(P) \otimes \pi_1^*(P) \to \pi_2^*(P)$$

of $U(1)$ bundles over $Y^{[3]}$.

2. *This multiplication is associative, that is, if we let $P_{(y_1,y_2)}$ denote the fibre of P over (y_1, y_2) then the following diagram commutes for all $(y_1, y_2, y_3, y_4) \in Y^{[4]}$:*

$$\begin{array}{ccc} P_{(y_1,y_2)} \otimes P_{(y_2,y_3)} \otimes P_{(y_3,y_4)} & \to & P_{(y_1,y_3)} \otimes P_{(y_3,y_4)} \\ \downarrow & & \downarrow \\ P_{(y_1,y_2)} \otimes P_{(y_2,y_4)} & \to & P_{(y_1,y_4)} \end{array}$$

We remark that for any $(y_1, y_2, y_3) \in Y^{[3]}$ the bundle gerbe multiplication defines an isomorphism:

$$m\colon P_{(y_1,y_2)} \otimes P_{(y_2,y_3)} \to P_{(y_1,y_3)}$$

of $U(1)$ spaces.

We can show using the bundle gerbe multiplication that there are natural isomorphisms $P_{(y_1,y_2)} \cong P^*_{(y_2,y_1)}$ and $P_{(y,y)} \simeq U(1)$.

We can rephrase the existence and associativity of the bundle gerbe multiplication to an equivalent pair of conditions in the following way. The bundle gerbe multiplication gives rise to a section s of $\delta(P) \to Y^{[3]}$. Moreover $\delta(s)$ is a section of $\delta(\delta(P)) \to Y^{[4]}$. But $\delta(\delta(P))$ is canonically trivial so it makes sense to ask that $\delta(s) = 1$. This is the condition of associativity. The family of spaces $\{Y^{[p]} \mid p = 1, 2, \dots\}$ is an example of a simplicial space (Dupont 1978) and by comparing to Brylinski and McLaughlin (1994) we see that a bundle gerbe is the same thing as a *simplicial line bundle* over this particular simplicial space.

Example 12.7 If we replace Y in the definition by $Y_{\mathcal{U}}$ for some open cover $\mathcal{U}$ of M we obtain the definition of gerbe given by Hitchin (2001) and by his student Chatterjee (1998). This consists of choosing an open cover $\mathcal{U}$ of M and a family of $U(1)$ bundles $P\colon U_\alpha \cap U_\beta$ such that over triple overlaps we have sections

$$s_{\alpha\beta\gamma} \in \Gamma(U_\alpha \cap U_\beta \cap U_\gamma \mid P_{\beta\gamma} \otimes P^*_{\alpha\gamma} \otimes P_{\alpha\beta})$$

and we require that $\delta(s) = 1$ in the appropriate way.

Example 12.8 The simplest example of a line bundle is given by the clutching construction on the two sphere S^2. If U_0 and U_1 are the open neighbourhoods of the north and south hemispheres we take the transition function $g\colon U_0 \cap U_1 \to U(1)$ to have winding number 1. As there are only two open sets there is no condition on triple overlaps and we obtain the $U(1)$ bundle over S^2 of chern class 1. In a similar fashion we can consider U_0 and U_1 to be open neighbourhoods of the north and south hemispheres of the three-sphere S^3. Their intersection is retractable to the two-sphere so we can choose over this the $U(1)$ bundle P of chern class 1. Again there are no additional conditions and we obtain the gerbe of degree 1 over S^3.

Example 12.9 Hitchin and Chatterjee also consider a gerbe as in Example 12.7 but with the added requirement that each $P_{\alpha\beta}$ is trivial in the form $P_{\alpha\beta} = U_\alpha \cap U_\beta \times U(1)$. Writing elements of the disjoint union $Y_{\mathcal{U}}^{[2]}$ as (α, β, x) where $x \in U_\alpha \cap U_\beta$ we see that the bundle gerbe multiplication must take the form

$$((\alpha, \beta, x), z) \otimes ((\beta, \gamma, x), w) \mapsto ((\alpha, \gamma, x), zw g_{\alpha\beta\gamma}(x))$$

for some $g_{\alpha\beta\gamma}\colon U_\alpha \cap U_\beta \cap U_\gamma \to U(1)$ and will be associative precisely when $g_{\alpha\beta\gamma}$ is a co-cycle.

We will refer to gerbes of the forms in Examples 12.7 and 12.9 as *Hitchin–Chatterjee gerbes*. The connection with bundle gerbes is simple. For clarity we define

Definition 12.2 *A bundle gerbe (P, Y) over M is called* local *if $Y = Y_{\mathcal{U}}$ for some open cover $\mathcal{U}$ of M.*

We then obviously have

Proposition 12.2 *A Hitchin–Chatterjee gerbe is the same thing as a local bundle gerbe.*

If (P, Y) is a bundle gerbe over M then associated to every point m of M we have a *groupoid* constructed as follows. The objects are the elements of the fibre Y_m and the morphisms between y_1 and y_2 in Y_m are $P_{(y_1,y_2)}$. Composition comes from the bundle gerbe multiplication. If we call a groupoid a $U(1)$ groupoid if it is transitive and the group of morphisms of a point is isomorphic to $U(1)$, then the algebraic conditions on the bundle gerbe (i.e. the multiplication and its associativity) are captured precisely by saying that a bundle gerbe is a bundle of $U(1)$ groupoids over M.

We now consider the properties given in Section 12.1 which we would like a 1-gerbe to satisfy and show how they are satisfied by bundle gerbes.

12.3.1 *Pullback*

If $f\colon N \to M$ then we can pullback $Y \to M$ to $f^*(Y) \to N$ with a map $\hat{f}\colon f^*(Y) \to Y$ covering f. There is an induced map $\hat{f}^{[2]}\colon f^*(Y)^{[2]} \to Y^{[2]}$. Let

$$f^*(P, Y) = (\hat{f}^{[2]*}(P), f^*(Y)).$$

To see this is a bundle gerbe notice that all this is doing is pulling back the $U(1)$ groupoid at $f(n) \in M$ and placing it at $n \in N$ so we have a bundle of $U(1)$ groupoids over N and thus a bundle gerbe.

12.3.2 *Dual and product*

If (P, Y) is a bundle gerbe then $(P, Y)^* = (P^*, Y)$ is also a bundle gerbe called the *dual* of (P, Y).

If (P, Y) and (Q, X) are bundle gerbes we can form the fibre product $Y \times_M X \to M$, a new surjective submersion and then define a $U(1)$ bundle

$$P \otimes Q \to (Y \times_M X)^{[2]} = Y^{[2]} \times_M X^{[2]}$$

by

$$(P \otimes Q)_{((y_1,x_1),(y_2,x_2))} = P_{(y_1,y_2)} \otimes Q_{(x_1,x_2)}.$$

We define $(P, Y) \otimes (Q, X) = (P \otimes Q, Y \times_M X)$.

12.3.3 *Characteristic class*

The characteristic class of a bundle gerbe is called the *Dixmier–Douady* class. We construct it as follows. Choose a good cover $\mathcal{U}$ of M (Bott and Tu 1982) with sections $s_\alpha\colon U_\alpha \to Y$. Then

$$(s_\alpha, s_\beta)\colon U_\alpha \cap U_\beta \to Y^{[2]}$$

is a section. Choose a section $\sigma_{\alpha\beta}$ of $P_{\alpha\beta} = (s_\alpha, s_\beta)^*(P)$. That is, some

$$\sigma_{\alpha\beta}\colon U_\alpha \cap U_\beta \to P$$

such that $\sigma_{\alpha\beta}(x) \in P_{(s_\alpha(x), s_\beta(x))}$. Over triple overlaps we have

$$m(\sigma_{\alpha\beta}(x), \sigma_{\beta\gamma}(x)) = g_{\alpha\beta\gamma}(x)\sigma_{\alpha\gamma}(x) \in P_{(s_\alpha(x), s_\gamma(x))}$$

for $g_{\alpha\beta\gamma}\colon U_\alpha \cap U_\beta \cap U_\gamma \to U(1)$. This defines a co-cycle which is the Dixmier–Douady class:

$$DD((P,Y)) = [g_{\alpha\beta\gamma}] \in H^2(M, U(1)) = H^3(M, \mathbb{Z}).$$

Example 12.10 If $\mathcal{U}$ is an open cover and $g_{\alpha\beta\gamma}$ a $U(1)$ co-cycle then we can build a Hitchin–Chatterjee gerbe or local bundle gerbe of the type considered in Example 12.9. It is easy to see that this has Dixmier–Douady class given by the Čech class $[g_{\alpha\beta\gamma}]$.

Notice that this example shows that every class in $H^3(M, \mathbb{Z})$ arises as the Dixmier–Douady class of some Hitchin–Chatterjee gerbe or of some (local) bundle gerbe.

It is straightforward to check that if $f\colon N \to M$ and (P,Y) is a bundle gerbe over M then $f^*(DD(P,Y)) = DD(f^*(P,Y))$. Moreover we have

1. $DD((P,Y)^*) = -DD(P,Y)$
2. $DD((P,Y) \otimes (Q,X)) = DD(P,Y) + DD(Q,X)$

We will defer the question of triviality of a bundle gerbe until the next section and consider next the notion of a connective structure on a bundle gerbe.

12.3.4 *Connective structure*

As $P \to Y^{[2]}$ is a $U(1)$ bundle we can pick a connection A. Call it a *bundle gerbe connection* if it respects the bundle gerbe multiplication. That is, if the section s of $\delta(P) \to Y^{[3]}$ satisfies $s^*(\delta(A)) = 0$, that is, if it is flat for $\delta(A)$. We would like bundle gerbe connections to exist. This is a straightforward consequence of the fact that the fundemental complex is exact. Indeed if A is any connection consider $s^*(\delta(A))$; we have $\delta(s^*(\delta(A))) = \delta(s)^*(\delta\delta(A)) = 0$ because $\delta\delta(A)$ is the flat connection on the canonically trivial bundle $\delta\delta(P)$. Hence there is a one-form a on $Y^{[2]}$ such that $\delta(a) = s^*(\delta(A))$ and thus $A - a$ is a bundle gerbe connection.

If A is a bundle gerbe connection then the curvature $F_A \in \Omega^2(Y^{[2]})$ satisfies $\delta(F_A) = 0$. From the exactness of the fundamental complex there must be an $f \in \Omega^2(Y)$ such that $F_A = \delta(f)$. As δ commutes with d we have $\delta(df) = d\delta(f) = dF_A = 0$. Hence $df = \pi^*(\omega)$ for some $\omega \in \Omega^3(M)$. So $\pi^*(d\omega) = d\pi^*(\omega) = ddf = 0$ and ω is closed. In fact it is a consequence of standard Čech de Rham theory that:

$$\left[\frac{1}{2\pi i}\omega\right] = r(DD(P,Y)) \in H^3(M, \mathbb{R}).$$

We call f a *curving* for A, the pair (A, f) a *connective structure* for (P,Y), and ω is called the *three-curvature* of the connective structure (A, f). In string theory the two-form f is called the B-field.

We can give a local description of the connective structure as follows. Assume we have an open cover $\mathcal{U}$ of M with local sections $s_\alpha \colon U_\alpha \to Y$ and sections over double-overlaps $\sigma_{\alpha\beta}$ of $(s_\alpha, s_\beta)^*(P) \to U_\alpha \cap U_\beta$. We define

$$A_{\alpha\beta} = (s_\alpha, s_\beta)^*(A) \in \Omega^1(U_\alpha \cap U_\beta)$$

and

$$f_\alpha = s_\alpha^*(f) \in \Omega^2(U_\alpha).$$

These satisfy

$$A_{\beta\gamma} - A_{\alpha\gamma} + A_{\alpha\beta} = g_{\alpha\beta\gamma}^{-1} dg_{\alpha\beta\gamma}$$

$$f_\beta - f_\alpha = df_{\alpha\beta}$$

and the three-curvature ω restricted to U_α is df_α.

Example 12.11 We can use this local description of the connective structure to calculate the Dixmier–Douady class of the Hitchin–Chatterjee gerbe on the three-sphere defined in Example 12.8. Stereographic projection from either pole identifies $S^3 - \{(1,0,0)\}$ and $S^3 - \{(-1,0,0)\}$ with $\mathbb{R}^3$ and maps the equator to the unit sphere $S^2 \subset \mathbb{R}^3$. Let U_0 and U_1 be the pre-images of the interior of a ball of radius 2 in $\mathbb{R}^3$ under both stereographic projections. We can identify $U_0 \cap U_1$ with $S^2 \times (-1, 1)$. Pull back the line bundle of chern class k on S^2, with connection A and curvature F, to $U_0 \cap U_1$. Because there are no triple overlaps this is a bundle gerbe connection. If we choose a partition of unity ψ_0 and ψ_1 for U_0 and U_1 then $f_0 = -\psi_1 F$ and $f_1 = \psi_0 F$ define two-forms on U_0 and U_1, respectively, satisfying $F = f_1 - f_0$ on $U_0 \cap U_1$. These two-forms define a curving for the bundle gerbe connection. The curvature is the globally defined three-form ω whose restriction to U_0 and U_1 is $-d\psi_1 \wedge F$ and $d\psi_0 \wedge F$, respectively. The integral of ω over the three-sphere reduces, by Stokes theorem, to the integral of F over the two-sphere. Hence this bundle gerbe has Dixmier–Douady class $k \in H^3(M, \mathbb{Z}) = \mathbb{Z}$.

Holonomy will need to wait until we have a considered the notion of triviality which we turn to now.

12.4 Triviality

Recall that a $U(1)$ bundle $P \to M$ is trivial if it is isomorphic to the bundle $M \times U(1)$ or, equivalently, has a global section. This occurs if and only if $P \to M$ has zero Chern class. If $s_a \colon U_\alpha \to P$ are local sections then P is determined by a transition function $g \colon U_\alpha \cap U_\beta \to U(1)$ given by $s_\alpha = s_\beta g_{\alpha\beta}$ and $P \to M$ is trivial if and only if there exist $h_\alpha \colon U_\alpha \to U(1)$ such that

$$g_{\alpha\beta} = h_\beta h_\alpha^{-1}.$$

In an analogous way Hitchin (2001) and Chatterjee (1998) define a gerbe $P_{\alpha\beta} \to U_\alpha \cap U_\beta$ to be trivial if there are $U(1)$ bundles $R_\alpha \to U_\alpha$ and isomorphisms $\phi_{\alpha\beta}\colon R_\alpha \otimes R_\beta^* \to P_{\alpha\beta}$ on double-overlaps in such a way that the multiplication becomes the obvious contraction

$$R_\alpha \otimes R_\beta^* \otimes R_\beta \otimes R_\gamma^* \to R_\alpha \otimes R_\gamma^*.$$

In the bundle gerbe formalism this idea takes the following form (Murray and Stevenson 2000). Let $R \to Y$ be a $U(1)$ bundle and let $\delta(R) \to Y^{[2]}$ be defined as above. Note that $\delta(R)$ has a natural associative bundle gerbe multiplication given by

$$\delta(R)_{(y_1,y_2)} \otimes \delta(R)_{(y_2,y_3)} = R_{y_1} \otimes R_{y_2}^* \otimes R_{y_2} \otimes R_{y_3}^* \simeq R_{y_1} \otimes R_{y_3}^* = \delta(R)_{(y_1,y_3)}.$$

Definition 12.3 *A bundle gerbe (P, Y) over M is called* trivial *if there is a $U(1)$ bundle $R \to Y$ such that (P, Y) is isomorphic to $(\delta(R), Y)$. In such a case we call a choice of R and the isomorphism $\delta(R) \simeq P$ a* trivialization *of (P, Y).*

Example 12.12 Let (P, Y) be a bundle gerbe and assume that $Y \to M$ admits a global section $s\colon M \to Y$. Define $R \to Y$ by $R_y = P_{(s(\pi(y)),y)}$. Then we have an isomorphism

$$\begin{aligned}
\delta(R)_{(y_1,y_2)} &= P_{(s(\pi(y_2)),y_2)} \otimes P^*_{(s(\pi(y_1)),y_1)} \\
&= P_{(s(\pi(y_2)),y_2)} \otimes P_{(y_1,s(\pi(y_1)))} \\
&\simeq P_{(y_1,y_2)}
\end{aligned}$$

using the bundle gerbe multiplication and the fact that $s(\pi(y_1)) = s(\pi(y_2))$. It is easy to check that this isomorphism preserves the respective bundle gerbe multiplications and we have shown that if Y admits a global section then any bundle gerbe (P, Y) is trivial. Notice that the converse is not true. Just take an open cover with more than one element and $g_{\alpha\beta\gamma} = 1$ to obtain a Hitchin–Chatterjee gerbe which has zero Dixmier–Douady class but for which $Y_{\mathcal{U}} \to M$ has no global section.

Consider the Dixmier–Douady class of $\delta(R)$. If $s_\alpha\colon U_\alpha \to Y$ are local sections for a good cover choose local sections η_α of $s_\alpha^*(R)$. Then we can take as local sections of $(s_\alpha, s_\beta)^*(\delta(R))$ the sections $\sigma_{\alpha\beta} = \eta_\alpha \otimes \eta_\beta^*$ and it follows that the corresponding $g_{\alpha\beta\gamma} = 1$ and $\delta(R)$ has Dixmier–Douady class equal to zero. The converse is also true. Consider a bundle gerbe with Dixmier–Douady class zero. So we have an open cover and $\sigma_{\alpha\beta}$ such that

$$g_{\alpha\beta\gamma} = h_{\beta\gamma} h_{\alpha\gamma}^{-1} h_{\alpha\beta}.$$

By replacing $\sigma_{\alpha\beta}$ by $\sigma_{\alpha\beta}/h_{\alpha\beta}$ we can assume that $g_{\alpha\beta\gamma} = 1$. Let $Y_\alpha = \pi^{-1}(U_\alpha)$ and define $R_\alpha \to Y_\alpha$ by letting the fibre of R_α over $y \in Y_\alpha$ be $P_{(y,s_\alpha(\pi(y)))}$. Construct an isomorphism $\chi_{\alpha\beta}(y)$ from the fibre of R_α over y to the fibre of

R_β over y by noting that

$$P_{(y,s_\alpha(\pi(y)))} = P_{(y,s_\beta(\pi(y)))} \otimes P_{(s_\alpha(\pi(y))s_\beta(\pi(y)))}$$

and using $\sigma_{\alpha\beta}(\pi(y)) \in P_{(s_\alpha(\pi(y))s_\beta(\pi(y)))}$. Because $\sigma_{\beta\gamma}\sigma_{\alpha\gamma}\sigma_{\alpha\beta} = 1$ we can show that $\chi_{\alpha\beta}(y) \circ \chi_{\beta\gamma}(y) = \chi_{\alpha\gamma}(y)$ and hence the R_α clutch together to form a global $U(1)$ bundle $R \to Y$. It is straightforward to check that $\delta(R) = P$.

Consider now a $U(1)$ bundle $R \to Y$ and assume that $\delta(R) \to Y^{[2]}$ has a section s with $\delta(s) = 1$ with respect to the canonical trivialization of $\delta(\delta(R)) \to Y^{[3]}$. The section s is called *descent data* for R and is equivalent to R being the pullback of a $U(1)$ bundle on M. Indeed s constitutes a family of isomorphisms

$$s(y_1, y_2) \colon R_{y_1} \to R_{y_2}$$

and $\delta(s) = 1$ is equivalent to $s(y_2, y_3) \circ s(y_1, y_2) = s(y_1, y_3)$ from which it is easy to define a bundle on M whose pullback is R.

Assume that a bundle gerbe (P, Y) over M is trivial and that R_1 and R_2 are two trivializations. Then we have $\delta(R_1) \simeq P \simeq \delta(R_2)$ and hence a section of $\delta\left(R_1^* \otimes R_2\right)$ which is descent data for $R_1^* \otimes R_2$. Thus $R_1 = R_2 \otimes \pi^*(Q)$ for some $U(1)$ bundle $Q \to M$. It is easy to show the converse that if R is a trivialization and $Q \to M$ a $U(1)$ bundle then $R \otimes \pi^*(Q)$ is another trivialization. We have now proved

Proposition 12.3 *Let (P, Y) be a bundle gerbe over M. Then*

1. *(P, Y) is trivial if and only if $DD(P, Y) = 0$.*
2. *If $DD(P, Y) = 0$ then any two trivializations of (P, Y) differ by a $U(1)$ bundle on M.*

This should be compared to the case of $U(1)$ bundles (0-gerbes) where two trivializations or sections of the bundle differ by a map into $U(1)$ which is a -1-gerbe. The general pattern is that we expect two trivializations of a p-gerbe to differ by a $(p-1)$-gerbe. Notice also that whereas any two trivial $U(1)$ bundles are isomorphic there are many trivial bundle gerbes which are not isomorphic. This leads us to the notion of stable isomorphism.

Definition 12.4 *If (P, Y) and (Q, X) are bundle gerbes over M we say they are* stably isomorphic *(Murray and Stevenson 2000) if $(P, Y)^* \otimes (Q, X)$ is trivial. A choice of a trivialization is called a* stable isomorphism *from (P, Y) to (Q, X).*

We have

Proposition 12.4 *Bundle gerbes (P, Y) and (Q, X) over M are stably isomorphic if and only if $DD(P, Y) = DD(Q, X)$. The Dixmier–Douady class defines a bijection between stable isomorphism classes of bundle gerbes on M and $H^3(M, \mathbb{Z})$.*

Proof. Bundle gerbes (P, Y) and (Q, X) over M are stably isomorphic if and only if $(P, Y)^* \otimes (Q, X)$ is trivial which occurs if and only if $-DD(P, Y) + D(Q, X) = 0$. We have already seen that every three class arises as the Dixmier–Douady class of some bundle gerbe on M. □

It follows that the correct notion of equivalence for bundle gerbes is stable isomorphism. It is actually possible to compose stable isomorphisms and the details are given in work of Stevenson (2000) where the structure of the two category of all bundles gerbes on M is discussed (see also Waldorf 2007).

Note that we also have

Proposition 12.5 *Every bundle gerbe is stably isomorphic to a Hitchin–Chatterjee gerbe.*

12.4.1 *Holonomy*

Consider now a bundle gerbe (P, Y) with connective structure (A, f) over a surface Σ. Because $H^3(\Sigma, \mathbb{Z}) = 0$ we know that (P, Y) is trivial. So there is a $U(1)$ bundle $R \to Y$ with $\delta(R) = P$. Choose a connection a for R and note that $\delta(a)$ is a connection for P using the isomorphism $\delta(R) = P$. But $\delta(\delta(a))$ is flat so $\delta(a)$ is a bundle gerbe connection. Hence $A = \delta(a) + \alpha$ for a one-form α on $Y^{[2]}$ with $\delta(\alpha) = 0$. Using the exactness of the fundamental complex we can solve $\alpha = \delta(\alpha')$ and hence show that $\delta(a + \alpha') = A$. So without loss of generality we can choose a connection a on R with $\delta(a) = A$. Consider the two-form $f - F_a$. This satisfies $\delta(f - F_a) = F_A - F_{\delta(a)} = 0$ so $f - F_A = \pi^*(\mu_a)$ for some two-form μ_a on Σ. Define the holonomy of (A, f) over Σ by

$$\mathrm{hol}((A, f), \Sigma) = \exp\left(\int_\Sigma \mu_a\right)$$

and note that this is independent of the choice of trivialization R and connection a. Indeed any two trivializations with connection will differ by a $U(1)$ bundle on M with connection and the corresponding μ_a will differ by the curvature of the connection on that $U(1)$ bundle. But the integral of the curvature of a $U(1)$ bundle over a closed surface is in $2\pi i\mathbb{Z}$ so the two definitions of holonomy agree.

If (P, Y) is a bundle gerbe with connective structure (A, f) on a general manifold M and $\Sigma \subset M$ is a submanifold we can pull (P, Y) and (A, f) back to Σ and define $\mathrm{hol}((A, f), \Sigma)$ as above. In this more general setting if $X \subset M$ is a three-dimensional submanifold with boundary ∂X also a submanifold of M we can trivialize (P, Y) over all of X and repeat the construction above. We then have $d\mu_a = \omega$, the three-curvature of (A, f), and thus

$$\mathrm{hol}((A, f), \partial \mathrm{X}) = \exp\left(\int_X \omega\right)$$

the final property in Section 12.1 which we wanted a 1-gerbe to satisfy.

Assume that we have a local description for (P, Y) and (A, f) as in Section 12.3.4 in terms of $g_{\alpha\beta\gamma}$, $A_{\alpha\beta}$, and f_α. Then there is a remarkable formula (Gawędzki and Reis 2002) for the holonomy, first proposed in Alvarez (1985) and also Gawędzki (1988) and subsequently derived by a number of authors, which can be described as follows. Choose a triangulation $\triangle$ of Σ and a map $\chi\colon \triangle \to I$ such that for any simplex $\sigma \in \triangle$ we have $\sigma \subset U_{\chi(\sigma)}$. Write σ^2, σ^1, and σ^0 for two-, one-, and zero-dimensional simplices, that is, faces, edges, and vertices, respectively. Then

$$\mathrm{hol}((A, f), \Sigma) =$$

$$\exp\left(\sum_{\sigma^2}\int_{\sigma^2} f_{\chi(\sigma^2)}\right)\exp\left(\sum_{\sigma^1\subset\sigma^2}\int_{\sigma^1} A_{\chi(\sigma^2)\chi(\sigma^1)}\right)\prod_{\sigma^0\subset\sigma^1\subset\sigma^2} g_{\chi(\sigma^2)\chi(\sigma^1)\chi(\sigma^0)}(\sigma^0).$$

12.4.2 *Obstructions to certain kinds of $Y \to M$*

When we consider the examples in the next section it will become apparent that they tend to cluster into two kinds: either $Y \to M$ has infinite-dimensional fibres or $Y \to M$ has discrete fibres as in the case of Hitchin–Chatterjee gerbes. There is a reason for this which is the following result:

Proposition 12.6 *Let (P, Y) be a bundle gerbe over M with $Y \to M$ a finite-dimensional fibration with both M and the fibres of $Y \to M$ one connected. Then the Dixmier–Douady class of (P, Y) is torsion.*

Proof. The proof uses the result from Gotay *et al.* (1983) which shows that if $Y \to M$ satisfies the hypothesis we have stated in the proposition and moreover there is a smooth two-form μ defined on the vertical tangent bundle of $Y \to M$ which is closed on each fibre then we can extend μ to a global, closed two-form on Y. To apply this choose a connection and curving f for (P, Y). If we restrict f to any fibre of $Y \to M$ it agrees with F and hence is closed in the fibre directions. So there exists a global two-form μ on $Y \to M$ which agrees with f in the vertical directions. Consider $\rho = f - \mu$. Both ρ and $d\rho$ are zero in the vertical directions so that $\rho = \pi^*(\chi)$ for some two-form χ on M. But then $\pi^*(d\rho) = d(f - \mu) = \pi^*(\omega)$ where ω is the three-curvature. Hence the image of the Dixmier–Douady class in real cohomology is zero so the Dixmier–Douady class is torsion. □

12.5 Examples of bundle gerbes

We expect to find bundle gerbes on manifolds M which have three-dimensional cohomology. There are two, related, cases we will discuss here. The first comes from lifting problems for principal bundles, the so-called *lifting bundle gerbe*, and the second is to take M a simple, compact Lie group.

12.5.1 *Lifting bundle gerbe*

Consider a central extension of Lie groups

$$0 \to U(1) \to \hat{\mathcal{G}} \xrightarrow{\rho} \mathcal{G} \to 0$$

so that $U(1)$ is the kernel of p and is in the centre of $\mathcal{G}$. If $Y \to M$ is a $\mathcal{G}$ principal bundle then we can ask if it lifts to a $\hat{\mathcal{G}}$ bundle $\hat{Y} \to M$. That is, can we find a $\hat{\mathcal{G}}$ bundle $\hat{Y} \to M$ and a bundle morphism $f\colon \hat{Y} \to Y$ such that $f(pg) = f(p)\rho(g)$ for all $p \in \hat{Y}$ and $g \in \hat{G}$? There is a well-known topological obstruction to the existence of $\hat{Y}$ which we can calculate as follows. Choose a good open cover $\mathcal{U}$ of M and local sections of Y that give rise to a transition function

$$g_{\alpha\beta}\colon U_\alpha \cap U_\beta \to G$$

in the usual way. Because these double overlaps are all contractible we can choose lifts of each $g_{\alpha\beta}$ to $\hat{g}_{\alpha\beta}\colon U_\alpha \cap U_\beta \to \hat{G}$. Note that

$$\epsilon_{\alpha\beta\gamma} = \hat{g}_{\beta\gamma}\hat{g}_{\alpha\gamma}^{-1}\hat{g}_{\alpha\beta}$$

takes values in $U(1)$. It is not difficult to show that $\epsilon_{\alpha\beta\gamma}$ is a $U(1)$ valued co-cycle and that the class

$$[\epsilon_{\alpha\beta\gamma}] \in H^2(M, U(1) \simeq H^3(M, \mathbb{Z})$$

vanishes if and only if P lifts to a $\hat{\mathcal{G}}$ bundle.

Given a $\mathcal{G}$ bundle $Y \to M$ there is a map $g\colon Y^{[2]} \to \mathcal{G}$ defined by $y_1 g(y_1, y_2) = y_2$. Notice that $\hat{\mathcal{G}} \to \mathcal{G}$ is a $U(1)$ bundle so we can pull it back to define a $U(1)$ bundle $Q \to Y^{[2]}$ whose fibres are cosets of $U(1)$ in $\hat{G}$ defined by

$$Q_{(y_1,y_2)} = U(1)g(y_1, y_2).$$

Because $g(y_1, y_2)g(y_2, y_3) = g(y_1, y_3)$ the product of an element in the coset containing $g(y_1, y_2)$ with an element in the coset containing $g(y_2, y_3)$ will be an element in the coset containing $g(y_1, y_3)$ which defines a bundle gerbe multiplication:

$$Q_{(y_1,y_2)} \otimes Q_{(y_2,y_3)} \to Q_{(y_1,y_3)}.$$

The bundle gerbe (Q, Y) is called the *lifting bundle gerbe* of Y (Murray 1996). It is easy to check that the lifting bundle gerbe has Dixmier–Douady class precisely the obstruction to lifting the bundle $Y \to M$. Indeed if we follow through the construction in Section 12.3.3 we find that $\hat{\sigma}_{\alpha\beta} = \hat{g}_{\alpha\beta}$. It follows from the discussion above that the lifting bundle gerbe is trivial if and only if the bundle $Y \to M$ lifts to $\hat{\mathcal{G}}$. In fact this follows directly because a lift $\hat{Y} \to Y$ will be a $U(1)$ bundle over Y and actually a trivialization of the lifting bundle gerbe defined above.

12.5.2 *Projective bundles*

Let H be a Hilbert space, possibly finite-dimensional. A Hilbert bundle with fibre H can be regarded locally as a collection of transition functions

$$g_{\alpha\beta}\colon U_\alpha \cap U_\beta \to U(H)$$

satisfying the co-cycle condition

$$g_{\beta\gamma} g_{\alpha\gamma}^{-1} g_{\alpha\beta} = 1.$$

In some situations we have slightly less than this, namely, a collection of transition functions

$$g_{\alpha\beta}\colon U_\alpha \cap U_\beta \to U(H)$$

satisfying

$$g_{\beta\gamma} g_{\alpha\gamma}^{-1} g_{\alpha\beta} = \epsilon_{\alpha\beta\gamma} 1_{U(H)}.$$

If we denote by $\rho\colon U(H) \to PU(H)$ the projection onto the projective unitary group we see that have

$$\rho(g_{\beta\gamma})\rho\left(g_{\alpha\gamma}^{-1}\right)\rho(g_{\alpha\beta}) = \rho(\epsilon_{\alpha\beta\gamma}1) = 1_{PU(H)}$$

so there is a well-defined bundle of projective spaces or a principal $PU(H)$ bundle. Given a projective bundle a natural question is to ask when it is the projectivization of a global Hilbert bundle. This is equivalent to lifting the $PU(H)$ bundle to $U(H)$. The obstruction to this lifting in the class is

$$[\epsilon_{\alpha\beta\gamma}] \in H^2(M, U(1))$$

arising from the sequence

$$0 \to U(1) \to U(H) \to PU(H) \to 0.$$

In the finite-dimensional class we can also consider the obstruction to lifting from $PU(n)$ to $SU(n)$ via the exact sequence

$$0 \to \mathbb{Z}_n \to SU(n) \to PU(n) \to 0.$$

In this case the lifting bundle gerbe is really a $\mathbb{Z}_n \subset U(1)$ bundle gerbe and the Dixmier–Douady class is a torsion class in the image of the Bockstein map

$$H^2(M, \mathbb{Z}_n) \to H^3(M, \mathbb{Z}).$$

For further details on $\mathbb{Z}_n$ bundle gerbes see Carey *et al.* (2000). Note finally that if $Y \to M$ is a $PU(n)$ bundle then the lifting bundle gerbe has finite-dimensional fibres so the fact that its Dixmier–Douady class is torsion is implied by Proposition 12.6.

12.5.3 *Bundle gerbes on Lie groups*

If G is a compact, simple, Lie group then $H^3(G, \mathbb{Z}) \simeq \mathbb{Z}$ so we expect to find bundle gerbes on G. There are, in fact, a number of constructions of bundle gerbes on G and we have seen one already for $SU(2) \simeq S^3$ in Section 12.3.2. For convenience let us fix an orientation of G and call the bundle gerbe over G of Dixmier–Douady class 1 the *basic bundle gerbe* on G.

Example 12.13 Let PG be the based path space of all smooth maps g of the interval $[0,1]$ into G with $g(0) = 1$. Let $\mathrm{ev}\colon \mathrm{PG} \to \mathrm{G}$ be the evaluation map $\mathrm{ev(g)} = \mathrm{g(1)}$. The kernel of ev is the space of based loops $\Omega(G)$, that is, all smooth maps $g\colon [0,1] \to G$ for which $g(0) = 1 = g(1)$. In the smooth Fréchet topology $PG \to G$ is a principal $\Omega(G)$ bundle. Moreover there is a well-known central extension (Pressley and Segal 1986)

$$0 \to U(1) \to \widehat{\Omega}(G) \to \Omega(G) \to 0$$

where $\hat{\Omega}(G)$ is the Kac–Moody group. Hence there is a corresponding lifting bundle gerbe over G which is the basic bundle gerbe. It is possible to give an explicit construction of this gerbe over G (see Murray 1996 and also Stevenson 2000).

If we wish to avoid the infinite-dimensional spaces there is a construction of the basic bundle gerbe over G in the simply connected case due to Meinrenken (2003) and see also Gawędzki and Reis (2004) for the non-simply connected case. This construction has disconnected fibres for $Y \to G$ and uses the standard structure theory of compact, simple, simply connected Lie groups.

Example 12.14 For $G = SU(n)$ there is a simple construction of the basic bundle gerbe due to Meinrenken (2003) (see also Mickelsson 2003) using an open cover of G which can be presented without the cover as follows. Let

$$Y = \{(g, z) \mid \det(g - z1) \neq 0\} \subset G \times U(1).$$

For convenience let us write an element $((g, z), (g, w)) \in Y^{[2]}$ as (g, w, z). If $u \in U(1)$ and $u \neq w$ and $u \neq z$ let us say that u is between w and z if an anticlockwise rotation of z into w passes through u. Then let $W_{(g,w,z)}$ be the sum of all the eigenspaces of g for eigenvalues between z and w and define $P_{(g,w,z)}$ to be the $U(1)$ frame bundle of $\det W_{(g,w,z)}$. To define the bundle gerbe product notice that if u is between w and z then

$$W_{(g,w,u)} \oplus W_{(g,u,z)} \oplus W_{(g,z,w)} = \mathbb{C}^n$$

so that

$$\det W_{(g,w,u)} \otimes \det W_{(g,u,z)} \otimes \det W_{(g,z,w)} = \mathbb{C}$$

and thus

$$\det W_{(g,w,u)} \otimes \det W_{(g,u,z)} = \det W^*_{(g,z,w)}.$$

Similarly

$$W_{(g,w,z)} \oplus W_{(g,z,w)} = \mathbb{C}^n$$

so that

$$\det W_{(g,w,z)} \otimes \det W_{(g,z,w)} = \mathbb{C}$$

and

$$\det W_{(g,w,z)} = \det W^*_{(g,z,w)}.$$

There are a number of other cases that can be dealt with in a similar fashion and putting all these facts together gives a bundle gerbe multiplication on $P \to Y^{[2]}$. A construction of the curving on (P, Y) appears in Murray and Stevenson (2008).

12.6 Applications of bundle gerbes

12.6.1 *Wess–Zumino–Witten term*

The Wess–Zumino–Witten term associates to a smooth map g of a surface Σ into a compact, simple, Lie group G an invariant $\Gamma(g) \in U(1)$. As noted by a number of authors (Carey *et al.* 2000; Hitchin 2002; Gawędzki and Reis 2002) this can be understood as the holonomy of a connection and curving on the basic gerbe on the group. The original definition of Witten (1984) is that we choose a three-manifold X with $\partial X = \Sigma$ and extend g to $\hat{g} \colon X \to G$. We then consider

$$\int_X \hat{g}^*(\omega)$$

where ω is a three-form on G representing a generator of $H^3(G, 2\pi i\mathbb{Z})$. If we choose a different extension $\tilde{g} \colon X \to G$ then the pair can be combined to define a map from the manifold $X \cup_\Sigma X$, formed by joining two copies of X (with opposite orientations) along Σ, into G. Call this map $\hat{g} \cup \tilde{g}$. Then

$$\int_X \hat{g}^*(\omega) - \int_X \tilde{g}^*(\omega) = \int_{X \cup_\Sigma X} (\hat{g} \cup \tilde{g})^*(\omega) \in 2\pi i\mathbb{Z}.$$

It follows that

$$\Gamma(g) = \exp\left(\int_X \hat{g}^*(\omega)\right) \in U(1)$$

depends only on g.

If we choose a suitable connection and curving (A, f) on the basic gerbe on G, so that it has curvature ω then we see that

$$\Gamma(g) = \mathrm{hol}(\Sigma, g^*(A, f)).$$

The gerbe approach to the Wess–Zumino–Witten term has two advantages. Firstly, it removes the topological restriction on M of two-connectedness necessary in Witten's definition so that the map g can be extended to the three-manifold X. Secondly, we can use the local formula for the holonomy given in Section 12.4.1. For details see Gawędzki and Reis (2002).

12.6.2 *Faddeev–Mickelsson anomaly*

We follow Carey and Murray (1996) and Segal (1985). Let X be a compact, Riemannian, spin, three-manifold and denote by $\mathcal{A}$ the space of connections on a complex vector bundle over M and by $\mathcal{G}$ the space of gauge transformations. For any $A \in \mathcal{A}$ the chiral Dirac operator D_A coupled to A has discrete spectrum. Let

$$Y = \{(A, t) \mid t \notin \operatorname{spec}(D_A)\}$$

be considered as a submersion over $\mathcal{A}$. Note that $\mathcal{G}$ acts on Y and $Y/\mathcal{G} \to \mathcal{A}/\mathcal{G}$ is another submersion. Following Segal (1985) for $(A, s) \in Y$ we can decompose the Hilbert space H of coupled spinors into a direct sum of eigenspaces of D_A for eigenvalues greater than s and a direct sum of eigenspaces of D_A for eigenvalues less than s. Denote these by $H^-_{(A,s)}$ and $H^+_{(A,s)}$, respectively. We can then form the Fock space

$$F_{(A,s)} = \bigwedge H^+_{(A,s)} \otimes \bigwedge \left(H^-_{(A,s)}\right)^*$$

which is a bundle $F \to Y$. Notice that $\mathcal{G}$ acts on F and gives rise to a bundle $F/\mathcal{G} \to Y/\mathcal{G}$.

If we choose another $t < s$ we have

$$H = H^-_{(A,t)} \oplus V_{(A,t,s)} \oplus H^+_{(A,s)}$$

where $V_{(A,t,s)}$ is the sum of all the eigenspaces of D_A for eigenvalues between t and s. If we use the canonical isomorphism

$$\bigwedge V^*_{(A,t,s)} \otimes \det V_{(A,t,s)} = \bigwedge V_{(A,t,s)}$$

then we can show that

$$F_{(A,s)} = F_{(A,t)} \otimes \det V_{(A,t,s)}.$$

It follows that the projective spaces of $F_{(A,s)}$ and $F_{(A,t)}$ are canonically isomorphic and descend to a projective bundle $\mathcal{P} \to \mathcal{A}$ and as $\mathcal{G}$ acts there is also a projective bundle $\mathcal{P}/\mathcal{G} \to \mathcal{A}/\mathcal{G}$. The question of interest (Segal 1985) is whether there is a Hilbert bundle $\mathcal{H} \to \mathcal{A}/\mathcal{G}$ whose projectivization is $\mathcal{P}/\mathcal{G}$. Notice that as $\mathcal{A}$ is contractible the answer to the equivalent question on $\mathcal{A}$ is clearly positive.

As noted in Section 12.5.2 we could give a bundle gerbe interpretation of this question via the lifting bundle gerbe for the central extension

$$0 \to U(1) \to PU(\tilde{\mathcal{H}}) \to U(\tilde{\mathcal{H}}) \to 0$$

for a suitable Hilbert space $\tilde{\mathcal{H}}$. However there is a more direct approach as follows. Let $P_{(A,s,t)}$ be the unitary frame bundle of $\det V_{(A,t,s)}$. Notice that if $r < s < t$ then

$$V_{(A,r,t)} \oplus V_{(A,t,s)} = V_{(A,r,s)}$$

so that

$$\det V_{(A,r,t)} \otimes \det V_{(A,t,s)} = \det V_{(A,r,s)}$$

gives rise to a bundle gerbe multiplication similar to the $SU(n)$ case in Example 12.14. Again $\mathcal{G}$ acts so this descends to a bundle gerbe on $\mathcal{A}/\mathcal{G}$. If this bundle gerbe is trivial then there is a line bundle $L \to Y$ such that

$$\det V_{(A,t,s)} = L_{(A,s)} \otimes L^*_{(A,t)}$$

and moreover as it is the bundle gerbe on $\mathcal{A}/\mathcal{G}$ which is trivial these are $\mathcal{G}$ equivariant isomorphisms. It follows that

$$F_{(A,s)} = F_{(A,t)} \otimes L_{(A,s)} \otimes L^*_{(A,t)}$$

or

$$F_{(A,s)} \otimes L^*_{(A,s)} = F_{(A,t)} \otimes L^*_{(A,t)},$$

and these are $\mathcal{G}$-equivariant isomorphisms. Hence $F_{(A,s)} \otimes L^*_{(A,s)}$ descends to a $\mathcal{G}$-equivariant Hilbert bundle on $\mathcal{A}$ whose projectivization is $\mathcal{P}$. It follows that there is a Hilbert bundle on $\mathcal{A}/\mathcal{G}$ whose projectivization is $\mathcal{P}/\mathcal{G}$.

12.6.3 *String structures*

In Killingback (1987), he introduced the notion of a *string structure*. Given a G bundle $Q \to X$ he considers the corresponding loop bundle $LQ \to LX$ which is an LG bundle. As noted above we have the Kac–Moody central extension

$$0 \to U(1) \to \widehat{LG} \to LG \to 0.$$

Killingback says that $Q \to X$ is *string* if $LQ \to LX$ lifts to $\widehat{LG}$, calls a choice of such a lift a *string structure*, and defines the obstruction class to the lift to be the *string class*. In Murray and Stevenson (2003) we consider the more general situation of an LG bundle on a manifold M and the question of whether it lifts to $\widehat{LG}$. There is an equivalence between LG bundles on M and G bundles on $S^1 \times M$ which is exploited to define a connection and curving on the lifting bundle gerbe associated to the LG bundle. This enables the derivation of a de Rham representative for the image of the string class in real cohomology. The same approach could be applied to the other level Kac–Moody central extensions.

12.7 Other matters

In the interests of brevity nothing has been said about the original approach to gerbes as sheaves of groupoids. Details are given in Brylinski (1993) and the relationship with bundle gerbes is discussed in Murray (1996) and in more detail in Stevenson (2000). In this same context I would like to thank Larry Breen for pointing out that in the work of Ulbrich (1990, 1991) the notion of cocycle bitorsors can be interpreted as a form of bundle gerbe.

In the definitions and theory above we could replace $U(1)$ by any abelian group H and there would only be obvious minor modifications such as the Dixmier–Douady class being in $H^2(M, H)$. If we want to replace H by a non-abelian group things become more difficult as we mentioned in the introduction because we cannot form the product of two H bundles if H is non-abelian. To get around this difficulty we need to replace principal bundles by principal bibundles which have a left and right group actions. The resulting theory becomes more complicated although closer to Giraud's original aim of understanding non-abelian cohomology. For details see Aschieri *et al.* (2005) and Breen and Messing (2005).

We have motivated gerbes by the idea of replacing the transition function $g_{\alpha\beta}$ by a $U(1)$ bundle $P_{\alpha\beta}$ on double-overlaps. It is natural to consider what happens if we take the next step and replace $P_{\alpha\beta}$ by a bundle gerbe on each double-overlap. This gives rise to the notion of bundle 2-gerbes whose characteristic class is a four class. In particular there is associated to any principal G bundle $P \to M$ a bundle 2-gerbe whose characteristic class is the pontrjagin class of P. For details of the theory see Stevenson (2000, 2004) and for an application to Chern–Simons theory see Carey *et al.* (2005).

It is clear that we can continue on in this fashion and consider bundle 2-gerbes on double-overlaps and more generally inductively use p-gerbes on double-overlaps to define $(p+1)$-gerbes. However the theory becomes increasingly complex for a reason that we have not paid much attention to in the discussion above. In the case of $U(1)$ bundles the transition function has to satisfy one condition, the co-cycle identity. In the case of gerbes the $U(1)$ bundles over double-overlaps satisfy two conditions, the existence of a bundle gerbe multiplication and its associativity. In the case of bundle 2-gerbes there are three conditions and the complexity continues to grow in this fashion.

Acknowledgements

The author acknowledges the support of the Australian Research Council and would like to thank the organizers of the conference for the invitation to participate. The referee is also thanked for useful remarks on the manuscript.

References

Alvarez, O. (1985). Topological quantization and cohomology. *Commun. Math. Phys.* **100**, 279–309.

Aschieri, P., Cantini, L., and Jurčo, B. (2005). Nonabelian bundle gerbes, their differential geometry and gauge theory. *Commun. Math. Phys.* **254**(2), 367–400.

Bott, R and Tu, L. W. (1982). *Differential Forms in Algebraic Topology*. Vol. 82 of Graduate Texts in Mathematics, Springer-Verlag, New York/Berlin.

Breen, L and Messing, W. (2005). Differential geometry of gerbes. *Adv. Math.* **198**(2), 732–846.

Brylinski, J.-L. (1993). *Loop Spaces, Characteristic Classes and Geometric Quantization*, Vol. **107** of Progress in Mathematics, Birkhäuser Boston Inc., Boston, MA.

Brylinski, J.-L. and McLaughlin, D. A. (1994). The geometry of degree-four characteristic classes and of line bundles on loop spaces. I. *Duke Math. J.* **75**(3), 603–38.

Carey, A. L. and Murray, M. K. (1996). Faddeev's anomaly and bundle gerbes. *Lett. Math. Phys.* **37**(1), 29–36.

Carey, A. L., Johnson, S., Murray, M. K., Stevenson, D., and Wang, B.-L. (2005). Bundle gerbes for Chern-Simons and Wess-Zumino-Witten theories. *Commun. Math. Phys.* **259**(3), 577–613.

Carey, A. L., Mickelsson, J., and Murray, M. K. (2000). Bundle gerbes applied to quantum field theory. *Rev. Math. Phys.* **12**(1), 65–90.

Chatterjee, D.S. (1998). On gerbs, PhD thesis, Trinity College, Cambridge, UK. Available at http://www.maths.ox.ac.uk/hitchin/hitchinstudents/chatterjee.pdf

Dupont, J. L. (1978). *Curvature and Characteristic Classes*. Vol. **640** of Lecture Notes in Mathematics, Springer–Verlag, Berlin, Germany.

Gawędzki, K. (1988). Topological actions in two-dimensional quantum field theories. In G. t'Hooft, A. Jaffe, G. Mack, P. K. Mitter, and R. Stora (eds.), *Nonperturbative Quantum Field Theories* Vol. **185** of NATO Series, pp. 101–41. Plenum Press New York/London.

Gawędzki, K. and Reis, N. (2002). WZW branes and gerbes. *Rev. Math. Phys.* **14**(12), 1281–334.

Gawędzki, K. and Reis, N. (2004). Basic gerbe over non-simply connected compact groups. *J. Geom. Phys.* **50**(1–4), 28–55.

Gotay, M. J., Lashof, R., Śniatycki, J., and Weinstein, A. (1983) Closed forms on symplectic fibre bundles. *Comment. Math. Helv.* **58**(4), 617–21.

Giraud, J. (1971). *Cohomologie non-abélienne.* Grundl. **179**, Springer Verlag, Berlin/Heidelberg/New York.

Hitchin, N. (2001). Lectures on special Lagrangian submanifolds. In *Winter School on Mirror Symmetry, Vector Bundles and Lagrangian Submanifolds (Cambridge, MA, 1999)* Vol. **23** of *AMS/IP Studies on Advanced Mathematics*, pp. 151–82, American Mathematical Society, Providence, RI.

Hitchin, N. (2002). The Wess-Zumino term for a harmonic map. *J. Reine Angew. Math.* **543**, 83–101.

Hitchin, N. (2003). What is a gerbe?, *Notices Am. Math. Soc.* **50**(2), 218–19.

Hitchin (2006). Instantons, Poisson structures and generalized Kähler geometry. *Commun. Math. Phys.* **265**(1), 131–64.

Killingback, T. (1987). World-sheet anomalies and loop geometry. *Nucl. Phys. B* **288**, 578–88.

Meinrenken., E. (2003). The basic gerbe over a compact simple Lie group. *Enseign. Math. (2)* **49**(3–4), 307–33. math.DG/0209194.

Mickelsson, J. (2003). Gerbes on quantum groups. math.DG/0308235.

Mickelsson, J. (2006). Gerbes and quantum field theory. math-ph/0603031. In J.-P. Francoise, G. L. Naber and T. S. Tsun (eds), Encyclopedia of Mathematical Physics, Elsevier Academic Press.

Murray, M. K. (1996). Bundle gerbes. *J. London Math. Soc. (2)* **54**(2), 403–16.

Murray, M. K. and Stevenson, D. (2000). Bundle gerbes: stable isomorphism and local theory. *J. London Math. Soc. (2)* **62**(3), 925–37.

Murray, M. K. and Stevenson, D. (2003). Higgs fields, bundle gerbes and string structures. *Commun. Math. Phys.* **243**(3), 541–55.

Murray, M. K. and Stevenson, D. (2008). The basic bundle gerbe on unitary groups. *J. Geometry Phys.* **58**(11), 1571–90.

Pressley, A. and Segal, G. (1986). *Loop Groups.* Oxford Mathematical Monographs, The Clarendon Press, Oxford.

Segal, G. (1985). Faddeev's anomaly in Gauss's Law. Preprint, Oxford University, Oxford.

Stevenson, D. (2000). The geometry of bundle gerbes, PhD thesis, University of Adelaide, Adelaide, Australia 2000. math.DG/0004117.

Stevenson, D. (2004). Bundle 2-gerbes. *Proc. London Math. Soc. (3)* **88**(2), 405–35.

Ulbrich, K.-H. (1990). On the correspondence between gerbes and bouquets. *Math. Proc. Camb. Philos. Soc.* **108**(1), 1–5.

Ulbrich, K.-H. (1991). On cocycle bitorsors and gerbes over a Grothendieck topos. *Math. Proc. Camb. Philos. Soc.* **110**(1), 49–55.

Waldorf, K. (2007). More morphisms between bundle gerbes. *Theory and Applications of Categories* **18**(9), 240–273.

Witten, E. (1984). Non-abelian bosonization in two dimensions. *Commun. Math. Phys.* **92**(4), 455–72.

XIII

PROJECTIVE LINKING AND BOUNDARIES OF POSITIVE HOLOMORPHIC CHAINS IN PROJECTIVE MANIFOLDS, PART I

F. Reese Harvey and H. Blaine Lawson, Jr.

Dedicated to Nigel Hitchin in celebration of his 60th birthday

13.1 Introduction

In 2000 H. Alexander and J. Wermer published the following result:

Theorem 13.1 (Alexander–Wermer) *Let Γ be a compact-oriented smooth submanifold of dimension $2p-1$ in $\mathbb{C}^n$. Then Γ bounds a positive holomorphic p-chain in $\mathbb{C}^n$ if and only if the linking number*

$$\mathrm{Link}(\Gamma, Z) \ \geq \ 0$$

for all canonically oriented algebraic subvarieties Z of codimension p in $\mathbb{C}^n - \Gamma$.

The *linking number* is an integer-valued topological invariant defined by the intersection $\mathrm{Link}(\Gamma, Z) \equiv N \bullet Z$ with any $2p$-chain N having $\partial N = \Gamma$ in $\mathbb{C}^n$ (see Section 13.3). A *positive holomorphic p-chain* is a finite sum of canonically oriented complex subvarieties of dimension p and finite volume in $\mathbb{C}^n - \Gamma$ (see Definition 13.7). Here the notion of boundary is taken in the sense of currents, that is, Stokes' theorem is satisfied. However, for smooth Γ there is boundary regularity almost everywhere, and if Γ is real analytic, one has complete boundary regularity in the sense of immersions (see Harvey and Lawson 1975).

The main point of this chapter is to formulate and prove an analogue of the Alexander–Wermer theorem for oriented (not necessarily connected) curves in a projective manifold. In the sequel we shall study the corresponding result for submanifolds of any odd dimension.

Before stating the main result we remark that a key ingredient in the proof of the Alexander–Wermer theorem is the following classical theorem and its generalizations (Wermer 1958):

Theorem 13.2 (Wermer) *Let $\Gamma \subset \mathbb{C}^n$ be a compact real analytic curve and denote by*

$$\widehat{\Gamma}_{\mathrm{poly}} = \{z \in \mathbb{C}^n \ : \ |p(z)| \leq \sup_{\Gamma} |p| \quad \textit{for all polynomials} \quad p\}$$

its polynomial hull. Then $\widehat{\Gamma}_{\text{poly}} - \Gamma$ *is a one-dimensional complex analytic subvariety of* $\mathbb{C}^n - \Gamma$.

For compact subsets K of complex projective n-space $\mathbb{P}^n$, the authors recently introduced the notion of the *projective hull*

$$\widehat{K}_{\text{proj}} = \{z \in \mathbb{P}^n : \exists C \text{ s.t. } \|\sigma_z\| \leq C^d \sup_K \|\sigma\| \quad \forall \sigma \in H^0(\mathbb{P}^n, \mathcal{O}(d)), d > 0\}$$

and defined K to be *stable* if the constant C can be chosen independently of the point $z \in \widehat{K}_{\text{proj}}$. A number of basic properties of $\widehat{K}_{\text{proj}}$ were established in Harvey and Lawson (2006*a*), and the following analogue of Wermer's theorem was proved in Harvey, Lawson, and Wermer 2008.

Theorem 13.3 (Harvey–Lawson–Wermer) *Let* $\Gamma \subset \mathbb{P}^n$ *be a stable real analytic curve. Then* $\widehat{\Gamma}_{\text{proj}} - \Gamma$ *is a one-dimensional complex analytic subvariety of* $\mathbb{P}^n - \Gamma$.

It is interesting to note that while Wermer's theorem holds for curves with only weak differentiability properties (see Alexander and Wermer 1998 or Dinh and Lawrence 2003 for an account), its projective analogue fails even for C^∞-curves. On the other hand there is much evidence for the following:

Conjecture 13.1 *Every real analytic curve in* $\mathbb{P}^n$ *is stable.*

This brings us to the notion of projective linking numbers. Suppose that $\Gamma \subset \mathbb{P}^n$ is a compact-oriented smooth curve, and let $Z \subset \mathbb{P}^n - \Gamma$ be an algebraic subvariety of codimension 1. The *projective linking number* of Γ with Z is defined to be

$$\text{Link}_{\mathbb{P}}(\Gamma, V) \;\equiv\; N \bullet Z - \deg(Z) \int_N \omega$$

where ω is the standard Kähler form on $\mathbb{P}^n$ and N is any integral 2-chain with $\partial N = \Gamma$ in $\mathbb{P}^n$. Here Z is given the canonical orientation, and $\bullet : H_2(\mathbb{P}^n, \Gamma) \times H_{2n-2}(\mathbb{P}^n - \Gamma) \to \mathbb{Z}$ is the topologically defined intersection pairing. This definition is independent of the choice of N (see Section 13.3). The associated *reduced linking number* is defined to be

$$\widetilde{\text{Link}}_{\mathbb{P}}(\Gamma, Z) \;\equiv\; \frac{1}{\deg(Z)} \text{Link}_{\mathbb{P}}(\Gamma, Z).$$

The basic result proved here is the following:

Theorem 13.4 *Let* Γ *be a oriented, stable, real analytic curve in* $\mathbb{P}^n$ *with a positive integer multiplicity on each component. Then the following are equivalent:*

1. Γ *is the boundary of a positive holomorphic* 1*-chain of mass* $\leq \Lambda$ *in* $\mathbb{P}^n$.
2. $\widetilde{\text{Link}}_{\mathbb{P}}(\Gamma, Z) \;\geq\; -\Lambda$ *for all algebraic hypersurfaces* Z *in* $\mathbb{P}^n - M$.

If Γ bounds any positive holomorphic 1-chain, then there is a unique such chain T_0 of least mass. (All others are obtained by adding algebraic 1-cycles to T_0.) Note that $\Lambda_0 \equiv M(T_0)$ is the smallest positive number such that (2) holds.

Corollary 13.1 *Let Γ be as in Theorem 13.4 and suppose T is a positive holomorphic 1-chain with $dT = \Gamma$. Then T is the unique holomorphic chain of least mass with $dT = \Gamma$ if and only if*

$$\inf_Z \left\{ \frac{T \bullet Z}{\deg Z} \right\} = 0$$

where the infimum is taken over all algebraic hypersurfaces in the complement $\mathbb{P}^n - \Gamma$.

Condition (2) in Theorem 13.4 has several equivalent formulations. The first is in terms of projective winding numbers. Given a holomorphic section $\sigma \in H^0(\mathbb{P}^n, \mathcal{O}(d))$, the *projective winding number* of σ on Γ is defined as the integral

$$\mathrm{Wind}_{\mathbb{P}}(\Gamma, \sigma) \equiv \int_\Gamma d^C \log\|\sigma\|,$$

and we set

$$\widetilde{\mathrm{Wind}}_{\mathbb{P}}(\Gamma, \sigma) \equiv \frac{1}{d}\mathrm{Wind}_{\mathbb{P}}(\Gamma, \sigma).$$

Another formulation involves the cone $PSH_\omega(\mathbb{P}^n)$ of *quasi-plurisubharmonic functions.* These are the upper semi-continuous functions $f : \mathbb{P}^n \to [-\infty, \infty)$ for which $dd^C f + \omega$ is a positive (1,1)-current on $\mathbb{P}^n$.

Proposition 13.1 *Let Γ be an oriented smooth curve in $\mathbb{P}^n$ with a positive integer multiplicity on each component. Then for any $\Lambda > 0$ the following are equivalent:*

1. $\widetilde{\mathrm{Link}}_{\mathbb{P}}(\Gamma, Z) \geq -\Lambda$ *for all algebraic hypersurfaces $Z \subset \mathbb{P}^n - \Gamma$.*
2. $\widetilde{\mathrm{Wind}}_{\mathbb{P}}(\Gamma, \sigma) \geq -\Lambda$ *for all holomorphic sections σ of $\mathcal{O}(d)$, and all $d > 0$.*
3. $\int_\Gamma d^C u \geq -\Lambda$ *for all $u \in PSH_\omega(\mathbb{P}^n)$.*

Any smooth curve Γ in $\mathbb{P}^n$ lies in some affine chart $\mathbb{C}^n \subset \mathbb{P}^n$, and it is natural to ask for a reformulation of condition (1) in terms of the conventional linking numbers of Γ with algebraic hypersurfaces in that chart. This is done explicitly in Theorem 13.4.

The results above extend to any projective manifold X. Given a very ample Hermitian line bundle λ on X there are intrinsically defined *λ-linking numbers* $\mathrm{Link}_\lambda(\Gamma, Z)$, *$\lambda$-winding numbers* $\mathrm{Wind}_\lambda(\Gamma, \sigma)$ for $\sigma \in H^0(X, \lambda^d)$, and $\omega = c_1(\lambda)$*-quasi-plurisubharmonic functions* $PSH_\omega(X)$. With these notions,

Proposition 13.1 and Theorem 13.4 carry over to X. This is done in Section 13.7.

Theorem 13.4 leads to the following interesting result:

Theorem 13.5 *Let $\gamma \subset \mathbb{P}^n$ be a finite disjoint union of real analytic curves and assume γ is stable. Then a class $\tau \in H_2(\mathbb{P}^n, \gamma; \mathbb{Z})$ is represented by a positive holomorphic chain with boundary on γ if and only if*

$$\tau \bullet u \;\geq\; 0$$

for all $u \in H_{2n-2}(\mathbb{P}^n - \gamma; \mathbb{Z})$ represented by positive algebraic hypersurfaces in $\mathbb{P}^n - \gamma$.

This result expands to a duality between the cones of relative and absolute classes which are representable by positive holomorphic chains (Harvey and Lawson 2006*d*).

The arguments given in Section 13.6 show that there is even more evidence for the following:

Conjecture 13.2 *Every oriented, real analytic curve in $\mathbb{P}^n$ which satisfies the equivalent conditions of Proposition 13.1 is stable.*

In Harvey and Lawson 2006*c* we prove that if Conjecture 13.2 holds for curves in $\mathbb{P}^2$, then all the results above continue to hold for real analytic Γ of any odd dimension $2p-1$. (No stability hypothesis is needed.)

Remark 13.1 There are several quite different characterizations of the boundaries of general (i.e. not necessarily positive) holomorphic chains in projective and certain quasi-projective manifolds (see, e.g., Harvey and Lawson 1977, 2005, Dolbeault 1983, and Dolbeault and Henkin 1994, 1997).

Note. To keep formulas simple throughout the chapter we adopt the convention that

$$d^C = \frac{i}{2\pi}(\bar{\partial} - \partial).$$

13.2 Projective hulls

In this section we recall the definition and basic properties of the projective hull introduced in Harvey and Lawson (2006*a*). This material is not really necessary for reading the rest of the chapter.

Let $\mathcal{O}(1) \to \mathbb{P}^n$ denote the holomorphic line bundle of Chern class 1 endowed with the standard unitary-invariant metric, and let $\mathcal{O}(d)$ be its dth tensor power with the induced tensor product metric.

Definition 13.1 *Let $K \subset \mathbb{P}^n$ be a compact subset of complex projective n-space. A point $x \in \mathbb{P}^n$ belongs to the projective hull of K if there exists a constant*

$C = C(x)$ such that

$$\|\sigma(x)\| \;\le\; C^d \sup_K \|\sigma\| \tag{13.1}$$

for all global sections $\sigma \in H^0(\mathbb{P}^n, \mathcal{O}(d))$ and all $d > 0$. This set of points is denoted $\widehat{K}$.

The set $\widehat{K}$ is independent of the choice of metric on $\mathcal{O}(1)$.

The projective hull possesses interesting properties. It and its generalizations function in projective and Kähler manifolds much as the polynomial hull and its generalizations function in affine and Stein manifolds. The following were established in Harvey and Lawson (2006*a*):

1. If $Y \subset \mathbb{P}^n$ is an algebraic subvariety and $K \subset Y$, then $\widehat{K} \subset Y$. That is, $\widehat{K}$ is contained in the Zariski closure of K. Furthermore, if $Y \subset \mathbb{P}^n$ is a projective manifold and $\lambda = \mathcal{O}(1)\big|_Y$, the λ-projective hull of $K \subset Y$, defined as in (13.1) with $\mathcal{O}(1)$ replaced by λ, agrees with $\widehat{K}$. The same is true of λ^k for any k.
2. If K is contained in an affine open subset $\Omega \subset \mathbb{P}^n$, then $(\widehat{K})_{\text{poly},\Omega} \subseteq \widehat{K}$.
3. If $K = \partial C$, where C is a holomorphic curve with boundary in $\mathbb{P}^n$, then $C \subseteq \widehat{K}$.
4. $\{\widehat{K}\}^- - K$ satisfies the *maximum modulus principle* for holomorphic functions on open subsets of $\mathbb{P}^n - K$. ($\{\widehat{K}\}^-$ denotes the closure of $\widehat{K}$.)
5. $\{\widehat{K}\}^- - K$ is 1-pseudoconcave in the sense of Dinh and Lawrence (2003). In particular, for any open subset $U \subset \mathbb{P}^n - K$, if the Hausdorff 2-measure $\mathcal{H}^2(\widehat{K}^- \cap U) < \infty$, then $\widehat{K}^- \cap U$ is a *complex analytic subvariety* of dimension 1 in U.
6. If K is a real analytic curve, then the Hausdorff dimension of $\widehat{K}$ is 2.
7. $\widehat{K}$ is pluripolar if and only if K is pluripolar.[1] Thus there exist smooth closed curves $\Gamma \subset \mathbb{P}^2$ with $\widehat{\Gamma} = \mathbb{P}^2$. However, real analytic curves are always pluripolar.

The projective hull has simple characterizations in both affine and homogeneous coordinates. For example, if $\pi : \mathbb{C}^{n+1} - \{0\} \longrightarrow \mathbb{P}^n$ is the standard projection, and we set $S(K) \equiv \pi^{-1}(K) \cap S^{2n+1}$, then

$$\widehat{K} = \pi\left\{\widehat{S(K)}_{\text{poly}} - \{0\}\right\}$$

where $\widehat{S(K)}_{\text{poly}}$ is the polynomial hull of $S(K)$ in $\mathbb{C}^{n+1}$.

There is also a *best constant function* $C : \widehat{K} \to \mathbb{R}^+$ defined at x to be the least $C = C(x)$ for which the defining property (13.1) holds. For $x \in \widehat{K}$, the set $\pi^{-1}(x) \cap \widehat{S(K)}_{\text{poly}}$ is a disk of radius $\rho(x) = 1/C(x)$. One deduces that $\widehat{K}$ is compact if C is bounded.

[1] A set K is pluripolar if it is locally contained in the $-\infty$ set of a plurisubharmonic function.

Definition 13.2 *A compact subset $K \subset \mathbb{P}^n$ is called stable if $C : \widehat{K} \to \mathbb{R}^+$ is bounded.*

Combining a classical argument of E. Bishop with (5) and (6) above gives the following:

Theorem 13.6 (Harvey, Lawson, and Wermer) *Let $\Gamma \subset \mathbb{P}^n$ be a compact stable real analytic curve (not necessarily connected). Then $\widehat{\Gamma} - \Gamma$ is a one-dimensional complex analytic subvariety of $\mathbb{P}^n - \Gamma$.*

As mentioned in the introduction, there is much evidence for the conjecture that any compact real analytic curve in $\mathbb{P}^n$ is stable, and therefore the stability hypothesis could be removed from Theorem 13.6. Interestingly, the conclusion of Theorem 13.6 fails to hold in general for smooth curves (see Harvey and Lawson 2006*a*, Section 4), but may hold for smooth curves which are pluripolar or, say, quasi-analytic.

Remark 13.2 The close parallel between polynomial and projective hulls is signaled by the existence of a *projective Gelfand transformation*, whose relation to the classical Gelfand transform is analogous to the relation between Proj of a graded ring and Spec of a ring in modern algebraic geometry. To any Banach graded algebra $A_* = \bigoplus_{d \geq 0} A_d$ (a normed graded algebra which is a direct sum of Banach spaces) one can associate a topological space X_{A_*} and a Hermitian line bundle $\lambda_{A_*} \to X_{A_*}$ with the property that A_* embeds as a closed subalgebra

$$A_* \subseteq \bigoplus_{d \geq 0} \Gamma_{\text{cont}}\left(X_{A_*}, \lambda^d_{A_*}\right)$$

of the algebra of continuous sections of powers of λ with the sup-norm. When $K \subset \mathbb{P}^n$ is a compact subset and $A_d = H^0(\mathbb{P}^n, \mathcal{O}(d))\big|_K$ with the sup-norm on K, there is a natual homeomorphism

$$\widehat{K} \cong X_{A_*} \quad \text{and} \quad \mathcal{O}(1) \cong \lambda_{A_*}.$$

This parallels the affine case where, for a compact subset $K \subset \mathbb{C}^n$, the Gelfand spectrum of the closure of the polynomials on K in the sup-norm corresponds to the polynomial hull of K.

Furthermore, when A_* is finitely generated, the space X_{A_*} can be realized, essentially uniquely, by a subset $X_{A_*} \subset \mathbb{P}^n$ with $\lambda_{A_*} \cong \mathcal{O}(1)$ and

$$\widehat{X}_{A_*} = X_{A_*}.$$

This parallels the classical correspondence between finitely generated Banach algebras and polynomially convex subsets of $\mathbb{C}^n$. Details are given in Harvey and Lawson (2006*a*).

13.3 Projective linking and projective winding numbers

In this section we introduce the notion of projective linking numbers and projective winding numbers for oriented curves in $\mathbb{P}^n$.

Let $M = M^{2p-1} \subset \mathbb{P}^n$ be a compact oriented submanifold of dimension $2p-1$, and recall the *intersection pairing*

$$\bullet : H_{2p}(\mathbb{P}^n, M; \mathbb{Z}) \times H_{2(n-p)}(\mathbb{P}^n - M; \mathbb{Z}) \longrightarrow \mathbb{Z} \tag{13.2}$$

which under Alexander duality corresponds to the Kronecker pairing:

$$\kappa : H^{2(n-p)}(\mathbb{P}^n - M; \mathbb{Z}) \times H_{2(n-p)}(\mathbb{P}^n - M; \mathbb{Z}) \longrightarrow \mathbb{Z}.$$

When homology classes are represented by cycles which intersect transversally in regular points, the map (13.2) is given by the usual algebraic intersection number.

Definition 13.3 *Fix M as above and let $Z \subset \mathbb{P}^n - M$ be an algebraic subvariety of dimension $n-p$. Then the projective linking number of M and Z is defined as follows. Choose a $2p$-chain N in $\mathbb{P}^n$ with $dN = M$ and set*

$$\mathrm{Link}_{\mathbb{P}}(M, Z) \equiv N \bullet Z - \deg(Z) \int_N \omega^p \tag{13.3}$$

where ω is the standard Kähler form on $\mathbb{P}^n$. This definition extends by linearity to algebraic $(n-p)$-cycles $Z = \sum_j n_j Z_j$ supported in $\mathbb{P}^n - M$.

Lemma 13.1 *The linking number* $\mathrm{Link}_{\mathbb{P}}(M, Z)$ *is independent of the choice of the cobounding chain N.*

Proof. Let N' be another choice and set $W = N - N'$. Then $dW = 0$ and

$$W \bullet Z - \deg(Z) \int_W \omega^p = \deg(W) \cdot \deg(Z) - \deg(Z) \cdot \deg(W) = 0.$$

□

For now we shall be concerned with the case $p = 1$. Here there is a naturally related notion of projective winding number defined as follows:

Definition 13.4 *Suppose $\Gamma \subset \mathbb{P}^n$ is a smooth, closed, oriented curve, and let σ be a holomorphic section of $\mathcal{O}(\ell)$ over $\mathbb{P}^n$ which does not vanish on Γ. Then the projective winding number of σ on Γ is defined to be*

$$\mathrm{Wind}_{\mathbb{P}}(\Gamma, \sigma) \equiv \int_\Gamma d^C \log\|\sigma\| \tag{13.4}$$

Proposition 13.2 *Let Γ and σ be as in Definition 13.4, and let Z be the divisor of σ. Then*

$$\mathrm{Wind}_{\mathbb{P}}(\Gamma, \sigma) = \mathrm{Link}_{\mathbb{P}}(\Gamma, Z).$$

Proof. We recall the fundamental formula

$$dd^C \log\|\sigma\| = Z - \ell\omega \tag{13.5}$$

which follows by writing $\|\sigma\| = \rho|\sigma|$ in a local holomorphic trivialization of $\mathcal{O}(\ell)$ and applying the Chern and Poincaré–Lelong formulas: $dd^C \log\rho = -\ell\omega$ and $dd^C \log|\sigma| = \mathrm{Div}(\sigma) = Z$. We now write $\Gamma = \partial N$ for a rectifiable 2-current N in $\mathbb{P}^n$ and note that

$$\int_\Gamma d^C \log\|\sigma\| = \int_N dd^C \log\|\sigma\| = N \bullet Z - \deg(Z) \int_N \omega$$

by (13.5). $\square$

It will be useful in what follows to normalize these linking numbers:

Definition 13.5 *For Γ and Z as above, we define the reduced projective linking number to be*

$$\widetilde{\mathrm{Link}}_{\mathbb{P}}(\Gamma, Z) \;\equiv\; \frac{1}{\deg(Z)} \mathrm{Link}_{\mathbb{P}}(\Gamma, Z)$$

and the reduced projective winding number to be $\widetilde{\mathrm{Wind}}_{\mathbb{P}}(\Gamma, Z) \equiv \mathrm{Wind}_{\mathbb{P}}(\Gamma, Z)/\deg(Z)$.

Note that if $Z = \mathrm{Div}(\sigma)$ for $\sigma \in H^0(\mathbb{P}^n, \mathcal{O}(\ell))$, then by Proposition 13.2 we have

$$\widetilde{\mathrm{Link}}_{\mathbb{P}}(\Gamma, Z) = \int_\Gamma d^C \log\|\sigma\|^{\frac{1}{\ell}}. \tag{13.6}$$

Remark 13.3 (Relation to affine linking numbers) Note that a smooth curve $\Gamma \subset \mathbb{P}^n$ does not meet the generic hyperplane $\mathbb{P}^{n-1}$ and is therefore contained in the affine chart $\mathbb{C}^n = \mathbb{P}^n - \mathbb{P}^{n-1}$. Let N be an integral 2-chain with support in $\mathbb{C}^n$ and $dN = \Gamma$. Then

$$\mathrm{Link}_{\mathbb{P}}(\Gamma, \mathbb{P}^{n-1}) = N \bullet \mathbb{P}^{n-1} - \int_N \omega = -\int_N \omega, \tag{13.7}$$

and for any divisor Z of degree ℓ which does not meet Γ we have

$$\mathrm{Link}_{\mathbb{P}}(\Gamma, Z) - \mathrm{Link}_{\mathbb{P}}(\Gamma, \ell\mathbb{P}^{n-1}) = N \bullet Z = \mathrm{Link}_{\mathbb{C}^n}(\Gamma, Z),$$

the classical linking number of Γ and Z.

Remark 13.4 (Relation to sparks and differential characters) The de Rham–Federer approach to differential characters (cf. Gillet and Soulé 1989, Harris 1989, and Harvey, Lawson, and Zweck 2003) is built on objects called *sparks*. These are generalized differential forms (or currents) α which satisfy the equation

$$d\alpha = R - \phi$$

where R is rectifiable and ϕ is smooth. By (13.5) the generalized 1-form $d^C \log\|\sigma\|$ is such a creature. Its associated Cheeger–Simons character $[\alpha] : Z_1(\mathbb{P}^n) \longrightarrow \mathbb{R}/\mathbb{Z}$ on smooth 1-cycles is defined by the formula

$$[\alpha](\Gamma) \equiv \int_N \ell\omega \pmod{\mathbb{Z}}$$

where N is a 2-chain with $dN = \Gamma$. Thus we see that the "correction term" in the projective linking number is the value of a secondary invariant related to differential characters.

Remark 13.5 Note that by (13.5) the function $u \equiv \log\|\sigma\|^{\frac{1}{\ell}}$ which appears in (13.6) has the property that

$$dd^C u + \omega = \frac{1}{\ell} Z \geq 0 \tag{13.8}$$

by the fundamental equation (13.5). This leads us naturally to the next section.

13.4 Quasi-plurisubharmonic functions

The following concept, due to Demailly (1982) and developed systematically by Guedj and Zeriahi (2003), is central to the study of projective hulls, and it is intimately related to projective linking numbers.

Definition 13.6 *An upper semi-continuous function $u : X \longrightarrow [-\infty, \infty)$ on a Kähler manifold (X, ω) is quasi-plurisubharmonic (or ω-quasi-plurisubharmonic) if $u \not\equiv -\infty$and*

$$dd^C u + \omega \geq 0 \quad \text{on} \quad X. \tag{13.9}$$

The convex set of all such functions on X will be denoted by $PSH_\omega(X)$.

These functions enjoy many of the properties of classical plurisubharmonic functions and play an important role in understanding various capacities in projective space (cf. Guedj and Zeriahi 2003). One of the appealing geometric properties of this class is the following. Suppose ω is the curvature form of a holomorphic line bundle $\lambda \to X$ with Hermitian metric g. Then a smooth function $u : X \to \mathbb{R}$ is quasi-plurisubharmonic if and only if the Hermitian metric $e^u g$ has curvature ≥ 0 on X.

The quasi-plurisubharmonic functions are directly relevant to projective hulls, as the next result shows (cf. Guedj and Zeriahi 2003, proof of Theorem 4.2 and Theorem 13.8 below).

Theorem 13.7 *Let ω denote the standard Kähler form on $\mathbb{P}^n$. Then the projective hull $\widehat{K}$ of a compact subset $K \subset \mathbb{P}^n$, is exactly the subset of points*

$x \in \mathbb{P}^n$ for which there exists a constant $\Lambda = \Lambda(x)$ with

$$u(x) \ \leq \sup_K u + \Lambda \quad \textit{for all} \quad u \in PSH_\omega(X) \tag{13.10}$$

This enables one to generalize the notion of projective hull from projective algebraic manifolds to general Kähler manifolds.

Note that the least constant $\Lambda(x)$ for which (13.10) holds is exactly $\Lambda(x) = \log C(x)$ where $C(x)$ is the best constant function discussed in Section 13.2.

By considering functions of the form $u = \log\{\|\sigma\|^{\frac{1}{d}}\}$ for sections $\sigma \in H^0(\mathbb{P}^n, \mathcal{O}(d))$, one immediatly sees the necessity of the condition in Theorem 13.7. Sufficiency follows from the fact that such functions are the extreme points of the cone $PSH_\omega(\mathbb{P}^n)$.

One importance of quasi-plurisubharmonic functions is that they enable us to establish Poisson–Jensen measures for points in $\widehat{K}$ (Harvey and Lawson 2006*a*, Theorem 11.1).

13.5 Boundaries of positive holomorphic chains

Let Γ be a smooth, oriented, closed curve in $\mathbb{P}^n$. We recall that (even if Γ is only class C^1) any irreducible complex analytic subvariety $V \subset \mathbb{P}^n - \Gamma$ of dimension 1 has finite Hausdorff 2-measure and defines a current $[V]$ of dimension 2 in $\mathbb{P}^n$ by integration on the canonically oriented manifold of regular points. This current satisfies $d[V] = 0$ in $\mathbb{P}^n - \Gamma$ (see e.g. Harvey 1977).

Definition 13.7 *By a positive holomorphic 1-chain with boundary Γ we mean a finite sum $T = \sum_k n_k[V_k]$ where each $n_k \in \mathbb{Z}^+$ and each $V_k \subset \mathbb{P}^n - \Gamma$ is an irreducible subvariety of dimension 1, so that*

$$dT = \Gamma \qquad \text{(as currents on } \ \mathbb{P}^n).$$

We shall be interested in conditions on Γ which are necessary and sufficient for it to be such a boundary.

Proposition 13.3 *Let $\Gamma \subset \mathbb{P}^n$ be a smooth, oriented, closed curve (not necessarily connected) with a positive integer multiplicity on each component. Then the following are equivalent:*

1. $\widetilde{\text{Link}}(\Gamma, Z) \ \geq \ -\Lambda$ *for all algebraic hypersurfaces $Z \subset \mathbb{P}^n - \Gamma$.*
2. $\widetilde{\text{Wind}}(\Gamma, \sigma) \ \geq \ -\Lambda$ *for all $\sigma \in H^0(\mathbb{P}^n, \mathcal{O}(\ell)), \ell > 0$, with no zeros on Γ.*
3. $\int_\Gamma d^C u \ \geq \ -\Lambda$ *for all $u \in PSH_\omega(\mathbb{P}^n)$.*

Proof. The equivalence of (1) and (2) follows immediately from Proposition 13.2. That (3) $\Rightarrow$ (1) follows from (13.6) and (13.8). To see that (1) $\Rightarrow$ (3) we use the following non-trivial fact due essentially to Demailly (1982) and Guedj (1999). □

Proposition 13.4 *The functions of the form $\log(\|\sigma\|^{\frac{1}{\ell}})$ for $\sigma \in H^0(\mathbb{P}^n, \mathcal{O}(\ell))$, $\ell > 0$, are weakly dense in $PSH_\omega(\mathbb{P}^n)$ modulo the constant functions.*

Proof. Fix $u \in PSH_\omega(\mathbb{P}^n)$ and consider the positive (1,1)-current $T \equiv dd^C u + \omega$. By Demailly (1982) and Guedj (1999, Theorem 0.1), there exist sequences $\sigma_j \in H^0(\mathbb{P}^n, \mathcal{O}(\ell_j))$ and $N_j \in \mathbb{R}^+$, $j = 1, 2, 3, \dots$ such that

$$T = \lim_{j\to\infty} \frac{1}{N_j}\mathrm{Div}(\sigma_j) = \lim_{j\to\infty} \frac{\ell_j}{N_j}\left\{dd^C \log\left(\|\sigma_j\|^{\frac{1}{\ell_j}}\right) + \omega\right\}. \tag{13.11}$$

Since $T(\omega^{n-1}) = \int_{\mathbb{P}^n}(dd^C u \wedge \omega^{n-1} + \omega^n) = \int_{\mathbb{P}^n}\omega^n = 1 = \lim_{j\to\infty}\ell_j/N_j$ by (13.11), we may assume that $N_j = \ell_j$ for all j. Therefore, setting $u_j = dd^C\log\|\sigma_j\|^{\frac{1}{\ell_j}}$, we have by (13.11) that $\lim_j dd^C u_j = dd^C u$ and, in particular, $\lim_j \Delta u_j = \Delta u$. Renormalizing each u_j by an additive constant, we may also assume that $\lim_j \int u_j = \int u$.

Recall the formula $Id = H + G \circ \Delta$ on $C^\infty(X)$ where H is harmonic projection and G is the Green's function. This decomposition carries over to distributions $\mathcal{D}'(X)$ by adjoint. It follows that $u_j - \int u_j = G(\Delta u_j) \to G(\Delta u) = u - \int u$ and so $u_j \to u$ as claimed. □

Theorem 13.8 *Let Γ be as above. Suppose $\Gamma = dT$ where T is a positive holomorphic chain with mass $\mathbf{M}(T) \le \Lambda$. Then Γ satisfies the equivalent conditions (1), (2), and (3) of Proposition 13.3.*

Proof. Suppose $u \in PSH_\omega(\mathbb{P}^n)$. Since $dd^C u + \omega \ge 0$, we have

$$\begin{aligned}\int_\Gamma d^C u = \int_{dT} d^C u = T(dd^C u) = T(dd^C u + \omega) - T(\omega) \;\ge\; -T(\omega)\\ = -\mathbf{M}(T) \;\ge -\Lambda\end{aligned}$$

as asserted. □

13.6 Projective Alexander–Wermer theorem for curves

We now examine the question of whether the equivalent necessary conditions, given in Theorem 13.8, are in fact sufficient.

Theorem 13.9 *Let $\Gamma \subset \mathbb{P}^n$ be an embedded, oriented, real analytic closed curve, not necessarily connected and with a positive integer multiplicity on each component. Suppose the underlying curve* supp(Γ) *is stable. If Γ satisfies the equivalent conditions (1), (2), and (3) of Proposition 13.3, then there exists a positive holomorphic 1-chain T in $\mathbb{P}^n$ with $\mathbf{M}(T) \le \Lambda$ such that*

$$dT = \Gamma.$$

Note. When it is clear from context, we also use Γ to denote the underlying curve supp Γ.

Proof. Fix a compact set $K \subset \mathbb{P}^n$ and consider the sets

$$\mathcal{C}_K \;\equiv\; \{dd^C T \;:\; T \in \mathcal{P}_{1,1},\;\; \mathbf{M}(T) \le 1 \;\text{ and }\; \operatorname{supp} T \subset K\}$$

$$\mathcal{S}_K \equiv \{u \in C^\infty(\mathbb{P}^n) \; : \; dd^C u + \omega \geq 0 \;\text{ on } K\}$$

where $\mathcal{P}_{1,1} \subset \mathcal{E}_2(\mathbb{P}^n)$ denotes the convex cone of positive currents of bidimension (1,1) on $\mathbb{P}^n$. Note that $\mathcal{C}_K$ is a weakly closed convex subset which contains 0 (since the set of $T \in \mathcal{P}_{1,1}$ with $\mathbf{M}(T) \leq 1$ and supp $T \subset K$ is weakly compact).

Suppose that $\mathcal{K}$ is a weakly closed convex subset containing 0 in a topological vector space V. Then the *polar* of $\mathcal{K}$ is defined to be the subset of the dual space given by

$$\mathcal{K}^0 \equiv \{L \in V' : (L, v) \geq -1 \;\text{ for }\; v \in \mathcal{K}\}.$$

Similarly given a subset $\mathcal{L} \subset V'$ we define

$$\mathcal{L}^0 \equiv \{v \in V : (L, v) \geq -1 \;\text{ for }\; L \in \mathcal{L}\}.$$

The Bipolar Theorem (Schaefer 1999) states that

$$(\mathcal{K}^0)^0 = \mathcal{K}.$$

□

Proposition 13.5

$$\mathcal{S}_K = (\mathcal{C}_K)^0$$

Proof. Suppose that $u \in C^\infty(\mathbb{P}^n)$ satisfies

$$u(dd^C T) = T(dd^C u) \geq -1 \tag{13.12}$$

for all $T \in \mathcal{P}_{1,1}$ with $\mathbf{M}(T) \leq 1$ and supp $T \subset K$. Consider $T = \delta_x \xi$ where $x \in K$ and ξ is a positive (1,1)-vector at x with mass-norm $\|\xi\| = 1$. Then

$$T(dd^C u) = (dd^C u)(\xi) = (dd^C u + \omega)(\xi) - 1 \geq -1$$

by (13.12), and so $u \in \mathcal{S}_K$. This proves that $(\mathcal{C}_K)^0 \subset \mathcal{S}_K$.

For the converse, let $u \in \mathcal{S}_K$ and fix $T \in \mathcal{P}_{1,1}$ with $\mathbf{M}(T) \leq 1$ and supp $T \subset K$. Then

$$(dd^C T)(u) = T(dd^C u) = T(dd^C u + \omega) - T(\omega) \geq T(\omega) = -\mathbf{M}(T) \geq -1,$$

and so $u \in (\mathcal{C}_K)^0$. □

As an immediate consequence we have the following. If we set

$$\Lambda \cdot \mathcal{S}_K \equiv \{dd^C T : T \in \mathcal{P}_{1,1}, \mathbf{M}(T) \leq \Lambda \text{ and supp } T \subset K\}, \qquad \text{then}$$
$$(\Lambda \cdot \mathcal{S}_K)^0 = \left\{u \in C^\infty(\mathbb{P}^n) \; : \; dd^C u + \tfrac{1}{\Lambda}\omega \geq 0 \;\text{ on }\; K\right\} \tag{13.13}$$

Recall from Section 13.4 that for each $\Lambda \in \mathbb{R}$ we have the compact set

$$\widehat{\Gamma}(\Lambda) = \{x \in \mathbb{P}^n \ : \ u(x) \leq \sup_{\Gamma} u + \Lambda \ \ \forall u \in PSH_\omega(\mathbb{P}^n)\}$$

and that $\widehat{\Gamma} = \bigcup_\Lambda \widehat{\Gamma}(\Lambda)$. The following lemma is established in Harvey and Lawson (2006*a*, 18.7):

Lemma 13.2 *Let u be a C^∞ function which is defined and quasi-plurisubharmonic on a neighborhood of $\{\widehat{\Gamma}\}^-$. Fix $\Lambda > 0$. Then there is a C^∞ function $\widetilde{u}$ which is defined and quasi-plurisubharmonic on all of $\mathbb{P}^n$ and agrees with u on a neighborhood of $\widehat{\Gamma}(\Lambda)$.*

Here $\{\widehat{\Gamma}\}^-$ denotes the closure of $\widehat{\Gamma}$. By our stability assumption, $\widehat{\Gamma} = \widehat{\Gamma}(\Lambda)$ for some Λ, and therefore $\widehat{\Gamma} = \{\widehat{\Gamma}\}^-$. However, we shall keep the notation $\{\widehat{\Gamma}\}^-$, when appropriate, in order to prove the more general result mentioned in Remark 13.6 below.

The lemma above leads to the following:

Proposition 13.6 *If Γ satisfies condition (3) in Proposition 13.3, that is, if*

$$(-d^C\Gamma)(u) = \int_\Gamma d^C u \ \geq \ -\Lambda$$

for all $u \in PSH_\omega(\mathbb{P}^n)$, then there exists $T \in \mathcal{P}_{1,1}$ with $\mathbf{M}(T) \leq \Lambda$ and $\operatorname{supp} T \subset \{\widehat{\Gamma}\}^-$, such that

$$dd^C T = -d^C \Gamma \tag{13.14}$$

Proof. If this is not true, then it must fail on some compact neighborhood K of $\{\widehat{\Gamma}\}^-$. (Otherwise, there exists a sequence of positive currents $\{T_j\}_j$, satisfying (13.14), with $M(T_j) \leq \Lambda$ and $\operatorname{supp} T_j \subset K_j$ where K_j are compact neighborhoods squeezing down to $\{\widehat{\Gamma}\}^-$. By the standard compactness theorem for positive currents, there would then be a convergent subsequence whose limit T would satisfy the conclusion of Proposition 13.6.)

By (13.13) and the Bipolar Theorem, we then conclude that there is a smooth function u which is quasi-plurisubharmonic on K with $-d^C\Gamma(u) < -\Lambda$. Applying Lemma 13.2 contradicts the hypothesis. □

Now let T be the current given by Proposition 13.6. Let V denote the projective hull of Γ and recall from Harvey, Lawson, and Wermer (Theorem 4.1) that V has the following strong regularity. There exists a Riemann surface S with finitely many components, a compact region $W \subset S$ with real analytic boundary, and a holomorphic map $\rho : S \to \mathbb{P}^n$ which is generically injective and satisfies

1. $\rho(W) = V$.
2. ρ is an embedding on a tubular neighborhood of ∂W in S.
3. $\rho(\partial W)$ is a union of components of the support of Γ.

Let Σ denote an ϵ-tubular neighborhood of ∂W on S (with respect to some analytic metric), with ϵ chosen so that

4. ρ is injective on Σ.
5. $\rho(\partial^+\Sigma) \cap V = \emptyset$ where $\partial^+\Sigma$ denotes the "outer" boundary of Σ, that is, the union of components of $\partial\Sigma$ not contained in W.

Write $\Sigma = \Sigma^+ \cup \partial W \cup \Sigma^-$ where Σ^+ denotes the union of components of $\Sigma - \partial W$ which are not contained in W. Then we have $d[\Sigma^+] = [\partial^+\Sigma] - [\partial W]$ (with standard orientations coming from the domains). Hence we have

$$dd^C[\Sigma^+] = -d^C[\partial^+\Sigma] + d^C[\partial W].$$

Let $\nu : \Sigma \to \mathbb{Z}$ denote the locally constant, integer-valued function with the property that

$$\rho_* \left(\nu[\partial W]\right) = \Gamma.$$

Then

$$dd^C \rho_* \left(\nu[\Sigma^+]\right) = -d^C \rho_* \left(\nu[\partial^+\Sigma]\right) + d^C\Gamma \stackrel{\text{def}}{=} -d^C\Gamma^+ + d^C\Gamma. \tag{13.15}$$

We now define

$$T^+ = T + \rho_* \left(\nu[\Sigma^+]\right)$$

and note that by (13.14) we have

$$dd^C T^+ = -d^C\Gamma^+. \tag{13.16}$$

Now in the open set $\mathbb{P}^n - \Gamma^+$, the current T^+ is a positive (1,1)-current which is supported in the analytic subvariety $V^+ \equiv \rho(W^+) \equiv \rho(\Sigma^+ \cup \overline{W})$ and satisfies $dd^C T^+ = 0$. We note that $\rho : W^+ \to V^+$ is just the normalization of V^+. It follows from Harvey and Lawson (1983, Lemma 32) that there is a harmonic function $h : W \to \mathbb{R}^+$ so that

$$T^+ = \rho_*(h[W^+]) \quad \text{in} \quad \mathbb{P}^n - \Gamma^+.$$

This function is evidently constant ($= \nu$) in Σ^+, and hence h is constant on every component of W^+. Thus $T = \rho_*(h[W])$ is a positive holomorphic chain.

Now since $\operatorname{supp} dT \subset \operatorname{supp}\Gamma$ and $\operatorname{supp}\Gamma$ is a regular curve, it follows from the Federer flat support theorem (Federer 1969, Theorem 4.1.15) that dT is of the form $dT = \sum_k c_k[\Gamma_k]$ where the c_k's are constants and the Γ_k's are the oriented, connected components of $\operatorname{supp}\Gamma$. The current Γ itself has the form $\Gamma = \sum_k \nu_k[\Gamma_k]$ for integers ν_k. Since $dd^C T = -d^C\Gamma$, we have $d^C(dT - \Gamma) = \sum(c_k - \nu_k)d^C[\Gamma_k] = 0$. It follows that $c_k = \nu_k$ for all k, that is, $dT = \Gamma$ as claimed. □

Remark 13.6 Theorem 13.9 remains true if one replaces the second statement with the assumption that the closed projective hull $\{\widehat{\Gamma}\}^-$ of Γ is locally contained in a 1-dimensional complex subvariety at each point of $\{\widehat{\Gamma}\}^- - \Gamma$.

Theorem 13.9 can be restated in terms of the classical (affine) linking numbers:

Theorem 13.10 *Let $\Gamma \subset \mathbb{C}^n \subset \mathbb{P}^n$ be an oriented, closed curve as in Theorem 13.9. Then there exists a constant Λ_0 so that the classical linking number*

$$\mathrm{Link}_{\mathbb{C}^n}(\Gamma, Z) \;\geq\; -\Lambda_0 \deg Z$$

for all algebraic hypersurfaces in $\mathbb{C}^n - \Gamma$ if and only if Γ is the boundary of a positive holomorphic 1-chain T in $\mathbb{P}^n = \mathbb{C}^n \cup \mathbb{P}^{n-1}$ with (projective) mass

$$\mathbf{M}(T) \;\leq\; \Lambda_0 + \frac{1}{2}\int_\Gamma d^C \log(1 + \|z\|^2).$$

Proof. This follows from Remark 13.3 and the fact that the projective Kähler form $\omega = \frac{1}{2} dd^C \log(1 + \|z\|^2)$ in $\mathbb{C}^n$. □

Note. The case $\Lambda_0 = 0$ corresponds to the theorem of Alexander and Wermer (2000). A proof of the full Alexander–Wermer theorem (for C^1-curves with no stability assumption) in the spirit of the arguments above is given in Harvey and Lawson (2006*b*).

Note that the current of least mass among those provided by Theorem 13.9 is uniquely determined by Γ. All others differ from this one by adding a positive algebraic cycle.

Corollary 13.2 *Let Γ be as in Theorem 13.9 and suppose T is a positive holomorphic 1-chain with $dT = \Gamma$. Then T is the unique holomorphic chain of least mass with $dT = \Gamma$ if and only if*

$$\inf_Z \left\{ \frac{T \bullet Z}{\deg Z} \right\} = 0$$

where the infimum is taken over all algebraic hypersurfaces in the complement $\mathbb{P}^n - \Gamma$.

Proof. Suppose $\inf_Z\{T \bullet Z/\deg Z\} = c > 0$. Then $\widetilde{\mathrm{Link}}_{\mathbb{P}}(\Gamma, Z) = T \bullet Z/\deg(Z) - T(\omega) \geq c - T(\omega)$ for all positive divisors in $\mathbb{P}^n - \Gamma$. Hence, by Theorem 13.9 there exists a positive holomorphic chain T' with $dT' = \Gamma$ and $M(T') \leq M(T) - c$.

On the other hand suppose that T is not the unique positive holomorphic chain T_0 of least mass. Then $T = T_0 + W$ where W is a positive algebraic cycle, and one has $T \bullet Z/\deg Z = T_0 \bullet Z/\deg Z + W \bullet Z/\deg Z \geq W \bullet Z/\deg Z = \deg W$. □

13.7 Theorems for general projective manifolds

The results established above generalize from $\mathbb{P}^n$ to any projective manifold. Let X be a compact complex manifold with a positive holomorphic line bundle λ. Fix a Hermitian metric on λ with curvature form $\omega > 0$, and give X the Kähler metric associated to ω. Let Γ be a closed curve with integral weights as in Theorem 13.9, and assume $[\Gamma] = 0$ in $H_1(X;\mathbb{Z})$.

Definition 13.8 *Let $Z = \mathrm{Div}(\sigma)$ be the divisor of a holomorphic section $\sigma \in H^0(X, \mathcal{O}(\lambda^\ell))$ for some $\ell \geq 1$. If Z does not meet Γ, we can define the* linking number *and the* reduced linking number *by*

$$\mathrm{Link}_\lambda(\Gamma, Z) \;\equiv\; N \bullet Z - \ell \int_N \omega \qquad \text{and} \qquad \widetilde{\mathrm{Link}}_\lambda(\Gamma, Z) \;\equiv\; \frac{1}{\ell}\,\mathrm{Link}(\Gamma, Z),$$

respectively, where N is any 2-chain in Z with $dN = \Gamma$ and where the intersection pairing $\bullet$ is defined as in (13.2) with $\mathbb{P}^n$ replaced by X.

To see that this is well defined suppose that N' is another 2-chain with $dN' = \Gamma$. Then $(N - N') \bullet Z - \ell \int_{N-N'} \omega = (N - N') \bullet (Z - \ell[\omega]) = 0$ because $Z - \ell\omega$ is cohomologous to zero in X.

Definition 13.9 *The winding number of a section $\sigma \in H^0(X, \mathcal{O}(\lambda^\ell))$ with $\|\sigma\| > 0$ on Γ is defined to be*

$$\mathrm{Wind}_\lambda(\Gamma, \sigma) \;\equiv\; \int_\Gamma d^C \log\|\sigma\|.$$

The reduced winding number is

$$\widetilde{\mathrm{Wind}}_\lambda(\Gamma, \sigma) \;\equiv\; \frac{1}{\ell}\,\mathrm{Wind}_\lambda(\Gamma, \sigma) = \int_\Gamma d^C \log\|\sigma\|^{\frac{1}{\ell}}$$

From the Poincaré–Lelong equation

$$dd^C \log\|\sigma\| = \mathrm{Div}(\sigma) - \ell\omega \tag{13.17}$$

we see that

$$\mathrm{Wind}_\lambda(\Gamma, \sigma) = \mathrm{Link}_\lambda(\Gamma, \mathrm{Div}(\sigma)). \tag{13.18}$$

From (13.17) *we also see that* $\log\|\sigma\|^{\frac{1}{\ell}}$ *belongs to the class $PSH_\omega(X)$ of quasi-plurisubharmonic functions on X defined in* (13.9).

Proposition 13.7 *The following are equivalent:*

1. $\widetilde{\mathrm{Link}}_\lambda(\Gamma, Z) \;\geq\; -\Lambda$ *for all divisors Z of holomorphic sections of λ^ℓ, and all $\ell > 0$.*
2. $\widetilde{\mathrm{Wind}}_\lambda(\Gamma, \sigma) \;\geq\; -\Lambda$ *for all holomorphic sections σ of λ^ℓ, and all $\ell > 0$.*
3. $\int_\Gamma d^C u \;\geq\; -\Lambda$ *for all $u \in PSH_\omega(X)$.*

Proof. That (1) $\Leftrightarrow$ (2) $\Rightarrow$ (3) is clear. To see that (3) $\Rightarrow$ (2) we use results of Demailly. Fix $u \in PSH_\omega(X)$ and consider the positive closed (1,1)-current $T \equiv dd^C u + \omega$. Note that $[T] = [\omega] = c_1(\lambda) \in H^2(X;\mathbb{Z})$. It follows from Demailly

(1982) that T is the weak limit

$$T = \lim_{j\to\infty} \frac{1}{N_j}\mathrm{Div}(\sigma_j)$$

where $\sigma_j \in H^0(X, \mathcal{O}(\lambda^{\ell_j}))$ and $N_j > 0$. We can normalize this sequence by scalars so that $\mathbf{M}(\frac{1}{N_j}\mathrm{Div}(\sigma_j)) = \mathbf{M}(T)$ for all j. Set $\Omega = \frac{1}{(n-1)!}\omega^{n-1}$. Then $\mathbf{M}(T) = \mathbf{M}(\frac{1}{N_j}\mathrm{Div}(\sigma_j)) = (\frac{1}{N_j}\mathrm{Div}(\sigma_j), \Omega) = \frac{1}{N_j}(\ell_j\omega, \Omega) = \frac{\ell_j}{N_j}(\omega, \Omega) = \frac{\ell_j}{N_j}\mathbf{M}(T)$. Therefore, $\ell_j = N_j$ for all j. Since $\mathrm{Div}(\sigma_j) = dd^C\log\|\sigma_j\| + \ell_j\omega$, we conclude that

$$dd^C u = T - \omega = \lim_{j\to\infty}\left\{dd^C\log\|\sigma_j\|^{\frac{1}{\ell_j}} + \omega\right\} - \omega = \lim_{j\to\infty} dd^C\log\|\sigma_j\|^{\frac{1}{\ell_j}}.$$

The remainder of the argument replicates the one given for Proposition 13.4. □

For a compact subset $K \subset X$ the authors introduced the notion of the λ-hull $\widehat{K}_\lambda$ of K and showed that $\widehat{K}_\lambda = \widehat{K}$ for any embedding $X \hookrightarrow \mathbb{P}^N$ by sections of some power of λ. We shall say that K is *λ-stable* if $\widehat{K}_\lambda$ is compact.

Theorem 13.11 *Let $\Gamma = \sum_{a=1}^{M} m_\alpha \Gamma_\alpha$ be an embedded, oriented, real analytic closed curve with integer multiplicities in X, and assume Γ is λ-stable. Then there exists a positive holomorphic 1-chain T in X with $dT = \Gamma$ and $\mathbf{M}(T) \le \Lambda$ if and only if any of the equivalent conditions of Proposition 13.6 is satisfied.*

Proof. If $dT = \Gamma$ and $\mathbf{M}(T) \le \Lambda$, then (1) follows as in the proof of Theorem 13.8 above.

For the converse we recall from Harvey and Lawson (2006*a*, (4.4)) that the hull $\widehat{K}_\lambda$ of a compact subset $K \subset X$ consists exactly of the points x where the extremal function

$$\Lambda_K(x) \;\equiv\; \sup\{u(x) : u \in PSH_\omega(X) \text{ and } \; u \le 0 \text{ on } X\}$$

is finite. This enables one to directly carry through the arguments for Theorem 13.9 in this case. □

13.8 Relative holomorphic cycles

Our main Theorem 13.9 has a nice reinterpretation in terms of the Alexander–Lefschetz duality pairing discussed in Section 13.3.

Theorem 13.12 *Let $\gamma \subset \mathbb{P}^n$ be a finite disjoint union of real analytic curves with γ-stable. Then a class $\tau \in H_2(\mathbb{P}^n, \gamma; \mathbb{Z})$ is represented by a positive holomorphic chain with boundary on γ if and only if*

$$\tau \bullet u \;\ge\; 0 \tag{13.19}$$

for all $u \in H_{2n-2}(\mathbb{P}^n - \gamma; \mathbb{Z})$ represented by positive algebraic hypersurfaces in $\mathbb{P}^n - \gamma$.

Proof. The implication $\Rightarrow$ is clear from the positivity of complex intersections. For the converse, consider the short exact sequence

$$0 \longrightarrow H_2(\mathbb{P}^n;\mathbb{Z}) \longrightarrow H_2(\mathbb{P}^n,\gamma;\mathbb{Z}) \stackrel{\delta}{\longrightarrow} H_1(\gamma;\mathbb{Z}) \longrightarrow 0.$$

Note that

$$\delta\tau = \sum_{k=1}^{\ell} m_k \overrightarrow{\gamma}_k \equiv \Gamma$$

where $\overrightarrow{\gamma}_1, ..., \overrightarrow{\gamma}_\ell$ are the connected components of γ with a chosen orientation and the m_k's are integers which we can assume to be positive. We are assuming that (13.19) holds for any positive algebraic class u. If $\Gamma = 0$, the desired conclusion is immediate, so we assume that $\Gamma \neq 0$. Now let$Z \subset \mathbb{P}^n - \gamma$ be any positive algebraic hypersurface, and note that

$$0 \leq \frac{\tau \bullet Z}{\deg Z} = \frac{\tau \bullet Z}{\deg Z} - \tau(\omega) + \tau(\omega) = \mathrm{Link}_{\mathbb{P}}(\Gamma, Z) + \tau(\omega).$$

Therefore, by Theorem 13.9, there exists a positive holomorphic 1-chain T with $dT = \Gamma$ and $M(T) \leq \tau(\omega)$. Note that $\delta([T] - \tau) = 0$ and $([T] - \tau)(\omega) \geq 0$. Hence, $[T] - \tau$ is a positive class in $H_2(\mathbb{P}^n;\mathbb{Z})$ and is represented by a positive algebraic 1-cycle, say W. Therefore, $\tau = [T + W]$ is represented by a positive holomorphic chain as claimed. □

This result leads to a nice duality between the cone in $H_2(\mathbb{P}^n,\gamma;\mathbb{Z})$ of those classes which are represented by positive holomorphic chains, and the cone in $H_{2n-2}(\mathbb{P}^n - \gamma;\mathbb{Z})$ of classes represented by positive algebraic hypersurfaces. This duality, in fact, extends to more general projective manifolds. All this is discussed in detail in Harvey and Lawson (2006*d*).

Acknowledgment

This chapter was partially supported by the NSF.

References

Alexander, H. and Wermer, J. (1998). *Several Complex Variables and Banach Algebras*, Springer-Verlag, New York.

Alexander, H. and Wermer, J. (2000). Linking numbers and boundaries of varieties. *Ann. Math.* **151**, 125–150.

Demailly, J.-P. (1982). Estimations L^2 pour l'opérateur $\overline{\partial}$ d'un fibré vectoriel holomorphe semi-positif au-dessus d'une variété kählérienne complète. *Ann. Sci. E. N. S.* **15**(4), 457–511.

Demailly, J.-P. (1982). Courants positifs extrêmaux et conjecture de Hodge. *Invent. Math.* **69**(4), 347–74.

Dinh ,T.-C. and Lawrence, M. (2003). Polynomial hulls and positive currents. *Ann. Fac. Sci. de Toulouse* **12**, 317–34.

Dolbeault, P. (1983). On holomorphic chains with given boundary in CP^n. *Springer Lect. Notes Math.* **1089**, 1135–40.

Dolbeault, P. and Henkin, G. (1994). Surfaces de Riemann de bord donné dans CP^n, pp. 163–87 in *Contributions to Complex Analysis and Analytic Geometry*, H. Skoda and J.-M. Trépreau (eds.), Aspects of Mathematics, Vol. **26**, Vieweg, Wiesbaden, Germany.

Dolbeault, P. and Henkin, G. (1997). Chaînes holomorphes de bord donné dans CP^n. *Bull. Soc. Math. France* **125**, 383–445.

Duval, J. and Sibony, N. (1995). Polynomial convexity, rational convexity and currents. *Duke Math. J.* **79**, 487–513.

Federer, H. (1969). *Geometric Measure Theory*, Springer–Verlag, New York.

Gillet H. and Soulé, C. (1989). Arithmetic chow groups and differential characters, pp. 30–68, in *Algebraic K-theory; Connections with Geometry and Topology*, J. F. Jardine and V. P. Snaith (eds.), Kluwer Academic Publishers Amsterdam.

Guedj, V. (1999). Approximation of currents on complex manifolds. *Math. Ann.* **313**, 437–474.

Guedj V. and Zeriahi, A. (2003). *Intrinsic capacities on compact Kähler manifolds*, Preprint, Université de Toulouse.

Harris, B. (1989). Differential characters and the Abel-Jacobi map, pp. 69–86, in Algebraic *K-theory; Connections with Geometry and Topology*, J. F. Jardine and V. P. Snaith (eds.), Kluwer Academic Publishers Amsterdam.

Harvey, F. R. (1977). Holomorphic chains and their boundaries, pp. 309–32, in *Several Complex Variables, Proceedings of Symposia in Pure Mathematics XXX Part 1*, A.M.S., Providence, RI.

Harvey, F. R. and Lawson, H. B. Jr., (1975). On boundaries of complex analytic varieties, I. *Ann. Math.* **102**, 223–90.

Harvey, F. R. and Lawson, H. B. Jr., (1977). On boundaries of complex analytic varieties, II. *Ann. Math.* **106**, 213–38.

Harvey, F. R. and Lawson, H. B. Jr., (1983). An intrinsic characterization of Kähler manifolds. *Invent. Math.* **74**, 169–98.

Harvey, F. R. and Lawson, H. B. Jr., (2005). Boundaries of varieties in projective manifolds. *J. Geom. Analysis* **14**(4), 673–95. ArXiv:math.CV/0512490.

Harvey, F. R. and Lawson, H. B. Jr., (2006*a*). Projective hulls and the projective Gelfand transformation. *Asian J. Math.* **10**(2), 279–318. ArXiv:math.CV/0510286.

Harvey, F. R. and Lawson, H. B. Jr., (2006*b*). Remarks on the Alexander-Wermer theorem for curves. *Matemática Contemporânea* **31**, 53–64. ArXiv: math.CV/0610611.

Harvey, F. R. and Lawson, H. B. Jr., (2006*c*). Projective linking and boundaries of holomorphic chains in projective manifolds, Part II, pp. 365–80, in Inspired by S. S. Chern, P. Griffiths (ed.), Nankai Tracts in Mathematics, Vol. **11**, World Scientific Publishing, Singapore.

Harvey, F. R. and Lawson, H. B. Jr., (2006*d*). Relative holomorphic cycles and duality, Preprint, Stony Brook.

Harvey, F. R. Lawson, H. B. Jr., and Zweck, J. (2003). A de Rham-Federer theory of differential characters and character duality. *Am. J. Math.* **125**, 791–847.

Harvey, F. R. Lawson, H. B. Jr., and Wermer, J. (2008). The projective hull of certain curves in C^2. Astérisque **322**, 241–54. ArXiv:math.CV/0611482.

Harvey, F. R. and Polking, J. (1975). Extending analytic objects. *Commun. Pure Appl. Math.* **28**, 701–27.

Schaefer, H. H. (1999). *Topological Vector Spaces*, Springer Verlag, New York.

Wermer, J. (1958). The hull of a curve in C^n. *Ann. Math.* **68**, 550–61.

Wermer, J. (2001). The argument principle and boundaries of analytic varieties. *Operator Theory: Adv. Appl.* **127**, 639–59.

XIV

SKYRMIONS AND NUCLEI

Nicholas S. Manton

Dedicated to Nigel Hitchin on the occasion of his 60th birthday

14.1 Skyrmions

T. H. R. Skyrme (1961, 1962) proposed that the interior of a nucleus is dominated by a non-linear semiclassical medium formed from the three pion fields, and he introduced the Skyrme model, a version of the Lorentz-invariant, non-linear sigma model, in which the pion fields $\boldsymbol{\pi} = (\pi_1, \pi_2, \pi_3)$ are combined into an $SU(2)$-valued scalar field

$$U = (1 - \boldsymbol{\pi} \cdot \boldsymbol{\pi})^{1/2} 1 + i\boldsymbol{\pi} \cdot \boldsymbol{\tau}\,, \tag{14.1}$$

where $\boldsymbol{\tau}$ are the Pauli matrices. There is an associated current, with spatial components $R_i = (\partial_i U)U^\dagger$. For static fields, the energy in the Skyrme model is given by

$$E = \frac{1}{12\pi^2} \int \left\{ -\frac{1}{2}\mathrm{Tr}(R_i R_i) - \frac{1}{16}\mathrm{Tr}([R_i, R_j][R_i, R_j]) + m^2 \mathrm{Tr}(1 - U) \right\} d^3x\,, \tag{14.2}$$

and the vacuum is $U = 1$. E is invariant under translations and rotations in $\mathbb{R}^3$ and also under $SO(3)$ isospin rotations given by the conjugation

$$U \mapsto \mathcal{O} U \mathcal{O}^\dagger, \quad \mathcal{O} \in SU(2)\,. \tag{14.3}$$

This rotates the pion fields among themselves. Stationary points of E satisfy the Skyrme field equation, and we shall mostly consider minima of E. (For a review, see Manton and Sutcliffe 2004.)

The expression (14.2) is in 'Skyrme units' and m is a dimensionless pion mass parameter. We will discuss below the calibration of the energy and length units by comparison with physical data. Traditionally m has been given a value of approximately 0.5 (Adkins and Nappi 1984), but recent work suggests a higher value, $m = 1$ (Battye *et al.* 2005; Battye and Sutcliffe 2006) or $m = 1.125$ (Manton and Wood 2006).

The model has a conserved, integer-valued topological charge B, the baryon number. This is the degree of the map $U : \mathbb{R}^3 \to SU(2)$, which is well-defined

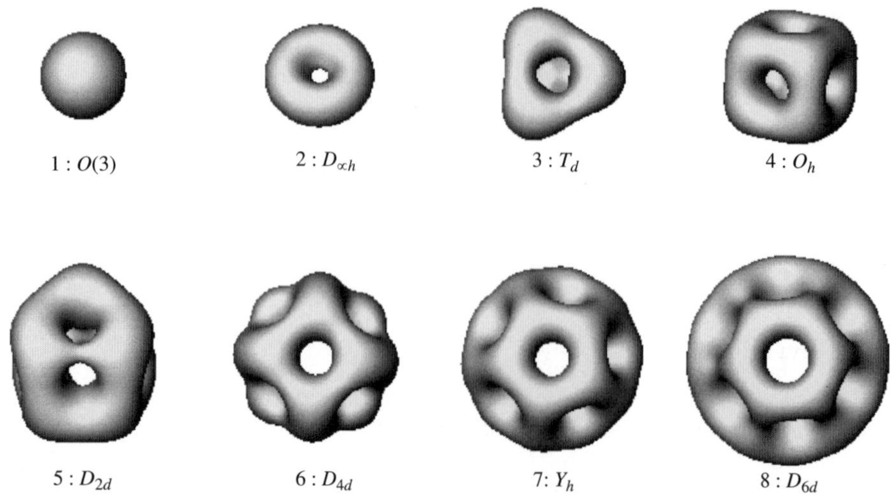

Figure 14.1 Skyrmions for $1 \leq B \leq 8$, with $m = 0$. A surface of constant baryon density is shown, together with the baryon number and symmetry.

because $U \to 1$ at spatial infinity. B is the integral of the baryon density

$$\mathcal{B} = -\frac{1}{24\pi^2}\epsilon_{ijk}\mathrm{Tr}(R_i R_j R_k), \tag{14.4}$$

which is proportional to the Jacobian of the map U.

In the above units there is the Faddeev–Bogomolny energy bound, $E \geq |B|$. The minimal energy solutions for each B are called Skyrmions. (More loosely, local minima and saddle points of E with nearby energies are also sometimes called Skyrmions.) Skyrme argued that these topological soliton solutions can be identified with nucleons and nuclei.

The $B = 1$ Skyrmion has the spherically symmetric, hedgehog form:

$$U(\mathbf{x}) = \exp\{if(r)\widehat{\mathbf{x}} \cdot \boldsymbol{\tau}\} = \cos f(r)1 + i \sin f(r)\widehat{\mathbf{x}} \cdot \boldsymbol{\tau}\,. \tag{14.5}$$

f is a radial profile function obeying an ordinary differential equation (ODE) with the boundary conditions $f(0) = \pi$ and $f(\infty) = 0$. Skyrmions with baryon numbers greater than 1 all have interesting shapes (see Figure 14.1); they are not spherical like the basic $B = 1$ Skyrmion. The $B = 2$ Skyrmion is toroidal, and the $B = 3$ Skyrmion tetrahedral. The $B = 4$ Skyrmion is cubic and can be obtained by bringing together two $B = 2$ toroids along their common axis. The $B = 6$ solution has D_{4d} symmetry and can be formed from three toroids stacked one above the other, and the $B = 7$ Skyrmion has icosahedral symmetry. The toroidal structure of the $B = 2$ Skyrmion has some phenomenological support from nuclear physics (Forest *et al.* 1996). Table 14.1 presents, for $m = 0$ and baryon numbers 1–8, the symmetries and energies of the Skyrmions, computed from numerical solutions of the field equation (Battye and Sutcliffe 1997, 2001, 2002).

Table 14.1. Symmetry K and energy per baryon E/B for numerically computed Skyrmions

B	K	E/B
1	$O(3)$	1.2322
2	$D_{\infty h}$	1.1791
3	T_d	1.1462
4	O_h	1.1201
5	D_{2d}	1.1172
6	D_{4d}	1.1079
7	Y_h	1.0947
8	D_{6d}	1.0960

$SU(2)$ Yang–Mills–Higgs monopoles, satisfying the Bogomolny equation, have been constructed with very similar symmetries (and baryon number replaced by monopole charge) (Hitchin *et al.* 1995), and there was, historically, an interesting interplay between the discovery of symmetric monopoles and symmetric Skyrmions.

When $m = 0$, the Skyrmions for baryon numbers from $B = 8$ up to $B = 22$ (Battye and Sutcliffe 2001, 2002), and various larger values of B (Battye *et al.* 2003) are hollow polyhedra, rather like carbon fullerenes. The baryon density is concentrated in a shell of roughly constant thickness, surrounding a region whose volume increases like $B^{3/2}$ where the energy and baryon density are very small. Such a hollow structure clearly disagrees with the rather uniform baryon density observed in the interior of real nuclei.

However, for baryon numbers $B \geq 8$, the hollow polyhedral Skyrmions do not remain stable when the pion mass parameter m is of order 1 (Battye and Sutcliffe 2005, 2006). This is because, in the interior of the hollow polyhedra, the Skyrme field is very close to $U = -1$ (i.e. the antipode to the vacuum value), and here the pion mass term gives the field a maximal potential energy density. Consequently, this interior region tends to pinch off and separate into smaller sub-regions.

The new, stable Skyrmions have a denser structure. Some of them are clusters of $B = 4$ Skyrmions; these are described in Section 14.3. Others, for $10 \leq B \leq 16$, have a planar character. It should be possible to interpret these solutions as fragments of an infinite crystalline sheet with hexagonal (or perhaps in some cases, square) symmetry. This would be a two-layer version of the solution presented in Battye and Sutcliffe (1998) (which by itself has the wrong boundary conditions).

14.2 Rational map ansatz

The observed similarities between Skyrmions and monopoles suggests there could be an approximate construction of Skyrmions from monopoles. As yet, there is no known direct transformation between the fields of a monopole and those of a Skyrmion, but there is an indirect transformation via rational maps between Riemann spheres. It is known that there is a precise one-to-one correspondence between charge N monopoles and degree N rational maps (we have in mind here the Jarvis 2000 maps); thus a Skyrme field constructed from a rational map is indirectly constructed from a monopole.

The rational map ansatz of Houghton *et al.* (1998) separates the angular from the radial dependence of the Skyrme field U. One introduces a complex (Riemann sphere) coordinate $z = \tan\frac{\theta}{2}\, e^{i\phi}$, where θ and ϕ are the usual spherical polar coordinates, and constructs the Skyrme field from a rational function of z:

$$R(z) = \frac{p(z)}{q(z)} \tag{14.6}$$

where p and q are polynomials with no common root, and a radial profile function $f(r)$ satisfying $f(0) = \pi$ and $f(\infty) = 0$. One should think of R as a smooth map from a two-sphere in space (at a given radius) to a two-sphere in the target $SU(2)$ (at a given distance from the identity). By standard stereographic projection, the point z corresponds to the Cartesian unit vector

$$\widehat{\mathbf{n}}_z = \frac{1}{1+|z|^2}(z+\bar{z},\, i(\bar{z}-z),\, 1-|z|^2)\,. \tag{14.7}$$

Similarly, an image point R can be expressed as a unit vector

$$\widehat{\mathbf{n}}_R = \frac{1}{1+|R|^2}(R+\bar{R}, i(\bar{R}-R), 1-|R|^2)\,. \tag{14.8}$$

The ansatz for the Skyrme field is then

$$U(r,z) = \exp(if(r)\widehat{\mathbf{n}}_{R(z)}\cdot\boldsymbol{\tau})\,, \tag{14.9}$$

generalizing the hedgehog formula (14.5). The ansatz is a suspension of the map $R : S^2 \to S^2$ to produce a map $U : \mathbb{R}^3 \to S^3$, the suspension points being the origin and infinity in $\mathbb{R}^3$, which are mapped to $U = -1$ and $U = 1$, respectively. The baryon number, B, of this Skyrme field equals the topological degree of the rational map $R : S^2 \to S^2$, which is the higher of the algebraic degrees of p and q.

An $SU(2)$ Möbius transformation on the domain S^2 of the rational map corresponds to a spatial rotation, whereas an $SU(2)$ Möbius transformation on the target S^2 corresponds to a rotation of $\widehat{\mathbf{n}}_R$, and hence to an isospin rotation of the Skyrme field. Thus if a rational map R has some symmetry (i.e. a rotation of the domain can be compensated by a rotation of the target), then the resulting Skyrme field has that symmetry (i.e. a spatial rotation can be compensated by an isospin rotation).

An important feature of the rational map ansatz is that when one substitutes it into the Skyrme energy function (14.2), the angular and radial parts decouple.

The energy simplifies to

$$E = \frac{1}{3\pi}\int_0^\infty \left(r^2 f'^2 + 2B(f'^2+1)\sin^2 f + \mathcal{I}\,\frac{\sin^4 f}{r^2} + 2m^2 r^2 (1-\cos f) \right) dr, \tag{14.10}$$

where $\mathcal{I}$ denotes the angular integral

$$\mathcal{I} = \frac{1}{4\pi}\int \left(\frac{1+|z|^2}{1+|R|^2}\left|\frac{dR}{dz}\right| \right)^4 \frac{2i\,dzd\bar{z}}{(1+|z|^2)^2}, \tag{14.11}$$

which only depends on the rational map R. $\mathcal{I}$ is an interesting function on the space of rational maps. To minimize the energy (for given B), it is sufficient to first minimize $\mathcal{I}$ with respect to the coefficients occurring in the rational map, and then to solve an ODE for $f(r)$ whose coefficients depend on the rational map only through B and the minimized $\mathcal{I}$. Optimal rational maps, and the associated profile functions, have been found for many values of B, and often have a high degree of symmetry. The optimized fields within the rational map ansatz are good approximations to Skyrmions, and they are also used as a starting point for a numerical relaxation to true Skyrmion solutions (which almost always have the same symmetry, but no exact separation of the angular and radial dependence of U).

The simplest degree 1 rational map is $R(z) = z$, which is spherically symmetric. The ansatz (14.9) then reduces to the hedgehog field (14.5). For $B = 2, 3, 4, 7$, the symmetries of the numerically computed Skyrmions are $D_{\infty h}, T_d, O_h, Y_h$, respectively. In each of these cases there is a unique rational map with this symmetry, up to rotations and isorotations, namely,

$$R(z) = z^2, \quad R(z) = \frac{z^3 - \sqrt{3}iz}{\sqrt{3}iz^2 - 1}, \quad R(z) = \frac{z^4 + 2\sqrt{3}iz^2 + 1}{z^4 - 2\sqrt{3}iz^2 + 1},$$
$$R(z) = \frac{z^7 - 7z^5 - 7z^2 - 1}{z^7 + 7z^5 - 7z^2 + 1}, \tag{14.12}$$

and these also minimize $\mathcal{I}$. For $B = 5$, 6, and 8, rational maps with dihedral symmetries are required, and these involve one or two coefficients that need to be determined numerically. Table 14.2 lists the energies of the approximate solutions obtained using the rational map ansatz, together with the values of $\mathcal{I}$, again for $m = 0$ (Battye and Sutcliffe 2001, 2002).

The Wronskian of a rational map $R(z) = p(z)/q(z)$ of degree B is the polynomial

$$W(z) = p'(z)q(z) - q'(z)p(z) \tag{14.13}$$

of degree $2B - 2$. Where W is zero, the derivative dR/dz is zero, so only the radial derivative of U is non-vanishing. The baryon density therefore vanishes along the entire radial half-line in the direction of a zero of W, and the energy density is also low. This explains why the Skyrmion baryon density contours

Table 14.2. Symmetry K, the value of the angular integral $\mathcal{I}$, and the energy per baryon E/B of the approximate Skyrmions obtained using the rational map ansatz

B	K	$\mathcal{I}$	E/B
1	$O(3)$	1.0	1.232
2	$D_{\infty h}$	5.8	1.208
3	T_d	13.6	1.184
4	O_h	20.7	1.137
5	D_{2d}	35.8	1.147
6	D_{4d}	50.8	1.137
7	Y_h	60.9	1.107
8	D_{6d}	85.6	1.118

look like polyhedra with holes in the directions given by the zeros of W, and why there are $2B-2$ such holes, precisely the structures seen in Figure 14.1.

As an example, the icosahedrally symmetric degree 7 map in (14.12) has Wronskian

$$W(z) = 28z(z^{10} + 11z^5 - 1), \tag{14.14}$$

which is proportional to one of the icosahedral Klein polynomials, and vanishes at the twelve face centres of a dodecahedron (including $z = \infty$).

The solutions we have described so far are for $m = 0$, but it is found that qualitatively similar solutions with 10–20% higher energy exist for m up to 1 and a little beyond, provided $B \leq 7$. There is, however, a qualitative change for $B > 7$, as we discuss next.

14.3 Skyrmions and α-particles

The ^{4}He nucleus, or α-particle, is particularly stable and can be regarded as a building block of larger nuclei. The α-particle model (Blatt and Weisskopf 1952; Brink *et al.* 1970; Wuosmaa *et al.* 1995; Von Oertzen *et al.* 2006) has considerable success describing the nuclei ^{8}Be, ^{12}C, ^{16}O, etc. with baryon numbers a multiple of four and having equal numbers of protons and neutrons, as 'molecules' of pointlike α-particles. For $m = 1$, new Skyrmion solutions with baryon number a multiple of four have recently been found (Battye *et al.* 2007), which make contact with the α-particle model. These new solutions are clusters of cubic $B = 4$ Skyrmions, and for $B \geq 12$ they are energetically more stable than the hollow polyhedral Skyrmions, the effect being marginal for $B = 8$.

14.3.1 $B = 8$

When $m = 0$, the $B = 8$ Skyrmion is a hollow polyhedron with D_{6h} symmetry, with no obvious relation to a pair of cubic $B = 4$ Skyrmions, and this solution persists at $m = 1$. However, motivated by the α-particle model, one might expect

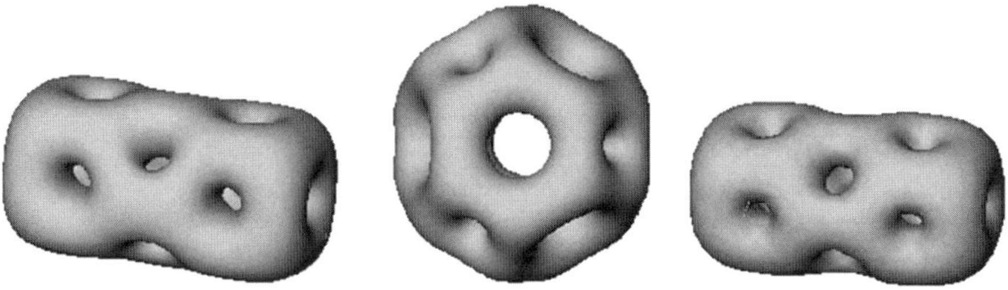

(a) (b) (c)

Figure 14.2 Baryon density contours for (a) two $B = 4$ cubes with one of the cubes rotated by 90° around the line joining them, (b) the $B = 8$ truncated octahedron, and (c) the relaxed $B = 8$ Skyrmion with $m = 1$.

that at $m = 1$, the lowest energy solution is a 'molecule' of two cubic, $B = 4$ Skyrmions. Two such Skyrmions, placed initially in the same orientation and next to each other, have a weak quadrupole–quadrupole attraction (Manton 1994; Manton *et al.* 2004). Because of a significant short-range octupole interaction in the single pion field component that has no quadrupole moment, it is best to also twist one cube by 90° relative to the other around the axis joining them (see Figure 14.2a). An alternative starting configuration has the shape of a truncated octahedron and is obtained using the rational map ansatz with a cubically symmetric degree 8 map (see Figure 14.2b).

Numerical relaxation from either starting point produces the stable solution displayed in Figure 14.2c, which has D_{4h} symmetry. For $m = 1$, the energy per baryon of this new Skyrmion and also of the old D_{6h}-symmetric Skyrmion is 1.294. The change of structure therefore has a marginal effect in this case, but one expects that for $m > 1$ and for larger B, clusters of $B = 4$ cubes will be the more stable solutions. Note that there is a clear attraction between $B = 4$ cubes, because the energy per baryon of the $B = 4$ cube is 1.307 for $m = 1$. Numerical errors are estimated as 0.5% or less.

14.3.2 $B = 12$

In the α-particle model, three α-particles form an equilateral triangle. This motivates the search for a triangular $B = 12$ solution in the Skyrme model, composed of three $B = 4$ cubes. A configuration with approximate D_{3h} symmetry can be obtained with each cube related to its neighbour by a spatial rotation through 120° followed by an isorotation by 120°. The isorotation cyclically permutes the values of the pion fields on the faces of the cube, so that these values match on touching faces, and the cubes attract. It is fairly easy to see that around the centre of the triangle the field has a winding equivalent to a single Skyrmion, and is unstable to this Skyrmion moving down and merging with the bottom face of the triangle; in fact, it fills a hole in the baryon density. This C_3-symmetric Skyrmion with $E/B = 1.288$ is presented in Figure 14.3.

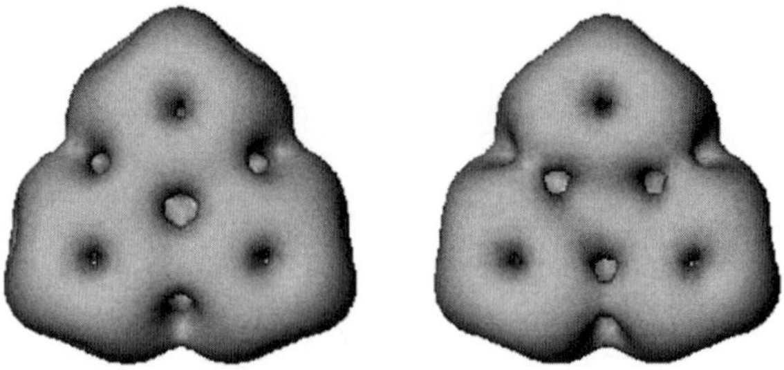

(a) (b)

Figure 14.3 Top and bottom views of the $B = 12$ Skyrmion with triangular symmetry.

The top view looks similar to the minimal energy $B = 11$ Skyrmion with $m = 0$, and suggests that the initial arrangement of three cubes can also be viewed as the $m = 1$ version of this $B = 11$ Skyrmion with a single Skyrmion placed inside at the origin. Such a field configuration can be constructed with exact D_{3h} symmetry using the double rational map ansatz (Manton and Piette 2001) (see Appendix). This involves a D_{3h}-symmetric outer map of degree 11, R^{out}, and a spherically symmetric degree 1 inner map, R^{in}, together with an overall radial profile function. The maps are

$$R^{\text{out}}(z) = \frac{z^2(1 + az^3 + bz^6 + cz^9)}{c + bz^3 + az^6 + z^9} \tag{14.15}$$

$$R^{\text{in}}(z) = -\frac{1}{z}, \tag{14.16}$$

where $a = -2.47$, $b = -0.84$, and $c = -0.13$. Note that the orientation of R^{in} has to be chosen compatibly with the D_{3h} symmetry of R^{out}. Numerically relaxing the field yields a solution with exact D_{3h} symmetry, but as we discussed above, this is only a saddle point and not a local energy minimum.

In Battye and Sutcliffe (2006) another $B = 12$ solution with C_3 symmetry was found, with $E/B = 1.289$. It is a general observation that rearrangements of clusters have only a tiny effect on the energy of a Skyrmion, so as B increases one expects an increasingly large number of local minima with extremely close energies.

Rearranged solutions are analogous to the rearrangements of the α-particles which model excited states of nuclei. An example is the Skyrme model analogue of the three α-particles in a chain configuration modelling the 7.65 MeV excited state of ^{12}C (Morinaga 1956; Friedrich *et al.* 1971). This is obtained from three $B = 4$ cubes placed next to each other in a line, with the middle cube twisted relative to the other two by 90° around the axis of the chain. The relaxed solution is displayed in Figure 14.4 and has $E/B = 1.285$. This is apparently the lowest energy of the $B = 12$ solutions, but note that the energy difference we are trying

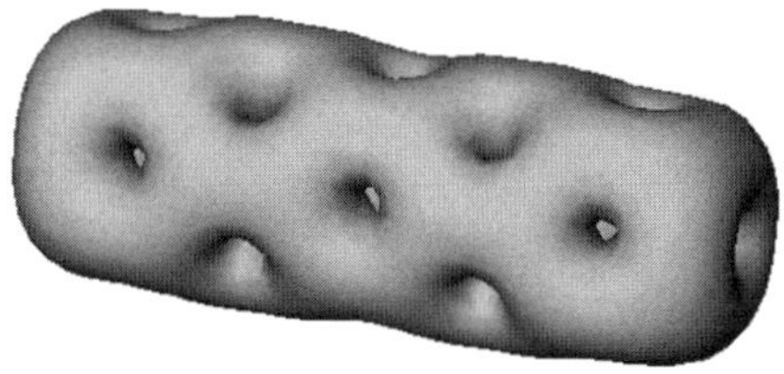

Figure 14.4 $B = 12$ Skyrmion formed from three cubes in a line, with the middle cube being rotated by 90° around the line of the cubes.

to understand, 7.65 MeV, is less than 0.1% of the total energy of a ^{12}C nucleus, smaller than the numerical errors in the Skyrmion energies.

14.3.3 $B = 16$

There is a tetrahedrally symmetric $B = 16$ solution which is an arrangement of four $B = 4$ cubes. It is created using the double rational map ansatz as a starting point, with an outer map of degree 12, R^{out}, and an inner map of degree 4, R^{in}, with compatible symmetries, and an overall radial profile function. There is a T_d-symmetric map R^{out}, and this can be combined with the O_h-symmetric map familiar from the $B = 4$ Skyrmion, giving T_d symmetry overall. The maps are

$$R^{\text{out}} = \frac{ap_+^3 + bp_-^3}{p_+^2 p_-} \tag{14.17}$$

$$R^{\text{in}} = \frac{p_+}{p_-}, \tag{14.18}$$

where $p_\pm(z) = z^4 \pm 2\sqrt{3}\, iz^2 + 1$, $a = -0.53$, and $b = 0.78$. Letting the field U relax, preserving the T_d symmetry, results in the solution displayed in Figure 14.5a, in which $U = -1$ at 16 points clustered into groups of four close to the centre of each cube. The solution has $E/B = 1.288$.

This tetrahedral solution is only a saddle point. It is energetically more favourable for the two cubes on a pair of opposite edges of the tetrahedron to open out, leading to the D_{2d} symmetric solution in Figure 14.5b, which has the slightly lower energy, $E/B = 1.284$. An α-particle molecule of similar shape has also been found, where it is termed a bent rhomb (Bauhoff *et al.* 1984).

A stable tetrahedral solution would be preferable, since the closed shell structure of ^{16}O is known to be compatible with clustering into a tetrahedral arrangement of four α-particles. Moreover, the ^{16}O ground state and the excited states at 6.1 MeV and 10.4 MeV, with spin/parity 0^+, 3^-, and 4^+, and some higher states, look convincingly like a rotational band for a tetrahedral intrinsic structure (Dennison 1954; Robson 1979).

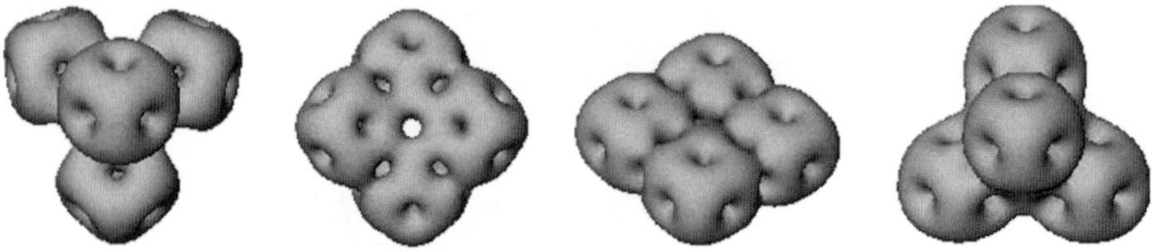

(a) (b) (c) (d)

Figure 14.5 $B = 16$ Skyrmions composed of four cubes: (a) Tetrahedral arrangement, (b) bent square, (c) flat square, and (d) another tetrahedral arrangement.

There is a further metastable solution of low energy, in which four $B = 4$ cubes all have the same orientation, and are connected together to form a flat square (see Figure 14.5c). The solution has $E/B = 1.293$.

Yet another tetrahedral saddle-point solution of four cubes exists (see Figure 14.5d). Its E/B is 1.295. A perturbation that breaks the tetrahedral symmetry results in the bent square solution of Figure 14.5b.

14.3.4 $B = 24$

Here one may begin with six $B = 4$ cubes on the vertices of an octahedron, all with the same spatial orientation (see Figure 14.6a). The cubes above and below the four in a square are given an isospin rotation by 180°, as this aids their attraction. The relaxed solution, shown in Figure 14.6b, resembles the $B = 16$ bent square Skyrmion of Figure 14.5b, joined with the $B = 8$ Skyrmion of Figure 14.2c. The fact that these $B = 16$ and $B = 8$ Skyrmions appear as substructures is further evidence that they are the minimal energy arrangements of $B = 4$ cubes. The energy per baryon of this $B = 24$ Skyrmion is 1.282.

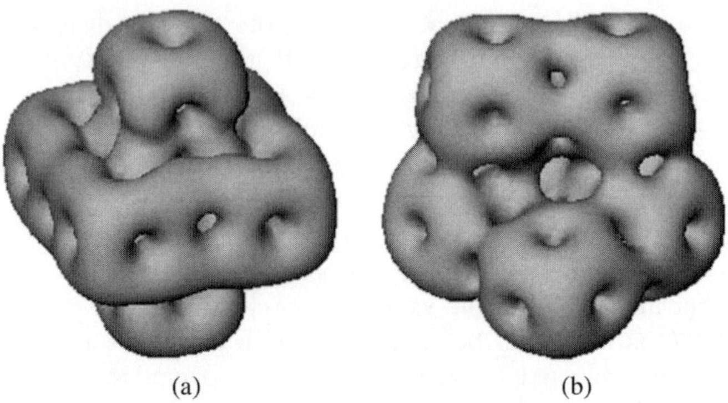

(a) (b)

Figure 14.6 (a) Initial configuration of six cubes in a square bipyramid arrangement and (b) the final relaxed $B = 24$ Skyrmion.

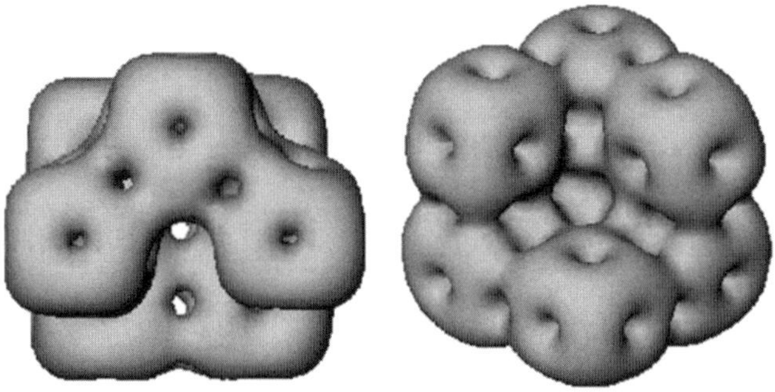

(a) (b)

Figure 14.7 (a) Initial configuration of four cubes in a square with three extra cubes placed above the square but not aligned with the cubes below and (b) the relaxed $B = 28$ Skyrmion.

14.3.5 $B = 28$

A starting configuration is shown in Figure 14.7a, with four cubes in a square and three more placed above, all with the same space and isospace orientations. Figure 14.7b displays the solution after relaxation. This $B = 28$ Skyrmion clearly resembles the cubic $B = 32$ Skyrmion (see below) with one $B = 4$ cube removed. The energy per baryon is 1.279, which is slightly lower than might have been expected given the energies for $B = 16$ and $B = 24$.

14.3.6 $B = 32$

Even for relatively small values of the pion mass parameter m, the $B = 32$ Skyrmion is cubic, and has lower energy than the minimal energy, hollow polyhedral structure (Battye and Sutcliffe 2006). The solution may be thought of as eight $B = 4$ cubic Skyrmions placed on the vertices of a cube, each with the same spatial and isospin orientations, and it may also be created by cutting out a cubic $B = 32$ chunk from the infinite, triply periodic Skyrme crystal (Baskerville 1996).

Alternatively, it may be obtained beginning with the double rational map ansatz. One places a $B = 4$ cube inside a $B = 28$ configuration with cubic symmetry using the maps

$$R^{\text{out}} = \frac{p_+(ap_+^6 + bp_+^3p_-^3 - p_-^6)}{p_-(p_+^6 - bp_+^3p_-^3 - ap_-^6)} \tag{14.19}$$

$$R^{\text{in}} = \frac{p_+}{p_-}, \tag{14.20}$$

where $a = 0.33$ and $b = 1.64$, and $p_\pm(z)$ are as before. This is displayed in Figure 14.8a. Numerical relaxation yields the solution in Figure 14.8b, which is the $B = 32$ Skyrmion, with $E/B = 1.274$.

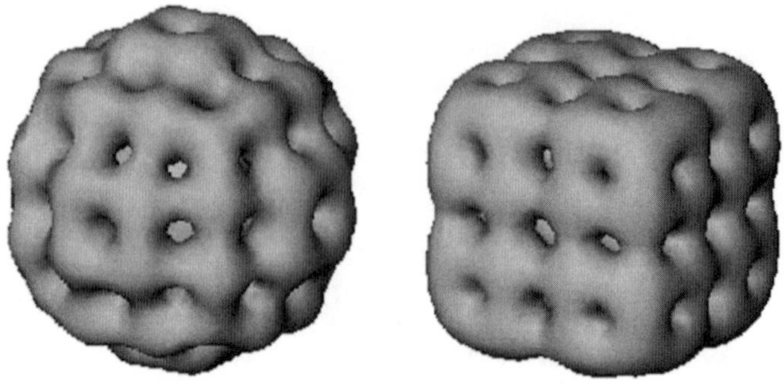

(a) (b)

Figure 14.8 (a) Initial condition of the cubic $B = 4$ Skyrmion inside a cubic $B = 28$ configuration and (b) the final relaxed $B = 32$ Skyrmion, which is a chunk of the Skyrme crystal.

Note that slicing the $B = 32$ Skyrmion in half produces the square $B = 16$ solution of Figure 14.5c, whereas if one removes every alternate $B = 4$ cube then the result is the unstable tetrahedral solution of Figure 14.5d.

Recall that a Skyrme field created using the rational map ansatz with a degree $\tilde{B}$ map has a polyhedral structure with $2\tilde{B} - 2$ holes in the baryon density. Figure 14.8b shows that the $B = 32$ crystal chunk contains 54 exterior holes and this corresponds to a degree 28 map. This motivates the above double rational map construction. The next crystal chunk with cubic symmetry contains 27 $B = 4$ cubes and therefore has $B = 108$. To create this from a triple rational map ansatz, wrapping a third shell around the $\tilde{B} = 4$ and $\tilde{B} = 28$ shells requires a map with degree 76. A degree 76 map has 150 holes, the correct number of exterior holes for the $B = 108$ crystal chunk. However, there is a six-parameter family of O_h-symmetric degree 76 maps and it has not yet been possible to obtain an appropriate member of this family with a suitable distribution of the 150 holes, either analytically or numerically.

14.4 Quantization and calibration

Quantization is a vital issue for Skyrmions, because Skyrmions are supposed to model physical nucleons (protons and neutrons) and nuclei, and a nucleon is a spin $\frac{1}{2}$ fermion. In the approach of Finkelstein and Rubinstein (1968), one quantizes a soliton as a fermion by lifting the classical field configuration space to its simply connected covering space. In the $SU(2)$ Skyrme model, this is a double cover for any value of B. States should be multiplied by a factor of -1 when acted upon by any operation corresponding to a circuit around a non-contractible loop in the configuration space. Equivalently, the wavefunction has opposite signs on the two points of the covering space that cover one point in the configuration space. Finkelstein and Rubinstein showed that the exchange of two $B = 1$ Skyrmions is a loop which is homotopic to a 2π rotation of one of the $B = 1$ Skyrmions, in agreement with the spin–statistics result, and it was verified by Williams (1970) that a 2π rotation of a $B = 1$ Skyrmion is a non-contractible loop, thus requiring the Skyrmion to be quantized as a fermion. This result was generalized by Giulini (1993), who showed that a 2π rotation of a Skyrmion of baryon number B is a non-contractible loop if B is odd and contractible if B is even.

A practical, approximate quantum theory of Skyrmions is achieved by a rigid body quantization of the spin and isospin rotations. Vibrational modes, whose excited states often have considerably higher energy, are ignored.

For the $B = 1$ Skyrmion, this quantization was carried out by Adkins, Nappi, and Witten (1983), who showed that the lowest energy states (compatible with the Finkelstein–Rubinstein [FR] constraints) have spin $\frac{1}{2}$, and may be identified with the proton/neutron isospin doublet. The next lowest states are identified with spin $\frac{3}{2}$ delta resonances. The masses of the nucleons and deltas were used to calibrate the Skyrme model. However, the delta is a broad resonance about 300 MeV above the nucleon ground state, and it strongly radiates pions. Related to this, it has been recently observed that the spin of the delta strongly deforms the $B = 1$ Skyrmion (Battye *et al.* 2005; Houghton and Magee 2006), and if this is taken into account, it has a rather complicated effect on the calibration of the model.

The Skyrmions with baryon numbers $B = 2, 3, 4$, and 6 have the right properties to model the deuteron ^{2}H, the isospin doublet ^{3}H/^{3}He, the α-particle ^{4}He, and the nucleus ^{6}Li (Braaten and Carson 1988; Carson 1991; Walhout 1992; Irwin 2000). In each case, the collective coordinate quantization is constrained by the symmetries of the classical solution, and the need to impose FR constraints. The resulting lowest energy states for $B = 2, 3, 4$, and 6 have spin/parity, respectively, $J^P = 1^+, \frac{1}{2}^+, 0^+$, and 1^+ (with isospin $T = 0$ for $B = 2, 4, 6$ and $T = \frac{1}{2}$ for $B = 3$), agreeing with those of real nuclei. However, in the traditional calibration of the model, the binding energies are too large and the sizes too small.

Collective coordinate quantization of the dodecahedral $B = 7$ Skyrmion leads to a lowest spin of $J = \frac{7}{2}$ when $T = \frac{1}{2}$ (Irwin 2000; Krusch 2003), disagreeing with the experimental value $J = \frac{3}{2}$ for the ground state of the isospin doublet ^{7}Li/^{7}Be. This suggests that the Skyrmion is too symmetric to model the ground state and it would be preferable if a less symmetric solution existed, which could have a larger classical energy, but be quantized with a lower spin (Manko and Manton 2007). Quantum states of the $B = 5$ Skyrmion also differ from those of ^{5}He/^{5}Li.

Finding the allowed spin and isospin states of Skyrmions has never been easy. There is a tricky interplay of the representations of the symmetry group of the Skyrmion and the FR constraints. Krusch (2003, 2006) has been able to simplify part of the calculation by exploiting the known topology of the space of rational maps (Segal 1979). Suppose a Skyrmion (approximately described by the rational map ansatz) is invariant under a combined rotation and isorotation. Krusch has found a simple formula for determining whether the FR factor associated with this symmetry operation is $+1$ or -1, in terms of the angles of rotation and the baryon number.

Recently, a recalibration of the Skyrme model around the properties of the ^{6}Li nucleus has been performed (Manton and Wood 2006). This is better suited to nuclear physics applications. The motivation for the choice of ^{6}Li is that it is a small nucleus of isospin zero, with a pair of levels that can reasonably be interpreted as a rotational band. Because the isospin is zero, the electric charge density is half the baryon density. The ground state of spin 1 and the first excited state of spin 3 are separated by just a few MeV, whereas the mass of the ^{6}Li nucleus is approximately 5600 MeV. The rotational motion is therefore quite non-relativistic, not leading to strong pion radiation, nor to significant Skyrmion deformation, which could affect the calibration. The $B = 6$ Skyrmion is also well-known, and can be approximated by the rational map ansatz, which is useful when estimating the energy and size. In the new calibration, the conversion from Skyrme units to physical energy and length units is fixed by fitting the ^{6}Li mass and charge radius. m is determined as usual from the physical pion mass, 138 MeV, but because of the change of units, m is roughly doubled and is now $m = 1.125$, which is in the range where the solutions described in Section 14.3, constructed from $B = 4$ cubes, are favoured.

In the new calibration, the masses of Skyrmions and nuclei are rather similar to those in the traditional calibration of Adkins and Nappi (1984), but the length scale is roughly doubled. This means that Skyrmion moments of inertia are four times larger, and hence the energy splitting of rotational levels is four times smaller. These moments of inertia (both rotational and isorotational) have been estimated for a range of baryon numbers, using the rational map ansatz where appropriate, and used to determine the energies of spin and isospin excited states for several nuclei (Manko *et al.* 2007). The results appear much closer to experimental nuclear energy levels than those found previously. For example, the spectrum of excited states of the double cube $B = 8$ Skyrmion matches well the experimental spectrum of ^{8}Be and its isobars ^{8}Li, ^{8}B, etc. There are also some detailed suggestions for a Skyrme model understanding of the physically observed spin states for $B = 5$ and $B = 7$, which, as mentioned above, have so far been problematic.

14.5 Conclusions

For small baryon numbers, the rational map ansatz gives good approximations to Skyrmions, some of which have the symmetries of Platonic solids. The ansatz has recently been shown to be helpful both in classifying the allowed spin and isospin states of quantized Skyrmions, and also for estimating the energies, radii, and moments of inertia of Skyrmions. The Skyrme model with pion mass parameter of order 1 has qualitatively new solutions for baryon numbers $B \geq 8$. These are not hollow polyhedra, the solutions found for $m = 0$ or $m = 0.528$ (the traditional value), but more dense structures often with clear clustering into $B = 4$ cubes, the Skyrme model analogue of α-particle molecules. This gives an improved fit of the model to nuclei like ^{8}Be and ^{12}C, provided the energy scale and especially the length scale of the model are recalibrated. The double rational map ansatz gives some insight into the shapes of Skyrmions for baryon numbers in the range $8 \leq B \leq 32$, and an improved version of this ansatz is proposed (see Appendix).

The interpretation of the Skyrmions discussed in this chapter has been that of nucleons and nuclei. However, there has been considerable interest in recent years in studying variants of non-linear scalar field theories in the context of condensed matter physics, the classic example being the continuum limit of the Heisenberg model of ferromagnets. The Skyrmions that occur in such theories are often localized in two dimensions, which means they can become extended vortex-like structures in a three-dimensional material. For a recent analysis of one model and further references, see Rössler *et al.* (2006). If the topology of the model is as in the original Skyrme model, then Skyrmions with three-dimensional localization should occur in the interior of the material, and it would be interesting if evidence could be found for the polyhedral structures of higher charge Skyrmions shown in Figure 14.1.

Appendix: Double rational map ansatz

Here, we briefly review the double rational map ansatz for Skyrmions (Manton and Piette 2001), which generalizes the original rational map ansatz (Houghton *et al.* 1998). This uses two rational maps $R^{\text{in}}(z)$ and $R^{\text{out}}(z)$, with a profile function $f(r)$ satisfying $f(0) = 2\pi$ and $f(\infty) = 0$. It is assumed that f decreases monotonically as r increases, passing through π at a radius r_0. The ansatz for the Skyrme field is again (14.9), with the understanding that for $r \leq r_0$, $R(z) = R^{\text{in}}(z)$, and for $r > r_0$, $R(z) = R^{\text{out}}(z)$. Notice that now $U = 1$ both at the origin and at spatial infinity. It can be shown that the total baryon number is the sum of the degrees of the maps R^{in} and R^{out}. The ansatz is optimized by adjusting the coefficients of both maps, allowing variations of r_0, and solving for $f(r)$. All this is quite hard, but easier if R^{in} and R^{out} share a substantial symmetry.

It has recently been realized that the double rational map ansatz is a special case of Skyrme's product ansatz (Skyrme 1962), in which Skyrme fields (often Skyrmion solutions, but not here) $U_1(\mathbf{x})$ with baryon number B_1 and $U_2(\mathbf{x})$ with baryon number B_2 are simply multiplied, giving the field $U(\mathbf{x}) = U_1(\mathbf{x})U_2(\mathbf{x})$ with baryon number $B_1 + B_2$. To express the double rational map ansatz in this way, define U_1 by the original rational map ansatz (14.9), with rational map R^{in} and profile function $f_1(r) = f(r) - \pi$ for $r \leq r_0$, $f_1(r) = 0$ for $r \geq r_0$, and define U_2 similarly, with rational map R^{out} and profile function $f_2(r) = \pi$ for $r \leq r_0$, $f_2(r) = f(r)$ for $r \geq r_0$. U_2 takes the value -1 in the region $r \leq r_0$, and this explains why f_1 differs by π from f.

A problem with the double rational map ansatz is that U takes the value -1 on the entire sphere at radius r_0, and true Skyrmions do not have this behaviour. But the formulation above suggests how this problem can be avoided. One simply needs to relax the non-overlapping character of the product of U_1 and U_2 by allowing both f_1 and f_2 to decrease freely from π at $r = 0$ to 0 at $r = \infty$, without further constraint at an intermediate radius r_0. One would expect f_1 to decrease more rapidly than f_2 if the Skyrmion has a genuine inner and outer structure. This version of the product ansatz preserves the joint rotational symmetries of U_1 and U_2, but not any inversion or reflection symmetries. As usual with the product ansatz, the field U is sensitive to the order in which U_1 and U_2 are multiplied, and the loss of reflection symmetry is related to this.

It is easy to verify that a small perturbation of the profiles f_1 and f_2 away from their initial, non-overlapping form will tend to reduce the energy. This is because the trajectory of the field U along a generic radial line will no longer pass through $U = -1$, but will rather take a short cut, which reduces the radial derivative of U without a significant increase in the angular derivatives, and also reduces the potential energy. The non-generic lines are those for which $R^{\text{in}}(z) = R^{\text{out}}(z)$, and there are B of these, counted with multiplicity. Therefore, $U = -1$ at B points, the number required topologically if they all have positive multiplicity. The points lie on these special lines and are all at the same distance from the origin. Further investigations are needed. With S. Krusch, the author is attempting to understand this improved double rational map ansatz more systematically, in order to find its optimal form. Alternatively, one could try *ad hoc* ansätze for f_1 and f_2, perhaps with one scale parameter each, and seek to minimize the energy of U numerically.

Acknowledgements

This chapter is largely based on Battye *et al.* (2007) and part of Manton and Sutcliffe (2004). I would like to acknowledge the contributions of Richard Battye and Paul Sutcliffe to these joint works, and especially thank them for all the figures that are reproduced here.

This is also an occasion to express my debt to Nigel Hitchin's mathematical insight. His work on monopoles, in particular, has inspired several of the developments discussed here.

References

Adkins, G. S. and Nappi, C. R. (1984). The Skyrme model with pion masses. *Nucl. Phys. B* **233**, 109.

Adkins, G. S., Nappi, C. R., and Witten, E. (1983). Static properties of nucleons in the Skyrme model. *Nucl. Phys. B* **228**, 552.

Baskerville, W. K. (1996). Making nuclei out of the Skyrme crystal. *Nucl. Phys. A* **596**, 611.

Battye, R. A. and Sutcliffe, P. M. (1997). Symmetric Skyrmions. *Phys. Rev. Lett.* **79**, 363.

Battye, R. A. and Sutcliffe, P. M. (1998). A Skyrme lattice with hexagonal symmetry. *Phys. Lett. B* **416**, 385.

Battye, R. A. and Sutcliffe, P. M. (2001). Solitonic fullerene structures in light atomic nuclei. *Phys. Rev. Lett.* **86**, 3989.

Battye, R. A. and Sutcliffe, P. M. (2002). Skyrmions, fullerenes and rational maps. *Rev. Math. Phys.* **14**, 29.

Battye, R. A. and Sutcliffe, P. M. (2005). Skyrmions and the pion mass. *Nucl. Phys. B* **705**, 384.

Battye, R. A. and Sutcliffe, P. M. (2006). Skyrmions with massive pions. *Phys. Rev. C* **73**, 055205.

Battye, R. A., Houghton, C. J., and Sutcliffe, P. M. (2003). Icosahedral Skyrmions. *J. Math. Phys.* **44**, 3543.

Battye, R. A., Krusch, S., and Sutcliffe, P. M. (2005). Spinning Skyrmions and the Skyrme parameters. *Phys. Lett. B* **626**, 120.

Battye, R. A., Manton, N. S., and Sutcliffe, P. M. (2007). Skyrmions and the α-particle model of nuclei. *Proc. R. Soc. A* **463**, 261.

Bauhoff, W., Schultheis, H., and Schultheis, R. (1984). Alpha cluster model and the spectrum of ^{16}O. *Phys. Rev. C* **29**, 1046.

Blatt, J. M. and Weisskopf, V. F. (1952). *Theoretical Nuclear Physics*. Wiley, New York, p. 292.

Braaten, E. and Carson, L. (1998). Deuteron as a toroidal Skyrmion. *Phys. Rev. D* **38**, 3525.

Brink, D. M., Friedrich, H., Weiguny, A., and Wong, C. W. (1970). Investigation of the alpha-particle model for light nuclei. *Phys. Lett. B* **33**, 143.

Carson, L. (1991). $B = 3$ nuclei as quantized multi-Skyrmions. *Phys. Rev. Lett.* **66**, 1406.

Dennison, D. M. (1954). Energy levels of the ^{16}O nucleus. *Phys. Rev.* **96**, 378.

Finkelstein, D. and Rubinstein, J. (1968). Connection between spin, statistics and kinks. *J. Math. Phys.* **9**, 1762.

Forest, J. L., Pandharipande, V. R., Pieper, S. C., Wiringa, R. B., Schiavilla, R., and Arriaga, A. (1996). Femtometer toroidal structures in nuclei. *Phys. Rev. C* **54**, 646.

Friedrich, H., Satpathy, L., and Weiguny, A. (1971). Why is there no rotational band based on the 7.65 MeV 0^+ state in ^{12}C?. *Phys. Lett. B* **36**, 189.

Giulini, D. (1993). On the possibility of spinorial quantization in the Skyrme model. *Mod. Phys. Lett. A* **8**, 1917.

Hitchin, N. J., Manton, N. S., and Murray, M. K. (1995). Symmetric monopoles. *Nonlinearity* **8**, 661.

Houghton, C. J. and Magee, S. (2006). A zero-mode quantization of the Skyrmion. *Phys. Lett. B* **632**, 593.

Houghton, C. J., Manton, N. S., and Sutcliffe, P. M. (1998). Rational maps, monopoles and Skyrmions. *Nucl. Phys. B* **510**, 507.

Irwin, P. (2000). Zero mode quantization of multi-Skyrmions. *Phys. Rev. D* **61**, 114024.

Jarvis, S. (2000). A rational map for Euclidean monopoles via radial scattering. *J. Reine Angew. Math.* **524**, 17.

Krusch, S. (2003). Homotopy of rational maps and the quantization of Skyrmions. *Ann. Phys.* **304**, 103.

Krusch, S. (2006). Finkelstein–Rubinstein constraints for the Skyrme model with pion masses. *Proc. R. Soc. A* **462**, 2001.

Manko, O. V. and Manton, N. S. (2007). On the spin of the $B = 7$ Skyrmion. *J. Phys. A* **40**, 3683.

Manko, O. V., Manton N. S., and Wood, S. W. (2007). Light nuclei as quantized Skyrmions. *Phys. Rev. C* **76**, 055203.

Manton, N. S. (1994). Skyrmions and their pion multipole moments. *Acta Phys. Pol. B* **25**, 1757.

Manton, N. S. and Piette, B. M. A. G. (2001). Understanding Skyrmions using rational maps. *Prog. Math.* **201**, 469.

Manton, N. and Sutcliffe, P. (2004). *Topological Solitons*. Cambridge University Press, Cambridge, UK, Chapter 9.

Manton, N. S. and Wood, S. W. (2006). Reparametrizing the Skyrme model using the lithium-6 nucleus. *Phys. Rev. D* **74**, 125017.

Manton, N. S., Schroers, B. J., and Singer, M. A. (2004). The interaction energy of well-separated Skyrme solitons. *Commun. Math. Phys.* **245**, 123.

Morinaga, H. (1956). Interpretation of some of the excited states of 4n self-conjugate nuclei. *Phys. Rev.* **101**, 254.

Robson, D. (1979). Evidence for the tetrahedral nature of ^{16}O. *Phys. Rev. Lett.* **42**, 876.

Rössler, U. K., Bogdanov, A. N., and Pfleiderer, C. (2006). Spontaneous Skyrmion ground states in magnetic metals. *Nature* **442**, 797.

Segal, G. (1979). The topology of the space of rational maps. *Acta Math.* **143**, 39.

Skyrme, T. H. R. (1961). A nonlinear field theory. *Proc. R. Soc. A* **260**, 127.

Skyrme, T. H. R. (1962). A unified field theory of mesons and baryons. *Nucl. Phys.* **31**, 556.

Von Oertzen, W., Freer, M., and Kanada-En'yo, Y. (2006). Nuclear clusters and nuclear molecules. *Phys. Rep.* **432**, 43.

Walhout, T. S. (1992). Quantizing the four-baryon Skyrmion. *Nucl. Phys. A* **547**, 423.

Williams, J. G. (1970). Topological analysis of a nonlinear field theory. *J. Math. Phys.* **11**, 2611.

Wuosmaa, A. H., Betts, R. R., Freer M., and Fulton, B. R. (1995). Recent advances in the study of nuclear clusters. *Ann. Rev. Nucl. Part. Sci.* **45**, 89.

XV

MIRROR SYMMETRY OF FOURIER–MUKAI TRANSFORMATION FOR ELLIPTIC CALABI–YAU MANIFOLDS

Naichung Conan Leung and Shing-Tung Yau

Dedicated to Nigel Hitchin on the occasion of his 60th birthday

15.1 Introduction

Mirror symmetry conjecture says that for any Calabi–Yau (CY) manifold M near the large complex/symplectic structure limit, there is another CY manifold X, called the *mirror manifold*, such that the B-model superstring theory on M is equivalent to the A-model superstring theory on X, and vice versa. Mathematically speaking, it roughly says that the complex geometry of M is equivalent to the symplectic geometry of X, and vice versa. It is conjectured (Strominger *et al.* 1996) that this duality can be realized as a Fourier-type transformation along fibers of special Lagrangian fibrations on M and X, called the *SYZ mirror transformation* $\mathcal{F}^{SYZ}$.

Suppose M has an elliptic fibration structure,

$$p : M \to S,$$

then there is another manifold W with a *dual* elliptic fibration over S. Since any elliptic curve is isomorphic to its dual, we have actually $M \cong W$ provided that p has irreducible fibers. There is also a Fourier–Mukai transformation (FM transform) $\mathcal{F}_{cx}^{FM}$ between the complex geometries of M and W. On the level of derived category of coherent sheaves, $\mathcal{F}_{cx}^{FM}$ is an equivalence of categories. On the level of cycles, this can be described as a *spectral cover construction* and it is a very powerful tool in the studies of holomorphic vector bundles over M.

In this chapter we address the following two questions: (i) What is the SYZ transform of the elliptic fibration structure on M? (ii) What is the SYZ transform of the FM transform $\mathcal{F}_{cx}^{FM}$?

The answer to the first question is a *twin Lagrangian fibration* structure on the mirror manifold X, coupled with a *superpotential*. To simplify the matter, we will ignore the superpotential in our present discussions. Similarly there is a twin Lagrangian fibration structure on the mirror manifold Y to W. We will explain several important properties of twin Lagrangian fibrations. In particular, we show that the twin Lagrangian fibration on Y is dual to the twin Lagrangian fibration on X. There is also an identification between X and Y, which is analogous to the identification between total spaces of dual elliptic fibrations M and W.

For the second question, the SYZ transform of $\mathcal{F}_{cx}^{FM}$ should be a symplectic FM transform $\mathcal{F}_{sym}^{FM}$ from X to Y. We will argue that this is actually the identity transformation! A naive explanation of this is because $\mathcal{F}_{sym}^{FM} = \mathcal{F}^{SYZ} \circ \mathcal{F}_{cx}^{FM} \circ \mathcal{F}^{SYZ}$ and each of the two $\mathcal{F}^{SYZ}$ transforms undo half of the complex FM transform.

The plan of the chapter is as follow: In Section 15.2 we review the SYZ mirror transformation and show that the mirror manifold to an elliptically fibered CY manifold has a twin Lagrangian fibration structure. In Section 15.3 we review the FM transform in complex geometry in general and also for elliptic manifolds. In Section 15.4 we first define the *symplectic* FM transform between Lagrangian cycles on X and Y. Then we define twin Lagrangian fibrations, give several examples of them, and study their basic properties. In Section 15.5 we show that the SYZ transformation of the complex FM transform between M and W is the symplectic FM transform between X and Y, which is actually the identity transformation.

15.2 Mirror symmetry and SYZ transformation

15.2.1 *Geometry of Calabi–Yau manifolds*

A real $2n$-dimensional Riemannian manifold M is a *CY manifold* if the holonomy group of its Levi-Civita connection is a subgroup of $SU(n)$. Equivalently, a CY manifold is a Kähler manifold with a parallel holomorphic volume form. A theorem (Yau 1978) of the second author says that any compact Kähler manifold with trivial canonical line bundle admits such a structure.

The complex geometry of M includes the study of (i) the moduli space of complex structures on M, (ii) complex submanifolds, holomorphic vector bundles, and Hermitian Yang–Mills metrics, and (iii) the derived category $D^b(M)$ of coherent sheaves on M. The symplectic geometry of M includes the study of (i) the moduli space of (complexified) symplectic structures on M, (ii) Lagrangian submanifolds and their intersection theory, and (iii) the Fukaya–Floer category $Fuk(M)$ of Lagrangian submanifolds in M. The complex geometry is more non-linear in nature, whereas the symplectic geometry requires the inclusion of quantum corrections, in which contributions from holomorphic curves in M needed to be included.

15.2.2 *Mirror symmetry conjectures*

Roughly speaking, the mirror symmetry conjecture says that for mirror CY manifolds M and X, the complex geometry of M is equivalent to the symplectic geometry of X and vice versa:

$$\text{Complex geometry}\,(M) \xleftrightarrow{\text{Mirror symmetry}} \text{Symplectic geometry}\,(X)\,.$$

This conjecture has far-reaching consequences in many different parts of mathematics and physics. For instance, (i) Candelas *et al.* (1991) studied the identification between the moduli of complex structures on M with the moduli of complexified symplectic structures on X and derived an amazing formula which enumerative the number of rational curves of each degree in the quintic CY threefold. This mirror formula has been proved mathematically by Liu, Lian, and the second author (1997) and Givental (1996) independently. (ii) A theorem of Donaldson (1985) and Uhlenbeck and the second author (1986) related the existence of Hermitian Yang–Mills metrics to the stability of holomorphic vector bundles. Thomas and the second author (2002) conjectured a mirror phenomenon to this for special Lagrangian submanifolds. (iii) Kontsevich's homological mirror conjecture (1995) identifies $D^b(M)$ with $Fuk(X)$.

In this chapter, we use the SYZ transform to study the mirror of an elliptic fibration structure on M and the FM transform associated to it.

15.2.3 *SYZ transform*

In order to explain the origin of this duality, Strominger, Yau, and Zaslow (1996) used physical reasonings to argue that (i) both M and X admit special Lagrangian tori fibrations with sections, which are fiberwise dual to each other. These are called SYZ fibrations. (ii) The equivalence between these two types of geometries is given by a geometric version of the Fourier–Mukai type transformation between M and X. We call this the SYZ transformation (see e.g. Leung 1998, 2000, 2005).

To see this, note that the manifold M itself is the moduli space of certain complex cycles in M, namely, points in M. Therefore M should also be the moduli space of certain special Lagrangian submanifolds with flat $U(1)$ bundles in X. These special Lagrangian submanifolds form a Lagrangian (tori) fibration on X since points form a fibration on M in a trivial way. By considering all those points in M which correspond to the same special Lagrangian torus in X but with different flat $U(1)$ connections, we know that M also admit a tori fibration which is naturally dual to the one on X. This is because the moduli space of flat $U(1)$ connections on a torus is naturally its dual torus.

$$\begin{array}{ccc} T & & T^* \\ \downarrow & & \downarrow \\ M & & X \\ \downarrow & & \downarrow \\ B & & B \end{array}$$

Dual Lagrangian fibrations on mirror manifolds

Similarly the mirror of the complex cycle M itself is a special Lagrangian section for $X \to B$. This heuristic reasoning is only expected to hold true asymptotically near the large complex structure limit (LCSL) (Strominger *et al.* 1996). We can apply this SYZ transform to other coherent sheaves on M. For example, suppose

L is a holomorphic line bundle on M, which restricts to a flat bundle on each fiber $T_b \subset M$ of the special Lagrangian fibration. Such a flat connection determines a point in the dual torus $T_b^* \subset X$. By varying $b \in B$, we obtain a Lagrangian section to $X \to B$, which is the mirror to L. This fiberwise Fourier transform forms the backbone of the *SYZ mirror transform.*

15.2.4 *Mirror of elliptic fibrations*

In this section we continue our reasonings to explain why the mirror manifold X to a CY manifold M with an elliptic fibration $p : M \to S$ admits another special Lagrangian fibration which is compatible with the original SYZ fibration on X. If the elliptic fibration on M has a section, then the corresponding special Lagrangian fibration on X also admits an appropriate section. Recall that the way we obtain the SYZ fibration $\pi : X \to B$ on X is by viewing the identity map $M \to M$ as a fibration on M by complex cycles and with a holomorphic section, both in trivial manners.

We consider the moduli space of coherent sheaves on M whose generic member has the same Hilbert polynomial as $\iota_* O_F$ where F is an elliptic fiber of $M \to S$ and $\iota : F \to M$ is the inclusion morphism. Geometrically speaking, this is a moduli space of elliptic curves in M together with holomorphic line bundles over them which are trivial topologically. This moduli space is nothing but the total space of the dual elliptic fibration $p' : W \to S$.

By the principle of mirror symmetry, we should view W also as the moduli space of certain Lagrangian cycles in the mirror manifold X. As $\dim_{\mathbb{C}} W = n$ and W consists of geometric cycles which foliate M, we can argue as before that X should have another Lagrangian fibration $p : X \to C$ such that W is also the moduli space of its Lagrangian fibers together with flat $U(1)$ connections over them. Similarly the section σ of the elliptic fibration $p : M \to S$ determines a Lagrangian section of $p : X \to C$:

$$\begin{array}{ccc} X & \xrightarrow{p} & C \\ \pi \downarrow & & \\ B & & \end{array}$$

Given any elliptic fiber F in M, there is an one complex parameter family of points in M which intersect F, namely, those points in F. Homologically speaking, suppose S is any coherent sheaf on M of the form $S = \iota_* L$ with $\iota : F \to M$ the natural inclusion and $L \in Pic^0(F) \simeq F^*$, the dual elliptic curve, then $Ext^*_{O_M}(\mathcal{S}, O_m) \; /\!= 0$ exactly whem$m \in F$. Translating this to the mirror side, given any Lagrangian fiber $\pi^{-1}(b)$, it should intersect an one real parameter family of $p^{-1}(c)$'s where $c \in C$.

On the other hand, given any point $m \in M$, there is a unique elliptic curve F that passes through m. But the coherent sheaves $S = \iota_* L$ with $L \in F^*$ also intersect m homologically and parametrized by an one complex parameter family, namely, by F^*. On the mirror side, it says that given any Lagrangian fiber $p^{-1}(c)$ to $p: X \to C$, it should intersect an one real parameter family of $\pi^{-1}(b)$'s where $b \in B$.

Furthermore if $m \in M$ and $S_1, S_2 \in W$ satisfying $Ext^*_{O_M}(\mathcal{S}_1, O_m) \neq 0 = Ext^*_{O_M}(\mathcal{S}_2, O_m)$, then $m \in F_1 = F_2$ with $F_i = \mathrm{Supp}\mathcal{S}_i$. On the mirror side, this means that if $\pi^{-1}(b) \cap p^{-1}(c_i) \neq 0$ for $i = 1, 2$ then $\pi(p^{-1}(c_1)) = \pi(p^{-1}(c_2))$. Similarly if $\pi^{-1}(b_i) \cap p^{-1}(c) \neq 0$ for $i = 1, 2$ then $p(\pi^{-1}(b_1)) = p(\pi^{-1}(b_2))$. Namely, $\pi(p^{-1}(c))$'s in B form a fibration over some space D, which is also the base space of a fibration on C given by $p(\pi^{-1}(b))$'s. That is,

$$\begin{array}{ccc} X & \to & C \\ \downarrow & & \downarrow \\ B & \to & D \end{array}$$

We might also expect that these fibrations on B and C over D are both affine fibrations. Such a structure on X will be called a twin Lagrangian fibration on X. On the first sight, it seems that these two Lagrangian fibrations on X are on equal footing. But these arguments are only valid outside singular fibers of $M \to S$. Recall that these two Lagrangian fibrations on X are mirror to holomorphic fibrations $id: M \to M$ and $p: M \to S$. The two base manifolds are quite different in nature: M is CY but S is not.

The Lagrangian fibers to $X \to C$ are mirror to the smooth elliptic curve fibers of $M \to S$. The situation near a singular elliptic curve fiber could be quite different. Their locus in S, called the discriminant locus $\mathcal{D}$, causes S fails to be CY because of the formula $K_S^{-1} = \frac{1}{12}\mathcal{D}$. In particular K_S^{-1} is an effective divisor on S, which is indeed ample in many cases, namely S, is a Fano manifold.

There is a version of the mirror symmetry conjecture for Fano manifolds, and their mirror involve Lagrangian fibrations together with a holomorphic function, called the *superpotential*. It is reasonable to expect that the Lagrangian fibration $X \to C$ should also interact with this superpotential corresponding to S. We hope to come back to further discuss this issue in the future.

15.2.4.1 *Large complex structure limits*

Mirror symmetry for M is expected to work well near the LCSL. In terms of Hodge theory, it means that M is a member of an one-parameter family of CY n-folds M_t with $0 < |t| < 1$ such that the monodromy operator $T: H^n(M) \to H^n(M)$ is of maximally unipotent, that is, $N = \log(I - T)$ satisfies $N^n \neq 0$ but $N^{n+1} = 0$. On the mirror side (Deligne 1997), this corresponds to the hard Lefschetz action $L = \wedge\omega_X : \oplus H^{p,p}(X) \to \oplus H^{p,p}(X)$ which satisfies $L^n \neq 0$ but $L^{n+1} = 0$.

We now assume that M has an elliptic fibration structure. In our above discussions, we need to require each member M_t in the family also have an elliptic fibration. When $n \geq 3$ the existence of an elliptic fibration is invariant under deformations of complex structures. Wilson (1997) showed that if M is a CY threefold then the existence of an elliptic fibration structure on M can be determined by cohomological conditions. When $n = 2$ existence of an elliptic fibration is not a deformation invariant property because of $H^{2,0}(M) \neq 0$. The existence of an elliptic fibration on a K3 surface M is equivalent to finding $f \in H^2(M, \mathbb{Z}) \cap H^{1,1}(M)$ satisfying $f^2 = 0$. On the mirror side, this corresponds to the fact that the ω_X-Lagrangian fibration $X \to C$ on X is indeed Lagrangian with respect to $t\omega_X$ for every $t > 0$.

15.2.4.2 *Calabi–Yau fibrations*

Some of the discussions in this section work for the mirror X of a CY n-fold M with a fibration by q-dimensional complex subvarieties. Every smooth fiber of such a fibration on M is automatically CY. In this case there will again be two Lagrangian fibrations on X with the property that each nonempty intersection of fibers of $X \to B$ and $X \to C$ should have codimension q in one of the fiber, and hence both.

We can also consider a CY manifold M with more than one fibration structures. For example, when M is a CY threefold with an elliptic fibration $M \to S$ over a Hirzebruch surface S. Then the $\mathbb{P}^1$-bundle structure on S gives a K3 fibration $M \to \mathbb{P}^1$ on M. In this circumstance, the mirror manifold X should admit three Lagrangian fibrations which are compatible to each other in certain large structure limit.

15.3 Fourier–Mukai transform and elliptic CY manifolds

15.3.1 *General Fourier–Mukai transform*

Suppose M and W are smooth projective varieties. Given any coherent sheaf $\mathcal{P}$ on $M \times W$ we can define a FM transform $\mathcal{F}$ between derived categories of coherent sheaves on M and W as follow (Orlov 1996):

$$\mathcal{F}_{cx}^{FM} : D^b(M) \to D^b(W)$$
$$\mathcal{F}_{cx}^{FM}(-) = R^{\bullet}p'_*\left(\mathcal{P} \otimes p^*(-)\right),$$

where $p : M \times W \to M$ and $p' : M \times W \to W$ are projection maps. It was originally introduced by Mukai in the situation when M and W are dual Abelian varieties and $\mathcal{P}$ is the Poincaré bundle. In this case $\mathcal{F}_{cx}^{FM}$ is an equivalence of derived categories. We will need to use $\mathcal{F}_{cx}^{FM}$ for families of Abelian varieties situations. In general the FM transform is a useful tool to verify equivalences of derived categories, for example, under flops or the McKay correspondence. On the other hand, a theorem of Orlov (1996) says that any triangle-preserving equivalence $\Phi : D^b(M) \to D^b(W)$ is given by a FM transform.

15.3.2 *Elliptic fibrations and their duals*

In this subsection we recall basic facts about elliptic fibrations (see Donagi 1997, 1998 and Friedman *et al.* 1997 for details). Suppose M is a CY manifold with an elliptic fibration $p : M \to S$ with section σ_M and with connected fibers. We denote its dual elliptic fibration as $p' : W \to S$, which is again an elliptic CY with a section σ_W. For example, a CY hypersurface in a Fano toric variety which is a $\mathbb{P}^2$-bundle always admits an elliptic fibration structure.

Recall that the Weierstrass model (e.g. Andreas *et al.* 2001) of a (reduced irreducible) elliptic curve T_τ in $\mathbb{P}^2$ is of the form $y^2z = 4x^3 - g_2xz^2 - g_3z^3$ with g_2, g_3 constants. The point $[0, 1, 0] \in \mathbb{P}^2$ is the origin of the elliptic curve T_τ. It is smooth if and only if $g_2^3 - 27g_3^2$ is nonzero. In our situation, we have a family of elliptic curves $p : M \to S$. We assume that if all fibers are isomorphic to reduced irreducible cubic plane curves and S is smooth, then we can express M in a similar fashion with g_2 and g_3 varying over S. To describe it, we let $L = R^1p_*O_M$ be a line bundle over S, which can also be identified as $O_M(-\sigma)|_\sigma$ where $\sigma = \sigma_M(S)$ is the given section. Then $g_2 \in H^0\left(S, L^{\otimes 4}\right)$ and $g_3 \in H^0\left(S, L^{\otimes 6}\right)$, and the Weierstrass model defines $M \subset \mathbb{P}\left(O_S \oplus L^{\otimes 2} \oplus L^{\otimes 3}\right)$. Furthermore the discriminant locus in S is the zero locus of the section $g_2^3 - 27g_3^2$ of the bundle $L^{\otimes 12}$.

The dual elliptic fibration W is the compactified relative Jacobian of M, namely, W parametrizes rank 1 torsion-free sheaves of degree zero on fibers of $p : M \to S$. There is a natural identification between M and W because every elliptic curve T_τ is canonically isomorphic to its Jacobian $Jac(T_\tau)$ given by $T_\tau \to Jac(T_\tau)$, $p \longmapsto O(p - p_0)$ where p_0 is the origin of T_τ.

15.3.3 *FM transform and spectral cover construction*

Given any elliptic curve T_τ, or more generally a principally polarized Abelian variety, the dual elliptic curve T_τ^* is its Jacobian, which parametrizes topologically trivial holomorphic line bundles on T_τ. Mukai (1987) used an analog of the Fourier transformation to define an equivalence of derived categories of coherent sheaves on T_τ and T_τ^*, $\mathcal{F}_{cx}^{FM} : D^b(T_\tau) \to D^b(T_\tau^*)$, called the *FM transform*. This can be generalized to the family version as follow: Consider the relative Poincaré line bundle, or more precisely a divisorial sheaf, $\mathcal{P}$ on $M \times_S W$ which is given by

$$\mathcal{P} = O(\Delta - \sigma_M \times W - M \times \sigma_W),$$

where Δ is the relative diagonal in $M \times_S W$ and σ_M (respectively, σ_W) is the section of $p : M \to S$ (respectively, $p' : W \to S$). We define the following Fourier-Mukai functor $\mathcal{F}_{cx}^{FM} : D^b(M) \to D^b(W)$ as $\mathcal{F}_{cx}^{FM}(-) = R^\bullet p'_*(\mathcal{P} \otimes p^*(-))$, where the one in Section 15.3.1 is a generalization of this. Then this can be proven to give an equivalence of derived categories (Bridgeland and Maciocia 2002; Hu *et al.* 2002; Donagi and Pantev 2008). Indeed

$$\mathcal{F}_{cx,\mathcal{P}^*\otimes K_S}^{FM}\left(\mathcal{F}_{cx,\mathcal{P}}^{FM}(S)\right) = S[-1]$$
$$\mathcal{F}_{cx,\mathcal{P}}^{FM}\left(\mathcal{F}_{cx,\mathcal{P}^*\otimes K_S}^{FM}(S)\right) = S[-1].$$

Besides working on the level of derived categories, we can also study the FM transform of a stable bundle, the so-called *spectral cover construction* (Donagi 1997, 1998; Friedman *et al.* 1997, 1999). The basic idea is any stable bundle over an elliptic curve is essentially a direct sum of line bundles. In the family situation, a stable bundle on M gives a multi-section for $p' : W \to S$, together with a line bundle over it. Such construction is important in describing the moduli space of stable bundles and it can also be generalized to construct principal G-bundles on M, which play an important role in the duality between F-theory and String theory (Friedman *et al.* 1997).

15.4 Symplectic FM transform and twin Lagrangian fibrations

15.4.1 *Symplectic Fourier–Mukai transform*

Definition 15.1 *A* Lagrangian cycle *in a symplectic manifold* (X, ω_X) *is a pair* $(L, \mathcal{L})$ *with* L *a Lagrangian submanifold in* X *and* $\mathcal{L}$ *a unitary flat line bundle over* L. *We denote* $\mathcal{C}(X)$ *the collection of Lagrangian cycles in* X.

Lagrangian cycles are the objects that form the sophisticated Fukaya–Floer category $Fuk(X)$, where morphisms are Floer homology groups which counts holomorphic disks bounding cycles of Lagrangian submanifolds. We could also generalize the notion of Lagrangian cycles to allow L to be a stratified Lagrangian submanifold and to allow $\mathcal{L}$ to be a higher rank flat bundle.

Suppose (X, ω_X) and (Y, ω_Y) are symplectic manifolds of dimensions $2m$ and $2n$, respectively. Then $(X \times Y, \omega_X - \omega_Y)$ is again a symplectic manifold. Given any Lagrangian cycle $(P, \mathcal{P}) \in \mathcal{C}(X \times Y)$ we can construct a Fourier-Mukai type transformation, or simply *symplectic FM transform*, defined as follow (Weinstein 1979):

$$\mathcal{F}_{sym}^{FM} : \mathcal{C}(X) \to \mathcal{C}(Y)$$
$$\mathcal{F}_{sym}^{FM}(L, \mathcal{L}) = \left(\hat{L}, \hat{\mathcal{L}}\right).$$

Claim 15.1 *Suppose that* $L \times Y$ *intersects* P *transversely. Then the projection of* $(L \times Y) \cap P \subset X \times Y$ *to* Y *is a Lagrangian (immersed) submanifold in* Y, *which we denote it as* $\hat{L}$.

Remark 15.1 If $L \times Y$ and P do not intersect transversely, then the image might have bigger dimension. But the vanishing of ω should still imply that it is a Lagrangian subspace, possibly singular.

Since $L \times Y$ intersects P transversely, for any point in L, by Darboux theorem, there exists a local coordinate $\{x^i, y^i\}_{i=1}^m$ centered at the point such that $\omega_X = \sum_{i=1}^m dx^i \wedge dy^i$ and L is given by $y^i = 0$ for all i. Suppose this point is also the X component of a transversal intersection point of $L \times Y$ and P. If we use an appropriate Darboux coordinates $\{u^k, v^k\}_{k=1}^n$ with $\omega_Y = \sum_{k=1}^n du^k \wedge dv^k$ and

P is locally determined by

$$P : x^i = x^i(u, y) \quad \text{and} \quad v^k = v^k(u, y) \qquad \text{for all } i, k.$$

Then $(L \times Y) \cap P \subset X \times Y$ is locally given by

$$(L \times Y) \cap P : \begin{cases} x^i = x^i(u, 0), \quad y^i = 0, \\ \\ u^k = u^k, \quad v^k = v^k(u, 0). \\ \text{(i.e. no restriction)} \end{cases}$$

Therefore its projection to Y becomes

$$\hat{L} : u^j = u^j \quad v^k = v^k(u, 0).$$

The Lagrangian condition for $P \subset X \times Y$ is

$$\frac{\partial x^i}{\partial y^j} + \frac{\partial x^j}{\partial y^i} = 0 \quad \text{and} \quad \frac{\partial v^k}{\partial u^l} + \frac{\partial v^l}{\partial u^k} = 0,$$

for any (u, y) and for all $i, j = 1, ..., m$ and $k, l = 1, ..., n$. By setting $y = 0$ for the second equation, this implies that $\hat{L}$ is a Lagrangian submanifold in Y.

To obtain the flat bundle $\hat{\mathcal{L}}$ over $\hat{L}$, in the generic case, we can identify $\hat{L} \subset Y$ with $(L \times Y) \cap P \subset X \times Y$. The flat bundle $\mathcal{L}$ on L induced one on $\hat{L}$ by pullback to $L \times Y$ and then restrict to $\hat{L}$. By restriction, $\mathcal{P}$ also determine a flat connection on $\hat{L}$. The tensor product of these two flat bundles is defined as $\hat{\mathcal{L}}$ over $\hat{L}$. More work will be needed to handle the pushforward in the non-generic case though.

Examples of Lagrangian cycles in products of symplectic manifolds include (i) the graph of any symplectic map $f : X \to Y$, that is, $f^*\omega_y = \omega_X$; (ii) if M (respectively N) is a Lagrangian submanifold in X (respectively Y), then $P = M \times N$ is obviously a Lagrangian submanifold in $X \times Y$; and (iii) let C be any coisotropic submanifold in (X, ω_X). It induces a canonical isotropic foliation on C and such that its leaf space $C/\sim$ has a natural symplectic structure, provided that it is smooth and Hausdorff. This is called the *symplectic reduction*. Then the natural inclusion $C \subset X \times (C/\sim)$ is a Lagrangian submanifold in the product symplectic manifold. Thus we can use this to obtain a symplectic FM transform $\mathcal{F}^{FM}_{sym} : \mathcal{C}(X) \to \mathcal{C}(C/\sim)$.

15.4.2 *Twin Lagrangian fibrations*

15.4.2.1 *Review of Lagrangian fibrations*

Suppose (X, ω) is a symplectic manifold with a Lagrangian fibration

$$\pi : X \to B$$

and with a Lagrangian section. Away from the singularities of π, we have an short exact sequence

$$0 \to T_{vert}X \to TX \to \pi^*TB \to 0,$$

where $T_{vert}X$ is the vertical tangent bundle. Using this and the Lagrangian condition, we have a canonical identification between $T_{vert}X$ and the pullback cotangent bundle:

$$T_{vert}X \cong \pi^* T^* B.$$

Therefore every cotangent vector at $b \in B$ determines a vector field on the fiber $X_b = \pi^{-1}(b)$. By integrating these constant vector fields on X_b, we obtain a natural *affine structure* on X_b. This implies that X_b must be an affine torus, that is, $X_b = T_p X_b / \Lambda_b \cong T^b$ for some lattice Λ_b in $T_p X_b = T_b^* B$, provided that π is proper. This lattice structure on T^*B in turn defines an affine structure on the base manifold B. Of course, this affine structure on B is only defined outside the discriminant locus, that is, those $b \in B$ with X_b singular.

Before we discuss twin Lagrangian fibrations in details, let us first explain the linear aspects of them.

15.4.2.2 *Linear algebra for twin Lagrangian fibrations*

Suppose $(V \simeq \mathbb{R}^{2n}, \omega)$ is a symplectic vector space and T_b and T_c are two Lagrangian subspaces in V. They give two Lagrangian fibrations $V \to B$ and $V \to C$ with $B = V/T_b$ and $C = V/T_c$. If B and C intersect transversely, then there is a natural isomorphism

$$C \cong B^*$$

given by the following composition

$$C \hookrightarrow V \overset{\omega}{\rightrightarrows} V^* \twoheadrightarrow B^*.$$

Suppose $T_{bc} := T_b \cap T_c$ has dimension $n - q$ and we write $D = V/(T_b + T_c)$ then we have the following diagram of affine morphisms

$$\begin{array}{ccc} V & \to & C \\ \downarrow & & \downarrow \\ B & \to & D \end{array}$$

The fibers of the columns (respectively, rows) are T_b and T_b/T_{bc} (respectively, T_c and T_c/T_{bc}). The Lagrangian conditions implies the existence of a natural isomorphism $T_c/T_{bc} \cong (T_b/T_{bc})^*$, namely, the two affine bundles $B \to D$ and $C \to D$ are fiberwise dual to each other.

The usual SYZ transform which switches the fibers of a Lagrangian fibration $V \to B$ to their duals will interchange complex geometry and symplectic geometry. In order to stay within the symplectic geometry, we should take the fiberwise dual to both Lagrangian fibrations $V \to B$ and $V \to C$.

Taking dual to both T_b and T_c has the same effect as taking dual to $(T_b + T_c)/T_{bc}$ while keeping T_{bc} fixed. This gives us the following new commutative diagram:

$$\begin{array}{ccc} U & \rightarrow & C' \\ \downarrow & & \downarrow \\ B' & \rightarrow & D \end{array}$$

Here B' (respectively C') is the total space of taking fiberwise dual to the fibration $B \rightarrow D$ (respectively $C \rightarrow D$). The fiber T'_b of $U \rightarrow B'$ is obtained by taking dual along the base of the fibration $T_b \rightarrow T_b/T_{bc}$. This is the same as taking fiberwise dual, up to conjugation with the duality of total spaces. The fiber of $B' \rightarrow D$ is $(T_c/T_{bc})^*$ and likewise for $C' \rightarrow D$.

A more intrinsic way to describe this double dual process is as follow: The fiber of $V \rightarrow D$ is a coisotropic subspace in V, say V_d. The symplectic reduction $V_d/\sim$ (see e.g. Weinstein 1979) is another symplectic vector space. Then U is obtained by replacing $V_d/\sim$ by its dual symplectic space from V.

In terms of an explicit coordinate system on V given by $\{x^i, x^\alpha, y_i, y_\alpha\}$ with $1 \leq i \leq n-q$ and $n-q+1 \leq \alpha \leq n$, then we have

$$\begin{array}{ccc} V & \rightarrow & C \\ \downarrow & & \downarrow \\ B & \rightarrow & D \end{array} \quad \text{with coordinates} \quad \begin{array}{ccc} \begin{matrix} y_i, y_\alpha \\ x^\alpha, x^i \end{matrix} & \rightarrow & \begin{matrix} y_\alpha \\ x^i \end{matrix} \\ \downarrow & & \downarrow \\ x^\alpha, x^i & \rightarrow & x^i \end{array}$$

and

$$\begin{array}{ccc} U & \rightarrow & C' \\ \downarrow & & \downarrow \\ B' & \rightarrow & D \end{array} \quad \text{with coordinates} \quad \begin{array}{ccc} \begin{matrix} y_i, y_\alpha^* \\ x^{\alpha *}, x^i \end{matrix} & \rightarrow & \begin{matrix} y_\alpha^* \\ x^i \end{matrix} \\ \downarrow & & \downarrow \\ x^{\alpha *}, x^i & \rightarrow & x^i \end{array}$$

The Lagrangian conditions actually imply that $B' \cong C$, $C' \cong B$, and $U \cong V$. We now return back to the general symplectic manifolds situation.

15.4.2.3 *Twin Lagrangian fibrations*

First we recall that every fiber of a Lagrangian fibration, say $\pi : X \rightarrow B$, has a natural affine structure.

Definition 15.2 *Let (X, ω) be a symplectic manifold of dimension $2n$ and $\pi : X \rightarrow B$ and $p : X \rightarrow C$ are two Lagrangian fibrations on X with Lagrangian sections. We call this a twin Lagrangian fibration of index q if for general $b \in B$ and $c \in C$ with $p^{-1}(c) \cap \pi^{-1}(b)$ nonempty, then it is an affine subspace of $\pi^{-1}(b)$ of codimension q. We denote such a structure as $B \overset{\pi}{\leftarrow} X \overset{p}{\rightarrow} C$.*

Since $\pi^{-1}(b)$ is the union of affine subspaces $p^{-1}(c) \cap \pi^{-1}(b)$, they form an affine foliation of $\pi^{-1}(b)$ of codimension q. In the above definition we assume that when $p^{-1}(c) \cap \pi^{-1}(b)$ is nonempty, then it is a codimension q affine submanifold of $\pi^{-1}(b)$. We did not assume that this affine structure is compatible with the one on $p^{-1}(c)$ which comes from the other Lagrangian fibration p. Nevertheless, under suitable assumptions, we will show that this is indeed the case and the above definition is symmetric with respect to B and C. Furthermore we have a commutative diagram of affine morphisms:

$$\begin{array}{ccc} X & \to & C \\ \downarrow & & \downarrow \\ B & \to & D \end{array}.$$

Claim 15.2 *Suppose that $B \overset{\pi}{\leftarrow} X \overset{p}{\rightarrow} C$ is a twin Lagrangian fibration. Then B admits a natural (singular) foliation whose leaves are q-dimensional subspaces $\pi(p^{-1}(c))$'s with $c \in C$.*

Proof. Since $p^{-1}(c) \cap \pi^{-1}(b)$ is always of codimension k in $p^{-1}(c)$ if nonempty, $\pi(p^{-1}(c))$ is a q dimensional subspace in B. Suppose b is a smooth point in $\pi(p^{-1}(c))$, we claim that its tangent space $T_b(\pi(p^{-1}(c))) \subset T_bB$ is independent of the choice of c. Assuming this, we obtain a q-dimensional (singular) distribution on B. Furthermore $\pi(p^{-1}(c))$'s with $c \in C$ are leaves of this distribution, thus we have the required foliation on B.

To prove the claim, we recall that $\pi^{-1}(b)$ has a natural affine structure and $\pi^{-1}(b)$ admits an affine foliation whose leaves are $p^{-1}(c) \cap \pi^{-1}(b)$ with $c \in C$, by the assumption of a twin Lagrangian fibration. The key observation is these imply that for any $x \in p^{-1}(c) \cap \pi^{-1}(b)$, under the natural identification $T_x^*(\pi^{-1}(b)) \simeq T_bB$, the conormal bundle of $p^{-1}(c) \cap \pi^{-1}(b)$ in $\pi^{-1}(b)$ at x is the same linear subspace in T_bB of dimension q. Furthermore this coincides with $\pi(T_x(p^{-1}(c)))$. Hence the result. □

We will assume that this foliation on B is indeed a fibration which we denote as

$$\overline{p} : B \to D.$$

As a corollary of the above claim, we have the following immediate result:

Corollary 15.1 *Suppose that $B \overset{\pi}{\leftarrow} X \overset{p}{\rightarrow} C$ is a Lagrangian twin fibration. Then we have a commutative diagram*

$$\begin{array}{ccc} X & \to & C \\ \downarrow & & \downarrow \\ B & \to & D \end{array}$$

where $\overline{p} : B \to D$ is the above fibration fibered by $\pi\left(p^{-1}(c)\right)$'s on B.

Claim 15.3 *Suppose that $B \overset{\pi}{\leftarrow} X \overset{p}{\rightarrow} C$ is a twin Lagrangian fibration with sections and π is a proper map. Then D has a natural affine structure and $\overline{p} : B \to D$ is an affine morphism.*

Proof. Since $\pi : X \to B$ is a Lagrangian fibration, its general fiber $\pi^{-1}(b)$ has a natural affine structure, thus as an affine manifold $\pi^{-1}(b) \simeq \mathbb{R}^n/\Lambda$ for some lattice $\Lambda \simeq \mathbb{Z}^r$ in T_b^*B. We have $r = n$ because π is proper, that is, $\pi^{-1}(b)$ is compact. This full rank lattice $\Lambda \subset T_b^*B$ determines an integral affine structure on B, away from the locus of singular fibers. Thus the proposition is equivalent to the fibration of $\pi(p^{-1}(c))$'s being an affine fibration on B.

From the proof of the previous proposition, we know that $p^{-1}(c) \cap \pi^{-1}(b)$ is an affine subtorus of $\pi^{-1}(b) \simeq T^n$. This implies that $T_b(p^{-1}(c)) \subset T_bB$ is an integral affine subspace. Because of the integral structure, these subspaces $T_b(p^{-1}(c))$'s are affinely equivalently to each other locally. Hence the foliation they determine is an affine foliation in B. In particular the leave space D inherits an affine structure such that $\overline{p} : B \to D$ is an affine morphism. Hence the result. □

This implies that, outside the singular locus of $\overline{p} : B \to D$, there exists a rank q vector bundle $\mathbb{R}^q \to E \xrightarrow{\varepsilon} D$ with affine gluing functions, together with a multi-section $E_{\mathbb{Z}}$ such that $\overline{p}$ is affine isomorphic to the projection $E/E_{\mathbb{Z}} \to D$. In particular $B = E/E_{\mathbb{Z}}$.

Given a general $x \in X$ with $b = \pi(x)$, $c = p(x)$, and $d = \overline{p}(b) = \overline{\pi}(b)$, we write $X_d = p^{-1}(c) \cap \pi^{-1}(b)$. From π and p both being Lagrangian fibrations on X, we have

$$0 \to N^*_{X_d/\pi^{-1}(b),x} \to T_bB \to T_x^*X_d \to 0$$

and

$$0 \to N^*_{X_d/p^{-1}(c),x} \to T_cC \to T_x^*X_d \to 0.$$

As we have shown earlier $N_{X_d/p^{-1}(c),x}$ can be identified with the fiber of $E \to D$ over $d \in D$, denoted as E_d. Moreover the exact sequence $0 \to N^*_{X_d/\pi^{-1}(b),x} \to T_bB \to T_x^*X_d \to 0$ is equivalent to $0 \to \varepsilon^*E \to TE \to \varepsilon^*TD \to 0$ for the vector bundle $\varepsilon : E \to D$. This implies that T_cC is naturally isomorphic to the tangent space of E^* at d.

Notice that the affine structure of the fibers of the Lagrangian fibration $p : X \to C$ is given by $T\left(p^{-1}(c)\right) \simeq p^*(T_c^*C)$. When p is proper, these data also determine an affine structure on C. Thus having such a natural identification between T_cC and $T_d(E^*)$, all the affine structures involved are compatible with each other. Such a claim can be checked by a direct diagram chasing method. Thus we have obtained the following result:

Claim 15.4 *Suppose $B \xleftarrow{\pi} X \xrightarrow{p} C$ is a twin Lagrangian fibration with π and p proper. Then $C \xleftarrow{p} X \xrightarrow{\pi} B$ is also a twin Lagrangian fibration and*

$$\begin{array}{ccc} X & \to & C \\ \downarrow & & \downarrow \\ B & \to & D \end{array}$$

is a commutative diagram of affine morphisms.

Similar to $B = E/E_{\mathbb{Z}}$ for a vector bundle $E \to D$, we have $C = F/F_{\mathbb{Z}}$ for another vector bundle $F \to D$ of rank q and multi-section $F_{\mathbb{Z}}$ of it. Furthermore the two bundles E and F over D are fiberwise dual to each other.

Similar discussions can be applied to Lagrangian sections to $\pi : X \to B$ and $p : X \to C$ and we can obtain affine sections to affine fibrations $B \to D$ and $C \to D$ and we can also prove that the following diagram of sections is commutative:

$$\begin{array}{ccc} X & \leftarrow & C \\ \uparrow & & \uparrow \\ B & \leftarrow & D \end{array}$$

Remark 15.2 The composition map $X \to D$ is a coisotropic fibration with fiber dimension equals $n + q$. If we apply the symplectic reduction on each coisotropic fiber, then we obtain a symplectic fibration over D which can be naturally identified with $B \times_D C = (E \oplus F) / (E_{\mathbb{Z}} \oplus F_{\mathbb{Z}})$.

Conversely, suppose that we have two proper Lagrangian fibrations $\pi : X \to B$ and $p : X \to C$ and a commutative diagram of maps

$$\begin{array}{ccc} X & \to & C \\ \downarrow & & \downarrow \\ B & \to & D_0 \end{array}$$

for some space D_0 satisfying for any $b \in B$ and $c \in C$ with the same image $d \in D_0$, then the preimage of d in X for the composition map is equal to $\pi^{-1}(b) \cap p^{-1}(c)$. We leave it as an exercise for readers to show that this gives a twin Lagrangian fibration structure on X.

15.4.3 *Dual twin Lagrangian fibration and its FM transform*

Suppose (X, ω) is a symplectic manifold with a twin Lagrangian fibration with sections:

$$\begin{array}{ccc} X & \to & C \\ \downarrow & & \downarrow \\ B & \to & D \end{array}$$

The dual of such a structure is obtained by taking the fiberwise dual to both fibrations $X \to B$ and $X \to C$ and we obtain the following commutative diagram:

$$\begin{array}{ccc} Y & \to & C' \\ \downarrow & & \downarrow \\ B' & \to & D \end{array}$$

Recall that a general fiber of $X \to D$ is a coisotropic submanifold in X which is a torus of dimension $n + q$, denoted T_d. Its symplectic reduction $(T_d/\sim)$ is a symplectic torus of dimension $2q$. We can view Y as being obtained from X by replacing those directions along the symplectic torus $(T_d/\sim)$ by the dual symplectic torus $(T_d/\sim)^*$. It is an important problem to describe Y near singular fibers. Such a Y is called the *dual twin Lagrangian fibration* to X. It is clear that the dual twin Lagrangian fibration to Y is X again.

As we have explained in the linear situation, the fibration $B' \to D$ (respectively $C' \to D$) is the dual fibration to $B \to D$ (respectively $C \to D$). Furthermore the Lagrangian conditions imply that there are natural identifications $B' \cong C$ and $C' \cong B$ and also

$$X \cong Y.$$

It is interesting to know whether this identification will continue to hold true if superpotentials are also included in our discussions.

Since $X \cong Y$, we choose the Lagrangian cycle $\left(P_{sym}^{FM}, \mathcal{P}_{sym}^{FM}\right)$ on $X \times Y$ given by the diagonal Lagrangian submanifold together with the trivial flat bundle over it. We call this the *Lagrangian Poincaré cycle* on $X \times Y$. The symplectic FM transform $\mathcal{F}_{sym}^{FM}$ for the twin Lagrangian fibration on X and its dual twin Lagrangian fibration on Y is defined using this Lagrangian cycle:

$$\mathcal{F}_{sym}^{FM} : \mathcal{C}(X) \to \mathcal{C}(Y)$$

$$\mathcal{F}_{sym}^{FM}(L, \mathcal{L}) = \left(\hat{L}, \hat{\mathcal{L}}\right).$$

Notice that this is actually an identity transformation.

15.4.4 *Examples of twin Lagrangian fibrations*

We are going to describe some examples of symplectic manifolds X with twin Lagrangian fibrations. First we notice that this property is stable under taking the product with another symplectic manifold with a Lagrangian fibration.

The trivial example is the product of a flat torus with its dual, $T \times T^*$. The complex projective space $\mathbb{CP}^n$ also has a twin Lagrangian fibration: its toric fibration

$$\mu : \mathbb{CP}^n \to \mathbb{R}^n$$

$$\mu[z_0, ..., z_n] = \left(\frac{|z_1|^2}{\sum_{j=0}^n |z_j|^2}, \frac{|z_2|^2}{\sum_{j=0}^n |z_j|^2}, ..., \frac{|z_n|^2}{\sum_{j=0}^n |z_j|^2}\right)$$

is a Lagrangian fibration. If we consider an automorphism of $\mathbb{CP}^n$ defined by $f([z_0, ..., z_n]) = [z_0 + z_1, z_0 - z_1, z_2, ..., z_n]$, then

$$\mu \circ f : \mathbb{CP}^n \to \mathbb{R}^n$$

gives another Lagrangian fibration on $\mathbb{CP}^n$. It is easy to check that the non-trivial intersection of any two generic fibers of μ and $\mu \circ f$ is an $(n-1)$-dimensional torus T^{n-1}. This example can be generalized to the Gelfand–Zeltin system on Grassmannians and partial flag varieties.

Taub-NUT example: Besides trivial examples of products of Lagrangian fibrations, we can write down explicit twin Lagrangian fibrations on four-dimensional hyperkähler manifolds X with S^1-symmetry. These spaces are classified by an positive integer n, denoted A_n, with explicit Taub-NUT metrics. Topologically A_n is the canonical resolution of $\mathbb{C}^2/\mathbb{Z}_{n+1}$ for a finite subgroup $\mathbb{Z}_{n+1} \subset SU(2)$. The hyperkähler moment map

$$\mu = (\mu_1, \mu_2, \mu_3) : X \to \operatorname{Im} \mathbb{H} = \mathbb{R}^3$$

is a singular S^1-fibration on X. For any unit vector $t \in S^2 \subset \mathbb{R}^3$, it determines a complex structure J_t on X. Furthermore the composition of μ with the projection to the orthogonal plane to t gives a Ω_{J_t}-complex Lagrangian fibration on X.

Corresponding to $i, j, k \in \operatorname{Im} \mathbb{H}$ we have complex structures I, J, K on X. Thus $(\mu_2, \mu_3) : X \to \mathbb{R}^2 = \mathbb{C}$ is a Ω_I-complex Lagrangian fibration with coordinate $(y, z) \in \mathbb{C}$. Similarly $(\mu_1, \mu_3) : X \to \mathbb{R}^2 = B$ is a Ω_J-complex Lagrangian fibration with coordinate $(x, z) \in B$. Then we have a ω_K twin Lagrangian fibration structure on X with D being the z-axis.

This construction can be easily generalized to any $4n$-dimensional hyperkähler manifold X with a tri-Hamiltonian T^n action with a hyperkähler moment map

$$\mu : X \to \mathbb{R}^n \otimes \mathbb{R}^3.$$

In the above example, the generic fiber of $X \to B$, or $X \to C$, is topologically a cylinder, thus noncompact. Therefore the base of the Lagrangian fibration only has a partial affine structure. For Lagrangian fibrations with compact fibers, there are usually singularities for the affine structures for the base spaces corresponding to singular fibers. When $\dim B = 2$, the affine structure near a generic singularity has been studied in Gross and Siebert (2003) and Kontsevich and Soibelman (2006). The monodromy across the "slit," namely, the positive x-axis, is given by $(x, y) \to (x, x + y)$. Near the singular point $0 \in B = \mathbb{R}^2$, the only affine fibration is given by horizontal lines. Other attempts to obtain fibrations on B will get overlapping fibers.

K3 surface with an elliptic fibration $M \to \mathbb{P}^1$ can be constructed as an anti-canonical divisor of a Fano toric threefold P_Δ which admits a toric $\mathbb{P}^2$-bundle structure, $\mathbb{P}^2 \to P_\Delta \to \mathbb{P}^1$. The mirror to M is another K3 surface X inside a Fano toric threefold P_∇ associated to a polytope which is a two-sided cone over a triangle.

Generically $X = \{f = 0\}$ admits a Lagrangian fibration over $B = \partial\nabla$, homeomorphic to the two-sphere S^2, with 24 singular fibers. B admits a natural affine structure with singularity along the base points of these singular fibers (see Gross and Siebert 2003 for its construction), which all lie on the edges of ∇, and their "slits" are also along these edges. We consider the restriction of the projection of $\mathbb{R}^3$ to the x-axis to $B = \partial\nabla$, the image is the interval D in $\mathbb{R}$ corresponding to the polytope of the base $\mathbb{P}^1$:

$$p : B \to D.$$

Since the slits for all the singular points in the middle triangle in $\partial\nabla$ are vertical with respect to p, we obtain an affine fibration around there. But p will cease to be an affine fibration around other singular points of B. However if we choose the defining function f for M appropriately, then we can arrange all other singular points to be far away from this middle triangle. Thus we obtain an affine fibration on a large portion of B. As a matter of fact, if we allow M to be singular, then we can make p to be an affine fibration on the whole $B\backslash\mathrm{Sing}\,(B)$. This picture can be generalized to certain higher dimensional manifolds, for instance CY hypersurfaces in toric varieties.

15.5 SYZ transformation of FM transform

Given an elliptically fibered CY manifold M, we argued in Section 15.2.4 that its mirror manifold X should admit a twin Lagrangian fibration (with superpotential which we neglect in our present discussions). In this section, we first argue that the mirror manifold to the dual elliptic fibration W to M is the dual twin Lagrangian fibration Y to X. Second, we will explain the mirror to the universal Poincaré sheaf that defines the FM transform between complex geometries of M and W is the universal Lagrangian Poincaré cycle that defines the FM transform between symplectic geometries of X and Y. Third, we show that these FM transforms commute with the SYZ transforms. Namely, the SYZ transform of the complex FM transform is the symplectic FM transform, which is actually an identity transformation.

15.5.1 *SYZ transform of dual elliptic fibrations*

As we mentioned above, we are going to argue that the following diagram commutes:

$$\begin{array}{ccccc} & M & \xleftrightarrow{\text{Mirror manifolds}} & X & \\ \begin{array}{c}\text{Dual elliptic}\\ \text{fibrations}\end{array} & \updownarrow & & \updownarrow & \begin{array}{c}\text{Dual twin Lagrangian}\\ \text{fibrations}\end{array} \\ & W & \xleftrightarrow{\text{Mirror manifolds}} & Y & \end{array}$$

Indeed our arguments only work in the large structure limit but we expect that a modified version of such will work in more general situations. To be precise, we are only dealing with the flat situation: We assume that M is CY n-fold with a q-dimensional Abelian varieties fibration $p : M \to S$ which is compatible with a Lagrangian fibration $\pi : M \to B$ in the sense that the restriction of π to any smooth Abelian variety fiber of p determines a Lagrangian fibration on the Abelian variety. We are going to impose a very restrictive assumption, which we expect to be the first-order approximation of what happens at the large structure limit. Namely, M is the product of an Abelian variety with S.

Let $z^1, ..., z^n$ be a local complex coordinate system on M such that $z = x + \sqrt{-1}y$ with $y^1, ..., y^n$ (respectively $x^1, ..., x^n$) be an affine coordinate system on fibers (respectively base) of the special Lagrangian fibration on $M \to B$. According to the SYZ proposal, the mirror manifold X to M is the total space of the dual Lagrangian torus fibration. For the affine coordinate system $y^1, ..., y^n$ on the torus fiber, we denote the dual coordinate system on its dual torus as $y_1, ..., y_n$. Thus X has a Darboux coordinate system given by $y_1, ..., y_n, x^1, ..., x^n$.

We assume the q-dimensional Abelian variety fibration $M \to S$ is compatible with the Lagrangian fibration structure in the sense that $z^{n-q+1}, ..., z^n$ gives a complex affine coordinate system on fibers to $M \to S$. In particular, the dual coordinate system on the dual Abelian variety is given by $z_{n-q+1}, ..., z_n$. Therefore such a coordinate system on M induces a similar coordinate system on the dual Abelian variety fibration $W \to S$, labeled $(z^1, ..., z^{n-q}, z_{n-q+1}, ..., z_n)$.

Note that on a fiber of the Abelian variety fibration $M \to S$, $y^{n-q+1}, ..., y^n$ (respectively, $x^{n-q+1}, ..., x^n$) are those coordinates belonging to Lagrangian fibers (respectively, base) for the special Lagrangian fibration on M. Since fibers are much smaller in scale when comparing to the base near the LCSL (Strominger *et al.* 1996), when we take the dual Abelian variety, $y_{n-q+1}, ..., y_n$ will become much larger in scale when comparing to $x_{n-q+1}, ..., x_n$. As a consequences, the special Lagrangian fiber (respectively, base) coordinates for the CY manifold Y are $y^1, ..., y^{n-q}, x_{n-q+1}, ..., x_n$ (respectively, $x^1, ..., x^{n-q}, y_{n-q+1}, ..., y_n$). Therefore the special Lagrangian fibration on the mirror manifold Y to W has fiber (respectively, base) coordinates as $y_1, ..., y_{n-q}, x^{n-q+1}, ..., x^n$ (respectively, $x^1, ..., x^{n-q}, y_{n-q+1}, ..., y_n$).

Next we need to check that Y is indeed the total space of the dual twin Lagrangian fibration to X. For the Abelian variety fiber in M, with coordinates $x^{n-q+1}, ...x^n, y^{n-q+1}, ...y^n$, its mirror Lagrangian cycle in X has coordinates $x^{n-q+1}, ...x^n, y_1, ...y_{n-q}$ and it is the fiber of the other Lagrangian fibration on X. Similarly, for the Abelian variety fibers in W, with coordinates $x_{n-q+1}, ...x_n, y_{n-q+1}, ...y_n$, its mirror Lagrangian cycle in Y has coordinates $y_1, ...y_{n-q}, y_{n-q+1}, ...y_n$ and it is the fiber of the other Lagrangian fibration on Y. It is now clear that X and Y are dual twin Lagrangian fibrations. All these fiber/base coordinates systems are summarized in the following diagram:

$$
\begin{array}{ccc}
M : \begin{array}{cc} y^i & y^\alpha \\ x^i & x^\alpha \end{array} & \leftrightarrow & X : \begin{array}{cc} y_\alpha & y_i \\ x^i & x^\alpha \end{array} \\
\updownarrow & & \updownarrow \\
W : \begin{array}{cc} y^i & x_\alpha \\ x^i & y_\alpha \end{array} & \leftrightarrow & Y : \begin{array}{cc} x^\alpha & y_i \\ x^i & y_\alpha \end{array}
\end{array}
$$

Our conventions are $i = 1, ..., n-q$ and $\alpha = n-q+1, ..., n$. For each space in the above diagram, the coordinates in the top (respectively, bottom) row are for the SYZ special Lagrangian fibers (respectively, base). Also the coordinates in the left (respectively, right) column are for the fibers (respectively, base) of the Abelian variety fibrations for M or W and the other Lagrangian fibrations for X or Y.

15.5.2 *SYZ transform of the universal Poincaré sheaf*

Recall that the FM transform on derived categories of manifolds with elliptic fibrations is given by

$$FM : D^b(M) \to D^b(W)$$
$$FM(-) = R^\bullet p'_*(\mathcal{P}_{cx} \otimes p^*(-)),$$

where $\mathcal{P}_{cx}$ is a coherent sheaf on $M \times W$ with support

$$\mathrm{Supp}(\mathcal{P}_{cx}) = M \times_S W \subset M \times W,$$

and

$$\mathcal{P}_{cx} = O(\Delta - \sigma_M \times W - M \times \sigma_W),$$

with Δ the relative diagonal in $M \times_S W$.

We note that $M \times W$ is again a CY manifold with its mirror manifold being $X \times Y$. We are going to describe the SYZ transform of $\mathcal{P}_{cx}$ from $D^b(M \times W)$ to $Fuk(X \times Y)$. We will apply the transformation on the level of object, as proposed in (Strominger *et al.* 1996).

Claim 15.5 *Corresponding to the mirror symmetry between CY $(2n)$-folds $M \times W$ and $X \times Y$*

$$\begin{matrix}\text{Complex}\\ \text{geometry}\end{matrix}(M \times W)\ \underset{\longleftrightarrow}{\text{SYZ transform}}\ \begin{matrix}\text{Symplectic}\\ \text{geometry}\end{matrix}(X \times Y)$$

the mirror of the coherent sheaf $\mathcal{P}_{cx}$ on $M \times W$ is the diagonal Lagrangian cycle $(P_{sym}, \mathcal{P}_{sym}) \in C(X \times Y)$ as described in Section 15.4.3.

Proof. We will continue to use the same coordinate systems as before. The Poincaré sheaf $\mathcal{P}_{cx}$ for the dual Abelian variety fibrations $M \to S$ and $W \to S$ has support $M \times_S W \subset M \times W$ which has coordinates $\{x^i, y^i, x^\alpha, y^\alpha, x_\alpha, y_\alpha\}_{i,\alpha}$ (where x^i and y^i are diagonal coordinates in the product space). Over $M \times_S W$, $\mathcal{P}_{cx}$ is a complex line bundle, indeed only a divisorial sheaf, with an $U(1)$ connection:

$$D_{cx}^{FM} = d + \sqrt{-1}\textstyle\sum_\alpha (x_\alpha dx^\alpha + x^\alpha dx_\alpha - y_\alpha dy^\alpha - y^\alpha dy_\alpha).$$

The reason that the signs for terms involving x and y are different is the following: For Abelian varieties, the dual *complex* coordinates are z^α and z_α, which induces the dual coordinates for x^α, y^α as $x_\alpha, -y_\alpha$ because $\mathrm{Re}(z^\alpha z_\alpha) = x^\alpha x_\alpha - y^\alpha y_\alpha$.

To apply the SYZ mirror transform to $\mathcal{P}_{cx}$ from the special Lagrangian fibration $M \times W \to B \times B^*$ to the one $X \times Y \to B \times B^*$, we first need to describe the Poincaré bundle $\mathcal{P}^{SYZ}$ over

$$(M \times W) \underset{B \times B^*}{\times} (X \times Y) \subset (M \times W) \times (X \times Y).$$

To describe the coordinates on this space, we need to rename the (x, y) coordinates on W and Y to (u, v) coordinates. That is,

$$\begin{array}{ccc} W & & Y \\ v^i \; u_\alpha & \underset{\longleftrightarrow}{\text{SYZ mirror}} & u^\alpha \; v_i \\ u^i \; v_\alpha & & u^i \; v_\alpha \end{array}$$

Now the universal $U(1)$ connection on the line bundle

$$\mathbb{C} \to \mathcal{P}^{SYZ} \to (M \times W) \underset{B \times B^*}{\times} (X \times Y)$$

is given by

$$\begin{aligned} \mathcal{D}^{SYZ} &= d + \sqrt{-1} \sum \left(y^i dy_i + y_i dy^i + y^\alpha dy_\alpha + y_\alpha dy^\alpha \right) \\ &\quad - \sqrt{-1} \sum \left(v^i dv_i + v_i dv^i + u_\alpha du^\alpha + u^\alpha du_\alpha \right). \end{aligned}$$

Note that we have used different signs for the SYZ transform between M and X and SYZ transform between W and Y.

In this newly named coordinates on W, the universal connection on the Poincaré bundle

$$\mathbb{C} \to \mathcal{P}_{cx}^{FM} \to M \times_S W$$

is given by

$$\mathcal{D}_{cx}^{FM} = d - \sqrt{-1} \sum_\alpha \left(u_\alpha dx^\alpha + x^\alpha du_\alpha - v_\alpha dy^\alpha - y^\alpha dv_\alpha \right).$$

To apply the SYZ transform, we need to first pullback the bundle $\mathcal{P}_{cx}^{FM}$ (with its connection $\mathcal{D}_{cx}^{FM}$) from $M \times_S W \subset M \times W$ to $(M \times W) \times_{B \times B^*} (X \times Y) \subset (M \times W) \times (X \times Y)$ and tensors it with $\mathcal{P}^{SYZ}$ (with its connection $\mathcal{D}^{SYZ}$). Then we pushforward along the projective map $(M \times W) \times_{B \times B^*} (X \times Y) \to X \times Y$.

We can separate our discussions into two parts: (i) perform the SYZ transform along those directions with indexes $i = 1, ..., n - q$ and (ii) perform the SYZ transform along those directions with indexes $\alpha = n - q + 1, ..., n$.

Part (i) with $i = 1, ..., n - q$: This part is easy because the Poincaré bundle for the FM transform does not involve here. In fact it is the trivial line bundle over $\{u^i = x^i\} \cap \{v^i = y^i\} \subseteq M \times W$. In these coordinates, the SYZ Poincaré bundle $\mathcal{P}^{SYZ}$ has support $(M \times W) \times_{B \times B^*} (X \times Y) \subset (M \times W) \times (X \times Y)$, which in our coordinate systems means the x^i's coordinates for M and X are the same and similarly the u^i's coordinates for W and Y are the same:

$$\begin{aligned}\mathcal{D}^{SYZ} &= d + \sqrt{-1}\textstyle\sum_i y^i dy_i + y_i dy^i - v^i dv_i - v_i dv^i \\ &= d + \sqrt{-1}\textstyle\sum_i y^i\left(dy_i - dv_i\right) + \left(y_i - v_i\right) dy^i.\end{aligned}$$

To pushforward to $X \times Y$, we integrate along y^i's directions. Along these directions, the restriction of the above connection is $d + \sqrt{-1}\sum_i \left(y_i - v_i\right) dy^i$, which has a (unique up to scaling) flat section precisely when $y_i - v_i = 0$ for all i. When this happens, we also have $y^i\left(dy_i - dv_i\right) = 0$. Hence the SYZ transform of $\mathcal{P}_{cx}^{FM}$ is given by the trivial bundle over $\left\{u^i = x^i \text{ and } y_i = v_i \text{ for all } i\right\} \subset X \times Y$. It is a Lagrangian submanifold in $X \times Y$ with the symplectic form $\omega_{X\times Y} = \Sigma dx^i \wedge dy_i + \Sigma du^i \wedge dv_i$.

Part (ii) with $\alpha = n - q + 1, ..., n$: In these coordinates, the support for the SYZ Poincaré bundle $\mathcal{P}^{SYZ}$ imposes a constraint which says that the x^α's coordinates for M and X are the same and the v_α's coordinates for W and Y are the same. Now we need to integrate the directions y^α's and u_α's. First we rearrange the terms

$$\begin{aligned}\mathcal{D}_{cx}^{FM} \otimes \mathcal{D}^{SYZ} &= d - \sqrt{-1}\textstyle\sum_\alpha \left(u_\alpha dx^\alpha + x^\alpha du_\alpha - v_\alpha dy^\alpha - y^\alpha dv_\alpha\right) \\ &\quad + \sqrt{-1}\textstyle\sum_\alpha \left(y^\alpha dy_\alpha + y_\alpha dy^\alpha - u_\alpha du^\alpha - u^\alpha du_\alpha\right) \\ &= d + \sqrt{-1}\textstyle\sum_\alpha \left(\left(-v_\alpha + y_\alpha\right) dy^\alpha + \left(x^\alpha - u^\alpha\right) du_\alpha\right) \\ &\quad + \sqrt{-1}\textstyle\sum_\alpha \left(u_\alpha dx^\alpha - y^\alpha dv_\alpha + y^\alpha dy_\alpha - u_\alpha du^\alpha\right).\end{aligned}$$

When we integrate along the y^α's and u_α's directions, the terms $u_\alpha dx^\alpha$, $y^\alpha dv_\alpha$, $y^\alpha dy_\alpha$, and $u_\alpha du^\alpha$ has non-trivial Fourier modes in these variables and therefore they contribute zero to $\mathcal{P}_{sym}$, the mirror cycle of $\mathcal{P}_{cx}$.

When we pushforward along y^α's direction, the restriction of the above connection becomes $d + \sqrt{-1}\sum_\alpha \left(-v_\alpha + y_\alpha\right) dy^\alpha$ along any fiber and it admits a (unique up to scaling) parallel section precisely when $y_\alpha = v_\alpha$ for all α. Similarly we obtain $u^\alpha = x^\alpha$ for all α when we pushforward along u_α's directions. Namely, the support of the mirror object to $\mathcal{P}_{cx}^{FM}$ is given by

$$\{y_\alpha = v_\alpha \text{ and } x^\alpha = u^\alpha \text{ for all } \alpha\} = P_{sym} \subset X \times Y.$$

Namely, this is precisely given by the diagonal Lagrangian submanifold as in the example in Section 15.4.3.

There is no remaining component of $\mathcal{D}_{cx}^{FM} \otimes \mathcal{D}^{SYZ}$ and therefore the unitary flat connection on $\mathcal{P}_{sym}$ is trivial.

By putting parts (i) and (ii) together, we have shown that the mirror transformation of the complex Poincaré cycle $\left(P_{cx}^{FM} = M \times_S W, \mathcal{P}_{cx}^{FM}\right) \in \mathcal{C}\left(M \times W\right)$ is the Lagrangian Poincaré cycle $\left(P_{sym}^{FM}, \mathcal{P}_{sym}^{FM}\right) \in \mathcal{C}\left(X \times Y\right)$ in the flat limit. Hence the claim. □

15.5.3 *SYZ transforms commute with FM transforms*

Recall from the last section that when $\mathcal{P}_{cx}^{FM}$ is the Poincaré sheaf for the FM transform between the elliptic CY manifold M and its dual elliptic CY manifold W, then its mirror $\mathcal{P}_{sym}^{FM}$ is the diagonal Lagrangian submanifold in $X \times Y$. Thus it defines a symplectic FM transform $\mathcal{F}_{sym}^{FM}$ from X to Y. We claim that the complex/symplectic FM transforms commute with the SYZ transforms:

$$\mathcal{F}_{sym}^{FM} \circ \mathcal{F}^{SYZ} = \mathcal{F}^{SYZ} \circ \mathcal{F}_{cx}^{FM}$$

as depicted in the following diagram:

$$\begin{array}{ccc}
\begin{matrix}\text{Complex}\\ \text{geometry}\end{matrix}\ (M) & \overset{\mathcal{F}^{SYZ}}{\longleftrightarrow} & \begin{matrix}\text{Symplectic}\\ \text{geometry}\end{matrix}\ (X) \\
\mathcal{F}_{cx}^{FM} \updownarrow & & \updownarrow \mathcal{F}_{sym}^{FM} \\
\begin{matrix}\text{Complex}\\ \text{geometry}\end{matrix}\ (W) & \overset{\mathcal{F}^{SYZ}}{\longleftrightarrow} & \begin{matrix}\text{Symplectic}\\ \text{geometry}\end{matrix}\ (Y)
\end{array}$$

Since $\mathcal{F}_{sym}^{FM}$ is essentially an identity transformation, we can regard the FM transform for elliptic CY as a *square* of the SYZ transforms! Note that even though one can identify M with W and X with Y, nevertheless, $\mathcal{F}_{(M,X)}^{SYZ}$ and $\mathcal{F}_{(W,Y)}^{SYZ}$ do not correspond to each other under these identifications of spaces. For instances, the whole manifold M (respectively, W) transforms to the zero section of the SYZ fibration on X (respectively, Y). However, they are sections of two different fibrations on $X \simeq Y$ as it admits a twin Lagrangian fibration structure.

In fact our claim follows from a more general statement: If M and W are two CY manifolds, possibly of different dimensions and X and Y are their mirror manifolds, respectively. Let $\mathcal{P}_{cx}$ be any complex cycle in $M \times W$ and $\mathcal{P}_{sym}$ be its mirror Lagrangian cycle in $X \times Y$. They define general complex/symplectic FM transforms $\mathcal{F}_{cx}^{FM}$ and $\mathcal{F}_{sym}^{FM}$, respectively. Then these FM transforms commute with the SYZ transforms.

The key point is the SYZ transform is an involution, that is, $\mathcal{F}^{SYZ} \circ \mathcal{F}^{SYZ} = id$. Suppose $\mathcal{S}$ is a complex cycle in M, we want to show that

$$\mathcal{F}_{sym}^{FM}\left(\mathcal{F}_{(M,X)}^{SYZ}(\mathcal{S})\right) = \mathcal{F}_{(W,X)}^{SYZ}\left(\mathcal{F}_{cx}^{FM}(\mathcal{S})\right).$$

Equivalently,

$$(\pi_X)_* (\pi_M)_* \mathcal{S}_M \otimes \mathcal{P}_{(M,X)}^{SYZ} \otimes \mathcal{P}_{sym}^{FM} = (\pi_W)_* (\pi_M)_* \mathcal{S}_M \otimes \mathcal{P}_{cx}^{FM} \otimes \mathcal{P}_{(W,Y)}^{SYZ}.$$

Therefore, it suffices to prove that

$$(\pi_X)_* \mathcal{P}_{(M,X)}^{SYZ} \otimes \mathcal{P}_{sym}^{FM} = (\pi_W)_* \mathcal{P}_{cx}^{FM} \otimes \mathcal{P}_{(W,Y)}^{SYZ},$$

over $M \times Y$. However,

$$\begin{aligned}\mathcal{P}_{sym}^{FM} &= (\pi_M)_* (\pi_W)_* \mathcal{P}_{cx}^{FM} \otimes \mathcal{P}_{(M\times W, X\times Y)}^{SYZ} \\ &= (\pi_M)_* (\pi_W)_* \mathcal{P}_{cx}^{FM} \otimes \mathcal{P}_{(M,X)}^{SYZ} \otimes \mathcal{P}_{(W,Y)}^{SYZ}.\end{aligned}$$

Hence

$$\begin{aligned}&(\pi_X)_* \mathcal{P}_{(M,X)}^{SYZ} \otimes \mathcal{P}_{sym}^{FM} \\ &= (\pi_X)_* \mathcal{P}_{(M,X)}^{SYZ} \otimes \left[(\pi_M)_* (\pi_W)_* \mathcal{P}_{cx}^{FM} \otimes \mathcal{P}_{(M,X)}^{SYZ} \otimes \mathcal{P}_{(W,Y)}^{SYZ}\right] \\ &= (\pi_W)_* \mathcal{P}_{cx}^{FM} \otimes \mathcal{P}_{(W,Y)}^{SYZ} \otimes \left[(\pi_M)_* (\pi_X)_* \mathcal{P}_{(M,X)}^{SYZ} \otimes \mathcal{P}_{(M,X)}^{SYZ}\right] \\ &= (\pi_W)_* \mathcal{P}_{cx}^{FM} \otimes \mathcal{P}_{(W,Y)}^{SYZ}.\end{aligned}$$

The last equality holds because SYZ transforms are involutive. Hence the result.

15.6 Conclusions and discussions

In this chapter we have introduced the notion of a twin Lagrangian fibration and explained several of its properties. We argued via the SYZ proposal that the mirror manifold of an elliptic CY manifold should admit such a structure, possibly coupled with a non-trivial superpotential which we have not fully understood yet.

An important tool in the study of the complex geometry of elliptic manifolds is the FM transform. We argued that under the SYZ transform, this FM transform will become the identity transformation for the symplectic geometry between dual twin Lagrangian fibrations. Even though we are mostly interested in the elliptic fibration situation, these arguments can be applied to Abelian varieties fibrations, and possibly to general CY fibrations with suitable adjustments.

We could also study mirror of the symplectic geometry of the elliptic CY manifolds. For example, let ω_M (respectively, ω_S) be any Kähler form on M (respectively, S). The pullback of ω_S under the elliptic fibration $M \to S$, denoted as ω_S again, is only nef but not ample. Obviously it satisfies $\omega_S^{n-1} \neq 0$ and $\omega_S^n = 0$. For any $t > 0$, $\omega_{M,t} = t\omega_M + (1-t)\,\omega_S$ is always a Kähler form on M. In particular it satisfies $\omega_{M,t}^n \neq 0$ and $\omega_{M,t}^{n+1} = 0$.

On the mirror side, these correspond to a two-parameter family of complex structures on X where (i) the generic monodromy is of maximally unipotent and the corresponding vanishing cycles are the Lagrangian fibers of $X \to B$; and (ii) a special monodromy T_0 for this family has the property $N_0^{n-1} \neq 0$ but $N_0^n = 0$ where $N_0 = \log(I - T_0)$ and the corresponding vanishing cycles are the $(n+1)$-dimensional coisotropic fibers of the composition map $X \to B \to D$.

Acknowledgments

The second author has known Nigel for almost forty years, having met him at the Institute of Advanced Study. In those days, we had many dinners together, and I (S.-T. Yau) always enjoyed talking to him. His many original ideas in mathematics influenced me tremendously. This is especially the case for K3 surfaces, in which his deep insight affected my thinking of the Calabi–Yau conjecture.

The research of N.C. Leung is partially supported by RGC Earmarked grants of Hong Kong. The work of S.-T. Yau is supported in part by NSF grants DMS-0354737, DMS-0306600, and DMS-0628341.

References

Andreas, B., Curio, G., Ruiperez, D. H. and Yau, S.-T. (2001). Fibrewise T-duality for D-branes on elliptic Calabi-Yau. *J. High Energy Phys.* No. **3**, Paper 20, 13 pp.

Bridgeland, T. and Maciocia, A. (2002). Fourier-Mukai transforms for K3 and elliptic fibrations. *J. Algebraic Geom.* **11**(4), 629–57.

Candelas, P., de la Ossa, X. C., Green, P. S. and Parkes, L. (1991). A pair of Calabi-Yau manifolds as an exactly soluble superconformal theory. *Nucl. Phys. B* **359**, 21–74.

Deligne, P. (1997). Local behavior of Hodge structures at infinity. In *Mirror Symmetry*, II, pp. 683–699, AMS/IP Studies in Advanced Mathematics, Vol. 1, American Mathematical Society, Providence, RI.

Donaldson, S. K. (1985). Anti self-dual Yang-Mills connections over complex algebraic surfaces and stable vector bundles. *Proc. London Math. Soc. (3)* **50**(1), 1–26.

Donagi, R. (1998). Taniguchi lectures on principal bundles on elliptic fibrations. In *Integrable Systems and Algebraic Geometry (Kobe/Kyoto, 1997)*, pp. 33–46, World Scientific Publishing, River Edge, NJ.

Donagi, R. (1997). Principal bundles on elliptic fibrations. *Asian J. Math.* **1**(2), 214–23.

Donagi R. and Pantev, T. (2008). Torus fibrations, gerbes, and duality. With an appendix by Dmitry Arinkin. *Mem. Amer. Math.* Soc. **193**(901), vi–90.

Friedman, R., Morgan, J. and Witten, E. (1997). Vector bundles and F theory. *Commun. Math. Phys.* **187**(3), 679–743.

Friedman, R., Morgan, J. and Witten, E. (1999). Vector bundles over elliptic fibrations. *J. Algebraic Geom.* **8**(2), 279–401.

Givental, A. (1996). Equivariant Gromov-Witten invariants. *Int. Math. Res. Notices*, No. **13**, 613–63.

Gross M. and Siebert, B. (2003). Affine manifolds, log structures, and mirror symmetry. *Turkish J. Math.* **27**(1), 33–60.

Hu, Y., Liu, C. H. and Yau, S.-T. (2002). Toric morphisms and fibrations of toric Calabi-Yau hypersurfaces. *Adv. Theor. Math. Phys.* **6**(3), 457–506.

Kontsevich, M. (1995). Homological algebra of mirror symmetry. In *Proceedings of the 1994 International Congress of Mathematicians*, Vol. I, Birkäuser, Zürich, p. 120.

Kontsevich, M. and Soibelman, Y. (2006). Affine structures and non-Archimedean analytic spaces. In *The Unity of Mathematics*, pp. 321–385, Progress of Mathematics, Vol. 244, Birkhauser Boston, Boston, MA.

Leung, N. C. (2005). Mirror symmetry without correction. *Commun. Anal. Geom.* **13**(2), 287–331.

Leung N. C. and Vafa, C. (1998). Branes and toric geometry. *Adv. Theor. Math. Phys.* **2**, 91–118.

Leung, N. C., Yau, S.-T. and Zaslow, E. (2000). From special Lagrangian to Hermitian-Yang-Mills via Fourier-Mukai transform. *Adv. Theor. Math. Phys.* **4**(6), 1319–41.

Lian, B. H., Liu, K. and Yau, S. T. (1997). Mirror principle I. *Asian J. Math.* **1**(4), 729–63.

Mukai, S. (1987). Fourier functor and its application to the moduli of bundles on an abelian variety. In *Algebraic Geometry, Sendai, 1985*, pp. 515–50, Advanced Studies in Pure Mathematics, Vol. 10, North-Holland, Amsterdam, the Netherlands.

Orlov, D. O. (1996). Equivalences of derived categories and K3 surfaces, Preprint. alg.geom/9606006.

Strominger, A., Yau, S.-T. and Zaslow, E. (1996). Mirror symmetry is T-duality. *Nucl. Phys. B* **479**, 243–59.

Thomas, R. and Yau, S. T. (2002). Special Lagrangians, stable bundles and mean curvature flow. *Commun. Anal. Geom.* **10**(5), 1075–113.

Uhlenbeck, K. and Yau, S. T. (1986). On the existence of Hermitian Yang-Mills connections in stable vector bundles. *Commun. Pure Appl. Math.* **39**, 257–93.

Weinstein, A. (1979). Lectures on symplectic manifolds. Corrected reprint. *CBMS Regional Conference Series in Mathematics*, Vol. 29, American Mathematical Society, Providence, RI.

Wilson, P. M. H. (1997). The role of c_2 in Calabi-Yau classification: a preliminary survey. In *Mirror Symmetry II*, pp. 381–92, AMS Studies in Advanced Mathematics, Vol. 1, American Mathematical Society, Providence, RI.

Yau, S.-T. (1978). On the Ricci curvature of a compact Kahler manifold and the complex Monge-Ampere equation. I. *Commun. Pure Appl. Math.* **31**(3), 339–411.

XVI

S-DUALITY IN HYPERKÄHLER HODGE THEORY

Tamás Hausel

To Nigel Hitchin for his 60th birthday

16.1 Introduction

In this chapter we survey the motivations, related results, and progress made towards the following problem, raised by Hitchin in 1995:

Problem 16.1 *What is the space of L^2 harmonic forms on the moduli space of Higgs bundles on a Riemann surface?*

The moduli space $\mathcal{M}^d_{Dol}(SL_n)$ of stable rank n Higgs bundles with fixed determinant of degree d on a Riemann surface was introduced and studied in Hitchin (1987), Nitsure (1991), and Simpson (1991). The Betti numbers of this space for $n = 2$ were determined in Hitchin (1987*b*) while for $n = 3$ in Gothen (1994). The above problem raised two new directions to study. First is the Riemannian geometry of $\mathcal{M}^d_{Dol}(SL_n)$, or more precisely the asymptotics of the natural hyperkähler metric, and its connection with Hodge theory. The second one, which can be considered the topological side of Problem 16.1, is to determine the intersection form on the middle-dimensional compactly supported cohomology of $\mathcal{M}^d_{Dol}(SL_n)$. While the first question seems still out of reach, although we will report on some modest progress below, the second is more approachable and we offer a conjecture at the end of this survey.

Problem 16.1 was motivated by S-duality conjectures emerging from the string theory literature about Hodge theory on certain hyperkähler moduli spaces, which are close relatives of $\mathcal{M}^d_{Dol}(SL_n)$.

In the physics literature S-duality stands for a *strong–weak duality* between two quantum field theories. The interest from the physics point of view is that it gives a tool to study physical theories with a large coupling constant via a conjectured equivalence with a theory with a small coupling constant where perturbative methods give a good understanding. The S-duality conjecture relevant for us is based on the Montonen–Olive electromagnetic duality proposal from 1977 in four-dimensional Yang–Mills theory (Montonen and Olive 1977). It was noted in (Witten and Olive 1978) that this duality proposal is more likely to hold in a supersymmetric version of the theory, and in Osborn (1979) it was argued that $N = 4$ supersymmetry is a good candidate. Hyperkähler Hodge theory is relevant in $N = 4$ supersymmetry as the space of differential forms on a hyperkähler manifold possesses an action of the $N = 4$ supersymmetry algebra via the various operators in hyperkähler Hodge theory.

In this chapter our interest lies in the mathematical predictions of such S-duality conjectures in physics. Sen (1994), using S-duality arguments in $N = 4$ supersymmetric Yang–Mills theory, predicted the dimension of the spaces $\mathcal{H}^d\left(\widetilde{M_k^0}\right)$ of L^2 harmonic d-forms on the universal cover $\widetilde{M_k^0}$ of the hyperkähler moduli space M_k^0 of certain $SU(2)$ magnetic monopoles on $\mathbb{R}^3$. In the interpretation of Sen (1994) the L^2 harmonic forms on $\widetilde{M_k^0}$ can be thought of as bound states of the theory, and the conjectured S-duality implies an action of $SL(2,\mathbb{Z})$ on $\bigoplus_k \mathcal{H}^*\left(\widetilde{M_k^0}\right)$. By further physical arguments Sen managed to predict this representation of $SL(2,\mathbb{Z})$ completely, implying the following:

Conjecture 16.1 *The dimension of the space of L^2 harmonic forms on $\widetilde{M_k^0}$ is*

$$\dim\left(\mathcal{H}^d\left(\widetilde{M_k^0}\right)\right) = \begin{cases} 0 & d \neq mid \\ \phi(k) & d = mid, \end{cases}$$

where $\phi(k) = \sum_{i=1}^k \delta_{1(i,k)}$ is the Euler ϕ function, and $mid = 2k-2$ is half of the dimension of $\widetilde{M_k^0}$.

Similar S-duality arguments led Vafa and Witten (1994) to get a conjecture on the space of L^2 harmonic forms on a certain smooth completion M_ϕ^{k,c_1}, constructed in Kronheimer (1990) and Nakajima (1998), of the moduli space of $U(n)$ Yang–Mills instantons of first Chern class c_1, energy k, and framing ϕ on one of Kronheimer's ALE spaces, which are four-dimensional complete hyperkähler manifolds, with an asymptotically locally Euclidean metric.

Conjecture 16.2 *The dimension of the space of L^2 harmonic forms on M_ϕ^{k,c_1} is*

$$\dim\left(\mathcal{H}^d\left(M_\phi^{k,c_1}\right)\right) = \begin{cases} 0 & d \neq mid \\ \dim\left(\operatorname{im}\left(H^{mid}_{cpt}\left(M_\phi^{k,c_1}\right) \to H^{mid}\left(M_\phi^{k,c_1}\right)\right)\right) & d = mid, \end{cases}$$

where mid now denotes half of the dimension of M_ϕ^{k,c_1}.

Vafa and Witten (1994) further argue that Conjecture 16.2 implies, via the work of Nakajima (1998) and Kac (1990), that

$$Z_\phi(q) = \sum_{c_1,k} q^{k-c/24} \dim\left(\mathcal{H}^{mid}\left(M_\phi^{k,c_1}\right)\right) \tag{16.1}$$

is a modular form, which, as was speculated in Vafa and Witten (1994), might be a consequence of S-duality.

This chapter will introduce the reader to various mathematical aspects of these three problems and offer mathematical techniques and results relating to them.

16.2 Hyperkähler quotients

A Riemannian manifold (M, g) is hyperkähler if it is Kähler with respect to three integrable complex structures $I, J, K \in \Gamma(\mathrm{End}(TM))$, which satisfy $I^2 = J^2 = K^2 = IJK = -1$, with Kähler forms ω_I, ω_J, and ω_K. Known compact examples are scarce (see e.g. Joyce 2000, section 7). Non-compact complete examples however are much more abundant. This is mostly because there is a widely applicable[1] *hyperkähler quotient construction*, due to Hitchin *et al.* (1987). The construction itself is an elegant quaternionization of the Marsden–Weinstein symplectic (or more precisely Kähler) quotient construction (see Mumford *et al.* 1994, chapter 8 for an introduction for the latter).

Let $\mathbb{M}$ be a hyperkähler manifold, $\mathcal{G}$ a Lie group, with Lie algebra $\mathfrak{g}$, and assume $\mathcal{G}$ acts on $\mathbb{M}$ preserving the hyperkähler structure (i.e. it acts by triholomorphic isometries). Let us further assume that we have moment maps $\mu_I : \mathbb{M} \to \mathfrak{g}^*$, $\mu_J : \mathbb{M} \to \mathfrak{g}^*$, and $\mu_K : \mathbb{M} \to \mathfrak{g}^*$ with respect to the symplectic forms ω_I, ω_J, and ω_K, respectively. We combine them into a single hyperkähler moment map:

$$\mu_{\mathbb{H}} = (\mu_I, \mu_J, \mu_K) : \mathbb{M} \to \mathbb{R}^3 \otimes \mathfrak{g}^*.$$

One takes $\xi \in \mathbb{R}^3 \otimes (\mathfrak{g}^*)^G$ and constructs the *hyperkähler quotient* at level ξ by

$$\mathbb{M}////_\xi \mathcal{G} := \mu_{\mathbb{H}}^{-1}(\xi)/\mathcal{G}.$$

The main result of Hitchin *et al.* (1987) is that the natural Riemannian metric on the smooth points of this quotient is hyperkähler.

Now we list three important examples of this construction, where the original hyperkähler manifold $\mathbb{M}$ and Lie group $\mathcal{G}$ are both infinite dimensional.

16.2.1 *Moduli of Yang–Mills instantons on* $\mathbb{R}^4$

Here we follow Hitchin (1987*a*, I example 3.6), compare also with Atiyah (1978).

Let G be a compact connected Lie group, which will be $U(n)$ or $SU(n)$ in this chapter. Let $P \to \mathbb{R}^4$ be a G-principal bundle over $\mathbb{R}^4$. Let $\mathbb{M}$ be the space of G-connections A on P of class C^∞, such that the energy

$$\left| \int_{\mathbb{R}^4} \mathrm{Tr}(F_A \wedge *F_A) \right| < \infty$$

is finite. Write

$$A = A_1 dx_1 + A_2 dx_2 + A_3 dx_3 + A_4 dx_4$$

in a fixed gauge, where $A_i \in \Omega^0(\mathbb{R}^4, \mathrm{ad}(P))$. Let $\mathcal{G} = \Omega(\mathbb{R}^4, Ad(P))$ be the gauge group of P. An element $g \in \mathcal{G}$ acts on $A \in \mathbb{M}$ by the formula $g(A) = g^{-1}Ag + g^{-1}dg$, preserving the hyperkähler structure. One finds that the hyperkähler

[1] Some colleagues even suggest, due to the success of this construction, that HyperKähLeR is in fact just a pronouncable version of the acronym HKLR.

moment map equation

$$\mu_{\mathbb{H}}(A) = 0 \Leftrightarrow F_A = *F_A$$

is just the self-dual Yang–Mills equation. Define the hyperkähler quotient $\mathcal{M}(\mathbb{R}^4, P) = \mu_{\mathbb{H}}^{-1}(0)/\mathcal{G}$, the moduli space of finite-energy self-dual Yang–Mills instantons on P. By its construction it has a natural hyperkähler metric.

A similar construction (Kronheimer and Nakajima 1990) for $G = U(n)$ yields a hyperkähler metric on moduli spaces of $U(n)$ Yang–Mills instantons on certain four-dimensional complete hyperkähler manifolds, the ALE spaces of Kronheimer (1989). These moduli spaces will have natural completions and various components of them will be the spaces M_ϕ^{k,c_1} which were mentioned in the introduction. They will resurface later as examples for Nakajima quiver varieties.

16.2.2 *Moduli space of magnetic monopoles on* $\mathbb{R}^3$

The following construction can be considered as a dimensional reduction of the previous example. Here we follow Hitchin (1987*a*, I example 3.5) and Atiyah and Hitchin (1988).

Assume that $G = SU(2)$ and the matrices A_i are independent of x_4. Then we have

$$A = A_1 dx_1 + A_2 dx_2 + A_3 dx_3$$

a connection on $\mathbb{R}^3$ and $A_4 = \phi \in \Omega^0(\mathbb{R}^3, \mathrm{ad}(P))$ becomes the *Higgs field.* The gauge group now is $\mathcal{G} = \Omega(\mathbb{R}^3, Ad(P))$ and $\mathbb{M}$ is the space of configurations (A, ϕ) satisfying certain boundary conditions. (The boundary condition is chosen to ensure finite energy.) The gauge group $\mathcal{G}$ acts on $\mathbb{M}$ by gauge transformations, preserving the natural hyperkähler metric on $\mathbb{M}$. The corresponding hyperkähler moment map equation

$$\mu_{\mathbb{H}}(A, \phi) = 0 \Leftrightarrow F_A = *d_A\phi$$

is equivalent to the Bogomolny equation.

Now by construction $M = \mu_{\mathbb{H}}^{-1}(0)/\mathcal{G}$, the moduli space of magnetic monopoles on $\mathbb{R}^3$ has a natural hyperkähler metric. It has infinitely many components $M = \cup_{k=1}^{\infty} M_k$ labeled by the magnetic charge k of the monopole.

M_k is acted upon by $\mathbb{R}^3$ by translations and by $U(1)$ by rotating the phase of the monopole. The quotient M_k^0 is still a smooth complete hyperkähler manifold of dimension $4k - 4$, with fundamental group $\mathbb{Z}_k$. We will denote by $\widetilde{M_k^0}$ its universal cover. In Atiyah and Hitchin (1985) they find the hyperkähler metric explicitly on the four-manifold M_2^0 and subsequently describe the scattering of two monopoles.

16.2.3 *Hitchin moduli space*

This example can be considered as a two-dimensional reduction of Section 16.2.1. We follow Hitchin (1987*b*, section 1; 1987*a* I example 3.3).

Now we assume that $G = U(n)$ and the matrices A_i in Section 16.2.1 are independent of x_3, x_4. We have now the connection $A = A_1 dx_1 + A_2 dx_2$ on the $U(n)$ principal bundle P on $\mathbb{R}^2$. We introduce $\Phi = (A_3 - A_4 i)dz \in \Omega^{1,0}(\mathbb{R}^2, \mathrm{ad}(P) \otimes \mathbb{C})$ the *complex Higgs field*. The gauge group now is $\mathcal{G} = \Omega(\mathbb{R}^2, Ad(P))$, which acts by gauge transformations on the space $\mathbb{M}$ of configurations (A, Φ) preserving the natural hyperkähler metric on $\mathbb{M}$. The moment map equations

$$\mu_{\mathbb{H}}(A, \Phi) = 0 \Leftrightarrow \begin{array}{c} F(A) = -[\Phi, \Phi^*], \\ d''_A \Phi = 0 \end{array}$$

are then equivalent with Hitchin's self-duality equations. There are no solutions of finite energy on $\mathbb{R}^2$, but as the equations are conformally invariant, we can replace $\mathbb{R}^2$ with a genus g compact Riemann surface C in the above definitions, and define $\mathcal{M}(C, P) = \mu_{\mathbb{H}}^{-1}(0)/\mathcal{G}$, the Hitchin moduli space, which has a natural hyperkähler metric by construction. There are different ways to think about this space with the different complex structures, which will be explained in Section 16.5.2.

16.3 Hodge theory

16.3.1 L^2 *harmonic forms on complete manifolds*

Let M be a complete Riemannian manifold of dimension n. We say that a smooth differential k-form $\alpha \in \Omega^k(M)$ is harmonic if and only if $d\alpha = d{*}\alpha = 0$, where $* : \Omega^k(M) \to \Omega^{n-k}(M)$ is the Hodge star operator. It is L^2 if and only if

$$\int_M \alpha \wedge *\alpha < \infty.$$

We denote by $\mathcal{H}^*(M)$ the space of L^2 harmonic forms.

A fundamental theorem of Hodge theory is the Hodge (orthogonal) decomposition theorem of de Rham (1984, section 32 theorem 24, Section 35 theorem 26):

$$\Omega^*_{L^2} = \overline{d\left(\Omega^*_{cpt}\right)} \oplus \mathcal{H}^* \oplus \overline{\delta\left(\Omega^*_{cpt}\right)}, \tag{16.2}$$

where δ is the adjoint of d. When M is compact this implies the celebrated Hodge theorem, which says that $\mathcal{H}^*(M) \cong H^*(M)$, that is, that there is a unique harmonic representative in every de Rham cohomology class. When M is non-compact we only have a topological lower bound. Namely, the Hodge decomposition theorem implies that the composite map

$$H^*_{cpt}(M) \to \mathcal{H}^*(M) \to H^*(M)$$

is just the forgetful map. (In the compact case these maps are isomorphisms, which gives the Hodge theorem mentioned above.) Thus

$$\operatorname{im}(H^*_{cpt}(M) \to H^*(M)) \tag{16.3}$$

is a "topological lower bound" for $\mathcal{H}^*(M)$. By Poincaré duality the map $H^*_{cpt}(M) \to H^*(M)$ is equivalent with the intersection pairing on $H^*_{cpt}(M)$.

In the cases most relevant for us M will be a hyperkähler manifold (sometimes orbifold) so $\dim(M) = 4k$ and we will additionally have $H^i(M) = 0$ for $i > 2k$. Therefore the possible non-trivial image in $\operatorname{im}(H^*_{cpt}(M) \to H^*(M))$ will be concentrated in the middle $2k$ dimension. (We will use the notation $mid = \dim(M)/2$ for the middle dimension of a manifold.) For such a hyperkähler manifold we denote

$$\begin{aligned}\chi_{L^2}(M) &= \dim\left(\operatorname{im}\left(H^{mid}_{cpt}(M) \to H^{mid}(M)\right)\right)\\ &= \dim\left(\operatorname{im}\left(H^*_{cpt}(M) \to H^*(M)\right)\right)\end{aligned} \tag{16.4}$$

the dimension of this image. $\chi_{L^2}(M)$ can be thought of either as a "topological lower bound" for $\dim(\mathcal{H}^*(M))$ or the Euler characteristic of topological L^2 cohomology.

16.3.2 *Results on L^2 harmonic forms*

There were few general theorems on describing $\mathcal{H}^*(M)$ for a non-compact complete manifold M (see however Hausel *et al.* 2004, introduction for an overview). It was thus a surprising development when Sen (1994), using arguments from S-duality, managed to predict the dimension of L^2 harmonic forms on $\widetilde{M}^k_0$ as was explained in Conjecture 16.1 in the Introduction. In particular, according to Sen's Conjecture 16.1 the space $\mathcal{H}^2(\widetilde{M}^0_2)$ should be one dimensional. Using the explicit description of Atiyah and Hitchin (1985) of the metric on $\mathcal{M}^0_2$ in Sen (1994) he was able to find an explicit L^2 harmonic two-form, called the *Sen two-form*, on $\widetilde{M}^0_2$. This was perhaps the strongest mathematical support exhibited for Conjecture 16.1 in Sen (1994).

More general mathematical support for Conjecture 16.1 came in 1996. Segal and Selby (1996) showed that the intersection form on $H^{mid}_{cpt}(\widetilde{M}^0_k)$ is definite. Moreover they obtained for the topological lower bound (16.3) for $\mathcal{H}^{mid}(\widetilde{M}^0_k)$

$$\chi_{L^2}(\widetilde{M}^0_k) = \dim(H^{mid}(\widetilde{M}^0_k)) = \phi(k).$$

This agrees with the predicted dimension of $\mathcal{H}^{mid}(\widetilde{M}^0_k)$ in Sen's Conjecture 16.1.

Motivated by Problem 16.1 and Segal–Selby's topological lower bound for Conjecture 16.1, the author calculated in Hausel (1998) that the intersection pairing on the g-dimensional space $H^{mid}_{cpt}(\mathcal{M}^1_{Dol}(SL_2))$ is trivial, in other words

$$\chi_{L^2}\left(\mathcal{M}^1_{Dol}(SL_2)\right) = 0 \tag{16.5}$$

for $g > 1$. This thus gave the surprising result that there are no L^2 harmonic forms on $\mathcal{M}^1_{Dol}(SL_2)$ plainly by topological reasons. The technique used in the proof of (16.5) was imitating Kirwan's proof (1992) of Mumford's conjecture on the cohomology ring of the moduli space of stable rank 2 bundles of degree 1 on the Riemann surface C. Therefore the extension of (16.5) to higher rank Higgs bundle moduli spaces $\mathcal{M}^d_{Dol}(SL_n)$ was not straightforward.

The next advance towards Sen's Conjecture 16.1 came in 2000. Hitchin (2000) showed that $\mathcal{H}^d(M) = 0$ unless $d = \dim(M)/2$ for a complete hyperkähler manifold M of linear growth. Examples include all our hyperkähler quotients discussed in this chapter. The proofs in Hitchin (2000) use techniques inspired by Jost and Zuo's extension (2000) of ideas of Gromov (1991). It is interesting to note that some of the proofs in Hitchin (2000) also exploit the operators in hyperkähler Hodge theory, which are relevant in $N = 4$ supersymmetry. Using the symmetries of the Atiyah–Hitchin metric (Hitchin 2000) proves Sen's conjecture for $k = 2$, that up to a scalar the only L^2 harmonic form on $\widetilde{M}^0_2$ is Sen's two-form.

A more topological approach was introduced in Hausel *et al.* (2004). Hausel *et al.* (2004) proves for fibered boundary manifolds M

$$\mathcal{H}^{mid}(M) \cong \operatorname{im}\left(IH^{mid}_{\underline{m}}(\overline{M}) \to IH^{mid}_{\bar{m}}(\overline{M})\right), \tag{16.6}$$

where $\overline{M}$ is a certain compactification of M, dictated by the asymptotics of the fibered boundary metric on M. Moreover $IH^{mid}_{\underline{m}}(\overline{M})$ denotes the intersection cohomology in dimension $mid = \dim(M)/2$ with lower middle perversity $\underline{m}$ and $IH^{mid}_{\bar{m}}(\overline{M})$ denotes the intersection cohomology in the middle dimension with upper middle perversity $\bar{m}$ of the possibly badly singular (i.e. not necessarily a Witt space) compactification $\overline{M}$. To illustrate (16.6) we take the compactification of $\widetilde{M}^0_2$, which happens to be the smooth space $\mathbb{CP}^2$ (with the non-standard orientation), where the above cohomologies in (16.6) all coincide, giving $\mathcal{H}^2(\widetilde{M}^0_2) \cong H^2(\mathbb{CP}^2)$. This provides a topological explanation for the existence and uniqueness of the Sen two-form.

The assumption that the metric is fibered boundary in Hausel *et al.* (2004) is fairly restrictive. Among hyperkähler quotients only a few examples satisfy this property (see the discussion in Hausel *et al.* 2004, section 7). Examples include all ALE gravitational instantons of Kronheimer (1989) and all known ALF (see Cherkis and Kapustin 1999) and some ALG gravitational instantons (see Cherkis and Kapustin 2002). In general our hyperkähler quotients have some kind of stratified asymptotic behaviour at infinity. For example, the metric on M^0_k is fibered boundary only when $k = 2$, for higher k it is known to behave differently at different regions of infinity. The first result which could handle Hodge theory on Riemannian manifolds with such a stratified behaviour at infinity appeared recently in a work by Carron. It proves for a QALE space M that:

$$\mathcal{H}^{mid}(M) \cong \operatorname{im}\left(H^{mid}_{cpt}(M) \to H^{mid}(M)\right).$$

A QALE space (Joyce 2000, section 9) by definition is a certain Calabi–Yau metric on a crepant resolution of $\mathbb{C}^k/\Gamma$, where $\Gamma \subset SU(k)$ is a finite subgroup. The asymptotics of the metric on such a QALE space is reminiscent to the asymptotics of the natural hyperkähler metric on M_ϕ^{k,c_1} appearing in the Vafa–Witten Conjecture 16.2. It is thus reasonable to hope that the Vafa–Witten Conjecture 16.2 will be decided soon.

As there have been extensive studies starting with Gibbons and Manton (1995) and more recently Bielawski (2008) on the asymptotics of the Riemannian metric on M_k^0, it is conceivable that we will have a precise understanding of the asymptotic behaviour of this metric, and in turn the Hodge theory of L^2 harmonic forms on $\widetilde{M}_k^0$, perhaps extending techniques from Carron. Thus one may be optimistic that Sen's Conjecture 16.1 will be decided in the foreseeable future.

Finally, one must admit that the description of the asymptotics of the metric at infinity on $\mathcal{M}^d_{Dol}(SL_n)$ is still lacking, thus calculation of $\mathcal{H}^*\left(\mathcal{M}^d_{Dol}(SL_n)\right)$ is presently hopeless. The topological side of Problem 16.1, that is, to determine $\chi_{L^2}\left(\mathcal{M}^d_{Dol}(SL_n)\right)$, when $(d,n)=1$, is more reasonable. After introducing a new arithmetic technique to study Hodge structures on the cohomology of our hyperkähler manifolds, we will be able to offer a general conjecture on the intersection form on Higgs moduli spaces, in particular that (16.5) holds for any n.

16.4 Mixed Hodge theory

As explained above there have been some limited successes of calculating $\mathcal{H}^*(M)$ for a hyperkähler quotient and understanding its relation to the cohomology $H^*(M)$ or more generally the cohomology of an appropriate compactification $H^*(\bar{M})$. Another extension of Hodge theory yields some different and in some ways more detailed insight into the cohomology of our hyperkähler quotients. This technique is Deligne's mixed Hodge structure on the cohomology of any complex algebraic variety. Instead of the global analysis on the Riemannian geometry of the complex algebraic variety it will relate to the arithmetic of the variety over finite fields.

16.4.1 *Mixed Hodge structure of Deligne*

Motivated by the (then still unproven) Weil conjectures and Grothendieck's "yoga of weights", which drew cohomological conclusions about complex varieties from the truth of those conjectures, Deligne (1971, 1974) proved the existence of mixed Hodge structures on the cohomology $H^*(M,\mathbb{Q})$ of a complex algebraic variety M. Here we give a quick introduction, for more details see Hausel and Rodriguez-Villegas (section 2.2) and the references therein. Deligne's mixed Hodge structure entails two filtrations on the rational cohomology of M. The increasing weight filtration

$$0 = W_{-1} \subseteq W_0 \subseteq \cdots \subseteq W_{2j} = H^j(X,\mathbb{Q})$$

and a decreasing Hodge filtration

$$H^j(X,\mathbb{C}) = F^0 \supseteq F^1 \supseteq \cdots \supseteq F^m \supseteq F^{m+1} = 0.$$

We can define mixed Hodge numbers obtained from these two filtrations by the following formula:

$$h^{p,q;j}(X) := \dim_{\mathbb{C}} \left(Gr^F_p Gr^W_{p+q} H^j(X)_{\mathbb{C}} \right). \tag{16.7}$$

From these numbers we form

$$H(M;x,y,t) = \sum_{p,q,k} h^{p,q;k}(M) x^p y^q t^k,$$

the *mixed Hodge polynomial*. By virtue of its definition it has the property that the specialization

$$P(M;t) = H(M;1,1,t)$$

gives the *Poincaré polynomial* of M. When M is smooth of dimension n we take another specialization

$$E(M;x,y) := x^n y^n H(1/x, 1/y, -1), \tag{16.8}$$

the so-called *E-polynomial* of a smooth variety M.

Deligne's construction of mixed Hodge structure is complex geometrical: for a smooth variety M it is defined by the log geometry of a compactification $\overline{M}$ with normal crossing divisors. In particular a global analytical description, like the Hodge theory of harmonic forms on a smooth complex projective manifold, of the mixed Hodge structure on a smooth variety is missing, which causes some difficulty in finding the meaning of mixed Hodge numbers in physical contexts (see the remark after Conjecture 16.3).

16.4.2 *Arithmetic and topological content of the E-polynomial*

The connection of the E-polynomial to the arithmetic of the variety is provided by the following theorem of Katz (Hausel and Rodriguez-Villegas, Appendix). Here we give an informal version of Katz's result for precise formulation (see Hausel and Rodriguez-Villegas 2008, theorem 6.1.2(3), theorem 2.1.8):

Theorem 16.1 *Let M be a smooth quasi-projective variety defined over $\mathbb{Z}$ (i.e. given by equations with integer coefficients). Assume that the number of points of M over a finite field $\mathbb{F}_q$, that is,*

$$E(q) := \#\{M(\mathbb{F}_q)\}$$

is a polynomial in q. Then the E-polynomial can be obtained from the count polynomial as follows:

$$E(M;x,y) = E(xy).$$

This theorem is especially useful when we further have $h^{p,q;k}(M) = 0$ unless $p + q = k$. In this case we say that the mixed Hodge structure on $H^*(M)$ is *pure*. In this case

$$H(M; x, y, t) = (xyt^2)^n E\left(\frac{-1}{xt}, \frac{-1}{yt}\right)$$

and so the Poincaré polynomial can be recovered from the E-polynomial as follows:

$$P(M; t) = H(M; 1, 1, t) = t^{2n} E\left(\frac{-1}{t}, \frac{-1}{t}\right).$$

Examples of varieties with pure MHS on their cohomology include smooth projective varieties (in this case we get the traditional Hodge structure, which is by definition pure), the moduli space of Higgs bundles $\mathcal{M}_{Dol}$, the moduli space of flat connections $\mathcal{M}_{DR}$ on a Riemann surface, and Nakajima's quiver varieties.

In general we can define the *pure part* of $H(M; x, y, t)$ as

$$PH(M; x, y) = \mathrm{Coeff}_{T^0}\left(H(M; xT, yT, tT^{-1})\right).$$

More generally we can define the *pure part* of the cohomology of M as

$$PH^*(M) := W_n H^n(M) \subset H^*(M),$$

which is a subring $PH^*(M) \subset H^*(M)$ of the cohomology of M. For a smooth M, the pure part of $H^*(M)$ is always the image of the cohomology of a smooth compactification (see Deligne 1971, corollaire 3.2.17). It is in fact this result which can be used to show that the spaces mentioned in the previous paragraph have pure mixed Hodge structure. That is, one can prove that they admit a smooth compactification which surjects on cohomology. Prototypes of such compactifications were constructed in Simpson (1997) for $\mathcal{M}_{DR}$ and in Hausel (1998) for $\mathcal{M}_{Dol}$.

16.5 Applications of mixed Hodge theory

Using the method sketched in the previous section the strongest results on cohomology can be achieved when the variety has a pure MHS on its cohomology, consequently the E-polynomial determines the mixed Hodge polynomial, and additionally it has polynomial-count so that Theorem 16.1 gives an arithmetic way to determine the E-polynomial. This is the case for Nakajima quiver varieties, where our method gives complete results.

16.5.1 *Nakajima quiver varieties*

Nakajima quiver varieties are constructed (Nakajima 1998) by a finite-dimensional hyperkähler quotient construction. Here we review the affine algebraic-geometric version of this construction.

Let Γ be a quiver (oriented graph) with vertex set $I = \{1, \ldots, n\}$ and edges $E \subset I \times I$. Let

$$\mathbf{v} = (\mathbf{v}_1, \ldots, \mathbf{v}_n), \quad \mathbf{w} = (\mathbf{w}_1, \ldots, \mathbf{w}_n) \in \mathbb{N}^I$$

be two-dimensional vectors and V_i and W_i corresponding complex vector spaces, that is, $\dim(V_i) = \mathbf{v}_i$ and $\dim(W_i) = \mathbf{w}_i$. We define the vector spaces

$$\mathbb{V}_{\mathbf{v},\mathbf{w}} = \bigoplus_{a \in E} \mathrm{Hom}(V_{t(a)}, V_{h(a)}) \oplus \bigoplus_{i \in I} \mathrm{Hom}(V_i, W_i)$$

of framed representations of the quiver Γ , and the action

$$\rho : GL(\mathbf{v}) := \prod_{i \in I} GL(V_i) \to GL(\mathbb{V}_{\mathbf{v}}),$$

with derivative

$$\varrho : \mathfrak{gl}(\mathbf{v}) := \prod_{i \in I} \mathfrak{gl}(V_i) \to \mathfrak{gl}(\mathbb{V}_{\mathbf{v}}).$$

The complex moment map

$$\mu : \mathbb{V}_{\mathbf{v},\mathbf{w}} \times \mathbb{V}^*_{\mathbf{v},\mathbf{w}} \to \mathfrak{gl}^*_{\mathbf{v}}$$

of ρ is given at $X \in \mathfrak{gl}_{\mathbf{v}}$ by

$$\langle \mu(v, w), X \rangle = \langle \varrho(X) v, w \rangle. \tag{16.9}$$

For $\xi = 1_{\mathbf{v}} \in \mathfrak{gl}(\mathbf{v})^{GL(\mathbf{v})}$ we define the (always smooth) Nakajima quiver variety by

$$\mathcal{M}(\mathbf{v}, \mathbf{w}) = \mu^{-1}(\xi) // GL(\mathbf{v}) = \mathrm{Spec}\left(\mathbb{C}[\mu^{-1}(\xi)]^{\mathrm{GL}(\mathbf{v})}\right)$$

as an affine GIT quotient. Alternatively one can construct the manifold underlying $\mathcal{M}(\mathbf{v}, \mathbf{w})$ as a hyperkähler quotient of $\mathbb{V}_{\mathbf{v},\mathbf{w}} \times \mathbb{V}^*_{\mathbf{v},\mathbf{w}}$ by the maximal compact subgroup $U(\mathbf{v}) \subset GL(\mathbf{v})$. This shows that $\mathcal{M}(\mathbf{v}, \mathbf{w})$ possesses a hyperkähler metric. The holomorphic symplectic quotient we presented above is the one where the arithmetic technique of Section 16.4 is applicable. Before we explain that, let us recall the following fundamental theorem of Nakajima (1998) about the cohomology of these Nakajima quiver varieties:

Theorem 16.2 *Assume that the quiver Γ has no edge-loops. Then there is an irreducible representation of the Kac–Moody algebra $\mathfrak{g}(\Gamma)$ of highest weight $\mathbf{w}$ on $\oplus_{\mathbf{v}} H^{mid}(\mathcal{M}(\mathbf{v}, \mathbf{w}))$. In particular the Weyl–Kac character formula gives the middle Betti numbers of Nakajima quiver varieties. Furthermore the intersection form on $H^{mid}_c(\mathcal{M}(\mathbf{v}, \mathbf{w}))$ is definite, thus $\chi_{L^2}(\mathcal{M}(\mathbf{v}, \mathbf{w}))$ equals the middle Betti number of $\mathcal{M}(\mathbf{v}, \mathbf{w})$.*

Remark 16.1 When Γ is an affine Dynkin diagram $\mathcal{M}(\mathbf{v}, \mathbf{w})$ could be identified with one of the spaces M^{k,c_1}_{ϕ} of certain Yang–Mills instantons on a ALE space X_Γ. In Kac (1990) he explains that the Weyl–Kac character formula for an affine Dynkin diagram has certain modular properties. This was the line of argument in Vafa and Witten (1994) that (16.1) is a modular form provided Conjecture 16.2 holds.

In Hausel (2006) a simple Fourier transform technique was found to enumerate the rational points of $\mathcal{M}(\mathbf{v},\mathbf{w})$ over a finite field $\mathbb{F}_q$. The corresponding count function $E(q)$ turned out to be polynomial, and as the mixed Hodge structure is pure on $H^*(\mathcal{M}(\mathbf{v},\mathbf{w}))$ the technique of Section 16.4 applies in its full strength to give a formula for the Betti numbers of the varieties $\mathcal{M}(\mathbf{v},\mathbf{w})$. The result is the following formula from Hausel (2006):

Theorem 16.3 *For any quiver Γ, the Betti numbers of the Nakajima quiver varieties are given by the following generating function, with the notation as in Hausel (2006, theorem 3):*

$$\sum_{\mathbf{v}\in\mathbb{N}^I} P_t(\mathcal{M}(\mathbf{v},\mathbf{w}))t^{-d(\mathbf{v},\mathbf{w})}T^{\mathbf{v}} = \frac{\displaystyle\sum_{\mathbf{v}\in\mathbb{N}^I} T^{\mathbf{v}} \sum_{\lambda\in\mathcal{P}(\mathbf{v})} \frac{\left(\prod_{(i,j)\in E} t^{-2\langle\lambda^i,\lambda^j\rangle}\right)\left(\prod_{i\in I} t^{-2\langle\lambda^i,(1^{\mathbf{w}_i})\rangle}\right)}{\prod_{i\in I}\left(t^{-2\langle\lambda^i,\lambda^i\rangle}\prod_k\prod_{j=1}^{m_k(\lambda^i)}(1-t^{2j})\right)}}{\displaystyle\sum_{\mathbf{v}\in\mathbb{N}^I} T^{\mathbf{v}} \sum_{\lambda\in\mathcal{P}(\mathbf{v})} \frac{\prod_{(i,j)\in E} t^{-2\langle\lambda^i,\lambda^j\rangle}}{\prod_{i\in I}\left(t^{-2\langle\lambda^i,\lambda^i\rangle}\prod_k\prod_{j=1}^{m_k(\lambda^i)}(1-t^{2j})\right)}}, \tag{16.10}$$

Remark 16.2 When Γ has no edge-loops Nakajima's Theorem 16.1 implies that the right-hand side of (16.10) is a deformation of the Weyl–Kac character formula. Simple reasoning gives the same result about the denominator of the right-hand side of (16.10) and the Kac denominator. Moreover, Kac's denominator formula and Hua's formula (2000, theorem 4.9) expressing the denominator of (16.10) as an infinite product imply a conjecture of Kac (cf. Hua 2000, corollary 4.10). Namely, if $A_\Gamma(\mathbf{v},q)$ denotes the number of absolutely indecomposable representations of Γ of dimension vector $\mathbf{v}$ over the finite field $\mathbb{F}_q$, then it turns out to be a polynomial in q and Kac's conjecture 1 (1983) says that the constant coefficient

$$A_\Gamma(\mathbf{v},0) = m_{\mathbf{v}} \tag{16.11}$$

equals with the multiplicity of the weight $\mathbf{v}$ in the Kac–Moody algebra $\mathfrak{g}(\Gamma)$. This can be proved, as sketched above and announced in Hausel (2006), to be a consequence of (16.10) and the above-mentioned results of Nakajima and Hua.

Remark 16.3 When the quiver is affine ADE and the RHS becomes an infinite product (indications that this can happen are the infinite product in Hausel 2006, section 3 and the infinite products in the recent Sasaki) we could get an alternative proof of the modularity of (16.1) in the Vafa–Witten S-duality conjecture.

In the remaining part of this survey we will motivate and study another application of the technique in Section 16.4, which will be less powerful as the mixed Hodge structure will fail to be pure, but will also open new interesting directions by the study of this more complicated mixed Hodge structure.

16.5.2 *Spaces diffeomorphic to the Hitchin moduli space* $\mathcal{M}(C, P_{U(n)})$

Among the spaces discussed in this chapter it is the Hitchin moduli space $\mathcal{M}(C, P_{U(n)})$ as defined in Section 16.2.2 which exhibits perhaps the most plentiful structures many of which are rooted in its hyperkähler quotient origin. In particular there are three distinct complex algebraic variety structures on $\mathcal{M}(C, P_{U(n)})$. These can be thought of (Simpson 1997) as the three types of non-Abelian (first) cohomology: Dolbeault, De Rham, and Betti, of the Riemann surface C. The survey paper Hausel (2005) gives a quick introduction to these spaces and some of the cohomological implications to be discussed below.

In this chapter the ground field is always $\mathbb{C}$ unless otherwise indicated. Following Hitchin (1987*b*) and Simpson (1997) we define a component of the twisted $GL_n = GL_n(\mathbb{C})$ Dolbeault cohomology of C as

$$\mathcal{M}^d_{Dol}(GL_n) := \left\{ \begin{matrix} \text{Moduli space of semistable rank } n \\ \text{degree } d \text{ Hitchin pairs on } C \end{matrix} \right\}$$

the GL_n De Rham cohomology as

$$\mathcal{M}^d_{DR}(GL_n) := \left\{ \begin{matrix} \text{Moduli space of flat } GL_n\text{-connections} \\ \text{on } C \setminus \{p\}, \text{ with holonomy } e^{\frac{2\pi i d}{n}} Id \text{ around } p \end{matrix} \right\}$$

and the GL_n Betti cohomology

$$\mathcal{M}^d_B(GL_n) := \Big\{ A_1, B_1, \ldots, A_g, B_g \in GL_n | \\ A_1^{-1}B_1^{-1}A_1B_1 \cdots A_g^{-1}B_g^{-1}A_gB_g = e^{\frac{2\pi i d}{n}} Id \Big\} //GL_n$$

as a twisted GL_n character variety of C.

When $d = 0$ these three varieties are diffeomorphic to the Hitchin moduli space $\mathcal{M}(C, P_{U(n)})$. However we prefer to consider the twisted versions, when $(d, n) = 1$, because then all the varieties are smooth. In this case these three varieties are all diffeomorphic to a twisted version $\mathcal{M}^d(C, P_{U(n)})$ of Hitchin moduli space and so to each other. The mixed Hodge structure is pure on $H^*\left(\mathcal{M}^d_{Dol}(GL_n)\right)$ and $H^*\left(\mathcal{M}^d_{DR}(GL_n)\right)$, while it is not pure on $H^*\left(\mathcal{M}^d_B(GL_n)\right)$. As the mixed Hodge structures are different on $H^*\left(\mathcal{M}^d_{DR}(GL_n)\right)$ and $H^*\left(\mathcal{M}^d_B(GL_n)\right)$, the spaces $M^d_{DR}(GL_n)$ and $\mathcal{M}^d_B(GL_n)$ cannot be isomorphic as complex algebraic varieties. Nevertheless as complex analytic manifolds the Riemann–Hilbert monodromy map

$$\mathcal{M}^d_{DR}(GL_n) \overset{RH}{\to} \mathcal{M}^d_B(GL_n) \tag{16.12}$$

sending a flat connection to its holonomy gives an isomorphism.

We will also consider the varieties $\mathcal{M}^d_{Dol}(SL_n)$, $\mathcal{M}^d_{DR}(SL_n)$, and $\mathcal{M}^0_B(SL_n)$, which can be defined by replacing GL_n with SL_n in the above definitions. Moreover $\mathcal{M}^0_{Dol}(GL_1)$, $\mathcal{M}^0_{DR}(GL_1)$, and $\mathcal{M}^0_B(GL_1)$ turn out to be Abelian groups. Then $\mathcal{M}^0_{Dol}(GL_1)$, $\mathcal{M}^0_{DR}(GL_1)$, and $\mathcal{M}^0_B(GL_1)$ will act on $\mathcal{M}^d_{Dol}(GL_n)$, $\mathcal{M}^d_{DR}(GL_n)$, and $\mathcal{M}^d_B(GL_n)$, respectively, by an appropriate form of tensorization. Finally we denote the corresponding (affine GIT) quotients by $\mathcal{M}^d_{Dol}(PGL_n)$, $\mathcal{M}^d_{DR}(PGL_n)$ and $\mathcal{M}^d_B(PGL_n)$. In our case, when $(d,n)=1$, they will turn out to be orbifolds. For more details on the construction of these varieties see Hausel (2005).

In the next section we explain the original motivation to consider the E-polynomials of these three complex algebraic varieties. The motivation is mirror symmetry, and most probably the same S-duality we discussed in the Introduction in connection with the Hodge cohomology of the moduli spaces of Yang–Mills instantons in four dimension and magnetic monopoles in three. S-duality ideas relating to mirror symmetry for Hitchin spaces have appeared in the physics literature (Bershadsky *et al.* 1995; Kapustin and Witten 2007).

16.5.3 *Topological mirror test*

For our mathematical considerations the relationship to mirror symmetry stems from the following observation of Hausel and Thaddeus (2003). It uses the famous *Hitchin map* (Hitchin 1987c), which makes the moduli space of Higgs bundles $\mathcal{M}_{Dol}$ into a completely integrable Hamiltonian system, so that the generic fibers are Abelian varieties.

Theorem 16.4 *In the following diagram*

$$\begin{array}{ccc} \mathcal{M}^d_{Dol}(PGL_n) & & \mathcal{M}^d_{Dol}(SL_n) \\ \downarrow{\scriptstyle \chi_{PGL_n}} & & \downarrow{\scriptstyle \chi_{SL_n}} \\ \mathcal{H}_{PGL_n} & \cong & \mathcal{H}_{SL_n} \end{array}$$

the generic fibers of the Hitchin maps χ_{PGL_n} and χ_{SL_n} are dual Abelian varieties.

Remark 16.4 If we change complex structures and consider $\mathcal{M}^d_{DR}(PGL_n)$ and $\mathcal{M}^d_{DR}(SL_n)$, then the Hitchin map on them becomes special Lagrangian fibrations, and consequently the pair of $\mathcal{M}^d_{DR}(PGL_n)$ and $\mathcal{M}^d_{DR}(SL_n)$ satisfies the requirements of the SYZ construction (Strominger *et al.* 1996) for a pair of mirror symmetric Calabi–Yau manifolds (see Hausel and Thaddeus 2001, 2003 for more details).

This motivates the calculation of Hodge numbers of $\mathcal{M}^d_{DR}(PGL_n)$ and $\mathcal{M}^d_{DR}(SL_n)$ to see if there is any relationship between them, which one would expect in mirror symmetry. Based on calculations in the $n=2,3$ cases Hausel and Thaddeus (2003) proposed

Conjecture 16.3 *For all $d, e \in \mathbb{Z}$, satisfying $(d,n) = (e,n) = 1$,*

$$E_{st}^{B^e}\left(x, y; \mathcal{M}_{DR}^d(SL_n)\right) = E_{st}^{\hat{B}^d}\left(x, y; \mathcal{M}_{DR}^e(PGL_n)\right),$$

where B^e and $\hat{B}^d$ are certain gerbes on the corresponding Hitchin spaces and the E-polynomials above are stringy E-polynomials for orbifolds twisted by the relevant gerbe as defined in Hausel and Thaddeus (2003).

Morally, this conjecture should be related to the S-duality considerations of Kapustin and Witten (2007) and in turn to the geometric Langlands programme of Beilinson and Drinfeld (1995). However the lack of global analytical interpretation of the mixed Hodge numbers (16.7) appearing in Conjecture 16.3 prevents a straightforward physical interpretation. Nevertheless the agreement of certain Hodge numbers for Hitchin spaces for Langlands dual groups is an interesting direction from a purely mathematical point of view. In particular, if we change our focus from $\mathcal{M}_{DR}$ and $\mathcal{M}_{Dol}$ to $\mathcal{M}_B$ we will uncover some surprising connections to the representation theory of finite groups of Lie type.

16.5.4 *Mirror symmetry for finite groups of Lie type*

As $\mathcal{M}_{DR}$ and $\mathcal{M}_B$ are complex analytically identical via the Riemann–Hilbert map (16.12), the complex analytical structure of dual special Lagrangian fibrations of Theorem 16.4 are present on the pair $\mathcal{M}_B^d(SL_n)$ and $\mathcal{M}_B^e(PGL_n)$. We might as well try to think of this pair as mirror symmetric in the SYZ picture. The mixed Hodge numbers of $\mathcal{M}_B$ are however different from the mixed Hodge numbers of $\mathcal{M}_{DR}$ so the corresponding topological mirror test (Hausel 2005) will also be different from Conjecture 16.3:

Conjecture 16.4 *For all $d, e \in \mathbb{Z}$, satisfying $(d,n) = (e,n) = 1$,*

$$E_{st}^{B^e}\left(x, y, \mathcal{M}_B^d(SL_n)\right) = E_{st}^{\hat{B}^d}\left(x, y, \mathcal{M}_B^e(PGL_n)\right).$$

For this conjecture however there is a powerful arithmetic method to calculate these E-polynomials. Using this technique we have already managed to check this conjecture (Hausel 2005) when n is a prime and $n = 4$. This arithmetic method is based on the technique explained in Section 16.4 and the following character formula from Hausel and Rodriguez-Villegas (2008):

Theorem 16.5 *Let $G = SL_n$ or GL_n, let $G(\mathbb{F}_q)$ be the corresponding finite group of Lie type*

$$E\left(\sqrt{q}, \sqrt{q}, \mathcal{M}_B^d(G)\right) = \#\left\{\mathcal{M}_B^d(G(\mathbb{F}_q))\right\} = \sum_{\chi \in Irr(G(\mathbb{F}_q))} \frac{|G(\mathbb{F}_q)|^{2g-2}}{\chi(1)^{2g-1}} \chi\left(\xi_n^d\right),$$

where the sum is over all irreducible characters of the finite group of Lie type $G(\mathbb{F}_q)$.

This character formula combined with Conjecture 16.4 implies certain relationships between the character tables of $PGL_n(\mathbb{F}_q)$ and $SL_n(\mathbb{F}_q)$. An intriguing way to formulate it is to say that *certain differences between the character tables of $PGL_n(\mathbb{F}_q)$ and its Langlands dual $SL_n(\mathbb{F}_q)$ are governed by mirror symmetry.* This kind of consideration could be interesting because the character tables of $PGL_n(\mathbb{F}_q)$ or more generally those of $GL_n(\mathbb{F}_q)$ have been known for a long time starting with the work of Green (1955), while the character tables of $SL_n(\mathbb{F}_q)$ have just recently been completed (Bonnafé 2006; Shoji 2006). It is especially enjoyable to follow the effect of the mirror symmetry proposal of Conjecture 16.4 by comparing the character tables of $GL_2(\mathbb{F}_q)$ and $SL_2(\mathbb{F}_q)$ first calculated a hundred years ago by Jordan 1907 and Schur 1907.

16.5.5 *Conjectural answer*

Finally, we can put all our observations and conjectures together to state a conjectural answer to the topological side of Problem 16.1.

As we already noted the mixed Hodge structure on $H^*(\mathcal{M}_B)$ is not pure. Therefore we are losing information by considering only $E(\mathcal{M}_B; x, y)$. It turns out that it is interesting to consider the full mixed Hodge polynomial $H(\mathcal{M}_B; x, y, t)$. When $n = 2$ it can be calculated via the explicit description of $H^*(\mathcal{M}_B)$ in Hausel and Thaddeus (2003). We get Hausel and Rodriguez-Villegas (Theorem 1.1.3):

$$\begin{aligned}
&H(\mathcal{M}_B(PGL_2); x, y, t)\\
&\quad = \frac{(q^2t^3+1)^{2g}}{(q^2t^2-1)(q^2t^4-1)} + \frac{q^{2g-2}t^{4g-4}(q^2t+1)^{2g}}{(q^2-1)(q^2t^2-1)} - \frac{1}{2}\frac{q^{2g-2}t^{4g-4}(qt+1)^{2g}}{(qt^2-1)(q-1)}\\
&\qquad - \frac{1}{2}\frac{q^{2g-2}t^{4g-4}(qt-1)^{2g}}{(q+1)(qt^2+1)},
\end{aligned}$$

where $q = xy$ and the four terms correspond to the four types of irreducible characters of $GL(2, \mathbb{F}_q)$. When $g = 3$ this equals

$$\begin{aligned}
&t^{12}q^{12} + t^{12}q^{10} + 6\,t^{11}q^{10} + t^{12}q^8 + t^{10}q^{10} + 6\,t^{11}q^8 + 16\,t^{10}q^8 + 6\,t^9q^8 + t^{10}q^6\\
&\quad + t^8q^8 + 26\,t^9q^6 + 16\,t^8q^6 + 6\,t^7q^6 + t^8q^4 + t^6q^6 + 6\,t^7q^4 + 16\,t^6q^4\\
&\quad + 6\,t^5q^4 + t^4q^4 + t^4q^2 + 6\,t^3q^2 + t^2q^2 + 1.
\end{aligned}$$

In particular we see that the pure part is $1 + q^2t^4 + q^4t^8$. These terms correspond to the cohomology classes 1, β, and β^2, and the term q^6t^{12} is not present because by the Newstead relation $\beta^g = \beta^3 = 0$ holds (Hausel and Thaddeus 2003). In particular there is no pure part in the middle = 12-dimensional cohomology. The same argument holds for all g, which shows that there is no pure part in the middle-dimensional cohomology of $\mathcal{M}_B^1(PGL_2)$. It is however easy to see that the intersection form on middle cohomology can only be non-trivial on the pure part and so this implies Hausel and Rodriguez-Villegas (2008, Corollary 5.4.1):

Corollary 16.1 *The intersection form on $H^*_{cpt}\left(\mathcal{M}_B^1(PGL_2)\right)$ is trivial.*

This gives an alternative proof of (16.5) as the equation

$$\chi_{L^2}\left(\mathcal{M}_B^1(SL_2)\right) = \chi_{L^2}\left(\mathcal{M}_B^1(PGL_2)\right)$$

is easy to prove. Moreover this approach is more promising to generalize for any n. We will offer a conjecture about the pure part of the cohomology of $\mathcal{M}_B^d(PGL_n)$ below and in turn that will yield a conjecture for the intersection form on the middle-dimensional compactly supported cohomology, answering the topological side of Problem 16.1.

To state our conjecture in its full generality we introduce character varieties on Riemann surfaces with k punctures and parabolic type $\mu = (\mu^1, \ldots, \mu^k)$ at the punctures, where μ^i is a partition of n. In other words we fix semisimple conjugacy classes $\mathcal{C}_1, \ldots, \mathcal{C}_k \subset GL_n$, which are generic and have type μ (in other words μ^i_j is the multiplicity of the jth eigenvalue of a matrix in $\mathcal{C}_i$). One can prove as in Hausel *et al.* (2008, lemma 2.1.2) that there exists generic semisimple conjugacy classes for every type $\mu = (\mu^1, \ldots, \mu^k)$. For a generic $\{\mathcal{C}_1, \ldots, \mathcal{C}_k\}$ of type μ we define

$$\begin{aligned}\mathcal{M}_B^\mu := \{A_1, B_1, \ldots, A_g, B_g \in GL_n, C_1 \in \mathcal{C}_1, \ldots, C_k \in \mathcal{C}_k | \\ [A_1, B_1] \cdots [A_g, B_g] C_1 \cdots C_k = I_n\} // GL_n\end{aligned}$$

as an affine GIT quotient by the diagonal adjoint action of GL_n. The generic choice of the semisimple conjugacy classes implies that $\mathcal{M}_B^\mu$ is smooth. The torus GL_1^{2g} acts on $\mathcal{M}_B^\mu$ by multiplying the matrices A_i and B_i by a scalar. We can define the quotient

$$\bar{\mathcal{M}}_B^\mu := \mathcal{M}_B^\mu // GL_1^{2g}$$

as the corresponding PGL_n character variety. The variety $\bar{\mathcal{M}}_B^\mu$ is an orbifold.

By studying the Riemann–Hilbert map on the level of cohomologies we are led (Hausel, in preparation) to consider the comet-shaped quiver Γ associated to g and μ. Namely, we can put g loops on a central vertex, and k legs of length $l(\mu^j)$. We also equip Γ with a dimension vector $\mathbf{v}$, which has dimension $\sum_{i=1}^{l} \mu_i^j$ at the lth vertex on the ith leg. Consider now the number $A_\Gamma(q, \mathbf{v})$ of absolutely indecomposable representations of Γ of dimension $\mathbf{v}$ over the finite field $\mathbb{F}_q$. Kac (1983, proposition 1.15) proved that $A_\Gamma(q, \mathbf{v})$ is a polynomial in q with integer coefficients. We have the following conjecture from Hausel (in preparation):

Conjecture 16.5 *The pure part of the cohomology of $\bar{\mathcal{M}}_B^\mu$ is given by*

$$PH\left(\bar{\mathcal{M}}_B^\mu, x, y\right) = (xy)^{d_\mu/2} A_\Gamma(\mathbf{v}, 1/(xy)),$$

where $(\Gamma, \mathbf{v})$ is the star-shaped quiver and dimension vector given by the parabolic type μ, and d_μ is the dimension of $\mathcal{M}_B^\mu$.

This conjecture gives a cohomological interpretation of $A_\Gamma(\mathbf{v}, q)$ and in particular implies that it has non-negative coefficients confirming (Kac 1983, conjecture 2) in the case when Γ is comet-shaped. When μ is indivisible Conjecture 16.5 can be proved to follow from the master conjecture in Hausel *et al.* (in preparation), which expresses the mixed Hodge polynomials of all the character varieties $\bar{\mathcal{M}}_B^\mu$ as a generating function generalizing the Cauchy formula for Macdonald polynomials. It also has the following consequence on the topological L^2 cohomology $\chi_{L^2}\left(\bar{\mathcal{M}}_B^\mu\right)$ of (16.4).

Conjecture 16.6 *The topological L^2 cohomology of the manifold $\bar{\mathcal{M}}_B^\mu$ is given by*

$$\chi_{L^2}(\bar{\mathcal{M}}_B^\mu) = 0, \ \ \textit{when } g > 1 \tag{16.13}$$

$$\chi_{L^2}(\bar{\mathcal{M}}_B^\mu) = 1, \ \ \textit{when } g = 1 \tag{16.14}$$

$$\chi_{L^2}(\bar{\mathcal{M}}_B^\mu) = m_{\mathbf{v}}, \ \ \textit{when } g = 0, \tag{16.15}$$

where $m_{\mathbf{v}}$ is the multiplicity of the weight $\mathbf{v}$ in the Kac–Moody algebra $\mathfrak{g}(\Gamma)$, which are encoded by the Kac denominator formula for the star-shaped quiver Γ.

When $g > 1$ and the parabolic type is $\mu = ((n))$, that is, we have only one puncture with central conjugacy class, then one can identify $\bar{\mathcal{M}}_B^\mu = \mathcal{M}_B^d(PGL_n)$, with some d such that $(d,n) = 1$. In this case (16.13) says that

$$\chi_{L^2}\left(\mathcal{M}_B^d(PGL_n)\right) = 0,$$

which appeared as (Hausel and Rodriguez-Villegas Conjecture 4.5.1). It follows from the mirror symmetry Conjecture 16.3 that

$$H_{cpt}^{mid}\left(\mathcal{M}_B^d(SL_n)\right) \cong H_{cpt}^{mid}\left(\mathcal{M}_B^d(PGL_n)\right)$$

and then the intersection forms also agree. This and (16.15) then imply that (16.5) holds for any n, that is, that the intersection form on the compactly supported cohomology of $\mathcal{M}_B^d(SL_n)$ is trivial. This gives a conjectural answer to the topological side of Problem 16.1.

When $g = 1$ the conjectured (16.14) follows from Conjecture 16.5 and the observation that the coefficient of q in the A-polynomial $A_\Gamma(q)$ for a $g = 1$ comet-shaped quiver Γ is always 1.

When $g = 0$ the varieties $\mathcal{M}_B^\mu = \bar{\mathcal{M}}_B^\mu$ coincide. Conjecture 16.5 then implies that

$$\chi_{L^2}\left(\mathcal{M}_B^\mu\right) = A_\Gamma(\mathbf{v}, 0).$$

Conjecture (16.15) is a combination of this and the equality $A_\Gamma(\mathbf{v}, 0) = m_{\mathbf{v}}$, that is, Kac's conjecture 1, in Kac (1983), which, as discussed in Remark 16.2, follows from Theorem 16.3.

Finally one can define $\bar{\mathcal{M}}_{Dol}^\mu$ the moduli space of stable parabolic PGL_n-Higgs bundles with quasi-parabolic type $\mu^j \in \mathcal{P}(n)$ and generic weights at the jth puncture on the Riemann surface (Boden and Yokogawa 1996, García-Prada *et al.* 2007). Then one can prove that $\bar{\mathcal{M}}_B^\mu$ is diffeomorphic to $\bar{\mathcal{M}}_{Dol}^\mu$. Thus Conjecture 16.6 also calculates the intersection form on the compactly supported cohomology of the moduli space $\bar{\mathcal{M}}^{\mu}_{Dol}$ of stable parabolic PGL_n-Higgs bundles of any rank.

Example 16.1 Consider the genus 0 Riemann surface $\mathbb{P}^1$ with four punctures. Consider the moduli space $\mathcal{M}_{toy}$ of stable rank 2 parabolic Higgs bundles on $\mathbb{P}^1$, with generic parabolic weights on the full parabolic flag at the punctures (see Boden and Yokogawa 1996). This is a complex surface and the intersection form on $H^2_c(\mathcal{M}_{toy})$ was discussed in Hausel (1998, example 2 for theorem 7.13). $H^2_c(\mathcal{M}_{toy})$ is five-dimensional but $\chi_{L^2}(\mathcal{M}_{toy})$ is only four. (The cohomology class of the generic fiber of the Hitchin map is the one in the kernel.)

$\mathcal{M}_{toy}$ is diffeomorphic to the character variety $\bar{\mathcal{M}}^{\mu}_{B}$ where $g = 0$ and $\mu = ((1,1),(1,1),(1,1),(1,1))$. Thus by Conjecture 16.6 we should be able to calculate $\chi_{L^2}\left(\bar{\mathcal{M}}^{\mu}_{B}\right)$ in terms of the representation theory of the corresponding quiver Γ. The corresponding quiver Γ in this case will be the affine $\tilde{D}_4$ Dynkin diagram, with $\mathbf{v} = (2,1,1,1,1)$ the minimal positive imaginary root. Its multiplicity $m_{\mathbf{v}}$ in the affine Kac–Moody algebra associated to Γ is known to be 4. Alternatively it is known (Kac 1983 example b to conjecture 2) that $A_\Gamma(\mathbf{v}, q) = q + 4$, which by (16.11) gives $m_{\mathbf{v}} = 4$. Thus indeed $\chi_{L^2}\left(\mathcal{M}^{\mu}_{B}\right) = m_{\mathbf{v}} = 4$ checking (16.15) in this case via Hausel (1998, Example 2 for Theorem 7.13).

Acknowledgements

This chapter is a write-up of the author's talk at the Geometry Conference in Honour of Nigel Hitchin in Madrid in September 2006. Problem 16.1 was raised by Nigel Hitchin in 1995, then the author's PhD supervisor, as a project for the author's PhD thesis. This chapter would like to show the impact of this modest-looking question on the author's subsequent research. The author's research has been supported by a Royal Society University Research Fellowship, NSF grant DMS-0604775 and an Alfred Sloan Fellowship. The visit to Madrid was supported by a Royal Society International Joint Project between the United Kingdom and Spain.

References

Atiyah, M. F. (1978). Geometry of Yang-Mills fields. In *Mathematical Problems in Theoretical Physics (Proceeding of the International Conference, University of Rome, Rome, 1977)*, Vol. 80 of *Lecture Notes in Physics.* pp. 216–21. Springer, Berlin, Germany.

Atiyah, M. F. and Hitchin, N. J. (1985). Low energy scattering of nonabelian monopoles. *Phys. Lett. A* **107**(1), 21–5.

Atiyah, M. F. and Hitchin, N. J. (1988). *The Geometry and Dynamics of Magnetic Monopoles.* M. B. Porter Lectures. Princeton University Press, Princeton, NJ.

Beilinson, A. and Drinfeld, V. Quantization of Hitchin's integrable system and Hecke eigensheaves. (384 pages, unpublished preprint, ca. 1995).

Bershadsky, M., Johansen, A., Shadov, V. and Vafa, C. (1995). Topological reduction of 4D SYM to 2D σ-models. *Nuclear Phys. B* **448**(1–2), 166–86.

Bielawski, R. (2008). Monopoles and clusters. arXiv:hep-th/0702190. *Commun. Math. Phys.* **284**, 675–712.

Boden, H. U. and Yokogawa, K. (1996). Moduli spaces of parabolic Higgs bundles and parabolic $K(D)$ pairs over smooth curves. I. *Int. J. Math.* **7**(5), 573–98.

Bonnafé, C. (2006). Sur les caractères des groupes réductifs finis à centre non connexe: applications aux groupes spéciaux linéaires et unitaires. *Astérisque* **(306)**, vi, 165.

Carron, G. Cohomologie L^2 des variétés QALE. arXiv: math.DG/0501290.

Cherkis, S. A. and Kapustin, A. (1999). Singular monopoles and gravitational instantons. *Commun. Math. Phys.* **203**(3), 713–28.

Cherkis S. A. and Kapustin, A. (2002). Hyper-Kähler metrics from periodic monopoles. *Phys. Rev. D (3)* **65**(8), 084015, 10.

de Rham, G. (1984). *Differentiable Manifolds*, Vol. 266 of *Grundlehren der Mathematischen Wissenschaften*. Springer-Verlag, Berlin, Germany.

Deligne, P. (1971). Théorie de Hodge. II. *Inst. Hautes Études Sci. Publ. Math.* **(40)**, 5–57.

Deligne, P. (1974). Théorie de Hodge. III. *Inst. Hautes Études Sci. Publ. Math.* (**44**), 5–77.

García-Prada, O., Gothen, P. and Muñoz, V. (2007). Betti numbers of the moduli space of rank 3 parabolic Higgs bundles. *Memoirs Am. Math. Soc.* **187**(879), 80.

Gibbons, G. W. and Manton, N. S. (1995). The moduli space metric for well-separated BPS monopoles. *Phys. Lett. B* **356**(1), 32–8.

Gothen, P. B. (1994). The Betti numbers of the moduli space of stable rank 3 Higgs bundles on a Riemann surface. *Int. J. Math.* **5**(6), 861–75.

Green, J. A. (1955). The characters of the finite general linear groups. *Trans. Am. Math. Soc.* **80**, 402–47.

Gromov, M. (1991). Kähler hyperbolicity and L_2-Hodge theory. *J. Differential Geom.* **33**(1), 263–92.

Hausel, T. Arithmetic harmonic analysis, Macdonald polynomials and the topology of the Riemann-Hilbert monodromy map (in preparation).

Hausel, T. (1998). Compactification of moduli of Higgs bundles. *J. Reine Angew. Math.* **503**, 169–92.

Hausel, T. (1998). Vanishing of intersection numbers on the moduli space of Higgs bundles. *Adv. Theor. Math. Phys.* **2**(5), 1011–40.

Hausel, T. (2005). Mirror symmetry and Langlands duality in the non-abelian Hodge theory of a curve. In *Geometric Methods in Algebra and Number Theory*, Vol. 235 of *Progress in Mathematics*. pp. 193–217. Birkhäuser Boston, Boston, MA.

Hausel, T. (2006). Betti numbers of holomorphic symplectic quotients via arithmetic Fourier transform. *Proc. Natl. Acad. Sci. USA* **103**(16), 6120–4.

Hausel, T., Hunsicker, E. and Mazzeo, R. (2004). Hodge cohomology of gravitational instantons. *Duke Math. J.* **122**(3), 485–548.

Hausel, T., Letellier, E. and Rodriguez-Villegas, F. Arithmetic harmonic analysis on character and quiver varieties, preprint arXiv:0810.2076.

Hausel, T. and Rodriguez-Villegas, F. (2008). Mixed Hodge polynomials of character varieties. *Invent. Math.* **174**(3), 555–624. arXiv:math.AG/0612668.

Hausel, T. and Thaddeus, M. (2001). Examples of mirror partners arising from integrable systems. *C. R. Acad. Sci. Paris Sér. I Math.* **333**(4), 313–18.

Hausel, T. and Thaddeus, M. (2003). Mirror symmetry, Langlands duality, and the Hitchin system. *Invent. Math.* **153**(1), 197–229.

Hausel, T. and Thaddeus, M. (2003). Relations in the cohomology ring of the moduli space of rank 2 Higgs bundles. *J. Am. Math. Soc.* **16**(2), 303–27.

Hitchin, N. J. (1987*a*). *Monopoles, Minimal Surfaces and Algebraic Curves*, Vol. 105 of *Séminaire de Mathématiques Supérieures.* Presses de l'Université de Montréal, Montreal, QC.

Hitchin, N. J. (1987*b*). The self-duality equations on a Riemann surface. *Proc. London Math. Soc. (3)* **55**(1), 59–126.

Hitchin, N. J. (1987*c*). Stable bundles and integrable systems. *Duke Math. J.* **54**(1), 91–114.

Hitchin, N. J. (2000). L^2-cohomology of hyperkähler quotients. *Commun. Math. Phys.* **211**(1), 153–65.

Hitchin, N. J., Karlhede, A., Lindström, U. and Roček, M. (1987). Hyper-Kähler metrics and supersymmetry. *Commun. Math. Phys.* **108**(4), 535–89.

Hua, J. (2000). Counting representations of quivers over finite fields. *J. Algebra* **226**(2), 1011–33.

Jordan, H. E. (1907). Group-characters of various types of linear groups. *Am. J. Math.* **29**(4), 387–405.

Jost, J. and Zuo, K. (2000). Vanishing theorems for L^2-cohomology on infinite coverings of compact Kähler manifolds and applications in algebraic geometry. *Commun. Anal. Geom.* **8**(1), 1–30.

Joyce, D. D. (2000). *Compact Manifolds with Special Holonomy.* Oxford Mathematical Monographs. Oxford University Press, Oxford.

Kac, V. G. (1983). Root systems, representations of quivers and invariant theory. In *Invariant Theory (Montecatini, 1982)*, Vol. 996 of *Lecture Notes in Mathematics.* pp. 74–108. Springer, Berlin, Germany.

Kac, V. G. (1990). *Infinite-Dimensional Lie Algebras.* 3rd edition. Cambridge University Press, Cambridge.

Kapustin, A. and Witten, E. (2007). Electric-magnetic duality and the geometric Langlands program. *Commun. Number Theory Phys.* **1**(1), 1–236. arXiv:hep-th/0604151.

Kirwan, F. (1992). The cohomology rings of moduli spaces of bundles over Riemann surfaces. *J. Am. Math. Soc.* **5**(4), 853–906.

Kronheimer, P. B. (1989). The construction of ALE spaces as hyper-Kähler quotients. *J. Differential Geom.* **29**(3), 665–83.

Kronheimer, P. B. and Nakajima, H. (1990). Yang-Mills instantons on ALE gravitational instantons. *Math. Ann.* **288**(2), 263–307.

Montonen, C. and Olive, C. D. I. (1977). Magnetic monopoles as gauge particles? *Phys. Lett. B* **B72**, 117–20.

Mumford, D., Fogarty, J. and Kirwan, F. (1994). *Geometric Invariant Theory.* Vol. 34 of *Ergebnisse der Mathematik und ihrer Grenzgebiete (2)*, 3rd edition. Springer-Verlag, Berlin, Germany.

Nakajima, H. (1998). Quiver varieties and Kac-Moody algebras. *Duke Math. J.* **91**(3), 515–60.

Nitsure, N. (1991). Moduli space of semistable pairs on a curve. *Proc. London Math. Soc. (3)* **62**(2), 275–300.

Osborn, H. (1979). Topological charges for N=4 supersymmetric gauge theories and monopoles of spin 1. *Phys. Lett. B* **83**, 321.

Sasaki, T. $O(-2)$ blow-up formula via instanton calculus on $\widehat{\mathbb{C}^2/\mathbb{Z}_2}$ and Weil conjecture. arXiv:hep-th/0603162.

Schur, I. (1907). Untersuchungen über die darstellung der endlichen gruppen durch gebrochene lineare substitutionen. *J. Reine Angew. Math.* **132**, 85.

Segal, G. and Selby, A. (1996). The cohomology of the space of magnetic monopoles. *Commun. Math. Phys.* **177**(3), 775–87.

Sen, A. (1994). Dyon-monopole bound states, self-dual harmonic forms on the multi-monopole moduli space, and SL(2, Z) invariance in string theory. *Phys. Lett. B* **329**(2–3), 217–21.

Shoji, T. (2006). Lusztig's conjecture for finite special linear groups. *Represent. Theory* **10**, 164–222 (electronic).

Simpson, C. (1997). The Hodge filtration on nonabelian cohomology. In *Algebraic Geometry—Santa Cruz 1995*, Vol. 62 of *Proceedings of the Symposium on Pure Mathematics*. pp. 217–81. American Mathematical Society, Providence, RI.

Simpson, C. T. (1991). Nonabelian Hodge theory. In *Proceedings of the International Congress of Mathematicians, Vols. I, II (Kyoto, 1990)*, pp. 747–56. Mathematical Society Japan, Tokyo, Japan.

Strominger, A., Yau, S.-T. and Zaslow, E. (1996). Mirror symmetry is T-duality. *Nuclear Phys. B* **479**(1–2), 243–59.

Vafa, C. and Witten, E. (1994). A strong coupling test of S-duality. *Nuclear Phys. B* **431**(1–2), 3–77.

Witten, E. and Olive, D. (1978). Supersymmetry algebras that include topological charges. *Phys. Lett. B* **78**, 97.

XVII

NON-EMBEDDING AND NON-EXTENSION RESULTS IN SPECIAL HOLONOMY

Robert L. Bryant

Dedicated to Nigel Hitchin with great admiration on the occasion of his 60th birthday

17.1 Introduction

In the early analyses of metrics with special holonomy in dimensions 7 and 8, particularly in regards to their existence and generality, heavy use was made of the Cartan–Kähler theorem, essentially because the analyses were reduced to the study of overdetermined PDE systems whose natures were complicated by their diffeomorphism invariance. The Cartan–Kähler theory is well suited for the study of such systems and the local properties of their solutions. However, the Cartan–Kähler theory is not particularly well suited for studies of global problems for two reasons: first, it is an approach to PDE that relies entirely on the local solvability of initial value problems and, second, the Cartan–Kähler theory is only applicable in the real-analytic category.

Nevertheless, when there are no other adequate methods for analyzing the local generality of such systems, the Cartan–Kähler theory is a useful tool and it has the effect of focusing attention on the initial value problem as an interesting problem in its own right. The point of this chapter is to clarify some of the existence issues involved in applying the initial value problem to the problem of constructing metrics with special holonomy. In particular, the role of the assumption of real-analyticity will be discussed and examples will be constructed to show that one cannot generally avoid such assumptions in the initial value formulations of these problems.

The general approach can be outlined as follows: As is well known (cf. Bryant 1987), the problem of understanding the local Riemannian metrics in dimension n whose holonomy is contained in a specified connected group $H \subset SO(n)$ is essentially equivalent to the problem of understanding the H-structures in dimension n whose intrinsic torsion vanishes, or, equivalently, that are parallel with respect to the Levi-Civita connection of the Riemannian metric associated to the underlying $SO(n)$-structure. In this chapter, an n-manifold M endowed with an H-structure $B \to M$ with vanishing intrinsic torsion will be said to be an H*-manifold.*[1] In particular, an H-manifold is a manifold M endowed with an H-structure B that is flat to first order.

It frequently happens (as it does for all of the cases to be considered here) that H acts transitively on S^{n-1}, with stabilizer subgroup $K \subset H$, where

$$K = \big(\{1\}\times SO(n-1)\big) \cap H.$$

In this case, an oriented hypersurface $N \subset M$ in an H-manifold M inherits, in a natural way, a K-structure $B' \to N$. Typically, this K-structure will not, itself, be torsion-free (unless N is a totally geodesic hypersurface in M), but will satisfy some weaker condition on its intrinsic torsion, essentially that its intrinsic torsion can be expressed in terms of the second fundamental form of N as a submanifold of M.[2] The problem then becomes to determine whether these weaker conditions on a given K-structure $B' \to N^{n-1}$ are sufficient to imply that (N, B') can be induced by immersion into an H-manifold (M, B).

It turns out that the concept of real-analyticity has an important bearing on this question. As was pointed out in DeTurck and Kazdan (1981), a metric g on M can only be real-analytic with respect to at most one real-analytic structure on M: one considers the (non-maximal) atlas $\mathcal{H}(g)$ that consists of the g-harmonic coordinate charts. If the overlaps of $\mathcal{H}(g)$ are not real-analytic, then g is not real-analytic with respect to any real-analytic structure on M, while, if the overlaps of $\mathcal{H}(g)$ are real-analytic, it lies in a unique maximal real-analytic atlas $\mathcal{A}(g)$, that is, a real-analytic structure on M, and this is the only real-analytic structure on M with respect to which g could be real-analytic. Because $H \subset O(n)$, an H-structure $B \to M$ induces a canonical "underlying" metric g_B on M. In particular, if the H-structure $B \to M$ is real-analytic with respect to some real-analytic structure $\mathcal{A}$, then $\mathcal{H}(g_B)$ must be a real-analytic atlas and $\mathcal{A}$ must equal $\mathcal{A}(g_B)$. Thus, it makes sense to say that an H-structure is real-analytic, meaning that $\mathcal{H}(g_B)$ is a real-analytic atlas *and* that B is real-analytic with respect to the underlying real-analytic structure $\mathcal{A}(g_B)$.

In the three cases to be considered in this chapter, in which H is one of $SU(2) \subset SO(4)$, $G_2 \subset SO(7)$, or $Spin(7) \subset SO(8)$, it will be shown that the weaker intrinsic torsion conditions on a hypersurface structure *are* sufficient to induce an embedding (in fact, essentially a unique one) *provided that the given* K*-structure is real-analytic.* It will also be shown, in each case, that there are K-structures $B' \to N^{n-1}$ that satisfy these weaker intrinsic torsion conditions that, nevertheless, *cannot* be induced by an immersion into an H-manifold. Of course, such structures are not real-analytic.

[1] While, strictly speaking, an H-manifold is a pair (M, B), it is common to refer to M as an H-manifold when the torsion-free H-structure B can be inferred from context.

[2] This observation that the intrinsic torsion of the induced K-structure on N can be expressed in terms of the second fundamental form can be found, for example, in Conti and Salamon (2007).

The existence of the desired embedding in the analytic case is a consequence of the Cartan–Kähler theorem and, for the cases of $G_2 \subset SO(7)$ and $Spin(7) \subset SO(8)$, is implicit in the original analyses of G_2 and $Spin(7)$ structures to be found in Bryant (1987). The case of $SU(2) \subset SO(4)$ appears not to have been treated in this manner before and will be discussed in detail in this chapter.[3]

The examples constructed below, showing that existence can fail when one does not have real-analyticity, appear to be new.

There is another interpretation of the initial value problem that has been considered by a number of authors, in particular, Hitchin (2001): The idea of a "flow" of K-structures that gives rise to a torsion-free H-structure. Let M be a connected n-manifold endowed with an H-structure and let g be the underlying metric. Let $N \subset M$ be an embedded, normally oriented hypersurface, and let $r : M \to [0,\infty)$ be the distance (in the metric g) from the hypersurface N. As is well known, there is an open neighborhood $U \subset M$ of N on which there is a smooth function $t : U \to \mathbb{R}$ satisfying $|t| = r$ and $dt \neq 0$ on$U$ as well as the condition that its gradient along N is the specified oriented normal. There is then a well-defined smooth embedding $(t, f) : U \to \mathbb{R} \times N$, where, for $p \in U$, the function $f(p)$ is the closest point of N to p. In this way, U can be identified with an open neighborhood of $\{0\} \times N$ in $\mathbb{R} \times N$ and, in particular, each level set of t in U can be identified with an open subset of N. (When N is compact, there will be an $\epsilon > 0$ such that the level sets $t = c$ for $|c| < \epsilon$ will be diffeomorphic to N.) Thus, at least locally, one can think of the H-structure on U as a one-parameter family of K-structures on N. When one imposes the condition that the H-structure on $U \subset \mathbb{R} \times N$ be torsion-free, this can be expressed as a first-order initial value problem with the given K-structure on $\{0\} \times N$ as the initial value. This first-order initial value problem is sometimes described as a "flow," but this can be misleading, especially if it causes one to think in terms of parabolic or hyperbolic PDE.

Indeed, the character of this PDE problem is more like that of trying to use the Cauchy–Riemann equations $u_x = v_y$ and $v_x = -u_y$ to extend a complex-valued function defined on the imaginary axis $x = 0$ to a holomorphic function on a neighborhood of the imaginary axis. One knows that a necessary and sufficient condition for being able to do this is that the given function on the imaginary axis must be real-analytic. In the cases to be considered in this chapter, the requirements are not this strong since there will be cases in which the initial K-structure is not real-analytic and yet a solution of the initial value problem will exist. However, as will be seen, real-analyticity is a sufficient condition.

For background on the use of exterior differential systems in this chapter, the reader might consult Bryant *et al.* (1991).

I have included the case of $SU(2)$-structures on four-manifolds because of its historical interest (it was the first case of special holonomy to be analyzed) and

[3] These three cases do not exhaust the possibilities; for the example of $SU(3) \subset \mathrm{SO}(6)$, see Conti and Salamon (2007).

because the algebra is simpler. Also, because other approaches, based on the existence of local holomorphic coordinates, have been employed in this case, there is an instructive comparison to be made between those methods and the Cartan–Kähler approach. For this reason, I go into the $SU(2)$-case in some detail. I hope that the reader will find this as interesting as I have.

17.2 Beginnings

17.2.1 *Holonomy*

Let (M^n, g) be a connected Riemannian n-manifold.

17.2.1.1 *Parallel transport*

To g, one associates its Levi-Civita connection ∇, which defines, for a piecewise-C^1 curve $\gamma : [0,1] \to M$, a parallel transport

$$P^{\nabla}_{\gamma} : T_{\gamma(0)}M \to T_{\gamma(1)}M, \tag{17.1}$$

which is a linear g-isometry between the two tangent spaces.

17.2.1.2 *Group structure*

In Schouten (1918), he considered the set

$$H_x = \left\{ P^{\nabla}_{\gamma} \mid \gamma(0) = \gamma(1) = x \right\} \subseteq O(T_xM) \tag{17.2}$$

and called its dimension the number of *degrees of freedom* of g.

It is easy to establish the identities

$$P^{\nabla}_{\bar{\gamma}} = \left(P^{\nabla}_{\gamma}\right)^{-1} \qquad \text{and} \qquad P^{\nabla}_{\gamma_2 * \gamma_1} = P^{\nabla}_{\gamma_2} \circ P^{\nabla}_{\gamma_1} \tag{17.3}$$

where $\bar{\gamma}$ is the reverse of γ and $\gamma_2 * \gamma_1$ is the concatenation of paths γ_1 and γ_2 satisfying $\gamma_1(1) = \gamma_2(0)$.

Consequently, $H_x \subset O(T_xM)$ is a subgroup and

$$H_{\gamma(1)} = P^{\nabla}_{\gamma} \; H_{\gamma(0)} \; \left(P^{\nabla}_{\gamma}\right)^{-1}. \tag{17.4}$$

In particular, fixing a linear isometry $u : T_xM \to \mathbb{R}^n$, the conjugacy class of $H_u = uH_xu^{-1}$ in $O(n)$ is well-defined, independent of the choice of $x \in M$ or the isometry $u : T_xM \to \mathbb{R}^n$. By abuse of terminology, we say that H is the *holonomy* of the metric g if $H \subset O(n)$ is a group conjugate to some (and hence any) of the groups H_u.

For later reference, if $u : T_xM \to \mathbb{R}^n$ is fixed, we let

$$B_u = \left\{ u \circ P^{\nabla}_{\gamma} \mid \gamma(1) = x \right\}. \tag{17.5}$$

This B_u is an H_u-subbundle of the orthonormal coframe bundle of g, that is, it is an H_u-structure on M. By its very construction, it is invariant under ∇-parallel translation and, since ∇ is torsion-free, it follows that this H_u-structure admits a torsion-free compatible connection.

Conversely, let $H \subset O(n)$ be a subgroup and let $B \to M$ be an H-structure on M. If $B \to M$ admits a compatible, torsion-free connection ∇, then B is said to be *torsion-free.* In this case, ∇ (necessarily unique) is the Levi-Civita connection of the underlying metric g on M and B is invariant under ∇-parallel translation, implying that $H_u \subset H$ for all $u \in B$. Thus, finding torsion-free H-structures on M provides a way to find metrics on M whose holonomy lies in H.

In this chapter, I will use the term *H-manifold* to denote an n-manifold M^n endowed with a torsion-free H-structure $B \to M$. (The intended embedding $H \subset O(n)$ is to be understood from context.)

17.2.1.3 *Cartan's early results*

In Cartan (1925), he made the following statements:

1. H_x is a Lie subgroup of $O(T_xM)$, connected if M is simply connected.
2. If H_x acts reducibly on T_xM, then g is locally a product metric.

Cartan's first statement was eventually proved (to modern standards of rigor) by Borel and Lichnerowicz (1952) and the second statement was globalized and proved by G. de Rham.

17.2.2 *A non-trivial case*

In dimensions 2 and 3, the above facts suffice to determine the possible holonomy groups of simply connected manifolds.

Cartan (1925)[4] studied the first non-trivial case, namely, $n = 4$ and $H_x \simeq SU(2)$, and observed that such metrics g

1. Have vanishing Ricci tensor
2. Are what we now call "self-dual"
3. Locally (modulo diffeomorphism) depend on two functions of three variables

While the first two observations are matters of calculation and/or definition, the third observation is non-trivial. However, Cartan gave no indication of his proof and, to my knowledge, never returned to this example again.

While I cannot be sure, I believe that it is likely that Cartan had a proof in mind along the following lines.[5]

The associated $SU(2)$-structure $B \to M$ of such a metric g satisfies structure equations of the form

$$
\begin{aligned}
\mathrm{d}\omega &= -\,\theta \wedge \omega, \\
\mathrm{d}\theta &= -\,\theta \wedge \theta + R(\omega \wedge \omega), \\
\mathrm{d}R &= -\,\theta.R + R'(\omega),
\end{aligned}
\tag{17.6}
$$

[4] Especially note chapter VII, section II.

[5] He had developed all of the tools necessary for this proof in his famous series of papers on pseudo-groups and the equivalence problem and, using those results, it would have been a simple observation for him.

where

1. The tautological one-form ω takes values in $\mathbb{R}^4$
2. The connection one-form θ takes values in $\mathfrak{su}(2) \subset \mathfrak{so}(4)$
3. The curvature function R takes values in W_4, the five-dimensional (real) irreducible representation of $SU(2)$ that lies in $\mathrm{Hom}\big(\Lambda^2(\mathbb{R}^4), \mathfrak{su}(2)\big)$
4. The derived curvature function R' takes values in V_5, the six-dimensional complex irreducible representation of $SU(2)$ that lies in $\mathrm{Hom}(\mathbb{R}^4, W_4)$

Calculation shows that the subspace V_5 is an involutive tableau in $\mathrm{Hom}(\mathbb{R}^4, W_4)$, with character sequence $(s_1, s_2, s_3, s_4) = (5, 5, 2, 0)$. Since the last non-zero character of this tableau is $s_3 = 2$, Cartan's generalization of the third fundamental theorem of Lie applies to the structure equation (17.6) to yield his third observation above.

17.2.3 *Hyper-Kähler viewpoint*

Riemannian manifolds (M^4, g) with

$$H_x \simeq SU(2) \subset SO(4)$$

are nowadays said to be *hyper-Kähler* and we understand them as special cases of Kähler manifolds. In fact, using our understanding of complex and Kähler geometry, we now arrive at Cartan's result by a somewhat different route.

Because the subgroup $SU(2) \subset SO(4)$ acts trivially on the space of self-dual two-forms on $\mathbb{R}^4$, when (M^4, g) has $H_x \simeq SU(2)$, then there exist three g-parallel (and, hence, closed) self-dual two-forms on M, say Υ_1, Υ_2, and Υ_3, such that

$$\Upsilon_i \wedge \Upsilon_j = 2\delta_{ij}\, \mathrm{d}V_g\,. \tag{17.7}$$

Conversely, a triple $\Upsilon = (\Upsilon_1, \Upsilon_2, \Upsilon_3)$ of closed two-forms on M that satisfies

$$\Upsilon_1{}^2 = \Upsilon_2{}^2 = \Upsilon_3{}^2 \neq 0 \quad \text{while} \quad \Upsilon_2 \wedge \Upsilon_3 = \Upsilon_3 \wedge \Upsilon_1 = \Upsilon_1 \wedge \Upsilon_2 = 0 \tag{17.8}$$

is easily shown to be g-parallel and self-dual with respect to a unique metric g on M for which (17.7) holds.

Suppose given a triple $\Upsilon = (\Upsilon_1, \Upsilon_2, \Upsilon_3)$ of two-forms on M satisfying (17.8). The complex-valued two-form $\Omega_1 = \Upsilon_2 + \mathrm{i}\,\Upsilon_3$ satisfies $\Omega_1{}^2 = 0$ because of (17.8) and hence is of type $(2,0)$ with respect to a unique almost-complex structure J_1 on M. Again appealing to (17.8), one sees that $\Upsilon_1 \wedge \Omega_1 = 0$, implying that the two-form Υ_1 is of type $(1,1)$ with respect to J_1. Moreover, since $\Upsilon_1^2 = \frac{1}{2}\big(\Omega_1 \wedge \overline{\Omega_1}\big)$, the two-form Υ_1 is either J_1-positive or J_1-negative. In the former case, one says that Υ satisfying (17.8) is *positive*, in the latter case, one says that Υ is *negative*. Clearly, by replacing Υ_1 by $-\Upsilon_1$, one can convert any positive Υ into a negative Υ and *vice versa*.

Thus, a positive triple Υ defines a unique $SU(2)$-structure $B_\Upsilon \to M$ for which the component forms Υ_i are a positive, orthonormal basis for the self-dual two-forms preserved by the $SU(2)$-structure.

Now, given a positive triple $(\Upsilon_1, \Upsilon_2, \Upsilon_3)$ of *closed* two-forms on M satisfying (17.8), one can prove (using the Newlander–Nirenberg theorem) that each point of M lies in a local coordinate chart $z = (z^1, z^2) : U \to \mathbb{C}^2$ for which there exists a real-analytic function $\phi : z(U) \to \mathbb{R}$ so that

$$\Upsilon_2 + \mathrm{i}\,\Upsilon_3 = \mathrm{d}z^1 \wedge \mathrm{d}z^2 \quad \text{and} \quad \Upsilon_1 = \tfrac{1}{2}\mathrm{i}\,\partial\bar{\partial}\phi, \tag{17.9}$$

where ϕ satisfies the elliptic Monge–Ampère equation

$$\det\left(\frac{\partial^2\phi}{\partial z^i \partial \bar{z}^j}\right) = 1 \tag{17.10}$$

and the strict pseudoconvexity condition with respect to z given by

$$\left(\frac{\partial^2\phi}{\partial z^i \partial \bar{z}^j}\right) > 0. \tag{17.11}$$

In fact, if one fixes an open set $U \subset \mathbb{C}^2$ and chooses a smooth, strictly pseudoconvex function $\phi : U \to \mathbb{R}$ satisfying (17.10), then the formulae (17.9) define a positive triple $(\Upsilon_1, \Upsilon_2, \Upsilon_3)$ on U consisting of g_ϕ-parallel, self-dual two-forms where

$$g_\phi = \frac{\partial^2\phi}{\partial z^i \partial \bar{z}^j}\,\mathrm{d}z^i \circ \mathrm{d}\bar{z}^j. \tag{17.12}$$

Thus, the holonomy of g_ϕ is (conjugate to) a subgroup of $SU(2)$ and it is not difficult to show that, for a generic strictly pseudoconvex ϕ satisfying (17.10), the holonomy of g_ϕ is (conjugate to) the full $SU(2)$.

Note that, because (17.10) is an analytic elliptic PDE at a strictly pseudoconvex solution ϕ, all of its strictly pseudoconvex solutions are real-analytic. In particular, a Riemannian metric g on M^4 with holonomy $SU(2)$ is real-analytic in harmonic coordinates (as would have followed anyway from its Ricci-flatness and a result of DeTurck and Kazdan 1981).

Now, one does not normally think of solving an elliptic PDE by an initial value problem, but, of course, in the analytic category, there is nothing wrong with such a procedure. Indeed, the Cauchy–Kovalewskaya theorem implies that, in this particular case, one can specify ϕ and its normal derivative along a hypersurface, say, $\mathrm{Im}(z^2) = 0$, as essentially arbitrary real-analytic functions (subject only to an open condition that guarantees the strict pseudoconvexity of the resulting solution) and thereby determine a unique strictly pseudoconvex solution of (17.10). Thus, in this sense, one sees, again, that the "general" metric with holonomy in $SU(2)$ modulo diffeomorphism depends on two functions of three variables, in agreement with Cartan's claim.

17.3 Hyper-Kähler four-manifolds

17.3.1 *An exterior differential systems proof*

One can use the characterization of $SU(2)$ as the stabilizer of three two-forms in dimension 4 as the basis of another analysis of the existence problem via an EDS (exterior differential system).[6]

Let M^4 be an analytic manifold and let $\boldsymbol{\Upsilon}$ be the tautological two-form on $\Lambda^2(T^*M)$. Let

$$X^{17} \subset \left(\Lambda^2(T^*M)\right)^3 \tag{17.13}$$

be the submanifold consisting of positive triples $(\beta_1, \beta_2, \beta_3) \in \Lambda^2(T_x^*M)$ such that

$${\beta_1}^2 = {\beta_2}^2 = {\beta_3}^2 \neq 0 \quad \text{and} \quad \beta_1 \wedge \beta_2 = \beta_3 \wedge \beta_1 = \beta_2 \wedge \beta_3 = 0. \tag{17.14}$$

Let $\pi_i : X \to \Lambda^2(T^*M)$ for $1 \le i \le 3$ denote the projections onto the three factors. The pullbacks $\boldsymbol{\Upsilon}_i = \pi_i^*(\boldsymbol{\Upsilon})$ define an EDS on X

$$\mathcal{I} = \{\mathrm{d}\boldsymbol{\Upsilon}_1, \mathrm{d}\boldsymbol{\Upsilon}_2, \mathrm{d}\boldsymbol{\Upsilon}_3\}. \tag{17.15}$$

The basepoint projection $\pi : X \to M$ makes X into a bundle over M whose fibers are diffeomorphic to the thirteen-dimensional homogeneous space $GL(4,\mathbb{R})/SU(2)$ and whose sections $\sigma : M \to X$ correspond to the $SU(2)$-structures on M.

An $\mathcal{I}$-integral manifold $Y^4 \subset X$ transverse to $\pi : X \to M$ then represents a choice of three closed two-forms Υ_i on an open subset $U \subset M$ that satisfy the algebra conditions needed to define an $SU(2)$-structure on U.

Calculation shows that $\mathcal{I}$ is involutive, with character sequence

$$(s_0, s_1, s_2, s_3, s_4) = (0, 0, 3, 6, 4). \tag{17.16}$$

In particular, a three-dimensional real-analytic $\mathcal{I}$-integral manifold $P \subset X$ that is transverse to the fibers of π can be "thickened" to a four-dimensional integral manifold that is transverse to the fibers of π. This "thickening" will not be unique, however, because of the invariance of the ideal $\mathcal{I}$ under the obvious action induced by the diffeomorphisms of M.

17.3.2 *A sharper result*

Suppose that (M^4, g) has holonomy $SU(2)$ and let Υ_i be three g-parallel two-forms on M satisfying

$$\Upsilon_i \wedge \Upsilon_j = 2\delta_{ij}\, \mathrm{d}V_g \tag{17.17}$$

such that $\Upsilon = (\Upsilon_1, \Upsilon_2, \Upsilon_3)$ is positive.

[6] The argument to be given in this subsection for $SU(2) \subset SO(4)$ is the analog of the arguments given in Bryant (1987) for the cases $G_2 \subset SO(7)$ and $\mathrm{Spin}(7) \subset SO(8)$. I omit the calculations in this case, since they are straightforward.

If $N^3 \subset M$ is an oriented hypersurface, with oriented normal $\mathbf{n}$, then there is a coframing η of N defined by

$$\eta = \begin{pmatrix} \eta_1 \\ \eta_2 \\ \eta_3 \end{pmatrix} = \begin{pmatrix} \mathbf{n} \lrcorner \Upsilon_1 \\ \mathbf{n} \lrcorner \Upsilon_2 \\ \mathbf{n} \lrcorner \Upsilon_3 \end{pmatrix} \tag{17.18}$$

and (because of the positivity condition) it satisfies

$$N^* \begin{pmatrix} \Upsilon_1 \\ \Upsilon_2 \\ \Upsilon_3 \end{pmatrix} = \begin{pmatrix} \eta_2{\wedge}\eta_3 \\ \eta_3{\wedge}\eta_1 \\ \eta_1{\wedge}\eta_2 \end{pmatrix} = *_\eta \eta. \tag{17.19}$$

where $*_\eta$ is the Hodge star associated to the metric $g_\eta = {\eta_1}^2 + {\eta_2}^2 + {\eta_3}^2$ and orientation $\eta_1{\wedge}\eta_2{\wedge}\eta_3 > 0$.

In particular, note that the coframing η is not arbitrary, but satisfies the system of three first-order PDE:

$$\mathrm{d}(*_\eta \eta) = N^* \mathrm{d}\Upsilon = 0. \tag{17.20}$$

An alternative manifestation of the involutivity of the system $\mathcal{I}$ that is better adapted to the initial value problem then becomes the following existence and uniqueness result:

Theorem 17.1 *Let η be a real-analytic coframing of N such that $\mathrm{d}(*_\eta \eta) = 0$. There exists an essentially unique embedding of N into a $SU(2)$-holonomy manifold (M^4, g) that induces the given coframing η in the above manner.*

Remark 17.1 (Essential uniqueness) The meaning of this term is as follows: If N can be embedded into two different $SU(2)$-manifolds (M, Υ) and $(\bar{M}, \bar{\Upsilon})$ in such a way that both embeddings induce the same coframing η on N by the above pullback formula, then there are open neighborhoods $U \subset M$ and $\bar{U} \subset \bar{M}$ of the respective images of N in the two manifolds and a diffeomorphism $f : U \to \bar{U}$ that is the identity on the image of N and that satisfies $f^*\bar{\Upsilon} = \Upsilon$.

Proof. Write $\mathrm{d}\eta = -\theta{\wedge}\eta$ where $\theta = -{}^t\theta$. (This θ exists and is unique by the fundamental lemma of Riemannian geometry.)

On $N \times GL(3, \mathbb{R})$ define[7]

$$\boldsymbol{\omega} = g^{-1}\eta \quad \text{and} \quad \gamma = g^{-1}\mathrm{d}g + g^{-1}\theta g, \tag{17.21}$$

so that $\mathrm{d}\boldsymbol{\omega} = -\gamma{\wedge}\boldsymbol{\omega}$. On $X = \mathbb{R} \times N \times GL(3, \mathbb{R})$ define the three two-forms

$$\begin{pmatrix} \boldsymbol{\Upsilon}_1 \\ \boldsymbol{\Upsilon}_2 \\ \boldsymbol{\Upsilon}_3 \end{pmatrix} = \begin{pmatrix} \mathrm{d}t{\wedge}\boldsymbol{\omega}_1 + \boldsymbol{\omega}_2{\wedge}\boldsymbol{\omega}_3 \\ \mathrm{d}t{\wedge}\boldsymbol{\omega}_2 + \boldsymbol{\omega}_3{\wedge}\boldsymbol{\omega}_1 \\ \mathrm{d}t{\wedge}\boldsymbol{\omega}_3 + \boldsymbol{\omega}_1{\wedge}\boldsymbol{\omega}_2 \end{pmatrix} = \mathrm{d}t \wedge \boldsymbol{\omega} + *_\omega \boldsymbol{\omega}. \tag{17.22}$$

[7] In the following formulae, I regard $g : N \times GL(3, \mathbb{R}) \to GL(3, \mathbb{R})$ as the projection onto the second factor. Also, to save writing, I will write η for $\pi_1^*\eta$ where $\pi_1 : N \times GL(3, \mathbb{R}) \to N$ is the projection onto the first factor.

Let $\mathcal{I}$ be the ideal on X generated by $\{d\boldsymbol{\Upsilon}_1, d\boldsymbol{\Upsilon}_2, d\boldsymbol{\Upsilon}_3\}$. One calculates that

$$d\boldsymbol{\Upsilon} = \left({}^t\gamma - (\operatorname{tr}\gamma)I_3\right) \wedge *_\omega \boldsymbol{\omega} + \gamma \wedge \boldsymbol{\omega} \wedge dt. \tag{17.23}$$

Consequently, $\mathcal{I}$ is involutive, with characters $(s_0, s_1, s_2, s_3, s_4) = (0, 0, 3, 6, 0)$.

Since $d(*_\eta \eta) = 0$, the locus $L = \{0\} \times N \times \{I_3\} \subset X$ is a regular, real-analytic integral manifold of the real-analytic ideal $\mathcal{I}$. (Note that L is just a copy of N.)

By the Cartan–Kähler theorem, L lies in a unique four-dimensional $\mathcal{I}$-integral manifold $M \subset X$. The $\boldsymbol{\Upsilon}_i$ thus pull back to M to be closed and to define the desired $SU(2)$-structure forms Υ_i on M inducing η on $L = N$. □

It is natural to ask whether it is necessary to assume that η be real-analytic for the conclusion of Theorem 17.1. The following result shows that one cannot weaken this assumption to "smooth" and still get the same conclusion:

Theorem 17.2 *If η is a coframing on N^3 that is not real-analytic in any local coordinate system and*

$$d(*_\eta \eta) = 0 \quad \textit{and} \quad *_\eta({}^t\eta \wedge d\eta) = 2C \tag{17.24}$$

for some constant C, then η cannot be induced by immersing N into an $SU(2)$-manifold.

Smooth-but-not-real-analytic coframings η satisfying (17.24) *do exist locally.*

Proof. Suppose that $\Upsilon = (\Upsilon_1, \Upsilon_2, \Upsilon_3)$ is a positive triple of parallel, self-dual two-forms on an (M^4, g) with holonomy $SU(2)$ and let $N^3 \subset M$ be an oriented hypersurface.

Calculation yields that the induced co-closed coframing η on N satisfies

$$*_\eta({}^t\eta \wedge d\eta) = 6H \tag{17.25}$$

where H is the mean curvature of N in M.

Now, by DeTurck and Kazdan (1981), since g is Ricci-flat, it is real-analytic in g-harmonic coordinates. In particular, such coordinate systems can be used to define a real-analytic structure on M, which is the one that will be meant henceforth. In particular, since the forms Υ_i are g-parallel, they, too, are real-analytic with respect to this structure. If H is constant, then elliptic regularity implies that N must be a real-analytic hypersurface in M and hence η must also be real-analytic.

Thus, if η is a non-real-analytic coframing on N^3 that satisfies (17.24) for some constant C, then η cannot be induced on N by an embedding into an $SU(2)$-manifold.

To finish the proof, I will now show how to construct a coframing η on an open subset of $\mathbb{R}^3$ that is not real-analytic in any coordinate system and yet satisfies (17.24).

To begin, note that, if a coframing η on N^3 is real-analytic in any coordinate system at all, it will be real-analytic in η-*harmonic* coordinates, that is, local coordinates $x : U \to \mathbb{R}^3$ satisfying

$$\mathrm{d}\big(*_\eta \mathrm{d}x\big) = 0. \tag{17.26}$$

Now, fix a constant C and consider a coframing $\eta = h(x)^{-1}\,\mathrm{d}x$ on $U \subset \mathbb{R}^3$ where $h : U \to GL(3,\mathbb{R})$ is a mapping satisfying the first-order, quasi-linear system:

$$\mathrm{d}(*_\eta \eta) = 0, \quad *_\eta\big({}^t\eta \wedge \mathrm{d}\eta\big) = 2C, \quad \text{and} \quad \mathrm{d}\big(*_\eta \mathrm{d}x\big) = 0. \tag{17.27}$$

The system (17.27) consists of seven equations for the nine unknown entries of h.

Calculation shows this first-order system to be underdetermined elliptic (i.e. its symbol is surjective at every real covector ξ). By standard theory, it has smooth local solutions that are not real-analytic.

Taking a non-real-analytic solution h, the resulting η will not be real-analytic in the x-coordinates, which, by construction, are η-harmonic. Thus, such an η is not real-analytic in any local coordinate system. □

Remark 17.2 ("Flow" interpretation) The condition $\mathrm{d}(\mathrm{d}t{\wedge}\omega + *\omega) = 0$ has sometimes been described as an "$SU(2)$-flow" on coframings of N. In fact, this closure condition can be written in the "evolutionary" form

$$\frac{\mathrm{d}}{\mathrm{d}t}\,\omega = *_\omega(\mathrm{d}\omega) - \tfrac{1}{2}\,*_\omega\big({}^t\omega \wedge \mathrm{d}\omega\big)\,\omega. \tag{17.28}$$

By Theorem 17.1, if η on N^3 is real-analytic and satisfies $\mathrm{d}(*_\eta\eta) = 0$, then (17.28) has a solution in a neighborhood of $t = 0$ in $\mathbb{R} \times N$ that satisfies the initial condition

$$\omega_{|t=0} = \eta. \tag{17.29}$$

One does not normally think of evolution equations as having to have real-analytic initial data. However, Theorem 17.2 shows that some such regularity assumption must be made.[8]

Remark 17.3 (Co-closed coframings with specified metric) Given a Riemannian three-manifold (N, g), it is an interesting question as to when there exist (either locally or globally) a g-orthonormal coframing $\eta = (\eta_1, \eta_2, \eta_3)$ that is co-closed.[9]

One can formulate this as an EDS for the section $\eta : N \to B$ of the g-orthonormal coframe bundle $B \to N$. The natural EDS for this is not involutive, but a slight extension is and is worth describing here.

[8] On the other hand, the reader should be aware that real-analyticity, while sufficient, is certainly not necessary for the initial-value problem to have a (local) solution. For example, it suffices to let N be a smooth hypersurface in an $SU(2)$-manifold M^4 that is not real-analytic with respect to the real-analytic structure provided by the holomorphic coordinates of the underlying integrable complex structure. The induced co-closed coframing on N will then be non-real-analytic initial data for which the initial-value problem will have a (local) solution.

[9] Of course, *closed* g-orthonormal coframing exist locally if and only if g is flat.

Let ω_i and $\omega_{ij} = -\omega_{ji}$ (where $1 \le i, j \le 3$) be the tautological and Levi-Civita connection forms on B. In particular, g pulls back to B to be ${\omega_1}^2 + {\omega_2}^2 + {\omega_3}^2$ and these forms satisfy the structure equations:

$$d\omega_i = -\omega_{ij} \wedge \omega_j \quad \text{and} \quad d\omega_{ij} = -\omega_{ik} \wedge \omega_{kj} + \tfrac{1}{2} R_{ijkl}\, \omega_k \wedge \omega_l\,. \tag{17.30}$$

Because of their tautological reproducing property, one has $\eta^*(\omega_i) = \eta_i$. Consequently, the g-orthonormal coframing η is co-closed if and only if, when regarded as a section $\eta : N \to B$, it is an integral manifold of the EDS $\mathcal{I}_0$ generated by the three closed three-forms $\Upsilon_{ij} = d(\omega_i \wedge \omega_j)$ with $1 \le i < j \le 3$. Computation shows that this ideal has $(s_1, s_2, s_3) = (0, 2, 1)$, while an integral manifold on which $\Omega = \omega_1 \wedge \omega_2 \wedge \omega_3$ is non-vanishing (an obvious requirement for a section $\eta(M) \subset B$) must satisfy

$$\eta^* \omega_{ij} = \epsilon_{ijk} S_{kl}\, \eta^* \omega_l \tag{17.31}$$

where ϵ_{ijk} is fully antisymmetric in its indices, with $\epsilon_{123} = 1$, and $S_{kl} = S_{lk}$. In particular, the space of admissible integral elements of $(\mathcal{I}_0, \Omega)$ at each point of B has dimension $6 < s_1 + 2s_2 + 3s_3$. Hence, the system $(\mathcal{I}_0, \Omega)$ is not involutive.

However, (17.31) show that any integral of $(\mathcal{I}_0, \Omega)$ is also an integral of the two-form

$$\Upsilon = \omega_1 \wedge \omega_{23} + \omega_2 \wedge \omega_{31} + \omega_3 \wedge \omega_{12} = \tfrac{1}{2} \epsilon_{ijk}\, \omega_i \wedge \omega_{jk}\,. \tag{17.32}$$

Moreover, since

$$d(\omega_i \wedge \omega_j) = \epsilon_{ijk}\, \omega_k \wedge \Upsilon, \tag{17.33}$$

the differential ideal $\mathcal{I}$ generated by Υ contains $\mathcal{I}_0$. Calculation using (17.30) yields

$$2\, d\Upsilon = \epsilon_{ijk}\, \omega_i \wedge \omega_{lj} \wedge \omega_{lk} - R\,\Omega \tag{17.34}$$

where R is the scalar curvature of the metric g. Thus, the ideal $\mathcal{I}$ is generated algebraically by Υ and $d\Upsilon$. An integral element of $(\mathcal{I}, \Omega)$ is now cut out by equations of the form

$$\omega_{ij} - \epsilon_{ijk}\, S_{kl}\, \omega_l = 0 \tag{17.35}$$

where $S = (S_{kl})$ is a 3×3 symmetric matrix satisfying $\sigma_2(S) = \frac{1}{2} R$ (where $\sigma_2(S)$ is the second elementary function of the eigenvalues of S). The characteristic variety of such an integral element consists of the null (co-)vectors of the quadratic form

$$Q_S = \mathrm{tr}(S)\, g - S_{ij}\, \omega_i\, \omega_j\,. \tag{17.36}$$

In particular, except for the case $S = 0$ (which can only occur when $R = 0$), this integral element is Cartan-regular, with Cartan character sequence $(s_0, s_1, s_2, s_3) = (0, 1, 2, 0)$. In particular, the system $(\mathcal{I}, \Omega)$ is involutive, and, in the real-analytic case, the general integral depends on two functions of two variables.

Note, by the way, that when Q_S is positive (or negative) definite, the linearization around such a solution is elliptic and hence such co-closed coframings are as regular as the metric g. In particular, this always happens when $R > 0$, that is, when the scalar curvature is positive.[10] Thus, for a real-analytic metric with positive scalar curvature, all of its co-closed coframings are real-analytic.

17.4 G₂-Manifolds

For background on the group $G_2 \subset SO(7)$ and G_2-manifolds, the reader can consult Bryant (1987), Salamon (1989), Joyce (2000), and Bryant (2006). I will generally follow the notation in Bryant (2006).

The crucial point is that the group G_2 can be defined as the stabilizer in $GL(7,\mathbb{R})$ of the three-form

$$\phi = e^{123} + e^{145} + e^{167} + e^{246} - e^{257} - e^{347} - e^{356}. \tag{17.37}$$

where $e^{ijk} = e^i\wedge e^j\wedge e^k$ and the e^i are a basis of linear forms on $\mathbb{R}^7$. The Lie group G_2 is connected, has dimension 14, preserves the metric and orientation for which the e^i are an oriented orthonormal basis, and acts transitively on the unit sphere $S^6 \subset \mathbb{R}^7$. The G_2-stabilizer of e^1 is the subgroup $SU(3) \subset SO(6)$ that preserves the two-form $e^{23} + e^{45} + e^{67}$ and the three-form

$$e^{246} - e^{257} - e^{347} - e^{356} = \mathrm{Re}\big((e^2 + \mathrm{i}\, e^3) \wedge (e^4 + \mathrm{i}\, e^5) \wedge (e^6 + \mathrm{i}\, e^7)\big). \tag{17.38}$$

The $GL(7,\mathbb{R})$-orbit of ϕ in $\Lambda^3(\mathbb{R}^7)$ is an open (but not convex) cone $\Lambda^3_+(\mathbb{R}^7) \subset \Lambda^3(\mathbb{R}^7)$ that consists precisely of the three-forms on $\mathbb{R}^7$ whose stabilizers are isomorphic to G_2. Consequently, for any smooth seven-manifold M^7, there is a well-defined open subset $\Lambda^3_+(T^*M) \subset \Lambda^3(T^*M)$ consisting of the three-forms whose stabilizers are isomorphic to G_2.

A G_2-structure on M is thus specified by a three-form σ on M with the property that σ_x lies in $\Lambda^3_+(T^*_xM)$ for all $x \in M$. Such a three-form σ will be said to be *definite*. Explicitly, the G_2-structure $B = B_\sigma$ consists of the linear isomorphisms $u : T_x \to \mathbb{R}^7$ that satisfy $u^*\phi = \sigma_x$. Conversely, given a G_2-structure $B \to M$, there is a unique definite three-form σ on M such that $B = B_\sigma$. Put another way, the G_2-structure bundle $\mathcal{F}(M)/G_2$ is naturally identified with $\Lambda^3_+(M)$ by identifying $[u] = u \cdot G_2$ with $u^*\phi \in \Lambda^3_+(T^*_{\pi(u)}M)$ for all $u \in \mathcal{F}(M)$.

A $\sigma \in \Omega^3_+(M)$ determines a unique metric g_σ and orientation $*_\sigma$ by requiring that the corresponding coframings $u \in B_\sigma$ be oriented isometries.

[10] It is interesting to note that, when $R > 0$, the ideal $\mathcal{I}$ on the six-manifold B is algebraically equivalent at each point to the special Lagrangian ideal on $\mathbb{C}^3$.

It is a fact (Bryant 1987) that B_σ is torsion-free if and only if σ is g_σ-parallel, which, in turn, holds if and only if

$$\mathrm{d}\sigma = 0 \quad \text{and} \quad \mathrm{d}(*_\sigma \sigma) = 0. \tag{17.39}$$

Thus, a G_2-manifold can be regarded as a pair (M^7, σ) where $\sigma \in \Omega^3_+(M)$ satisfies the non-linear system of PDE (17.39).

By a theorem of Bonan (see Besse 1987, chapter X), for any G_2-manifold (M, σ), the associated metric g_σ has vanishing Ricci tensor. In particular, by a result of DeTurck and Kazdan (1981), the metric g_σ is real-analytic in g_σ-harmonic coordinates. Since σ is g_σ-parallel, it, too, must be real-analytic in g_σ-harmonic coordinates.

There is a natural differential ideal on $\Lambda^3_+(T^*M) = \mathcal{F}(M)/G_2$ defined as follows. For $[u] \in \mathcal{F}(M)/G_2$, define

$$\boldsymbol{\sigma}_{[u]} = \pi^*(u^*\phi) \quad \text{and} \quad \boldsymbol{\tau}_{[u]} = \pi^*\big(u^*(*_\phi\phi)\big), \tag{17.40}$$

where $\pi : \mathcal{F}(M)/G_2 \to M$ is the natural basepoint projection. Let $\mathcal{I}$ be the differential ideal on $\mathcal{F}(M)/G_2 = \Lambda^3_+(T^*M)$ generated by $\mathrm{d}\boldsymbol{\sigma}$ and $\mathrm{d}\boldsymbol{\tau}$. The following result is proved in Bryant (1987):

Theorem 17.3 *The ideal $\mathcal{I}$ on $\Lambda^3_+(T^*M)$ is involutive. A section $\sigma \in \Omega^3_+(M)$ is an integral of $\mathcal{I}$ only if it is g_σ-parallel. Modulo diffeomorphisms, the general $\mathcal{I}$-integral σ depends on 6 functions of 6 variables.*

17.4.1 *Hypersurfaces*

The group G_2 acts transitively on $S^6 \subset \mathbb{R}^7$, with stabilizer $SU(3)$. Hence, an oriented $N^6 \subset M$ inherits a canonical $SU(3)$-structure, which is determined by the $(1,1)$-form ω and $(3,0)$-form $\Omega = \phi + \mathrm{i}\,\psi$ defined by

$$\omega = \mathbf{n} \lrcorner \sigma \quad \text{and} \quad \Omega = \phi + \mathrm{i}\,\psi = N^*\sigma - \mathrm{i}\,(\mathbf{n} \lrcorner *_\sigma \sigma). \tag{17.41}$$

In fact, if one defines $f : \mathbb{R} \times N \to M$ by

$$f(t,p) = \exp_p\big(t\,\mathbf{n}(p)\big), \tag{17.42}$$

then

$$f^*\sigma = \mathrm{d}t \wedge \omega + \mathrm{Re}(\Omega) \quad \text{and} \quad f^*(*_\sigma\sigma) = \tfrac{1}{2}\,\omega^2 - \mathrm{d}t \wedge \mathrm{Im}(\Omega), \tag{17.43}$$

where, now, ω and Ω are forms on N that depend on t.

For each fixed $t = t_0$, the induced $SU(3)$-structure on N satisfies

$$\mathrm{d}\,\mathrm{Re}(\Omega) = \mathrm{d}(f^*_{t_0}\sigma) = 0 \quad \text{and} \quad \mathrm{d}(\tfrac{1}{2}\,\omega^2) = \mathrm{d}\big(f^*_{t_0}(*_\sigma\sigma)\big) = 0, \tag{17.44}$$

so these are necessary conditions on the $SU(3)$-structure on N that it be induced by immersion into a G_2-holonomy manifold M.

Theorem 17.4 *A real-analytic $SU(3)$-structure on N^6 is induced by embedding into a G_2-manifold if and only if its defining forms ω and Ω satisfy*

$$\mathrm{d}\,\mathrm{Re}(\Omega) = 0 \quad \textit{and} \quad \mathrm{d}\left(\tfrac{1}{2}\,\omega^2\right) = 0. \tag{17.45}$$

Proof. The necessity of (17.45) has already been demonstrated, so I will just prove the sufficiency.

Define a tautological two-form $\boldsymbol{\omega}$ and three-form $\boldsymbol{\Omega}$ on $\mathcal{F}(N)/SU(3)$ as follows. For a coframe $u : T_xN \to \mathbb{C}^3$, define these forms at $[u] = u \cdot SU(3) \in \mathcal{F}(N)/SU(3)$ by

$$\boldsymbol{\omega}_{[u]} = \pi^*\left(u^*\left(\tfrac{\mathrm{i}}{2}({}^t\mathrm{d}z \wedge \mathrm{d}\bar{z})\right)\right) \quad \text{and} \quad \boldsymbol{\Omega}_{[u]} = \pi^*\left(u^*(\mathrm{d}z^1 \wedge \mathrm{d}z^2 \wedge \mathrm{d}z^3)\right) \tag{17.46}$$

where $\pi : \mathcal{F}(N)/SU(3) \to N$ is the basepoint projection.

On $X = \mathbb{R} \times \mathcal{F}(N)/SU(3)$, consider the three-form and four-form defined by

$$\begin{aligned} \boldsymbol{\sigma} &= \mathrm{d}t \wedge \boldsymbol{\omega} + \mathrm{Re}(\boldsymbol{\Omega}) \\ \boldsymbol{\tau} &= \tfrac{1}{2}\,\boldsymbol{\omega}^2 - \mathrm{d}t \wedge \mathrm{Im}(\boldsymbol{\Omega}). \end{aligned} \tag{17.47}$$

Let $\mathcal{I}$ be the EDS generated by the closed four-form $\mathrm{d}\boldsymbol{\sigma}$ and five-form $\mathrm{d}\boldsymbol{\tau}$. Then the calculation used to prove Theorem 17.3 (see Bryant 1987) shows that $\mathcal{I}$ is involutive, with characters

$$(s_0, s_1, \ldots, s_7) = (0, 0, 0, 1, 4, 10, 13, 0). \tag{17.48}$$

Since $\mathrm{d}\big(\mathrm{Re}(\Omega)\big) = \mathrm{d}\left(\tfrac{1}{2}\,\omega^2\right) = 0$, the given $SU(3)$-structure on N defines a regular integral manifold $L \subset X$ of $\mathcal{I}$ lying in the hypersurface $t = 0$.

Since the given $SU(3)$-structure is assumed to be real-analytic, the system $\mathcal{I}$ and the integral manifold $L \subset X$ are real-analytic by constuction. Hence the Cartan–Kähler theorem can be applied to conclude that L lies in a unique $\mathcal{I}$-integral $M^7 \subset X$. The pullback of $\boldsymbol{\sigma}$ to M is then a closed definite three-form σ on M while the pullback of $\boldsymbol{\tau}$ to M is closed and equal to $*_\sigma\sigma$. Thus, (M, σ) is a G_2-manifold. By construction, N is imbedded into M as the locus $t = 0$ and the $SU(3)$-structure on N induced by σ is the given one. □

Theorem 17.5 *There exist non-real-analytic $SU(3)$-structures on N^6 whose associated forms (ω, Ω) satisfy* (17.45) *but that are not induced from an immersion into a G_2-manifold (M, σ).*

In fact, if a non-analytic $SU(3)$-structure satisfies (17.45) *and*

$$*\big(\omega \wedge \mathrm{d}\big(\mathrm{Im}(\Omega)\big)\big) = C \tag{17.49}$$

for some constant C, then it cannot be G_2-immersed.

Non-analytic $SU(3)$-structures satisfying (17.45) *and* (17.49) *do exist.*

Proof. When an $SU(3)$-structure on N^6 with defining forms (ω, Ω) is induced via a G_2-immersion $N^6 \hookrightarrow M^7$, the mean curvature H of N in M is given by

$$-12H = *\big(\omega \wedge \mathrm{d}\big(\mathrm{Im}(\Omega)\big)\big). \tag{17.50}$$

Thus, when the right-hand side of this equation is constant, it follows by elliptic regularity that N^6 is a real-analytic submanifold of the real-analytic (M^7, σ).

Thus, if the given $SU(3)$-structure on N satisfying (17.45) and (17.49) is not real-analytic, it cannot be induced by an embedding into a G_2-holonomy manifold M.

It remains to construct a non-analytic example satisfying (17.45) and (17.49). Here is why it is somewhat delicate: Since $\dim\big(GL(6,\mathbb{R})/SU(3)\big) = 28$, a choice of an $SU(3)$-structure (ω, Ω) on N^6 depends on 28 functions of 6 variables. Modulo diffeomorphisms, this leaves 22 functions of 6 variables. On the other hand, the equations

$$\mathrm{d}\big(\mathrm{Re}(\Omega)\big) = 0, \quad \mathrm{d}\big(\tfrac{1}{2}\,\omega^2\big) = 0, \quad \text{and} \quad *\,\big(\omega\wedge \mathrm{d}\big(\mathrm{Im}(\Omega)\big)\big) = C \tag{17.51}$$

constitute $15 + 6 + 1 = 22$ equations for the $SU(3)$-structure.

Thus, the equations to be solved are "more determined" than in the analogous $SU(2)$ case. Nevertheless, their diffeomorphism invariance still allows one to construct the desired example, as will now be shown.

Say that a three-form $\phi \in \Omega^3(N^6)$ is *elliptic* if, at each point, it is linearly equivalent to $\mathrm{Re}\big(\mathrm{d}z^1\wedge\mathrm{d}z^2\wedge\mathrm{d}z^3\big)$. This is a open pointwise condition on ϕ (i.e. it is *stable* in Hitchin's sense 2001): The elliptic three-forms are sections of an open subbundle $\Lambda^3_e(T^*N) \subset \Lambda^3(T^*N)$. I will denote the set of elliptic three-forms on N by $\Omega^3_e(N)$.

Fix an orientation of N^6. An elliptic $\phi \in \Omega^3_e(N)$ then defines a unique, orientation-preserving almost-complex structure J_ϕ on N^6 such that

$$\Omega_\phi = \phi + \mathrm{i}\, J^*_\phi(\phi) \tag{17.52}$$

is of J_ϕ-type $(3,0)$.

Now assume that $\phi \in \Omega^3_e(N)$ is closed. Then $\mathrm{d}\Omega_\phi$ is purely imaginary and yet must be a sum of terms of J_ϕ-type $(3,1)$ and $(2,2)$. Thus, $\mathrm{d}\Omega_\phi$ is purely of J_ϕ-type $(2,2)$.

Let $\Lambda^{1,1}_+(N, J_\phi)$ denote the set of real two-forms that are of J_ϕ-type $(1,1)$ and that are positive on all J_ϕ-complex lines. The squaring map

$$\sigma : \Lambda^{1,1}_+(N, J_\phi) \to \Lambda^{2,2}(N, J_\phi)$$

given by $\sigma(\omega) = \omega^2$ is a diffeomorphism onto the open set $\Lambda^{2,2}_+(N, J_\phi) \subset \Lambda^{2,2}(N, J_\phi)$ that consists of the real four-forms of J_ϕ-type $(2,2)$ that are positive on all J_ϕ-complex two-planes.

Now fix a constant $C \neq 0$. One sees from the above discussion that it is a C^1-open condition on ϕ that

$$\mathrm{d}\Omega_\phi = \tfrac{\mathrm{i}}{6} C\,(\omega_\phi)^2 \quad \text{for some} \quad \omega_\phi = \overline{\omega_\phi} \in \Omega^{1,1}_+(N, J_\phi). \tag{17.53}$$

Now, the pair $(\omega_\phi, \Omega_\phi)$ are the defining forms of an $SU(3)$-structure on N if and only if

$$\tfrac{1}{6}(\omega_\phi)^3 - \tfrac{1}{8}\mathrm{i}\,\Omega_\phi\wedge\overline{\Omega_\phi} = 0. \tag{17.54}$$

This is a single, first-order scalar equation on the closed three-form ϕ. It is easy to see that there are non-analytic solutions. (For example, if one starts with the standard structure induced on the six-sphere in flat $\mathbb{R}^7$ endowed with its flat G_2-structure, then small perturbations of the corresponding closed ϕ can be made that solve (17.54) but for which the induced $SU(3)$-structure has constant curvature on a proper open subset of S^6. Such an $SU(3)$-structure clearly cannot be real-analytic everywhere.)

Assuming (17.54) is satisfied, we have

$$\mathrm{d}(\mathrm{Re}\,\Omega_\phi) = \mathrm{d}\phi = 0, \tag{17.55}$$

and

$$\mathrm{d}\big(\tfrac{1}{2}(\omega_\phi)^2\big) = \mathrm{d}\big(-3\mathrm{i}\,\tfrac{1}{C}\,\mathrm{d}\Omega_\phi\big) = 0, \tag{17.56}$$

and finally,

$$*_\phi\big(\omega_\phi \wedge \mathrm{d}(\mathrm{Im}\,\Omega_\phi)\big) = *_\phi\big(\omega_\phi \wedge \tfrac{1}{6}C\,(\omega_\phi)^2\big) = C. \tag{17.57}$$

□

17.4.2 *Flow interpretation*

On $N^6 \times \mathbb{R}$, with (ω, Ω) defining an $SU(3)$-structure on N^6 depending on $t \in \mathbb{R}$, consider the equations

$$\mathrm{d}\big(\mathrm{d}t \wedge \omega + \mathrm{Re}(\Omega)\big) = 0 \quad \text{and} \quad \mathrm{d}\big(\tfrac{1}{2}\,\omega^2 - \mathrm{d}t \wedge \mathrm{Im}(\Omega)\big) = 0, \tag{17.58}$$

which assert that $\sigma = \mathrm{d}t{\wedge}\omega + \mathrm{Re}(\Omega)$, which is a definite three-form on $M = N^6 \times \mathbb{R}$, is both closed and co-closed (since $*_\sigma\sigma = \frac{1}{2}\,\omega^2 - \mathrm{d}t{\wedge}\,\mathrm{Im}(\Omega)$).

Think of Ω as $\phi + \mathrm{i}J^*_\phi(\phi)$, so that the $SU(3)$-structure is determined by (ω, ϕ) where $\phi = \mathrm{Re}(\Omega)$. The conditions (17.58) for fixed t are then

$$\mathrm{d}\phi = 0 \quad \text{and} \quad \mathrm{d}(\omega^2) = 0, \tag{17.59}$$

while the G_2-evolution equations implied by (17.58) for such (ω, ϕ) are then

$$\frac{\mathrm{d}}{\mathrm{d}t}(\phi) = \mathrm{d}\omega \quad \text{and} \quad \frac{\mathrm{d}}{\mathrm{d}t}(\omega) = -{L_\omega}^{-1}\left(\mathrm{d}\big(J^*_\phi(\phi)\big)\right), \tag{17.60}$$

where $L_\omega : \Omega^2(N) \to \Omega^4(N)$ is the invertible map $L_\omega(\beta) = \omega{\wedge}\beta$.

Theorems 17.4 and 17.5 show that solutions to the "G_2-flow" (17.60) do exist for analytic initial $SU(3)$-structures satisfying the closure conditions (17.59), but need not exist for non-analytic initial $SU(3)$-structures satisfying these closure conditions.

17.5 Spin(7)-manifolds

For background on the group $Spin(7) \subset SO(8)$ and $Spin(7)$-manifolds, the reader can consult Bryant (1987), Salamon (1989), and Joyce (2000). I will generally follow the notation in Bryant (1987).

The main point is that the group $Spin(7) \subset SO(8)$ is the $GL(8,\mathbb{R})$-stabilizer of the four-form $\Phi_0 \in \Lambda^4(\mathbb{R}^8)$, defined by

$$\Phi_0 = e^0 \wedge \phi + *_\phi \phi \tag{17.61}$$

where ϕ is defined by (17.37) and $e^0, \ldots, e^7$ is a basis of linear forms on $\mathbb{R}^8 = \mathbb{R} \oplus \mathbb{R}^7$. The group $Spin(7)$ is a connected Lie group of dimension 21 that is the double cover of $SO(7)$. It acts transitively on the unit sphere in $\mathbb{R}^8$ and the $Spin(7)$-stabilizer of e^0 is G_2. The $GL(8,\mathbb{R})$-orbit of Φ_0 in $\Lambda^4(\mathbb{R}^8)$ will be denoted by $\Lambda^4_s(\mathbb{R}^8)$. This orbit has dimension 43, so it is not open in $\Lambda^4(\mathbb{R}^8)$.

A $Spin(7)$-structure on M^8 is thus specified by a four-form $\Phi \in \Omega^4(M)$ that is linearly equivalent to Φ_0 at each point of M. The set of such four-forms will be denoted $\Omega^4_s(M)$. They are the sections of the bundle $\mathcal{F}(M)/Spin(7) \to M$, which has a natural embedding into $\Lambda^4(T^*M)$.

Given a four-form $\Phi \in \Omega^4_s(M)$, the corresponding $Spin(7)$-structure $B = B_\Phi$ is defined to be the set of coframings $u : T_xM \to \mathbb{R}^8$ that satisfy $u^*\Phi_0 = \Phi_x$. Conversely, every $Spin(7)$-structure $B \to M$ is of the form $B = B_\Phi$ for a unique four-form $\Phi \in \Omega^4_s(M)$.

In particular, each $\Phi \in \Omega^4_s(M)$ determines a metric g_Φ and orientation $*_\Phi$ on M by requiring that the elements $u \in B_\Phi$ be oriented isomorphisms.

It is a fact (Bryant 1987) that B_Φ is torsion-free if and only if Φ is g_Φ-parallel, which, in turn, holds if and only if

$$\mathrm{d}\Phi = 0. \tag{17.62}$$

Thus, a $Spin(7)$-manifold can be regarded as a pair (M^8, Φ) where $\Phi \in \Omega^4_s(M)$ satisfies the non-linear system of PDE (17.62).

By a theorem of Bonan (see Besse 1987, chapter X), for any $Spin(7)$-manifold (M, Φ), the associated metric g_Φ has vanishing Ricci tensor. In particular, by a result of DeTurck and Kazdan (1981), the metric g_Φ is real-analytic in g_Φ-harmonic coordinates. Since Φ is g_Φ-parallel, it, too, must be real-analytic in g_Φ-harmonic coordinates.

Define a four-form $\boldsymbol{\Phi}$ on $\mathcal{F}(M)/\mathrm{Spin}(7)$ by the following rule: For $u : T_xM \to \mathbb{R}^8$ and $[u] = u{\cdot}\mathrm{Spin}(7)$, set

$$\boldsymbol{\Phi}_{[u]} = \pi^*\left(u^*\Phi_0\right) \tag{17.63}$$

where $\pi : \mathcal{F}(M) \to M$ is the basepoint projection. Let $\mathcal{I}$ be the ideal generated by $\mathrm{d}\boldsymbol{\Phi}$ on $\mathcal{F}(M)/Spin(7)$. The following result is proved in Bryant (1987):

Theorem 17.6 *A section $\Phi \in \Omega^4_s(M)$ satisfies (17.62) if and only if it is an integral of $\mathcal{I}$. The ideal $\mathcal{I}$ is involutive. Modulo diffeomorphisms, the general $\mathcal{I}$-integral Φ depends on 12 functions of 7 variables.*

17.5.1 *Hypersurfaces*

Spin(7) acts transitively on S^7 and the stabilizer of a point is G_2. An oriented hypersurface $N^7 \subset M^8$ inherits a G_2-structure $\sigma \in \Omega^3_+(M)$ that is defined by the rule

$$\sigma = \mathbf{n} \lrcorner \Phi \tag{17.64}$$

where $\mathbf{n}$ is the oriented normal vector field along N. It also satisfies

$$*_\sigma \sigma = N^* \Phi. \tag{17.65}$$

The structure equations show that

$$*_\sigma \big(\sigma \wedge \mathrm{d}\sigma\big) = 28H \tag{17.66}$$

where H is the mean curvature of N in (M, g_Φ).

Theorem 17.7 *If* $\sigma \in \Omega^3_+(N^7)$ *is real-analytic and satisfies* $\mathrm{d}(*_\sigma\sigma) = 0$, *then* σ *is induced by an immersion of* N *into a Spin*(7)*-manifold* (M, Φ).

Proof. The argument in this case is entirely analogous to the $SU(2)$ and G_2 cases already treated.

Recall the definitions of $\boldsymbol{\sigma}$ and $\boldsymbol{\tau}$ on $\mathcal{F}(N^7)/G_2$ and, on $X = \mathbb{R} \times \mathcal{F}(N)/G_2$, define

$$\boldsymbol{\Phi} = \mathrm{d}t \wedge \boldsymbol{\sigma} + \boldsymbol{\tau}. \tag{17.67}$$

Let $\mathcal{I}$ be the ideal on $\mathbb{R} \times \mathcal{F}(N)/G_2$ generated by $\mathrm{d}\boldsymbol{\Phi}$. The same calculation used to prove Theorem 17.6 (see Bryant 1987) then yields that $\mathcal{I}$ is involutive, with character sequence

$$(s_0, s_1, \ldots, s_8) = (0, 0, 0, 0, 1, 4, 10, 20, 0). \tag{17.68}$$

Since $\mathrm{d}(*_\sigma\sigma) = 0$, the G_2-structure σ defines a regular $\mathcal{I}$-integral $L \subset X$ within the locus $t = 0$.

Since σ is assumed to be real-analytic, it follows that X, $\mathcal{I}$, and L are real-analytic with respect to the obvious induced analytic structures on the appropriate underlying manifolds. Thus, the Cartan–Kähler theorem applies to show that L lies in a unique $\mathcal{I}$-integral $M^8 \subset X$. The form $\boldsymbol{\Phi}$ then pulls back to M to be a closed $\Phi \in \Omega^4_s(M)$ which induces the given σ on $N \simeq L \subset M$ defined by $t = 0$. □

Theorem 17.8 *There exist non-real-analytic* G_2*-structures* $\sigma \in \Omega^3_+(N^7)$ *that satisfy*

$$\mathrm{d}(*_\sigma\sigma) = 0 \tag{17.69}$$

but that are not induced from a Spin(7)*-immersion.*

In fact, if a non-analytic G_2*-structure* σ *satisfies* (17.69) *and*

$$*_\sigma \big(\sigma \wedge \mathrm{d}\sigma\big) = C \tag{17.70}$$

where C is a constant, then it cannot be $Spin(7)$-immersed.

Non-analytic G_2-structures $\sigma \in \Omega^3_+(N^7)$ satisfying (17.69) and (17.70) do exist.

Proof. The first claim follows from the second and third.

If $\sigma \in \Omega^3_+(N^7)$ is induced from at $Spin(7)$-immersion $(N, \sigma) \hookrightarrow (M^8, \Phi)$ and has $*_\sigma(\sigma \wedge \mathrm{d}\sigma)$ equal to a constant, then the hypersurface $N \subset M$ has constant mean curvature. Since (M, g_Φ) is real-analytic, constant mean curvature hypersurfaces in M are also real-analytic, so it follows that σ must be real-analytic. This proves the second claim.

It remains now to verify the third claim by explaining how to construct non-analytic G_2-structures σ satisfying (17.69) and (17.70). To save space, I will only sketch the argument, the details of which are somewhat involved, though the idea is the same as for $SU(2)$:

If such a σ is to be real-analytic, it will have to be real-analytic in g_σ-harmonic coordinates. Now, the system of first-order equations

$$\mathrm{d}(*_\sigma \sigma) = 0, \quad *_\sigma(\sigma \wedge \mathrm{d}\sigma) = C, \quad \text{and} \quad \mathrm{d}(*_\sigma \mathrm{d}x) = 0$$

for $\sigma \in \Omega^3_+(\mathbb{R}^7)$ is only $21 + 1 + 7 = 29$ equations for 35 unknowns. This underdetermined system is not elliptic, but its symbol mapping has constant rank and it can be embedded into an appropriate sequence to show that it has non-real-analytic solutions. □

17.5.2 *Flow interpretation*

Finally, let us consider the "flow" interpretation. If $f : N \hookrightarrow M$ is an oriented smooth hypersurface in a $Spin(7)$-manifold (M, Φ) with oriented unit normal $\mathbf{n} : N \to TM$, then the normal exponential mapping can be used to embed a neighborhood U of N in M into $\mathbb{R} \times N$ in such a way that, on this neighborhood U, one can write

$$\Phi = \mathrm{d}t \wedge \sigma + *_\sigma \sigma \tag{17.71}$$

where $\sigma \in \Omega^3_+(N)$ now depends on t (in the case that N is noncompact, the domain of σ might depend on t). The closure of Φ implies that

$$\frac{\mathrm{d}}{\mathrm{d}t}(*_\sigma \sigma) = \mathrm{d}\sigma. \tag{17.72}$$

This equation is enough to determine the time derivative of σ as well because of the following observation. For any real vector space V of dimension 7, the map $\mathsf{S} : \Lambda^3_+(V^*) \to \Lambda^4(V^*)$ defined by $\mathsf{S}(\phi) = *_\phi \phi$ is a 2-to-1 smooth local diffeomorphism of $\Lambda^3_+(V^*)$ onto an open cone $\Lambda^4_+(V^*) \subset \Lambda^4(V^*)$. In fact, if $\mathsf{S}(\phi) = \mathsf{S}(\psi)$, then $\phi = \pm\psi$. Both ϕ and $-\phi$ are definite, but they each determine opposite orientations, that is, $*_{-\phi} 1 = - *_\phi 1$. In particular, if one fixes an orientation on V, then for any $\tau \in \Lambda^4_+(V^*)$, there is a unique element $\phi \in \Lambda^3_+(V^*)$ such that $*_\phi \phi = \tau$ and $*_\phi 1$ is a positive volume form on V. I will denote this element by $\mathsf{S}^{-1}(\tau) = \phi$. Using this notation and the assumption that N is oriented, the above equation can be written in the more obviously "evolutionary" form

$$\frac{\mathrm{d}}{\mathrm{d}t}(\tau) = \mathrm{d}\left(\mathsf{S}^{-1}(\tau)\right) \tag{17.73}$$

where

$$\Phi = \mathrm{d}t \wedge \mathsf{S}^{-1}(\tau) + \tau \tag{17.74}$$

with $\tau \in \Omega^4_+(N)$ depending on t.

The content of Theorems 17.7 and 17.8 is then that (17.73) has a solution when the initial τ_0 is closed and real-analytic, but need not have a solution for a non-analytic closed τ_0.

Acknowledgments

Thanks to Duke University for its support via a research grant and to the NSF for its support via DMS-0604195. Finally, I would like to thank the referee for a careful reading of the manuscript and for helpful suggestions regarding references.

References

Besse, A. (1987). Einstein manifolds. *Ergebnisse der Mathematik und ihrer Grenzgebiete (3)* [*Results in Mathematics and Related Areas (3)*]. 10. Springer-Verlag, Berlin, Germany. MR 0867684 (88f:53087).

Borel, A. and Lichnerowicz, A. (1952). Groupes d'holonomie des variétés riemanniennes. *C. R. Acad. Sci. Paris* **234**, 1835–37. MR 48133 (13,986b).

Bryant, R. (1987). Metrics with exceptional holonomy. *Ann. Math. (2)* **126**, 525–76. MR 916718 (1989b:53084).

Bryant, R. *et al.*, (1991). *Exterior Differential Systems.* MSRI Series **18**, Springer-Verlag, New York. MR 1083148 (1992h:58007).

Bryant, R. (2006). Some remarks on G_2-structures. *Proceedings of Gökova Geometry-Topology Conference 2005.* pp. 75–109, Gökova Geometry/Topology Conference (GGT), Gökova, MR 2282011 (2007k:53019) arXiv:math.DG/0305124.

Cartan, É. (1925). La géometrie des espaces de Riemann. *Mémorial des Sciences Mathematiques* Fasc. IX.

Conti, D. and Salamon, S. (2007). Generalized Killing spinors in dimension 5. *Trans. Am. Math. Soc.* **359**, 5319–43.

DeTurck, D. and Kazdan, J. (1981). Some regularity theorems in Riemannian geometry. *Ann. Sci. Éc. Norm. Sup.* **14**, 249–60. MR 644518 (83f:53018).

Hitchin, N. (2001). Stable forms and special metrics. *Global Differential Geometry: The Mathematical Legacy of Alfred Gray (Bilbao, 2000).* pp. 70–89, Contemporary Mathematics, Vol. **288**, American Mathematical Society, Providence, RI. MR 1871001 (2003f:53065).

Joyce, D. (2000). *Compact Manifolds with Special Holonomy.* Oxford Mathematical Monographs, Oxford University Press, Oxford. MR 1787733 (2001k:53093).

Salamon, S. (1989). *Riemannian Geometry and Holonomy Groups.* Pitman Research Notes in Mathematics, Vol. **201**. Longman, Harlow, UK. MR 1004008 (90g:53058).

Schouten, J. (1918). On the number of degrees of freedom of the geodetically moving systems. *Proc. Kon. Acad. Wet. Amsterdam* **21**, 607–13.

XVIII

BRANES ON POISSON VARIETIES

Marco Gualtieri

Dedicated to Nigel Hitchin on the occasion of his 60th birthday

18.1 Introduction

In this chapter we shall take a second look at a classical structure in differential and algebraic geometry, that of a holomorphic Poisson structure, which is a complex manifold with a holomorphic Poisson bracket on its sheaf of regular functions. The structure is determined, on a real smooth manifold M, by the choice of a pair (I, σ_I), where I is an integrable complex structure tensor and σ_I is a holomorphic Poisson tensor. We shall view (I, σ_I) not as we normally do but instead as a *generalized complex structure*, in the sense of Hitchin (2003). In so doing, we shall obtain a new notion of equivalence between the pairs (I, σ_I) which does not imply the holomorphic equivalence of the underlying complex structures.

In studying this equivalence relation we are naturally led to an unexpected connection to *generalized Kähler geometry*, as defined in Gualtieri (2004), and to a method for constructing certain examples of these structures which extends the recent work of Hitchin constructing bi-Hermitian metrics on Del Pezzo surfaces (Hitchin 2007); in particular we obtain similar families of bi-Hermitian metrics on all smooth Poisson Fano varieties, and in fact on any smooth Poisson variety admitting a positive Poisson line bundle. We therefore give an explicit construction of a subclass of the generalized Kähler structures proven to exist by the generalized Kähler stability theorem of Goto (2007).

In both these efforts we shall find it useful to introduce an extension of the notion of connection on a vector bundle, to allow differentiation not only in the tangent but also in the cotangent directions; we call such a structure a *generalized connection*. We also show that in the presence of a generalized metric, there is a canonical connection D which plays the role of the Levi-Civita connection in Kähler geometry: namely, we show that $(\mathcal{J}, G)$ is generalized Kähler if and only if $D\mathcal{J} = 0$.

In the final section we make some speculative comments concerning the relationship between generalized Kähler geometry and non-commutative geometry, a topic we hope to clarify in the future.

18.2 Gerbe trivializations

Let M be a manifold equipped with a $U(1)$ gerbe with connection (specifically, a gerbe with connective structure in the sense of Brylinski 1993). This determines canonically a *Courant algebroid* E over M, in the same way that a $U(1)$ principal bundle P determines an Atiyah Lie algebroid $E = TP/U(1)$ over M. See Ševera (1998–2000) and Hitchin (2006*a*) for details of this construction, and see Courant (1990), Roytenberg (1999), and Liu, Weinstein, and Xu (1997) for details concerning Courant algebroids; we review their main properties presently.

The Courant algebroid E is an extension of real vector bundles

$$0 \longrightarrow T^* \xrightarrow{\pi^*} E \xrightarrow{\pi} T \longrightarrow 0, \tag{18.1}$$

where T and T^* denote the tangent and cotangent bundles of M. Further, E is equipped with a nondegenerate symmetric bilinear form $\langle \cdot, \cdot \rangle$ of split signature, such that $\langle \pi^*\xi, a\rangle = \xi(\pi(a))$. Finally, there is a bilinear *Courant bracket* $[\cdot, \cdot]$ on $C^\infty(E)$ such that

- $[[a, b], c] = [a, [b, c]] - [b, [a, c]]$ (Jacobi identity)
- $[a, fb] = f[a, b] + (\pi(a)f)b$ (Leibniz rule)
- $\pi(a)\langle b, c\rangle = \langle [a, b], c\rangle + \langle b, [a, c]\rangle$ (invariance of bilinear form)
- $[a, a] = \pi^* d\langle a, a\rangle$ (skew-symmetry anomaly)

The choice of an isotropic complement to T^* in E is a contractible one, and so an isotropic splitting $s : T \longrightarrow E$ of the sequence (18.1) always exists. Each such splitting determines a closed three-form $H \in \Omega^3(M)$, given by

$$(i_X i_Y H)(Z) = \langle [s(X), s(Y)], s(Z)\rangle. \tag{18.2}$$

The cohomology class $[H]/2\pi \in H^3(M, \mathbb{R})$ is independent of the choice of splitting, and coincides with the image of the Dixmier–Douady class of the gerbe in real cohomology. Furthermore, $[H]$ classifies the Courant algebroid up to isomorphism, as shown by Ševera (1998–2000).

Courant algebroids may be naturally pulled back by the inclusion $S \subset M$ of a submanifold; as a bundle over S, the result is simply given by

$$E_S = \frac{\pi^{-1}(TS)}{\mathrm{Ann}(TS)}, \tag{18.3}$$

and its bracket and inner product are inherited in a straightforward manner.

A trivialization of the gerbe along S induces a Courant trivialization in the following sense:

Definition 18.1 *A* Courant trivialization *along S is an integrable isotropic splitting $s : TS \longrightarrow E_S$ of the pullback Courant algebroid. Such trivializations exist if and only if $\iota^*[H] = 0$ for $\iota : S \hookrightarrow M$ is the inclusion.*

Integrability is the requirement that the subbundle $s(TS) \subset E_S$ be closed under the Courant bracket. Integrable maximal isotropic subbundles of a Courant algebroid are called Dirac structures; therefore $s(TS)$ is simply a Dirac structure transverse to T^*S. As a result of a Courant trivialization, E_S is canonically isomorphic to $TS \oplus T^*S$ with its natural pairing and the bracket

$$[X + \xi, Y + \eta] = L_X(Y + \eta) - i_Y d\xi.$$

Now suppose that $S_0, S_1 \subset M$ are submanifolds with smooth intersection, and suppose we have gerbe trivializations on each of them. Then on $X = S_0 \cap S_1$ we obtain a pair of gerbe trivializations, which must differ by a line bundle L_{01} with $U(1)$ connection ∇_{01}. Let s_0, s_1 be the splittings of E_X determined by the two gerbe trivializations. Then $s_1 - s_0 : TX \longrightarrow T^*X$ is given by $X \mapsto i_X F_{01}$ where $F_{01} \in \Omega^2(M)$ is the curvature of ∇_{01}.

The notion of Courant trivialization provides a convenient way of characterizing isomorphisms of Courant algebroids, as in the following example. The notation $\overline{E}$ denotes the Courant algebroid E, equipped with the opposite bilinear form $-\langle\cdot,\cdot\rangle$.

Example 18.1 Let E_M, E_N be Courant algebroids over the manifolds M, N, respectively. They are isomorphic precisely if there is a Courant trivialization of the product Courant algebroid $\overline{E}_M \times E_N$ along the graph of a diffeomorphism $\varphi : M \longrightarrow N$ in the product $M \times N$.

18.3 Generalized connections

Let E be a Courant algebroid as in the previous section. In keeping with the notion that the Courant algebroid is an analogue of the tangent bundle, we have the following generalization of the usual notion of connection:

Definition 18.2 *A* generalized connection *on a vector bundle V is a first-order linear differential operator*

$$D : C^\infty(V) \longrightarrow C^\infty(E \otimes V)$$

such that $D(fs) = fDs + (\pi^* df) \otimes s$. Furthermore, if V has a Hermitian metric h, then D is unitary *when*

$$d(h(s,t)) = h(Ds,t) + h(s,Dt).$$

If $s : T \longrightarrow E$ is any splitting (not necessarily isotropic) of the Courant algebroid, then using the decomposition $E = s(T) \oplus T^* \cong T \oplus T^*$, we have

$$D = \nabla + \chi, \qquad (18.4)$$

where ∇ is a usual unitary connection and χ is a vector field with values in the bundle of skew-adjoint endomorphisms of V, that is, $\chi \in C^\infty(T \otimes \mathfrak{u}(V))$. The tensor χ is independent of the choice of splitting, and we note that if V is of rank 1, χ is simply a vector field on the manifold.

With respect to a different splitting s', such that

$$s' - s = \theta : T \longrightarrow T^*,$$

we obtain a different decomposition $D = \nabla' + \chi$, where $\nabla' = \nabla + \theta(\chi)$.

A generalized connection has a natural curvature operator: for $a, b \in C^\infty(E)$, we define

$$R(a,b) = [D_a, D_b] - D_{[a,b]} \in C^\infty(\mathfrak{u}(V)).$$

This becomes tensorial in a, b when restricted to a Dirac structure $L \subset E$:

$$R|_L \in C^\infty(\wedge^2 L^* \otimes \mathfrak{u}(V)).$$

If $L = T^*$, for example, we obtain a bivector with values in the skew-adjoint endomorphisms, $R|_{T^*} = [\chi, \chi]$.

The tensorial curvatures $R_{s'}, R_s$ associated to integrable splittings s, s' of E with $s' - s = F \in \Omega^2_{cl}(M)$ may be compared by projection to T:

$$R_{s'} = R_s + d^\nabla(a) + a \wedge a,$$

where $a = F(\chi)$. Therefore if V has rank 1, we have that $\chi = iX$ for a real vector field X, and

$$R_{s'} - R_s = di_X F.$$

In the particular case that we have a generalized connection D on E itself, it is natural to compare the connection derivative with the Courant bracket; we therefore introduce the *torsion* of D, and leave it as an exercise to verify it is well-defined.

Definition 18.3 *The torsion $T_D \in C^\infty(\wedge^2 E \otimes E)$ of a generalized connection D on E itself is defined by*

$$T(a,b,c) = \langle D_a b - D_b a - [a,b]_{sk}, c\rangle + \tfrac{1}{2}(\langle D_c a, b\rangle - \langle D_c b, a\rangle),$$

where $[a,b]_{sk} = \frac{1}{2}([a,b] - [b,a])$. If D preserves the canonical bilinear form $\langle \cdot, \cdot \rangle$ on E, then T_D is totally skew, that is, $T_D \in C^\infty(\wedge^3 E)$.

A generalized Riemannian metric on the Courant algebroid E is the choice of a maximal positive-definite subbundle $C_+ \subset E$; this reduces the $O(n,n)$ structure of E to $O(n) \times O(n)$, and defines a positive-definite metric on E:

$$G(\cdot,\cdot) = \langle \cdot, \cdot \rangle|_{C_+} - \langle \cdot, \cdot \rangle|_{C_-},$$

where $C_- = C_+^\perp$ is the orthogonal complement with respect to $\langle \cdot, \cdot \rangle$. We now describe a construction of a canonical connection associated to the choice of such a metric, inspired by calculations in Gualtieri (2004) and Hitchin (2006*a*) relating metric connections with skew torsion to the Courant bracket.

The G-orthogonal complement to T^* is an isotropic splitting $C_0 \subset E$ and we identify it with T, so that G induces a splitting $E = T \oplus T^*$. The Courant bracket in this splitting is

$$[X+\xi, Y+\eta]_H = [X,Y] + L_X\eta - i_Y d\xi + i_Y i_X H,$$

where $H \in \Omega^3_{cl}(M)$ is defined by (18.2). The splitting also defines an anti-orthogonal automorphism $C : E \longrightarrow E$ defined by $C(X+\xi) = X-\xi$, which satisfies $C(C_\pm) = C_\mp$. It also has the property, for $Z, W \in C^\infty(E)$:

$$[CZ, CW]_H = C([Z,W]_{-H}).$$

Theorem 18.1 *Let $C_+ \subset E$ be a maximal positive-definite subbundle, i.e. a generalized metric, as above. Let $C : E \longrightarrow E$ be the map defined above. Write $Z = Z_+ + Z_-$ for the orthogonal projections of $Z \in C^\infty(E)$ to $C_\pm$. Then the operator*

$$D_Z(W) = [Z_-, W_+]_+ + [Z_+, W_-]_- + [CZ_-, W_-]_- + [CZ_+, W_+]_+ \tag{18.5}$$

defines a generalized connection on E, preserving both $\langle\cdot,\cdot\rangle$ and the positive-definite metric G. We call this the generalized Bismut connection

Proof. Using the properties of the Courant bracket and the orthogonality $C_+ = C_-^\perp$, we have immediately the property $D_{fZ}W = fD_Z W$, for $f \in C^\infty(M)$. We also have

$$\begin{aligned} D_Z(fW) &= fD_Z(W) + (Z_-f)W_+ + (Z_+f)W_- + (Z_-f)W_- + (Z_+f)W_+ \\ &= fD_Z(W) + (Zf)W, \end{aligned}$$

proving that D is a generalized connection.

It is clear from (18.5) that $C_\pm$ are preserved by the connection, since $D_Z W$ has nonzero component in $C_\pm$ if and only if W does. Hence we obtain a decomposition

$$D = D^+ \oplus D^-,$$

where $D^\pm$ are generalized connection on $C_\pm$ respectively.

To prove that D preserves the canonical metric $\langle\cdot,\cdot\rangle$ as well as the metric G, we show that $D^\pm$ preserve the induced metrics on $C_\pm$. Let $V, W \in C^\infty(C_+)$, and $Z \in C^\infty(E)$. Then

$$\begin{aligned} Z_+\langle V, W\rangle &= (CZ_+)\langle V, W\rangle \\ &= -\langle [CZ_+, V], W]\rangle - \langle V, [CZ_+, W]\rangle \\ &= \langle D_{Z_+}V, W\rangle + \langle V, D_{Z_+}W\rangle \end{aligned}$$

Also, we have

$$\begin{aligned} Z_-\langle V, W\rangle &= \langle [Z_-, V], W\rangle + \langle V, [Z_-, W]\rangle \\ &= \langle D_{Z_-}V, W\rangle + \langle V, D_{Z_-}W\rangle. \end{aligned}$$

Summing these two results, we obtain that D^+ preserves the metric on C_+; the same argument holds for C_-, completing the proof. □

The generalized connections $D^\pm$ define tensors $\chi^\pm \in C^\infty(T \otimes C_\pm)$, via the decomposition (18.4). We see now that these vanish, since for $Z \in C^\infty(T^*)$ and $W \in C^\infty(C_\pm)$, we have

$$\begin{aligned}\chi_Z^\pm W &= D_Z W = [Z_\mp, W_\pm]_\pm + [CZ_\pm, W_\pm]_\pm \\ &= [Z_\mp + (CZ)_\mp, W_\pm]_\pm \\ &= 0,\end{aligned}$$

where we use the fact that $Z \in T^*$ if and only if $CZ = -Z$.

As a result of this, we conclude that $D^\pm$ may be viewed as the usual metric connections $\nabla^\pm$ on T, via the projection isomorphisms $\pi_\pm : C_\pm \longrightarrow T$, i.e.

$$D^\pm = \pi_\pm^{-1} \nabla^\pm \pi_\pm$$

The connections $\nabla^\pm$ may be described as follows:

$$\begin{aligned}\nabla_X^\pm Y &= 2\pi_\pm D_X^\pm Y_\pm \\ &= 4\pi_\pm D_{X_\mp}^\pm Y_\pm \\ &= 4\pi_\pm [X_\mp, Y_\pm]_\pm.\end{aligned}$$

We may easily compute the torsion $T^\pm$ of the connections $\nabla^\pm$, for vector fields X, Y, Z:

$$\begin{aligned}2g(T^+(X,Y),Z) &= \langle T^+(X,Y), Z_+\rangle \\ &= \langle 4\pi_+[X_-,Y_+]_+ - 4\pi_+[Y_-,X_+]_+ - \pi[X,Y], Z_+\rangle \\ &= \langle 2[X_-,Y_+]_+ + 2[X_+,Y_-]_+ - [X,Y] + i_Y i_X H, Z_+\rangle \\ &= 2H(X,Y,Z) - 2\langle [X_+ - X_-, Y_+ - Y_-], Z_+\rangle \\ &= 2H(X,Y,Z),\end{aligned}$$

by the fact that $[X_+ - X_-, Y_+ - Y_-] = 0$ since the Courant bracket vanishes on 1-forms. A similar calculation gives $g(T^-(X,Y),Z) = -H(X,Y,Z)$.

The above calculation shows that $\nabla^\pm$ coincide with the Bismut connections with totally skew torsion $\pm H$. In this way, we have essentially repeated the observation of (Hitchin 2006*a*) that $\nabla^\pm$ may be conveniently expressed in terms of the Courant bracket. To summarize, the generalized Bismut connection is essentially a usual connection on E which restricts to $C_\pm$ to give the Bismut connections with torsion $\pm H$.

Proposition 18.1 *The torsion T_D of the generalized Bismut connection lies in $\wedge^3 C_+ \oplus \wedge^3 C_- \subset \wedge^3 E$, and is given by*

$$T_D = 2(\pi_+^* H + \pi_-^* H).$$

Proof. First we show that $T(C_+, C_-, \cdot) = 0$, so that $T \in C^\infty(\wedge^3 C_+ \oplus \wedge^3 C_-)$. Let $x \in C^\infty(C_+)$, $y \in C^\infty(C_-)$ and $z \in C^\infty(E)$. Then

$$\begin{aligned} T_D(x, y, z) &= \langle D_x y - D_y x - [x, y], z\rangle \\ &= \langle [x, y]_- - [y, x]_+ - [x, y], z\rangle = 0, \end{aligned}$$

as required.

Now take $x, y, z \in C^\infty(C_+)$. Since $\chi_D = 0$, we have the identity

$$\begin{aligned} \langle D_z x, y\rangle - \langle D_z y, x\rangle &= \langle D_{Cz} x, y\rangle - \langle D_{Cz} y, x\rangle \\ &= \langle [Cz, x], y\rangle - \langle [Cz, y], x\rangle \\ &= \langle [x, y] - [y, x], Cz\rangle. \end{aligned}$$

Therefore the torsion is given by

$$\begin{aligned} T(x, y, z) &= \langle D_x y - D_y x - [x, y]_{sk}, z\rangle + \frac{1}{2}\langle [x, y] - [y, x], Cz\rangle \\ &= \langle D_x y - D_y x, z\rangle + \langle [x, y], Cz - z\rangle \\ &= 2g(\nabla^+_X Y - \nabla^+_Y X - [X, Y], Z) \\ &= 2H(X, Y, Z). \end{aligned}$$

A similar calculation for $x, y, z \in C^\infty(C_-)$ gives $T(x, y, z) = 2H(x, y, z)$ as well, yielding the result. $\square$

As we have explained, the generalized Bismut connection D is completely determined by a usual connection on $T \oplus T^*$. Using the fact that the Bismut connections satisfy $\nabla^\pm = \nabla \pm \frac{1}{2} g^{-1} H$ for ∇ the Levi-Civita connection, we may write D explicitly with respect to the splitting $E = T \oplus T^*$, and for $X \in C^\infty(T)$, as follows:

$$D_X = \begin{pmatrix} \nabla_X & \frac{1}{2} \wedge^2 g^{-1}(i_X H) \\ \frac{1}{2} i_X H & \nabla^*_X \end{pmatrix}$$

The significance of this connection in the context of generalized geometry was first understood and investigated by Ellwood. Here we simply view it as a generalized connection[1] mainly for the purpose of highlighting its properties and defining its torsion tensor.

18.4 Generalized holomorphic bundles and branes

Suppose now that we have a generalized complex structure $\mathcal{J}$ on (M, E), which is an orthogonal almost complex structure $\mathcal{J} : E \longrightarrow E$ whose $+i$ eigenbundle $L \subset E \otimes \mathbb{C}$ is closed under the Courant bracket (Hitchin 2003). We now

[1] The author thanks Yicao Wang for pointing this out and correcting an error in the previous version.

describe how the structures in the previous two sections may be made *compatible* with $\mathcal{J}$.

18.4.1 *Generalized holomorphic bundles*

The integrability of $\mathcal{J}$ guarantees that $L = \ker(\mathcal{J} - i1)$ is a complex Lie algebroid, with associated de Rham complex

$$C^\infty(\wedge^k L^*) \xrightarrow{d_L} C^\infty(\wedge^{k+1} L^*) \tag{18.6}$$

A complex vector bundle equipped with a flat L-connection is called a generalized holomorphic bundle (Gualtieri 2007). Therefore, generalized holomorphic bundles form a category of Lie algebroid representations in the sense of (Evens, Lu, and Weinstein 1999).

In the case that $\mathcal{J}$ is a usual complex structure, for instance, a generalized holomorphic bundle consists of a holomorphic bundle V, together with a holomorphic section $\Phi \in H^0(M, T_{1,0}M \otimes \mathrm{End}(V))$ satisfying

$$\Phi \wedge \Phi = 0 \in H^0(M, \wedge^2 T_{1,0}M \otimes \mathrm{End}(V).$$

Note that if M is holomorphic symplectic, then $T_{1,0}$ is isomorphic to $T^*_{1,0}$, and Φ may be viewed as a Higgs bundle, in the sense of Simpson (1992).

In the case that $\mathcal{J}$ is a symplectic structure, a generalized holomorphic bundle is simply a flat bundle.

Definition 18.4 *A unitary generalized connection D on a complex vector bundle V is compatible with $\mathcal{J}$ when its curvature along $L = \ker(\mathcal{J} - i1)$ is zero.*

It follows immediately that the restriction of D to L defines a flat L-module structure on V, making V a generalized holomorphic bundle. Conversely, suppose V is a $\mathcal{J}$-holomorphic bundle, that is, it is equipped with an L-connection as follows:

$$\overline{\partial} : C^\infty(V) \longrightarrow C^\infty(L^* \otimes V), \quad \overline{\partial}^2 = 0. \tag{18.7}$$

This operator has symbol sequence given by wedging with $\sigma(\xi) = \frac{1}{2}(1 + i\mathcal{J})\xi \in \overline{L}$, where we identify $L^* = \overline{L}$ using the metric on E.

Choosing a Hermitian metric h on the bundle V, so that $\overline{V} \simeq V^*$, we may view the complex conjugate of (18.7),

$$\partial : C^\infty(\overline{V}) \longrightarrow C^\infty(L \otimes \overline{V}),$$

as a L-connection on V^*; we then form the dual ∂^* of this partial connection. Finally we form the sum

$$D = \overline{\partial} + \partial^* : C^\infty(V) \longrightarrow C^\infty((\overline{L} \oplus L) \otimes V) = C^\infty(E \otimes V),$$

which has symbol $\sigma + \bar{\sigma} = \pi^*$. Hence it defines a generalized connection on V. We summarize the above in the following:

Proposition 18.2 *Let V be a complex vector bundle with $\mathcal{J}$-holomorphic structure given by $\overline{\partial}$, and choose a Hermitian metric on V. Then the operator*

$$D = \overline{\partial} + \partial^* : C^\infty(V) \longrightarrow C^\infty(E \otimes V)$$

is the unique unitary generalized connection extending $\overline{\partial}$.

When V is a line bundle, there is a useful formula for the generalized connection one-form in terms of a holomorphic trivialization, analogous to the Poincaré–Lelong formula for the Chern connection on a Hermitian holomorphic line bundle.

Proposition 18.3 (Generalized Poincaré–Lelong formula) *Let V be a generalized holomorphic Hermitian line bundle, and let $s \in C^\infty(V)$ be a holomorphic section. Where it is nonzero, it defines a trivialization of the unitary generalized connection, $D = d + i\mathcal{A}$, where*

$$\mathcal{A} = \mathcal{J} d \log |s| \in C^\infty(E).$$

Proof. Whenever s is nonzero, we have

$$D\frac{s}{|s|} = i\mathcal{A}\frac{s}{|s|}.$$

Taking the projection to $\overline{L}$, we obtain

$$d_L \log |s| = i\mathcal{A}^{0,1},$$

so that $\mathcal{A} = -i(d_L - d_{\overline{L}}) \log |s| = \mathcal{J} d \log |s|$, as required. □

In particular, if s is nonzero on an open dense set, then the vector field $\pi \mathcal{J} d \log |s| = X$ must extend to a smooth vector field on the whole of M, since $\pi(i\mathcal{A})$ coincides with $\chi \in C^\infty(T \otimes \mathfrak{u}(V))$, which is globally defined for any generalized connection. But the map $\pi\mathcal{J}|_{T^*} : T^* \longrightarrow T$ is actually a Poisson structure $Q \in C^\infty(\wedge^2 T)$ (see Gualtieri 2007 for a discussion of this fact), and hence s vanishes only along the zero locus of the Poisson structure Q, which is a strong constraint on any generalized holomorphic section.

The above proposition may be used, by invoking the local existence of non-vanishing holomorphic sections near points for which $\mathcal{J}$ is regular (i.e. Q has locally constant rank), to show that the vector field χ of any Hermitian $\mathcal{J}$-holomorphic line bundle must be a Poisson vector field. It therefore defines a characteristic class in the Poisson cohomology of Lichnerowicz (1977), which is the cohomology of the complex $(C^\infty(\wedge^\bullet T), d_Q)$, where $d_Q \Pi = [Q, \Pi]$ is the Schouten bracket with Q.

Corollary 18.1 *The real vector field $X = -i\chi$ associated to any Hermitian generalized holomorphic line bundle preserves the Poisson structure Q, that is, it is a Poisson vector field. Furthermore its Poisson cohomology class $[X] \in H^1_Q(M)$ is independent of the Hermitian metric.*

Proof. X is Poisson since, by the proposition, it is locally Hamiltonian on an open dense subset of M. Hence $L_X Q = 0$ everywhere. Rescaling the Hermitian metric by a positive function e^f, we obtain a new vector field $X' = X + \frac{1}{2}Qdf$, which differs from X by a global Hamiltonian vector field. Hence $[X] \in H^1_Q(M)$ is independent of the choice of Hermitian structure. □

We may also deduce this result from the more general fact that the tensor product of a L-module with a $\overline{L}$ module is a Poisson module for Q. (This is a direct consequence of the fact that the tensor product of the Dirac structures $L, \overline{L}$ is the Dirac structure associated to Q, shown in Gualtieri 2007.) For any generalized holomorphic line bundle V, therefore, the trivial bundle $V \otimes \overline{V}$ acquires a Q-module structure, and therefore, as described in Evens, Lu, and Weinstein (1999), a characteristic class in $H^1_Q(M)$.

There are always two natural $\mathcal{J}$-holomorphic line bundles on any generalized complex manifold: the trivial bundle, for which $\chi = 0$ (for the standard Hermitian structure), and the canonical line bundle $K_\mathcal{J}$ of pure spinors associated to the maximal isotropic subbundle $L \subset E \otimes \mathbb{C}$. Since $K_\mathcal{J} \otimes \overline{K_\mathcal{J}}$ is naturally the determinant line $\det T^*$, it follows that $[X] = [-i\chi]$ is actually the *modular class* of the Poisson structure Q, in the sense of Weinstein (1997).

18.4.2 *Generalized complex branes*

Suppose we have a submanifold $\iota : S \hookrightarrow M$ equipped with a Courant trivialization $s : TS \longrightarrow E_S$. The Dirac structure $s(TS) \subset E_S$ may be canonically lifted to a maximal isotropic subbundle of $\iota^* E$; this operation is called the push-forward of Dirac structures (Bursztyn and Radko 2003):

$$\tau_S := \iota_* s(TS) = \{e \in E \; : \; \pi(e) \in TS \text{ and } e + \mathrm{Ann}(TS) \in s(TS)\}.$$

Note that τ_S is an extension of the tangent bundle of S by its conormal bundle:

$$0 \longrightarrow N^*S \longrightarrow \tau_S \longrightarrow TS \longrightarrow 0.$$

In the presence of the generalized complex structure, there is a natural compatibility condition, as follows:

Definition 18.5 *A* generalized complex submanifold *is a trivialization of the Courant algebroid along a submanifold $\iota : S \hookrightarrow M$ which is compatible with the generalized complex structure $\mathcal{J}$, in the sense that*

$$\mathcal{J}\tau_S = \tau_S, \tag{18.8}$$

where $\tau_S = \iota_* s(TS) \subset \iota^* E$.

As shown in Gualtieri (2007), in the complex case (and for the trivial gerbe), generalized complex submanifolds correspond to holomorphic submanifolds equipped with unitary holomorphic line bundles, whereas in the symplectic case they correspond to Lagrangian submanifolds equipped with flat line bundles *or* the co-isotropic A-branes of Kapustin and Orlov (2003). A useful general example of a generalized complex submanifold is the graph of an isomorphism of generalized complex manifolds, as follows. The notation $\overline{\mathcal{J}}$ denotes the same endomorphism as $\mathcal{J}$ but in the opposite Courant algebroid $\overline{E}$.

Example 18.2 Let $(M, \mathcal{J}_M)$, $(N, \mathcal{J}_N)$ be generalized complex manifolds. They are isomorphic when there is a Courant algebroid isomorphism in the sense of Example 18.1 which is a generalized complex submanifold of the product $(M \times N, \overline{\mathcal{J}}_M \times \mathcal{J}_N)$.

In Gualtieri (2007), it is shown that in the eigenspace decomposition with respect to $\mathcal{J}$

$$\tau_S \otimes \mathbb{C} = \ell + \bar{\ell},$$

the $+i$ eigenbundle ℓ inherits a Lie bracket, by extending sections randomly to sections over M which remain $+i$ eigensections of $\mathcal{J}$, taking their Courant bracket and restricting to S. Thus ℓ becomes an elliptic complex Lie algebroid over S. Therefore there is a notion of flat ℓ-module, The resulting ℓ-modules are called branes in analogy to the physics literature.

Definition 18.6 *A* generalized complex brane *is a vector bundle with flat ℓ-connection, supported over a generalized complex submanifold.*

Remark 18.1 Just as for generalized holomorphic bundles, we may choose to represent branes using unitary connections with values in τ_S^*, that is, operators

$$D : C^\infty(V) \longrightarrow C^\infty(\tau_S^* \otimes V)$$

with symbol given by the inclusion $T^*S \subset \tau_S^*$, and with vanishing curvature along $\ell \subset \tau_S \otimes \mathbb{C}$.

For a usual complex structure, a brane consists of a holomorphic bundle V supported on a complex submanifold $S \subset M$ together with a choice of holomorphic section $\phi \in H^0(S, N_{1,0}S \otimes \mathrm{End}(V))$ satisfying

$$\phi \wedge \phi = 0 \in H^0(S, \wedge^2 N_{1,0}S \otimes \mathrm{End}(V)),$$

where $N_{1,0}S$ denotes the holomorphic normal bundle of S.

On the other hand, for a symplectic structure, branes are complex flat bundles if they are supported on Lagrangian submanifolds; they may also be supported on co-isotropic submanifolds with holomorphic structure transverse to the characteristic foliation (Kapustin and Orlov 2003, Gualtieri 2007), in which case they are transversally holomorphic bundles, flat along the leaves.

Example 18.3 (Higgs bundles) Let $C \subset X$ be a curve in a complex surface equipped with holomorphic symplectic form $\Omega \in H^0(X, \Omega^{2,0})$, for example, a K3 surface. Also, let $V \longrightarrow C$ be a Higgs bundle in the sense of Hitchin (1987), that is, a holomorphic bundle together with a Higgs field $\theta \in H^0(C, \Omega^1 \otimes \mathrm{End}(V))$. Since C is Lagrangian with respect to Ω, we have an isomorphism $T_{1,0}C \longrightarrow$

$N^*_{1,0}C$, so that we may form $\phi = \Omega^{-1}\theta \in H^0(C, N_{1,0} \otimes \mathrm{End}(V))$, making (V, ϕ) into a brane for the complex structure.

On the other hand, if (V, θ) is a stable Higgs bundle, then by the existence theorem (Hitchin 1987) for solutions to Hitchin's equations we obtain a complex flat connection ∇ on V, rendering (C, V, ∇) into a symplectic brane with respect to either the real or imaginary parts of Ω.

Example 18.4 Let V be a generalized holomorphic bundle, that is, a complex vector bundle equipped with a flat L-connection, where $L = \ker(\mathcal{J} - i1)$ for $\mathcal{J}$ a generalized complex structure. Then the pullback of V to any generalized complex submanifold $S \subset M$ defines a generalized complex brane, as can be seen easily from the inclusion $\ell \subset L$.

Another simple example of a generalized complex brane occurs when it is supported on an isomorphism of generalized complex manifolds, as in Example 18.16.

Proposition 18.4 *Let $S \subset M_0 \times M_1$ define an isomorphism of the generalized complex manifolds $(M_0, \mathcal{J}_0)$, $(M_1, \mathcal{J}_1)$. Then the Lie algebroid ℓ is isomorphic to both $L_i = \ker(\mathcal{J}_i - i1)$, so that branes on S may be identified with generalized holomorphic bundles on either manifold.*

Proof. Let $\pi_i : M_0 \times M_1 \longrightarrow M_i$ be the usual projection maps. The subbundle $\ell \subset (\pi_0^* L_0 \oplus \pi_1^* L_1)|_S$ is transverse to both $\pi_i^* L_i$, as we now show. If $x \in \ell \cap \pi_0^* L_0$, for instance, then $(\pi_1)_* x = \varphi_*((\pi_0)_* x)$, where φ is the diffeomorphism defining S. This implies $(\pi_i)_* x = 0$ and $x \in N^*S \otimes \mathbb{C}$, which clearly is transverse to $\pi_0^* L_0$. Hence $x = 0$, and similarly for L_1.

This transversality means that we have isomorphic bundle maps onto each factor:

$$\begin{array}{ccccc} L_0 & \xleftarrow{p_0} & \ell & \xrightarrow{p_1} & L_1 \\ \downarrow & & \downarrow & & \downarrow \\ M_0 & \xleftarrow{\pi_0} & S & \xrightarrow{\pi_1} & M_1 \end{array} \tag{18.9}$$

We now show that the projections p_0, p_1 are isomorphisms of Lie algebroids.

Given $Z \in C^\infty(S, \ell)$, let $X = p_0(Z)$ and $Y = p_1(Z)$. Then Z may be expressed as $Z = (\pi_0^* X + \pi_1^* Y)|_S$. Computing the bracket of Z, Z', we may use the given extensions to $M_0 \times M_1$ and compute their Courant bracket:

$$\begin{aligned} [Z, Z'] &= [\pi_0^* X + \pi_1^* Y, \pi_0^* X' + \pi_1^* Y']|_S \\ &= [\pi_0^* X, \pi_0^* X']|_S + [\pi_1^* Y, \pi_1^* Y']|_S \\ &= (\pi_0^*[X, X'] + \pi_1^*[Y, Y'])|_S, \end{aligned} \tag{18.10}$$

where we use the fact that sections pulled back from opposite factors M_0, M_1 Courant commute in the product. Applying the projections to the final formula, we obtain

$$p_0([Z, Z']) = [p_0(Z), p_0(Z')] \text{ and } p_1([Z, Z']) = [p_1(Z), p_1(Z')],$$

as required. □

We now describe the general form of a generalized complex brane when it is supported on the whole manifold M; these are usually called "space-filling branes." We first observe that the requirement that M be a generalized complex submanifold of itself places a very strong constraint on $\mathcal{J}$.

Proposition 18.5 *$(M, \mathcal{J})$ is a generalized complex submanifold of itself if and only if there exists an integrable isotropic splitting $E = T \oplus T^*$ of the Courant algebroid with respect to which $\mathcal{J}$ has the form*

$$\mathcal{J} = \begin{pmatrix} I & Q \\ & -I^* \end{pmatrix}, \tag{18.11}$$

where I is a usual complex structure on the manifold and $\sigma = P + iQ$, for $P = IQ$, is a holomorphic Poisson structure, that is, it satisfies $[\sigma, \sigma] = 0$.

Proof. Compatibility of the splitting with $\mathcal{J}$ forces $\mathcal{J}T = T$, which holds if and only if $\mathcal{J}$ is upper triangular, and the orthogonality of $\mathcal{J}$ together with the fact $\mathcal{J}^2 = -1$ guarantees that I is an almost complex structure and that Q is a bivector of type $(2,0) + (0,2)$. The $-i$ eigenbundle of $\mathcal{J}$ is then the direct sum of $T_{0,1}$ with the graph of $\sigma : T^*_{1,0} \longrightarrow T_{1,0}$. This is closed (involutive) for the Courant bracket if and only if $T_{0,1}$ is integrable and $[\sigma, \sigma] = 0$, as required. □

In the splitting $E = T \oplus T^*$ for which $\mathcal{J}$ has the form (18.11), we see that $\tau_S = TM$, and further that $\ell = T_{1,0}$, so that ℓ-modules are precisely holomorphic bundles with respect to the complex structure I.

18.5 Multiple branes and holomorphic Poisson varieties

Suppose that we have a Courant trivialization s making $(M, \mathcal{J})$ a generalized complex submanifold of itself, so that $E = T \oplus T^*$ and $\mathcal{J}$ has the form (18.11). Now we investigate the consequences of having a second trivialization s' which is also compatible with $\mathcal{J}$. Let $F \in \Omega^2_{cl}(M, \mathbb{R})$ be the two-form taking s to s'. By Proposition 18.5, and the fact that the Poisson structure Q is independent of splitting, we have

$$\begin{pmatrix} 1 & \\ -F & 1 \end{pmatrix} \begin{pmatrix} I & Q \\ & -I^* \end{pmatrix} \begin{pmatrix} 1 & \\ F & 1 \end{pmatrix} = \begin{pmatrix} J & Q \\ & -J^* \end{pmatrix}, \tag{18.12}$$

for a second complex structure J such that $\sigma' = JQ + iQ$ is holomorphic Poisson. In particular we note the important fact that a generalized complex structure may be expressed as a holomorphic Poisson structure *in several different ways*, and with respect to different underlying complex structures, depending on the choice of splitting. Equation (18.12) is equivalent to the conditions

$$\begin{cases} J - I = QF, \\ FJ + I^*F = 0. \end{cases} \tag{18.13}$$

Phrased as a single condition on F, we obtain the non-linear equation

$$FI + I^*F + FQF = 0, \tag{18.14}$$

which may be viewed as a deformation of the usual condition $FI + I^*F = 0$ that F be of type $(1,1)$ with respect to the complex structure. Equation (18.14) has been studied by Kapustin (2005), who showed that it corresponds to a non-commutative version of the $(1,1)$ condition via the Seiberg–Witten transform on tori. We take a different approach here, focusing rather on a groupoid interpretation of the equivalent system (18.13).

The set of compatible global Courant trivializations forms a groupoid; we may label each trivialization by the complex structure it induces on the base, and we see from (18.12) or (18.13) that if F_{IJ} takes I to J and F_{JK} takes J to another trivialization K, then $F_{IJ} + F_{JK}$ takes I to K.

Definition 18.7 *Fix a real manifold M with real Poisson structure Q. Let $\mathcal{G}$ be the groupoid whose objects are holomorphic Poisson structures (I_i, σ_i) on M with fixed imaginary part given by $\mathrm{Im}(\sigma_i) = Q$, and whose morphisms $\mathrm{Hom}(i,j)$ consist of real closed two-forms $F_{ij} \in \Omega^2_{cl}(M, \mathbb{R})$ such that*

$$\begin{cases} I_j - I_i = QF_{ij}, \\ F_{ij}I_j + I_i^*F_{ij} = 0. \end{cases} \tag{18.15}$$

The composition of morphisms is then simply addition of two-forms $F_{ij} + F_{jk}$. In keeping with the interpretation of F_{ij} as differences between gerbe trivializations, we could define $\mathrm{Hom}(i,j)$ to consist of unitary line bundles L_{ij} with curvature F_{ij}, such that composition of morphisms would coincide with tensor product.

Automorphisms of the Courant algebroid which fix $\mathcal{J}$ give rise to automorphisms of the groupoid of trivializations defined above; we describe these now. Orthogonal automorphisms of the standard Courant bracket on $T \oplus T^*$ consist of pairs $(\varphi, B) \in \mathrm{Diff}(M) \times \Omega^2_{cl}(M, \mathbb{R})$, which act on $T \oplus T^*$ via $X + \xi \mapsto \varphi_* X + (\varphi^{-1})^*\xi + i_{\varphi_* X}B$. Since our generalized complex structure has the form (18.11), we may easily determine its automorphism group.

Proposition 18.6 *The automorphism group* Aut($\mathcal{J}$) *of the generalized complex structure* (18.11) *is the set of pairs* $(\varphi, B) \in \mathrm{Diff}(M) \times \Omega^2_{cl}(M, \mathbb{R})$ *such that*

$$Q^\varphi = Q$$
$$I^\varphi - I = QB$$
$$BI^\varphi + I^* B = 0,$$

where $Q^\varphi = \varphi_* Q$ *and* $I^\varphi = \varphi_* I \varphi_*^{-1}$.

These automorphisms therefore act on the groupoid of global generalized complex submanifolds (18.15), sending $(I_i, \sigma_i) \mapsto (I_i^\varphi, (\varphi^{-1})^* \sigma_i)$ and $F_{ij} \mapsto (\varphi^{-1})^* F_{ij} + B$. Of course, we may wish to interpret B as the curvature of a unitary line bundle U, in which case it would act on the groupoid line bundles L_{ij} by tensor product $L_{ij} \mapsto (\varphi^{-1})^* L_{ij} \otimes U$.

Instead of viewing F_{ij} as the difference between two generalized complex submanifolds of $(M, \mathcal{J})$, we may interpret (18.12) as giving an isomorphism between two different generalized complex structures on $T \oplus T^*$. This rephrasing leads immediately to the following:

Proposition 18.7 *Let* (I_i, σ_i) *and* (I_j, σ_j) *be holomorphic Poisson structures on* M *with associated generalized complex structures* $\mathcal{J}_i, \mathcal{J}_j$ *on* $T \oplus T^*$ *via* (18.11), *let* $\mathrm{Im}(\sigma_i) = \mathrm{Im}(\sigma_j) = Q$, *and let* $F_{ij} \in \Omega^2_{cl}(M, \mathbb{R})$ *satisfy* (18.15). *Then the graph of* F_{ij} *over the diagonal* $\Delta \subset M \times M$ *defines a generalized complex submanifold of* $(M \times M, \overline{\mathcal{J}}_i \times \mathcal{J}_j)$, *yielding an isomorphism of generalized complex manifolds*

$$(M, \mathcal{J}_i) \xrightarrow{\cong} (M, \mathcal{J}_j). \tag{18.16}$$

In view of Proposition 18.4, this result implies that a morphism F_{ij} from (I_i, σ_i) to (I_j, σ_j) induces an equivalence between the categories of generalized holomorphic bundles associated to $\mathcal{J}_i, \mathcal{J}_j$. We now explain this equivalence explicitly, and its significance for the holomorphic Poisson structures involved.

Proposition 18.8 *Let* $\mathcal{J}$ *be of the form* (18.11), *for* I *is a complex structure and* $\sigma = P + iQ$ *is a holomorphic Poisson structure. Then a generalized holomorphic bundle is precisely a holomorphic Poisson module (Polishchuk 1997), that is, a holomorphic bundle* V *with an additional action of the structure sheaf on the sheaf of holomorphic sections, denoted* $\{f, s\}$, *satisfying*

$$\{f, gs\} = \{f, g\}s + g\{f, s\}, \tag{18.17}$$
$$\{\{f, g\}, s\} = \{f, \{g, s\}\} - \{g, \{f, s\}\}, \tag{18.18}$$

where $f, g \in \mathcal{O}$, $s \in \mathcal{O}(V)$, *and* $\{f, g\}$ *denotes the Poisson bracket induced by* σ.

Proof. Let $L = \ker(\mathcal{J} + i1)$, so that for $\mathcal{J}$ as in (18.11), we have

$$L = T_{0,1} \oplus \Gamma_\sigma, \tag{18.19}$$

where $\Gamma_\sigma = \{\xi + \sigma(\xi) \ : \ \xi \in T^*_{1,0}\}$. Let $\overline{\partial}_V : C^\infty(V) \longrightarrow C^\infty(L^* \otimes V)$ be a generalized holomorphic structure. Decomposing using (18.19) and identifying $\Gamma_\sigma = T^*_{1,0}$, we write $\overline{\partial}_V = \overline{\partial}' + \overline{\partial}''$, where $\overline{\partial}' : C^\infty(V) \longrightarrow C^\infty(T^*_{0,1} \otimes V)$ is a usual holomorphic structure and $\overline{\partial}'' : C^\infty(V) \longrightarrow C^\infty(T_{1,0} \otimes V)$ satisfies, for $f \in C^\infty(M, \mathbb{C})$ and $s \in C^\infty(V)$,

$$\overline{\partial}''(fs) = f\overline{\partial}'' s + Z_f \otimes s,$$

where $Z_f = \sigma(df)$ is the Hamiltonian vector field of f. This is equivalent to condition (18.17). Furthermore $\overline{\partial}_V^2 = 0$ implies that $\overline{\partial}''$ is holomorphic and defines a Poisson module structure via

$$\{f, s\} = \overline{\partial}''_{\partial f} s,$$

since $(\overline{\partial}'')^2 = 0$ implies $\overline{\partial}''_{\partial\{f,g\}} = [\overline{\partial}''_{\partial f}, \overline{\partial}''_{\partial g}]$, which is equivalent to (18.18), as required. □

Letting $L_k = \ker(\mathcal{J}_k + i1)$, we see from (18.12) that $\exp(F_{ij})$ takes L_i to L_j. Hence the map on generalized holomorphic bundles induced by the isomorphism (18.16) may be described as composition with $\exp(F_{ij})$

$$\overline{\partial}_i \longmapsto e^{F_{ij}} \circ \overline{\partial}_i.$$

This map may be made more explicit in terms of the associated generalized connections. Choose a Hermitian structure on the $\mathcal{J}_i$-holomorphic bundle (i.e. σ_i-Poisson module), and let $D = \nabla + \chi$ be the extension of $\overline{\partial}_i$ as in Proposition 18.2. Then F_{ij} acts on D via

$$\begin{cases} D \longmapsto D' = \nabla' + \chi, \\ \nabla' = \nabla + F_{ij}(\chi). \end{cases} \tag{18.20}$$

which then defines a σ_j-Poisson module. It is important to note that the σ_i-Poisson module, which is I_i-holomorphic, inherits via (18.20) a I_j-holomorphic structure, without the presence of any holomorphic map between (M, I_j) and (M, I_i).

Given this result, it is natural to ask how restrictive the condition of admitting a Poisson module structure actually is. The following is a simple result describing the complete obstruction to the existence of a Poisson module structure on a holomorphic line bundle.

Proposition 18.9 *Let M be a holomorphic Poisson manifold, and let V be a holomorphic line bundle on M. Then the Atiyah class of V, $\alpha \in H^1(T^*_{1,0})$, combines with the Poisson structure $\sigma \in H^0(\wedge^2 T_{1,0})$ to give the class $\sigma\alpha \in H^1(T_{1,0})$. If $\sigma\alpha = 0$, then there is a well-defined secondary characteristic class $f_\alpha \in H^2_\sigma(M)$ in Poisson cohomology. V admits a Poisson module structure if and only if both classes $\{\sigma\alpha, f_\alpha\}$ vanish. The space of Poisson module structures is affine, modeled on $H^1_\sigma(M)$.*

Proof. A Poisson module structure on V is a holomorphic differential operator $\partial : \mathcal{O}(V) \longrightarrow \mathcal{O}(T_{1,0} \otimes V)$ satisfying $\partial(fs) = f\partial s + Z_f \otimes s$, for $f \in \mathcal{O}$ and $s \in \mathcal{O}(V)$, where Z_f is the σ-Hamiltonian vector field of f, and such that the curvature vanishes. Let $\{U_i\}$ be an open cover of M and let $\{s_i \in \mathcal{O}(U_i, V)\}$ be a local trivialization of V such that $s_i = g_{ij}s_j$ for holomorphic transition functions g_{ij}; then

$$\partial s_i = X_i \otimes s_i, \tag{18.21}$$

where X_i are holomorphic Poisson vector fields (since $\partial_{d\{f,g\}} = [\partial_{df}, \partial_{dg}]$) such that

$$X_i - X_j = Z_{\log g_{ij}}. \tag{18.22}$$

The Hamiltonian vector fields $Z_{\log g_{ij}} = \sigma(d \log g_{ij})$ are a Čech representative for the image of the Atiyah class under σ. Therefore, (18.22) holds if and only if $\sigma\alpha = 0 \in H^1(T_{1,0})$. If $\sigma\alpha = 0$, then we may solve (18.22) for some holomorphic vector fields $\tilde{X}_i$. We can modify these by a global holomorphic vector field so that they are each Poisson if and only if the global bivector field f_σ defined by $f_\sigma|_{U_i} = [\tilde{X}_i, \sigma]$ vanishes in Poisson cohomology, that is, $f_\sigma = [Y, \sigma]$ for $Y \in H^0(T_{1,0})$, in which case $X_i = \tilde{X}_i - Y|_{U_i}$ defines a Poisson module structure as required.

Given any holomorphic Poisson vector field $Z \in H^1_\sigma(M)$ and Poisson module structure ∂, the sum $\partial + Z$ defines a new Poisson module structure. Conversely, two Poisson module structures ∂', ∂ must satisfy $\partial' - \partial \in H^1_\sigma(M)$, as claimed. □

It is remarked in (Polishchuk 1997) that the canonical line bundle K always admits a natural Poisson module structure for any holomorphic Poisson structure σ via the action, for $f \in \mathcal{O}$ and $\rho \in \mathcal{O}(K)$,

$$\{f, \rho\} = L_{Z_f}\rho.$$

Based on these considerations, we obtain the following example.

Example 18.5 Let $M = \mathbb{C}P^2$, equipped with a holomorphic Poisson structure $\sigma \in H^0(\mathcal{O}(3))$. Note that $H^1(T_{1,0}) = 0$. Then $K = \mathcal{O}(-3)$ is canonically a Poisson module, and since $\mathcal{O}(1)^{-3} = K$, we see that the obstruction f_σ from Proposition 18.9 must vanish for $\mathcal{O}(1)$ as well (note that $\dim H^2_\sigma(\mathbb{C}P^2) = 2$, so the obstruction space is nonzero). Hence all holomorphic line bundles $\mathcal{O}(k)$ admit Poisson module structures. If σ is generic, these Poisson module structures are unique, since $H^1_\sigma(M) = 0$, due to the fact that only the zero holomorphic vector field on $\mathbb{C}P^2$ stabilizes a smooth cubic curve.

We conclude this section with a simple example of a generalized complex manifold admitting multiple trivializations with *non-biholomorphic* induced complex structures.

Proposition 18.10 *Let $E_0 = E \times \mathbb{C}$, the trivial line bundle over an elliptic curve E, and let E_c, for $c \in \mathbb{R}$, be the alternative holomorphic structure on $E \times \mathbb{C}$ obtained by endowing the bundle $E \times \mathbb{C}$ with the holomorphic structure associated to the point $ic \in H^1(E, \mathcal{O}) = \mathbb{C}$. Then E_0 and E_c are diffeomorphic, non-biholomorphic complex manifolds. They are equipped with canonical holomorphic Poisson structures σ_0, σ_c vanishing to first order on the zero section, and furthermore (E_0, σ_0) and (E_c, σ_c) are isomorphic as generalized complex manifolds $\forall c \in \mathbb{R}$ (and hence have equivalent categories of Poisson modules).*

Proof. Represent E as $\mathbb{C}^*/\{z \mapsto \lambda z\}$ and let w be the linear coordinate on the fiber of $E \times \mathbb{C}$. Then the holomorphic structure E_c is given by the complex volume form

$$\Omega_c = \frac{dz}{z} \wedge \left(dw + icw \frac{d\bar{z}}{\bar{z}} \right),$$

and the holomorphic Poisson structure σ_c is given by

$$\sigma_c = \left(z\frac{\partial}{\partial z} + ic\bar{w}\frac{\partial}{\partial \bar{w}} \right) \wedge w\frac{\partial}{\partial w}.$$

The pure spinor corresponding to the generalized complex structure (E_c, σ_c) is

$$\rho_c = e^{\sigma_c}\Omega_c.$$

Now let $F_c = ic\frac{dz \wedge d\bar{z}}{z\bar{z}}$ be a real multiple of the volume form on E (which may be viewed as a curvature when $c \in 2\pi\mathbb{Z}$). Then we verify that

$$\begin{aligned} e^{F_c}e^{\sigma_0}\Omega_0 &= w + \tfrac{dz}{z} \wedge \left(dw + icw\tfrac{d\bar{z}}{\bar{z}}\right) \\ &= e^{\sigma_c}\Omega_c, \end{aligned}$$

showing that (E_0, σ_0) and (E_c, σ_c) are isomorphic as generalized complex manifolds. □

18.6 Relation to generalized Kähler geometry

A generalized Kähler structure is a pair $(\mathcal{J}_A, \mathcal{J}_B)$ of commuting generalized complex structures such that

$$G(\cdot, \cdot) = \langle \mathcal{J}_A \cdot, \mathcal{J}_B \cdot \rangle$$

is a generalized Riemannian metric.

In Gualtieri (2004) it is shown that the integrability of the pair $(\mathcal{J}_A, \mathcal{J}_B)$ is equivalent to the fact that the induced decomposition of the definite subspaces $C_\pm$ given by

$$C_\pm \otimes \mathbb{C} = L_\pm \oplus \overline{L_\pm},$$

where $L_\pm = \ker(\mathcal{J}_A - i1) \cap \ker(\mathcal{J}_B \mp i1)$, satisfies the condition that $L_\pm$ are each involutive. Using the canonical generalized connection D introduced in

Theorem 18.1, we provide the following equivalent description of generalized Kähler geometry:

Theorem 18.2 *Let G be a generalized metric and let $\mathcal{J}$ be a G-orthogonal almost generalized complex structure. Then $(\mathcal{J}, G)$ defines a generalized Kähler structure if and only if $D\mathcal{J} = 0$ and the torsion $T_D \in C^\infty(\wedge^3 E)$ is of type $(2,1) + (1,2)$ with respect to $\mathcal{J}$.*

Proof. We leave the forward direction to the reader. We show that if $D\mathcal{J} = 0$ (where, as usual, $(D_x\mathcal{J})y = D_x(\mathcal{J}y) - \mathcal{J}(D_x y)$) and the torsion is as above, then $\mathcal{J}$ is integrable as a generalized complex structure. Note that under these assumptions, the complementary generalized complex structure $\mathcal{J}' = G\mathcal{J}$ would also be covariant constant, and be compatible with the torsion as well, by Proposition 18.1. Therefore by the following argument $\mathcal{J}'$ is also integrable, and we obtain the result.

We compute the Nijenhuis tensor of $\mathcal{J}$, for $x, y, z \in C^\infty(E)$ (in the following, $[\cdot,\cdot]$ refers to the skew-symmetrized Courant bracket):

$$\begin{aligned}\langle N_{\mathcal{J}}(x,y), z\rangle &= \langle [\mathcal{J}x, \mathcal{J}y] - \mathcal{J}[\mathcal{J}x, y] - \mathcal{J}[x, \mathcal{J}y] - [x,y], z\rangle \\ &= \langle D_z(\mathcal{J}x), \mathcal{J}y\rangle - \langle D_z(\mathcal{J}y), \mathcal{J}x\rangle + \langle D_{\mathcal{J}z}(\mathcal{J}x), y\rangle - \langle D_{\mathcal{J}z}y, \mathcal{J}x\rangle \\ &\quad + \langle D_{\mathcal{J}z}x, \mathcal{J}y\rangle - \langle D_{\mathcal{J}z}(\mathcal{J}y), x\rangle - \langle D_z x, y\rangle + \langle D_z y, x\rangle \\ &\quad - T_D(\mathcal{J}x, \mathcal{J}y, z) - T_D(\mathcal{J}x, y, \mathcal{J}z) - T_D(x, \mathcal{J}y, \mathcal{J}z) - T(x,y,z).\end{aligned}$$

The first eight terms cancel since $D_x(\mathcal{J}y) = \mathcal{J}D_x y$, and the last four terms cancel since T_D is of type $(2,1) + (1,2)$. Therefore $\mathcal{J}$ is integrable, as claimed. □

We now explain that a solution to the system (18.13), if positive in a certain sense, gives rise to a generalized Kähler structure. When the Poisson structure Q vanishes, this result specializes to the fact that a positive holomorphic line bundle with Hermitian structure defines a Kähler structure.

Definition 18.8 *Let (I, J, Q, F) be a solution to the system* (18.13)*, that is, it defines two global Courant trivializations compatible with a generalized complex structure, separated by the two-form F. Then*

$$g = -\tfrac{1}{2}F(I + J)$$

is a symmetric tensor, and if it is positive-definite, we say that F is positive.

If F is positive, then (g, I), (g, J) are both Hermitian structures. Let $\omega_I = gI$, $\omega_J = gJ$ be their associated two-forms. Then we have the following:

Theorem 18.3 *Let (I, J, Q, F) be as above, and let F be positive. Then the pair*

$$\mathcal{J}_B = \frac{1}{2}\begin{pmatrix}1 & \\ b & 1\end{pmatrix}\begin{pmatrix}J+I & -(\omega_J^{-1}-\omega_I^{-1}) \\ \omega_J-\omega_I & -(J^*+I^*)\end{pmatrix}\begin{pmatrix}1 & \\ -b & 1\end{pmatrix},$$
$$\mathcal{J}_A = \frac{1}{2}\begin{pmatrix}1 & \\ b & 1\end{pmatrix}\begin{pmatrix}J-I & -(\omega_J^{-1}+\omega_I^{-1}) \\ \omega_J+\omega_I & -(J^*-I^*)\end{pmatrix}\begin{pmatrix}1 & \\ -b & 1\end{pmatrix} \tag{18.23}$$

defines a generalized Kähler structure on the standard Courant algebroid $(T \oplus T^*, [\cdot,\cdot]_0)$, for $b \in \Omega^2(M, \mathbb{R})$ given by

$$b = -\tfrac{1}{2}F(J-I).$$

Proof. It is easily verified that $\mathcal{J}_A^2 = \mathcal{J}_B^2 = -1$ and that $[\mathcal{J}_A, \mathcal{J}_B] = 0$. To show integrability, we first observe that $\mathcal{J}_A$ has the form of a pure symplectic structure; indeed, with the definitions above,

$$\mathcal{J}_A = \begin{pmatrix} & -F^{-1} \\ F & \end{pmatrix}.$$

We see therefore that $\mathcal{J}_A$ is integrable since $dF = 0$.

The structure $\mathcal{J}_B$ is also integrable, as follows. Let $L_B = \ker(\mathcal{J}_B - i)$ and Let $L_B = L_+ \oplus L_-$ be its decomposition into $\pm i$ eigenspaces for $\mathcal{J}_A$. Then

$$L_+ = \{X + (b+g)X \ : \ X \in T_J^{1,0}\}$$
$$L_- = \{X + (b-g)X \ : \ X \in T_I^{1,0}\}$$

It follows from the definitions of b, g that $b + g = -FJ$ whereas $b - g = FI$. As a result we have

$$L_+ = \{X - iFX \ : \ X \in T_J^{1,0}\}$$
$$L_- = \{X + iFX \ : \ X \in T_I^{1,0}\}$$

which are integrable precisely when $i_X i_Y dF = 0$ for all X, Y in $T_I^{1,0}$ or $T_J^{1,0}$. Of course this holds since F is closed. □

We note that the converse of this argument also holds; using the result from Gualtieri (2004) that any generalized Kähler structure has the form (18.23), we may show that any generalized Kähler structure $(\mathcal{J}_A, \mathcal{J}_B)$ with the property that $\mathcal{J}_A$ is symplectic gives rise to a solution to the system (18.13). More explicitly, given the bi-Hermitian data (g, I, J) we determine F via

$$F = -2g(I+J)^{-1},$$

where $(I + J)$ is invertible by the assumption on $\mathcal{J}_A$, and the Poisson structure Q is given by

$$Q = (J-I)F^{-1} = \tfrac{1}{2}[I, J]g^{-1}.$$

This is consistent with Hitchin's general observation (Hitchin 2006*b*) that $[I, J]g^{-1}$ defines a holomorphic Poisson structure for both I and J, for any generalized Kähler structure.

In fact, the interpretation of F_{ij} in Proposition 18.7 as defining a morphism between holomorphic Poisson structures allows us to view the generalized Kähler structure as a morphism between the holomorphic Poisson structures (I, σ_I), (J, σ_J). This point of view is related to the approach in Lindström *et al.* (2007). to defining a generalized Kähler potential, and may help to resolve the problems encountered there at non-regular points.

Given the equivalence between certain generalized Kähler structures and configurations of generalized complex submanifolds shown in this section, we may apply it to produce new examples of generalized Kähler structures, or indeed of configurations of branes. We do this in the following section.

18.7 Construction of generalized Kähler metrics

Given a generalized complex submanifold, it is natural to construct more by deformation; this is a familiar construction in symplectic geometry, where new Lagrangian submanifolds may be produced by applying Hamiltonian or symplectic diffeomorphisms. Therefore we would like to deform a given generalized complex submanifold by an automorphism of the underlying geometry, as described in Proposition 18.6. If the automorphism used is positive in the sense of Definition 18.8, then we will have constructed a generalized Kähler structure, by Theorem 18.3. This construction is inspired by a construction of Joyce contained in Apostolov, Gauduchon, and Grantcharov (1999), and its generalization by Hitchin (2006*b*) to the construction of generalized Kähler structures on Del Pezzo surfaces.

To reiterate, the goal of the construction is as follows: given a holomorphic Poisson structure (I, σ_I) on M, with real and imaginary parts $\sigma_I = P + iQ$, find a second complex structure J and a two-form F solving the system (18.13), that is,

$$\begin{cases} J - I = QF, \\ FJ + I^*F = 0. \end{cases} \tag{18.24}$$

We are particularly interested in the case where $g = -\frac{1}{2}F(I + J)$ is positive-definite, as this then defines a generalized Kähler structure, however the construction does not depend on it.

In this construction, the complex structure J will be obtained from I by flowing along a vector field; as a result, J will be biholomorphic to I. Also, we shall describe the construction in the case that F is the curvature of a unitary connection, although it will be clear that integrality of the form F is not required.

1. We begin with a Hermitian complex line bundle L over a compact complex manifold M; the two-form F solving (18.24) will be chosen from the

cohomology class $c_1(L)$. We first assume that L admits a holomorphic structure $\overline{\partial}_0$ with respect to the "initial" complex structure $I = I_0$. The associated Chern connection will be called ∇_0, and its curvature denoted F_0. Recall that ∇_0 is the unique Hermitian connection on L such that $\nabla_0^{0,1} = \overline{\partial}_0$.

2. We then assume that L admits the structure of a holomorphic Poisson module with respect to a holomorphic Poisson structure σ_I on M, which by Proposition 18.9 occurs if and only if $[\sigma_I F_0] \in H^1_{\overline{I}}(T_{1,0})$ vanishes and the secondary characteristic class in $H^2_{\sigma_I}(M)$ also vanishes. By Proposition 18.2, we construct the Hermitian generalized connection D associated to this generalized holomorphic structure, and decompose it according to the splitting $T \oplus T^*$:

$$D = \nabla_0 + iX,$$

where X is a real Q-Poisson vector field such that $\overline{\partial} X^{1,0} = \sigma_I F_0$, giving rise to the real equations

$$\begin{cases} L_X Q &= 0, \\ L_X I_0 &= QF_0. \end{cases} \tag{18.25}$$

3. Let φ_t be the time t flow of the vector field X. Then we may transport F_0 by the flow, yielding the cohomologous family of two-forms $F_t = \varphi^*_{-t} F_0$, which satisfies

$$\dot{F}_t = L_X F_t = d i_X F_t.$$

We may also transport I_0 by the flow, obtaining a family $I_t = I_0^{\varphi_t}$ satisfying

$$\dot{I}_t = L_X I_t = QF_t,$$

by (18.25). Note that F_t is type $(1,1)$ with respect to I_t. Also note that F_t is the curvature of the family of connections

$$\nabla_t = \nabla_0 + \int_0^t i_X F_s ds,$$

which are therefore the Chern connections associated to a family of holomorphic structures $\overline{\partial}_t$ on L, each holomorphic with respect to I_t.

4. We then compute the difference

$$\begin{aligned} I_t - I_0 &= \int_0^t QF_s ds \\ &= tQ\tfrac{1}{t} \int_0^t F_s ds \\ &= tQ\overline{F}_t, \end{aligned} \tag{18.26}$$

where $\overline{F}_t$ is the curvature of the average Chern connection on L:

$$\overline{\nabla}_t = \tfrac{1}{t}\int_0^t \nabla_s ds.$$

Setting $t = 1$ we obtain a solution to the first part of (18.24):

$$I_1 - I_0 = Q\overline{F}_1.$$

5. Observe that the second part of (18.24) is automatically satisfied: from (18.26) we have $I_t - I_0 = QG_t$, where

$$G_t = \int_0^t F_s ds.$$

For $t = 0$, the expression

$$G_t I_t + I_0^* G_t \tag{18.27}$$

vanishes, since $G_0 = 0$. Taking the time derivative, we obtain

$$\begin{aligned}\dot{G}_t I_t + G_t \dot{I}_t + I_0^* \dot{G}_t &= F_t I_t + G_t Q F_t + I_0^* F_t\\ &= -(I_t^* - I_0^*)F_t + G_t Q F_t\\ &= -G_t Q F_t + G_t Q F_t = 0.\end{aligned}$$

Therefore (18.27) vanishes for all t; since $\overline{F_t} = t^{-1}G_t$, we obtain the result.

6. *Positivity:* If F_0 is positive, that is, if the original line bundle L is positive, then $\overline{F}_t$ is positive for sufficiently small t. By (18.26), this gives a solution to the system (18.24) for the Poisson structure $t\sigma_I$ replacing σ_I.

We summarize the main result of this construction in the following:

Theorem 18.4 *Let L be a positive holomorphic line bundle with Poisson module structure over a compact complex manifold with holomorphic Poisson structure σ. Let (g_0, I_0) be the original Kähler structure it determines. Then the choice of Hermitian structure on L determines a canonical family of generalized Kähler structures $\{(g_t, I_t, I_0) : -\epsilon < t < \epsilon\}$ such that the complex structure I_t coincides with I_0 only along the vanishing locus of σ for t /=.0*

Example 18.6 One case where the existence of a positive Poisson module is guaranteed is in the case of a Fano Poisson manifold, since the anti-canonical bundle, which always admits a Poisson module structure, is positive. This extends the result of Hitchin (2007), who showed that all smooth Fano surfaces (the Del Pezzo surfaces) admit the families of generalized Kähler structures described here.

We remark finally upon the relation of our construction to Hitchin's result for Del Pezzo surfaces (Hitchin 2007). To obtain the family of generalized Kähler structures, he used a flow generated by a Poisson vector field X which he expressed as the Hamiltonian vector field of $\log|s|^2$, for s a holomorphic section of the anti-canonical bundle vanishing at the zero locus of the Poisson structure. From our point of view, he was making use of the generalized Poincaré–Lelong formula of Proposition (18.3), since in the two-dimensional case there is always a non-trivial generalized holomorphic section of the anti-canonical bundle of a Poisson surface, namely, the Poisson structure itself. However, in higher dimension, there is a dearth of global generalized holomorphic sections; indeed by Proposition (18.3), such a section (if generically nonzero) must vanish only along the zero locus of σ, which has codimension greater than one in general.

18.8 Relation to non-commutative algebraic geometry

Since the observation in Gualtieri (2004) that the deformation space of a complex manifold as a generalized complex manifold includes the "non-commutative" directions in $H^0(\wedge^2 T_{1,0})$, it was hoped that there might be a more precise relationship between generalized complex structures and non-commutativity. The presence of an underlying Poisson structure, for example, lends credence to this idea. In the realm of generalized Kähler four-manifolds, we have even more evidence in this direction, since, as observed originally in Apostolov, Gauduchon, and Grantcharov (1999), the locus where the bi-Hermitian complex structures (I_+, I_-) coincide is an anti-canonical divisor for both structures.

If smooth, each connected component of this coincidence locus is an elliptic curve C, and we may view it as embedded in two different complex manifolds $X_\pm = (M, I_\pm)$.

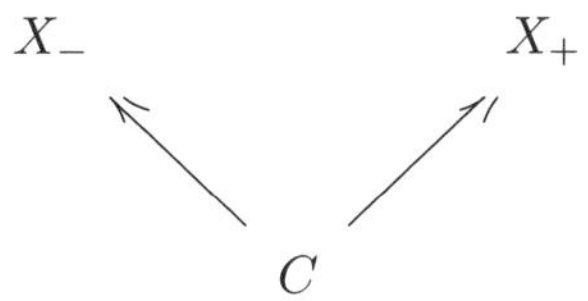

In the examples constructed in Section 18.7 and by Hitchin (2007), $X_\pm$ have natural holomorphic line bundles sitting over them, which we called L_0, L_1. Pulling them back to C, we obtain holomorphic line bundles $\mathcal{L}_0, \mathcal{L}_1$ over C. Furthermore, since the flow satisfied $L_X I_t = QF_t$, we know that the flow restricts to a holomorphic flow on C, the vanishing locus of Q. As a result, $\mathcal{L}_0, \mathcal{L}_1$ are related by an automorphism of C. This data $(C, \mathcal{L}_0, \mathcal{L}_1)$ is precisely what is used in the approach of Bondal and Polishchuk (1993) to the classification of $\mathbb{Z}$-algebras describing non-commutative projective surfaces.

In fact, in our construction, we produce an example of an automorphism $(\varphi, F) = (\varphi_1, \overline{F}_1)$ in the sense of Proposition 18.6. Therefore, we may apply it successively, producing an infinite family of generalized complex submanifolds with induced complex structures $\{I_k = I_0^{\varphi_1^k}\}$, each I_k separated from I_0 by the line bundle L^k with connection $\overline{\nabla}_k$, and all coinciding on the vanishing locus C of Q. As a result, we obtain an infinite family of embeddings

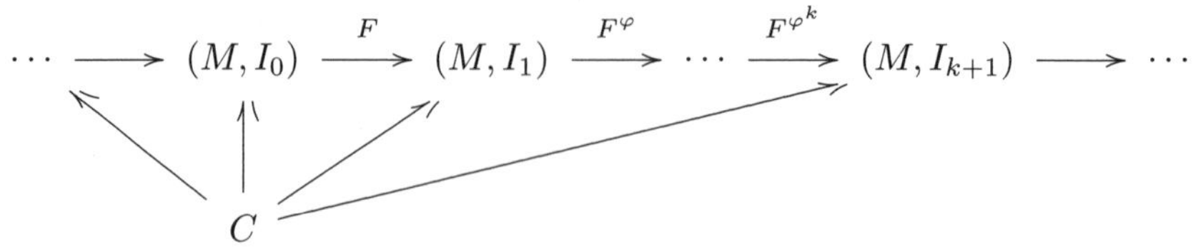

Where the arrows on the top row indicate morphisms in the sense of the groupoid of Definition 18.7. This may provide an alternative interpretation of Van den Bergh's construction of the twisted homogeneous coordinate ring (see Stafford and van den Bergh 2001): let $\mathcal{L} = L_0|_C$, and let $\mathcal{L}^{\varphi} = (\varphi^{-1})^*\mathcal{L}$. Then define the vector spaces

$$\mathrm{Hom}(i,j) = H^0(C, \mathcal{L}^{\varphi^i} \otimes \mathcal{L}^{\varphi^{i+1}} \otimes \cdots \otimes \mathcal{L}^{\varphi^{j-1}})$$

and define a $\mathbb{Z}^{>0}$-graded algebra structure on

$$A^{\bullet} = \bigoplus_{k \geq 0} \mathrm{Hom}(0,k), \tag{18.28}$$

via the multiplication, for $a \in A^p$ and $b \in A^q$:

$$a \cdot b = a \otimes b^{\varphi^p},$$

where we use the natural map $b \mapsto b^{\varphi^p}$ taking $\mathrm{Hom}(0,q) \longrightarrow \mathrm{Hom}(p,p+q)$, and the tensor product is viewed as a composition of morphisms.

Of course this is nothing but a recasting of the Van den Bergh construction; there is a sense in which it captures only certain morphisms between the generalized complex submanifolds, namely, those which are visible upon restriction to C. Though rare, there are sometimes generalized holomorphic sections of the bundles L^k supported over all of M. In some sense, these sections must be included in the morphism spaces as well.

For instance, performing our construction for $L = \mathcal{O}(1)$ over $\mathbb{C}P^2$, equipped with a holomorphic Poisson structure $\sigma \in H^0(\mathbb{C}P^2, \mathcal{O}(3))$ with smooth zero locus $\iota : C \hookrightarrow \mathbb{C}P^2$, the graded algebra (18.28) has linear growth instead of the quadratic growth needed to capture a full non-commutative deformation of the coordinate ring of $\mathbb{C}P^2$ (these are the Sklyanin algebras, classified by Artin, Tate and Van den Bergh 1991). It fails to include an additional generator in degree 3, as can be seen from the fact that the restriction map $H^0(\mathbb{C}P^2, \mathcal{O}(3)) \longrightarrow H^0(C, \iota^*\mathcal{O}(3))$ has one-dimensional kernel. However, it is important to note that neither $\mathcal{O}(1)$ nor $\mathcal{O}(2)$ has generalized holomorphic sections over $\mathbb{C}P^2$, while $\mathcal{O}(3)$ has a one-dimensional space of them. We end with this vague indication that the morphisms supported on C should be combined with those supported on the whole holomorphic Poisson manifold.

Acknowledgements

I would like to thank Nigel Hitchin for many illuminating discussions, and also the organizers of his birthday conference, especially Luis Álvarez-Cónsul and Oscar García-Prada, for a stimulating event. I would also like to thank Mike Artin, Gil Cavalcanti, Izzet Coskun, and Pavel Etingof for helpful discussions. Finally, I thank Yicao Wang for pointing out an error in a previous draft.

References

Apostolov, V., Gauduchon, P., and Grantcharov, G. (1999). Bihermitian structures on complex surfaces, *Proc. London Math. Soc.* **79**(3), 414–28.

Artin, M., Tate, J., and Van den Bergh, M. (1991). Modules over regular algebras of dimension 3, *Invent. Math.* **106**(2), 335–88.

Bondal, A. I. and Polishchuk, A. E. (1993). Homological properties of associative algebras: the method of helices, *Izv. Ross. Akad. Nauk Ser. Mat.* **57**(2), 3–50.

Brylinski, J.-L. (1993). *Loop Spaces, Characteristic Classes and Geometric Quantization*, Vol. 107 of *Progress in Mathematics*, Birkhäuser Boston Inc., Boston, MA.

Bursztyn, H. and Radko, O. (2003). Gauge equivalence of Dirac structures and symplectic groupoids, *Ann. Inst. Fourier (Grenoble)* **53**(1), 309–37.

Courant, T. (1990). Dirac manifolds, *Trans. Am. Math. Soc.* **319**, 631–61.

Ellwood, I. T. (2007). NS-NS fluxes in Hitchin's generalized geometry, *J. High Energy Phys.* **12**, 084, 1–24.

Evens, S., Lu, J.-H., and Weinstein, A. (1999). Transverse measures, the modular class and a cohomology pairing for Lie algebroids, *Q. J. Math. Oxford Ser. (2)* **50**(200), 417–36.

Goto, R. (2007). Deformations of generalized complex and generalized Kähler structures, arXiv:0705.2495v2.

Gualtieri, M. (2004). Generalized complex geometry, DPhil thesis, Oxford University, Oxford, math.DG/0401221.

Gualtieri, M. (2007). Generalized complex geometry, arXiv:math/0703298v2.

Hitchin, N. (1987). The self-duality equations on a Riemann surface, *Proc. London Math. Soc. (3)* **55**(1), 59–126.

Hitchin, N. (2003). Generalized Calabi-Yau manifolds, *Q. J. Math.* **54**, 281–308.

Hitchin, N. (2006*a*). Brackets, forms and invariant functionals, *Asian J. Math.* **10**(3), 541–60.

Hitchin, N. (2006*b*). Instantons, Poisson structures and generalized Kähler geometry, *Commun. Math. Phys.* **265**(1), 131–64.

Hitchin, N. (2007). Bihermitian metrics on Del Pezzo surfaces, *J. Symplectic Geom.* **5**, 1–7.

Kapustin, A. (2005). A-branes and noncommutative geometry, hep-th/0502212.

Kapustin, A. and Orlov, D. (2003). Remarks on A-branes, mirror symmetry, and the Fukaya category, *J. Geom. Phys.* **48**, 84.

Lichnerowicz, A. (1977). Les variétés de Poisson et leurs algèbres de Lie associées, *J. Diff. Geom.* **12**(2), 253–300.

Lindström, U., Roček, M., von Unge, R., and Zabzine, M. (2007). Generalized Kähler manifolds and off-shell supersymmetry, *Commun. Math. Phys.* **269**(3), 833–49.

Liu, Z.-J., Weinstein, A., and Xu, P. (1997). Manin triples for Lie bialgebroids, *J. Diff. Geom.* **45**, 547–74.

Polishchuk, A. (1997). Algebraic geometry of Poisson brackets, *J. Math. Sci. (New York)* **84**(5), 1413–44. Algebraic Geometry, 7.

Roytenberg, D. (1999). Courant algebroids, derived brackets and even symplectic supermanifolds, PhD thesis, University of California, Berkeley, CA.

Simpson, C. T. (1992). Higgs bundles and local systems, *Inst. Hautes Études Sci. Publ. Math.* (75), 5–95.

Stafford, J. T. and van den Bergh, M. (2001). Noncommutative curves and noncommutative surfaces, *Bull. Am. Math. Soc. (N.S.)* **38**(2), 171–216 (electronic).

Ševera, P. (1998–2000). Letters to A. Weinstein concerning Courant algebroids. Unpublished.

Weinstein, A. (1997). The modular automorphism group of a Poisson manifold, *J. Geom. Phys.* **23**, 379–94.

XIX

CONSISTENT ORIENTATION OF MODULI SPACES

Daniel S. Freed, Michael J. Hopkins, and Constantin Teleman

For Nigel, with admiration

In a series of papers by Freed, Hopkins, and Teleman (2003, 2005, 2007a) we develop the relationship between positive energy representations of the loop group of a compact Lie group G and the twisted equivariant K-theory $K_G^{\tau+\dim G}(G)$. Here G acts on itself by conjugation. The loop group representations depend on a choice of "level," and the twisting τ is derived from the level. For all levels the main theorem is an isomorphism of abelian groups, and for special transgressed levels it is an isomorphism of *rings*: the fusion ring of the loop group and $K_G^{\tau+\dim G}(G)$ as a ring. For G connected with $\pi_1 G$ torsionfree we prove in Freed *et al.* (2007*a*, section 4 and 2007*b*, section 7) that the ring $K_G^{\tau+\dim G}(G)$ is a quotient of the representation ring of G and we calculate it explicitly. In these cases it agrees with the fusion ring of the corresponding centrally extended loop group. We also treat $G = SO_3$ in Freed *et al.* (2007b (A.10)). In this chapter we explicate the multiplication on the twisted equivariant K-theory for an arbitrary compact Lie group G. We work purely in topology; loop groups do not appear. In fact, we construct a *Frobenius ring* structure on $K_G^{\tau+\dim G}(G)$. This is best expressed in the language of topological quantum field theory: we construct a two-dimensional TQFT over the integers in which the abelian group attached to the circle is $K_G^{\tau+\dim G}(G)$.

At first glance the ring structure seems apparent. The multiplication map $\mu : G \times G \to G$ induces a pushforward on K-theory: the Pontrjagin product. But in K-cohomology the pushforward is the wrong-way, or *umkehr*, map. Thus to define it we must K-orient the map μ. Furthermore, the twistings must be accounted for in the orientations. Finally, to ensure associativity we must consistently K-orient maps constructed from μ by iterated composition. For connected and simply connected groups there is essentially a unique choice, but in general one must work more. This orientation problem is neatly formulated in the language of topological quantum field theory. Cartesian products of G then appear as moduli spaces of flat connections on surfaces, and the maps along which we push forward are restriction maps of the connections to the boundary. What is required, then, is a consistent orientation of these moduli spaces and restriction maps. The existence of consistent orientations, which we prove in Theorem 19.1, is in some sense due to the Narasimhan–Seshadri theorem which identifies moduli spaces of flat connections with complex manifolds of stable bundles: complex manifolds carry a canonical orientation in K-theory. Our proof, though, uses only the much more simple linear statement that the symbol of the de Rham complex on a surface is the complexification of the symbol of the Dolbeault complex. As we explain in Section 19.1, which serves as a heuristic introduction and motivation, "consistent orientations on moduli spaces" is the topological analog of "consistent measures on spaces of fields" in quantum field theory. The latter is what one would like to construct in the path integral approach to quantum field theory.

Our topological construction, outlined in Section 19.3, proceeds via a universal orientation (Definition 19.1). The main observation is that the problem of consistent orientations is a bordism problem, and the relevant bordism groups are those constructed by Madsen and Tillmann (2001) in their formulation of the Mumford conjectures (see Madsen and Weiss 2005 and Galatius *et al.* 2009 for proofs and generalizations). A universal orientation induces a level (Definition 19.2). The map from universal orientations to levels is an isomorphism for simply connected and connected compact Lie groups G, but in general it may fail to be injective, surjective, or both. The theories we construct are parametrized by universal orientations, not by levels. It is interesting to ask whether universal orientations also appear in related topological and conformal field theories as a refinement of the level.

The two-dimensional TQFT we construct here is the dimensional reduction of three-dimensional Chern–Simons theory, refined to have base ring $\mathbb{Z}$ in place of $\mathbb{C}$. Our construction is *a priori* in the sense that the axioms of TQFT—the topological invariance and gluing laws—are deduced directly from the definition. By contrast, rigorous constructions of many other TQFTs, such as the Chern–Simons theory, proceed via *generators and relations.* Such constructions are based on general theorems which tell that these generators and relations generate a TQFT: gluing laws and topological invariance are satisfied. One can ask if there is an *a priori* topological construction of Chern–Simons theory using twisted K-theory. We do not know of one. In another direction we can extend TQFTs to lower dimension, so look for a theory in 0-1-2 dimensions which extends the 1-2 dimensional theory constructed here. Again, we do not know a construction of that extended theory.

Section 19.2 of this chapter is an exposition of twistings and orientation, beginning on familiar ground with densities in differential geometry. Section 19.4 briefly considers this TQFT for *families* of one- and two-manifolds. Our purpose is to highlight an extra twist which occurs: that theory is "anomalous."

As far as we know, the problem of consistently orienting moduli spaces first arises in work of Donaldson (Donaldson 1987; Donaldson and Kronheimer 1990). He works with anti-self-dual connections on a four-manifold and uses excision in index theory to relate all of the different moduli spaces. In both his situation and ours the moduli spaces in question sit inside infinite-dimensional function spaces, and the virtual tangent bundle to the moduli space extends to a virtual bundle on the function space. Thus it suffices to orient over the function space, and this becomes a universal problem. Presumably our methods apply to his situation as well, but we have not worked out the details.

It is a pleasure and an honor to dedicate this chapter to Nigel Hitchin. We greatly admire his mathematical taste, style, and influence. *¡Feliz cumpleaños y que cumplas muchos más!*

19.1 Push–pull construction of TQFT

19.1.1 *Quantum field theory*

The basic structure of an n-dimensional Euclidean quantum field theory may be axiomatized simply. Let $\mathcal{BR}iem_n$ be the bordism category whose objects are closed oriented $(n-1)$-dimensional Riemannian manifolds. A morphism $X : Y_0 \to Y_1$ is a compact-oriented n-dimensional Riemannian manifold X together with an orientation-preserving isometry of its boundary to the disjoint union $-Y_0 \sqcup Y_1$, where $-Y_0$ is the oppositely oriented manifold. We term Y_0 the *incoming* boundary and Y_1 the *outgoing* boundary. A quantum field theory is a functor from $\mathcal{BR}iem_n$ to the category of Hilbert spaces and trace class maps. The functoriality encodes the gluing law; there is also a symmetric monoidal structure which encodes the behavior under disjoint unions. There are many details and subtleties (e.g. see Chapter 9), but our concern is a simpler topological version. Thus we replace $\mathcal{BR}iem_n$ by the bordism category $\mathcal{BSO}_n$ of smooth-oriented manifolds and consider orientation-preserving diffeomorphisms in place of isometries. We define an n-dimensional *topological quantum field theory* (TQFT) to be a functor from $\mathcal{BSO}_n$ to the category of complex vector spaces. The functor is required to be monoidal: disjoint unions map to tensor products. The functoriality expresses the usual gluing law and the structure of the domain category $\mathcal{BSO}_n$ encodes the topological invariance. The example of interest here has an integral structure: the codomain is the category of abelian groups rather than complex vector spaces. The integrality reflects that the theory is a dimensional reduction (see Freed 2009 for a discussion).

Physicists often employ a path integral to construct a quantum field theory. Here is a cartoon version. To each manifold M is attached a space $\mathcal{F}_M$ of fields and so to a bordism $X : Y_0 \to Y_1$ a correspondence diagram:

$$\begin{array}{ccc} & \mathcal{F}_X & \\ {}^{s}\swarrow & & \searrow^{t} \\ \mathcal{F}_{Y_0} & & \mathcal{F}_{Y_1} \end{array} \tag{19.1}$$

in which s, t are restriction maps. The important property of fields is locality: in the diagram

$$\begin{array}{ccccccc} & & & \mathcal{F}_{X'\circ X} & & & \\ & & \stackrel{r}{\swarrow} & & \stackrel{r'}{\searrow} & & \\ & \mathcal{F}_X & & & & \mathcal{F}_{X'} & \\ \stackrel{s}{\swarrow} & & \stackrel{t}{\searrow} & & \stackrel{s'}{\swarrow} & & \stackrel{t'}{\searrow} \\ \mathcal{F}_{Y_0} & & & \mathcal{F}_{Y_1} & & & \mathcal{F}_{Y_2} \end{array} \tag{19.2}$$

the space of fields $\mathcal{F}_{X'\circ X}$ on the composition of bordisms $X : Y_0 \to Y_1$ and $X' : Y_1 \to Y_2$ is the fiber product of the maps t, s'. Fields are really infinite-dimensional *stacks*—for example, in gauge theories the gauge transformations act as morphisms of fields—and the maps and fiber products must be understood in that sense.

The backdrop for the path integral is measure theory. If there exist measures μ_X, μ_Y on the spaces $\mathcal{F}_X, \mathcal{F}_Y$ with appropriate gluing properties, then one can construct a quantum field theory. Namely, define the Hilbert space $\mathcal{H}_Y = L^2(\mathcal{F}_Y, \mu_Y)$ and the linear map attached to a bordism $X : Y_0 \to Y_1$ as the *push–pull*

$$Z_X = t_* \circ s^* : \mathcal{H}_{Y_0} \longrightarrow \mathcal{H}_{Y_1}.$$

The pushforward t_* is integration. Thus if $f \in L^2(\mathcal{F}_{Y_0}, \mu_{Y_0})$ and $g \in L^2(\mathcal{F}_{Y_1}, \mu_{Y_1})$, then

$$\langle \bar{g}, Z_X(f) \rangle_{\mathcal{H}_{Y_1}} = \int_{\mathcal{F}_X} \overline{g\big(t(\Phi)\big)}\, f\big(s(\Phi)\big)\, d\mu_X(\Phi).$$

One usually postulates an *action functional* $S_X : \mathcal{F}_X \to \mathbb{C}$ and a measure $\tilde{\mu}_X$ such that $\mu_X = e^{-S_X}\tilde{\mu}_X$ and the action satisfies the gluing law

$$S_{X'\circ X}(\Phi) = S_X\big(r(\Phi)\big) + S_{X'}\big(r'(\Phi)\big)$$

in (19.2). These measures have not been constructed in most examples of geometric interest.

19.1.2 *Topological construction*

Our idea is to replace the infinite-dimensional stack $\mathcal{F}_X$ by a finite-dimensional stack $\mathcal{M}_X \subset \mathcal{F}_X$ of solutions to a first-order partial differential equation and to shift from measure theory to algebraic topology. Examples of finite-dimensional moduli spaces $\mathcal{M}_X$ in supersymmetric field theory include anti-self-dual connections in four dimensions and holomorphic maps in two dimensions. From the physical point of view the differential equations are the BPS (Bogomolńyi-Prasad-Sommerfield) equations of supersymmetry; from a mathematical point of view they define the minima of a calculus of variations functional. In this chapter we consider pure gauge theories. Fix a compact Lie group G and for any manifold M let $\mathcal{F}_M$ denote the stack of G-connections on M. Define $\mathcal{M}_M$ as the stack of *flat* G-connections on M. If we choose a set $\{m_i\} \subset M$ of "basepoints," one for each component of M, then $\mathcal{M}_M$ is represented by the product of groupoids $\prod_i \bigl[\mathrm{Hom}\bigl(\pi_1(M, m_i), G\bigr)//G\bigr]$. A basic property of flat connections is the gluing law (see (19.2)).

Lemma 19.1 *Suppose $X : Y_0 \to Y_1$ and $X' : Y_1 \to Y_2$ are bordisms of smooth manifolds. Then $\mathcal{M}_{X' \circ X}$ is the fiber product of*

$$\begin{array}{ccc} \mathcal{M}_X & & \mathcal{M}_{X'} \\ & {\scriptstyle t}\searrow \quad \swarrow{\scriptstyle s'} & \\ & \mathcal{M}_Y & \end{array} \tag{19.3}$$

Roughly speaking, this says that given flat connections on X, X' and an isomorphism of their restrictions to Y, one can construct a flat connection on $X' \circ X$ and every flat connection on $X' \circ X$ comes this way.

Replace the infinite-dimensional correspondence diagram (19.1) with the finite-dimensional correspondence diagram of flat connections:

$$\begin{array}{ccc} & \mathcal{M}_X & \\ & {\scriptstyle s}\swarrow \quad \searrow{\scriptstyle t} & \\ \mathcal{M}_{Y_0} & & \mathcal{M}_{Y_1} \end{array} \tag{19.4}$$

Whereas the path integral linearizes (19.1) using measure theory, we propose instead to linearize (19.4) using algebraic topology. Let E be a generalized cohomology theory. To every closed $(n-1)$-manifold we assign the abelian group

$$A_Y = E^\bullet(\mathcal{M}_Y).$$

To a morphism $X : Y_0 \to Y_1$ we would like to attach a homomorphism $Z_X : A_{Y_0} \to A_{Y_1}$ defined as the push–pull

$$Z_X := t_* \circ s^* : E^\bullet(\mathcal{M}_{Y_0}) \longrightarrow E^\bullet(\mathcal{M}_{Y_1}) \tag{19.5}$$

in E-cohomology. Whereas the path integral requires *measures* consistent under gluing to define integration t_*, in our topological setting we require *orientations* of t consistent with gluing to define pushforward t_*. The *consistency* of orientations under gluing ensures that (19.5) defines a TQFT which satisfies the gluing law (functoriality).

This, then, is the goal of the chapter: we formulate the algebro-topological home for consistent orientations and study a particular example. Namely, specialize to $n = 2$ and require that the one-manifolds Y and two-manifolds X be oriented. In other words, the domain category of our TQFT is $\mathcal{BSO}_2$. For $Y = S^1$ the moduli stack of flat connections is the global quotient

$$\mathcal{M}_Y \cong G//G$$

of G by its adjoint action; the isomorphism is the holonomy of a flat connection around the circle. Take the cohomology theory E to be complex K-theory. The resulting two-dimensional TQFT on oriented manifolds is the dimensional reduction of three-dimensional Chern–Simons theory for the group G. In this case there is a map from consistent orientations to "levels" on G; the level is what is usually used to describe Chern–Simons theory. A two-dimensional TQFT on oriented manifolds determines a Frobenius ring and conversely. The Frobenius ring constructed here is the Verlinde ring attached to the loop group of G. The abelian group A_{S^1} is a *twisted* form of $K(G//G) = K_G(G)$ and its relation to positive energy representations of the loop group is developed in Freed *et al.* (2003, 2005, 2007*a*). In this chapter we describe a topological construction of the ring structure.

19.1.3 *Remarks*

- Let X be the "pair of pants" with the two legs incoming and the single waist outgoing. Then restriction to the outgoing boundary is the map $t : (G \times G)//G \to G//G$ induced by multiplication $\mu : G \times G \to G$. So $Z_X = t_* \circ s^*$, which defines the ring structure in a two-dimensional TQFT, is pushforward by multiplication on G. Therefore, we do construct the Pontrjagin product on $K_G^{\tau + \dim G}(G)$—here τ is the twisting and there is a degree shift as well—and have implicitly used an isomorphism of twistings $\mu^*\tau \to \tau \otimes 1 + 1 \otimes \tau$ which, since the TQFT guarantees an associative product, satisfies a compatibility condition for triple products. This isomorphism and compatibility are embedded in our consistent orientation construction.
- We do not use the theorem (Abrams 1996, Moore and Segal 2006) which constructs a two-dimensional TQFT from a Frobenius ring. Rather, our *a priori* construction manifestly produces a TQFT which satisfies the gluing law, and we deduce the Frobenius ring as a derived quantity.
- Three-dimensional Chern–Simons theory is defined on a bordism category of manifolds which carry an extra topological structure. For oriented manifolds this extra structure is described as a trivialization of p_1 or signature, or a certain sort of framing. (For spin manifolds it is described as a string structure or, since we are in sufficiently low dimensions, an ordinary framing.) The two-dimensional reduction constructed here factors through the bordism categories of oriented manifolds.

- The topological push–pull construction extends to families of bordisms parametrized by a base manifold S. A choice of consistent orientation determines this extension to a theory for families of manifolds, albeit an "anomalous" theory (see Section 19.4 for a discussion).
- The pushforward t_* is only defined if $t : \mathcal{M}_X \to \mathcal{M}_{Y_1}$ is a *representable* map of stacks, that is, only if the fibers of t are spaces—no automorphisms allowed. This happens only if each component of X has a nonempty outgoing boundary. Therefore, the push–pull construction only gives a partial TQFT. We complete to a full TQFT using the nondegeneracy of a certain bi-additive form (see Freed *et al.* 2003, section 17).
- As mentioned earlier, a standard TQFT is defined over the ring $\mathbb{C}$ whereas this theory, being a dimensional reduction of a three-dimensional theory, is defined over $\mathbb{Z}$. It is possible to go further and refine the push–pull construction to obtain a theory over K, where K is the K-theory ring spectrum (see Freed 2009 for further discussion).
- The theory constructed here has two tiers—it concerns one- and two-manifolds—so could be termed a "1-2 theory." Extensions to 0-1-2 theories, which have three tiers, are of great interest. The general structure of such theories has been much studied recently in various guises (Moore and Segal 2006; Costello 2007; Hopkins and Lurie in preparation). A theory defined down to points is completely local, and so ultimately has a simpler structure than less local theories. We do not know if the push–pull construction here can be extended to construct a 0-1-2 theory.

19.2 Orientation and twisting

19.2.1 *Ordinary cohomology*

The first example for a differential geometer is de Rham theory. Let M be a smooth manifold and suppose it has a dimension equal to n. An *orientation* on M, which is an orientation of the tangent bundle TM, enables integration

$$\int_M : \Omega^n_c(M) \longrightarrow \mathbb{R}$$

on forms of compact support. Absent an orientation we can integrate twisted forms, or densities. The *twisting* is defined as follows. For any real vector space V of dimension r let $\mathcal{B}(V)$ denote the $GL_r\mathbb{R}$-torsor of bases of V. There is an associated $\mathbb{Z}$-graded real line $\mathfrak{o}(V)$ of functions $f : \mathcal{B}(V) \to \mathbb{R}$ which satisfy $f(b \cdot A) = \operatorname{sign}\det A \cdot f(b)$ for $b \in \mathcal{B}(V)$, $A \in GL_r\mathbb{R}$; the degree of $\mathfrak{o}(V)$ is r. Applied fiberwise this construction yields a flat $\mathbb{Z}$-graded line bundle $\mathfrak{o}(V) \to M$ for a real vector bundle $V \to M$. There is a twisted de Rham complex

$$0 \longrightarrow \Omega^{\mathfrak{o}(V)-r}(M) \xrightarrow{d} \Omega^{\mathfrak{o}(V)-r+1}(M) \xrightarrow{d} \cdots \xrightarrow{d} \Omega^{\mathfrak{o}(V)}(M) \longrightarrow 0 \quad (19.6)$$

where $\Omega^{\mathfrak{o}(V)+q}(M)$ is the space of smooth sections of the *ungraded* vector bundle $\bigwedge^{r+q}T^*M \otimes \mathfrak{o}(V)$. The cohomology of (19.6) is the twisted de Rham cohomology $H_{dR}^{\mathfrak{o}(V)+\bullet}(M)$. Let $\mathfrak{o}(M) = \mathfrak{o}(TM)$. Then integration is a map

$$\int_M : \Omega_c^{\mathfrak{o}(M)}(M) \longrightarrow \mathbb{R}. \tag{19.7}$$

Notice this formulation notation works if M has several components of varying dimension: the degree of $\mathfrak{o}(M)$ is then the locally constant function $\dim M : M \to Z$.

A similar construction works in integer cohomology. If $\pi : V \to M$ is a real vector bundle over a space M (which need not be a manifold) we define $\mathfrak{o}(V)$ as the orientation double cover of M determined by V and introduce a $\mathbb{Z}$-grading according to the rank of V. (Note $\operatorname{rank} V : M \to \mathbb{Z}$ is a locally constant function.) There is an $\mathfrak{o}(V)$-twisted singular complex analogous to (19.6): cochains in this complex are cochains on the double cover which change sign under the deck transformation. The equivalence class of the twisting $\mathfrak{o}(V)$ is

$$\big[\mathfrak{o}(V)\big] = \big(\operatorname{rank} V, w_1(V)\big) \in H^0(M;\mathbb{Z}) \times H^1(M;\mathbb{Z}/2\mathbb{Z}),$$

where w_1 is the Stiefel–Whitney class. The relationship of the twisting to integration occurs in the Thom isomorphism. The Thom class $U \in H_{\mathrm{cv}}^{\pi^*\mathfrak{o}(V)}(V)$ lies in twisted cohomology with *compact vertical* support. Let $B(V)$ and $S(V)$ be the ball and sphere bundles relative to a metric on V, respectively. The *Thom space* M^V is the pair $\big(B(V), S(V)\big)$ or equivalently, assuming M is a CW complex, the quotient $B(V)/S(V)$. The composite

$$H^{-\mathfrak{o}(V)+\bullet}(M) \xrightarrow{\pi^*} H^{-\pi^*\mathfrak{o}(V)+\bullet}(V) \xrightarrow{U} H^\bullet(M^V)$$

is an isomorphism—the Thom isomorphism—a generalization of the suspension isomorphism. If M is a *compact* manifold and $i : M \hookrightarrow S^n$ a (Whitney) embedding with normal bundle $\nu \to M$, then the *Pontrjagin–Thom collapse* is the map $c : S^n \to M^\nu$ defined by identifying ν with a tubular neighborhood of M and sending the complement of $B(\nu)$ in S^n to the basepoint of M^ν. Integration is then the composite

$$H^{-\mathfrak{o}(\nu)+n}(M) \xrightarrow[\cong]{\text{Thom}} H^n(M^\nu) \xrightarrow{c^*} H^n(S^n) \xrightarrow[\cong]{\text{suspension}} \mathbb{Z}. \tag{19.8}$$

Twistings obey a Whitney sum formula: there is a natural isomorphism

$$\mathfrak{o}(V_1 \oplus V_2) \xrightarrow{\cong} \mathfrak{o}(V_1) + \mathfrak{o}(V_2).$$

Applied to $TM \oplus \nu = \underline{n}$, where $\underline{n}$ is the trivial bundle of rank n, we conclude that integration (19.8) is a map (compare (19.7))

$$H^{\mathfrak{o}(M)}(M) \longrightarrow \mathbb{Z}.$$

More generally, if $p : M \to N$ is a proper map there is a pushforward

$$p_* : H^{\mathfrak{o}(p)+\bullet}(M) \longrightarrow H^{\bullet}(N), \tag{19.9}$$

where $\mathfrak{o}(p) = \mathfrak{o}(M) - p^*\mathfrak{o}(N)$.

19.2.2 *K-theory*

This discussion applies to any multiplicative cohomology theory.[1] The only issue is to determine the twisting of a real vector bundle in that theory. For complex K-theory there are many possible models for the twisting τ_V of a vector bundle $V \to M$. In the Donovan–Karoubi picture (Donovan and Karoubi 1970) τ_V is represented by the bundle of complex $\mathbb{Z}/2\mathbb{Z}$-graded Clifford algebras defined by V. A bundle of algebras $A \to M$ of this type is considered trivial if $A = \mathrm{End}(W)$ for a $\mathbb{Z}/2\mathbb{Z}$-graded complex vector bundle $W \to M$, that is, if A is Morita equivalent to the trivial bundle of algebras $M \times \mathbb{C}$. The equivalence class of τ_V is

$$[\tau_V] = \big(\operatorname{rank} V, w_1(V), W_3(V)\big) \in H^0(M;\mathbb{Z}/2\mathbb{Z}) \times H^1(M;\mathbb{Z}/2\mathbb{Z}) \times H^3(M;\mathbb{Z}). \tag{19.10}$$

Only torsion classes in $H^3(M;\mathbb{Z})$ are realized by bundles of finite-dimensional algebras, but we have in mind a larger model which includes nontorsion classes. (Such models are developed in Atiyah and Segal 2006, Freed *et al.* 2007*a*, and Murray 1996 among other works.) There is a Whitney sum isomorphism

$$\tau_{V_1 \oplus V_2} \xrightarrow{\;\cong\;} \tau_{V_1} + \tau_{V_2}; \tag{19.11}$$

the sum of twistings is realized by the tensor product of algebras. A spin^c structure on V induces an *orientation*, that is, a Morita equivalence

$$\tau_{\underline{\operatorname{rank} V}} \xrightarrow{\;\cong\;} \tau_V. \tag{19.12}$$

An A-*twisted* vector bundle is a vector bundle with an A-module structure; it represents an element of twisted K-theory.

The Whitney formula (19.11) allows us to attach a twisting to any virtual real vector bundle: set

$$\tau_{-V} = -\tau_V. \tag{19.13}$$

Since the Thom space satisfies the stability condition $X^{V \oplus \underline{n}} \cong \Sigma^n X^V$, where Σ denotes suspension, there is also a Thom spectrum attached to any virtual vector bundle and a corresponding Thom isomorphism theorem. An orientation, which is an isomorphism as in (19.12), is equivalently a trivialization of the twisting attached to the reduced bundle $(V - \underline{\operatorname{rank} V})$.

[1] Also to a cohomology theory defined by a module over a ring spectrum.

Suppose τ is any twisting on a manifold N. We can put that extra twisting into the pushforward in K-theory associated to a proper map $p : M \to N$ (compare (19.9)):

$$p_* : K^{(\tau_p + p^*\tau) + \bullet}(M) \longrightarrow K^{\tau + \bullet}(N). \tag{19.14}$$

Here τ_p is the twisting $\tau_p = \tau_M - p^*\tau_N$ of the relative tangent bundle. In the next section we encounter a situation in which $\tau_p + p^*\tau$ is trivialized, and so construct a pushforward from untwisted K-theory to twisted K-theory (see (19.38)).

Twistings of $K^\bullet(\mathrm{pt})$ form a symmetric monoidal two-groupoid; its classifying space $\operatorname{Pic_g} K$ is thus an infinite loop space. The notation: $\operatorname{Pic} K$ is the classifying space of invertible K-modules and $\operatorname{Pic_g} K$ the subspace classifying certain "geometric" invertible K-modules including twisted forms of K-theory defined by real vector bundles. As a space there is a homotopy equivalence

$$\operatorname{Pic_g} K \sim K(\mathbb{Z}/2\mathbb{Z}, 0) \times K(\mathbb{Z}/2\mathbb{Z}, 1) \times K(\mathbb{Z}, 3) \tag{19.15}$$

with a product of Eilenberg–MacLane spaces, but the group structure on $\operatorname{Pic_g} K$ is not a product. The group of equivalence classes of twistings on M is the group of homotopy classes of maps $[M, \operatorname{Pic_g} K]$, which as a set is the product of cohomology groups in (19.10).

Let $pic_{\mathrm{g}}K$ denote the spectrum whose 0-space is $\operatorname{Pic_g} K$ and which is a Postnikov section of the real KO-theory spectrum: the "$\mathbb{Z}/2$, $\mathbb{Z}/2$, 0, $\mathbb{Z}$" bit of the "…, $\mathbb{Z}/2$, $\mathbb{Z}/2$, 0, $\mathbb{Z}$, 0, 0, 0, $\mathbb{Z}$, …" song. Thus the one-space $B \operatorname{Pic_g} K$ of the spectrum $pic_{\mathrm{g}}K$ is a Postnikov section of BO. Also, let ko denote the connective KO-theory spectrum. Its zero-space is the group completion of the classifying space of the symmetric monoidal category of real vector spaces of finite dimension (Segal 1977). The map which attaches a twisting of $K^\bullet(\mathrm{pt})$ to a real vector space, say via the Clifford algebra, induces a spectrum map

$$\tau : ko \longrightarrow pic_{\mathrm{g}}K. \tag{19.16}$$

Remark 19.1 If M is a smooth *stack*, presented as the groupoid of a Lie group acting on a smooth manifold, then its tangent space is presented as a graded vector bundle. An orientation of M is then an orientation of this graded bundle, viewed as a virtual bundle by taking the alternating sum of the homogeneous terms. The virtual tangent bundle is classified by a map $M \to ko$ whose composition with (19.16) gives the induced twisting of complex K-theory. We apply this in next section 19.3 to the moduli space of flat connections on a fixed oriented two-manifold. In that case the virtual tangent bundle is the index of an elliptic complex and the map $M \to ko$ is computed from the Atiyah–Singer index theorem (Atiyah and Singer 1971).

The $\mathbb{Z}$ part of $\operatorname{Pic_g} K$, the Eilenberg–MacLane space $K(\mathbb{Z}, 3)$, has an attractive geometric model: gerbes. Nigel Hitchin (2001) has developed beautiful applications of gerbes in differential geometry. There is a gerbe model for $\operatorname{Pic_g} K$ too—one need only add $\mathbb{Z}/2\mathbb{Z}$-gradings.

19.3 Universal orientations and consistent orientations

19.3.1 *Overview*

In this section, the heart of the chapter, we define universal orientations (Definition 19.1) and prove that they exist (Theorem 19.1). A universal orientation simultaneously orients the maps t in (19.4) along which we pushforward classes in twisted K-theory (see (19.40) for the precise push–pull maps in the theory). Universal orientations form a torsor for an abelian group (19.23). A universal orientation determines a level (Definition 19.2), which is the quantity typically used to label theories. The map (19.35) from universal orientations to levels is not an isomorphism in general.

We begin with a closed oriented surface X. The virtual tangent space to the stack $\mathcal{M}_X$ of flat G-connections on X is the index of a twisted de Rham complex (19.17), and we construct a universal symbol (19.20)—whence universal index—for this operator. A trivialization of the universal twisting (19.22) is a universal orientation, and it simultaneously orients the moduli stacks $\mathcal{M}_X$ for all closed oriented X.

For a surface X with (outgoing) boundary we must orient the restriction map $t: \mathcal{M}_X \to \mathcal{M}_{\partial X}$ on flat connections. It turns out that a universal orientation does this, simultaneously and coherently for all X, as expressed in the isomorphism (19.37). An important step in the argument is Lemma 19.4, which uses work of Atiyah and Bott (1964) to interpret the universal symbol in terms of standard local boundary conditions for the de Rham complex.

19.3.2 *Closed surfaces*

Fix a compact Lie group G with Lie algebra $\mathfrak{g}$. Let X be a *closed* oriented two-manifold and $\mathcal{M}_X$ the moduli stack of flat G-connections on X. A point of $\mathcal{M}_X$ is represented by a flat connection A on a principal bundle $P \to X$, and the tangent space to $\mathcal{M}_X$ at A by the deformation complex

$$0 \longrightarrow \Omega^0_X(\mathfrak{g}_P) \xrightarrow{d_A} \Omega^1_X(\mathfrak{g}_P) \xrightarrow{d_A} \Omega^2_X(\mathfrak{g}_P) \longrightarrow 0, \tag{19.17}$$

the de Rham complex with coefficients in the adjoint bundle associated to P. This is an elliptic complex. Its symbol σ satisfies the reality condition $\sigma(-\xi) = \sigma(\xi)$ for $\xi \in TX$, since (19.17) is a complex of real differential operators (Atiyah and Singer 1971). Recall that the symbol of any complex differential operator lies in $K^0_{\text{cv}}(TX) \cong K^0(X^{TX})$. The reality condition gives a lift $\sigma \in KR^0(X^{iTX})$, where the imaginary tangent bundle iTX carries the involution of complex conjugation (Atiyah 1966). Bott periodicity asserts that $V \oplus iV$ is canonically KR-oriented for any real vector bundle V with no degree shift. In the language of twistings of KR this means

$$\tau_V^{(KR)} + \tau_{iV}^{(KR)} = 0. \tag{19.18}$$

Therefore

$$KR^0(X^{iTX}) \xrightarrow[\cong]{\text{Thom}} KR^{-\tau^{(KR)}_{iTX}}(X) \xrightarrow[\cong]{(19.18)} KR^{\tau^{(KR)}_{TX}}(X) \xrightarrow[\cong]{(19.13)}$$

$$KR^{-\tau^{(KR)}_{-TX}}(X) \xrightarrow[\cong]{\text{Thom}} KO^0(X^{-TX}) \tag{19.19}$$

from which we locate the symbol $\sigma \in KO^0(X^{-TX})$. Note that by Atiyah duality $KO^0(X^{-TX}) \cong KO_0(X)$ and the KO-homology group is well-known to be the home of real elliptic operators.

Now (19.17) is a *universal* operator: its symbol is constructed from the exterior algebra of TX and the adjoint representation of G; it does not depend on details of the manifold X. Thus it is pulled back from a universal symbol. Let $V_n \to BSO_n$ denote the universal oriented n-plane bundle. The universal symbol lives on the bundle $V_2 \to BSO_2 \times BG$, and by (19.19) we identify it as an element

$$\sigma_{\text{univ}} \in KO^0(BSO_2^{-V_2} \wedge BG_+). \tag{19.20}$$

Here BG_+ is the space BG with a disjoint basepoint adjoined and "$\wedge$" is the smash product. Introduce the notation

$$MTSO_n = BSO_n^{-V_n}$$

for this Thom spectrum and so write

$$\sigma_{\text{univ}} \in KO^0(MTSO_2 \wedge BG_+).$$

If $f : X \to BSO_2 \times BG$ is a classifying map for TX and P, and $\tilde{f} : X^{-TX} \to MTSO_2 \wedge BG_+$ the induced map on Thom spectra, then $\sigma = \tilde{f}^*\sigma_{\text{univ}}$. It is in this sense that (19.20) is a *universal* symbol.

Remark 19.2 We digress to explain $MTSO_n$ in more detail. Let $Gr_n^+(\mathbb{R}^N)$ be the Grassmannian of oriented n-planes in $\mathbb{R}^N$ and

$$0 \to V_n \to \underline{N} \to Q_{N-n} \to 0$$

the exact sequence of real vector bundles over $Gr_n^+(\mathbb{R}^N)$ in which V_n is the universal subbundle and Q_{N-n} the universal quotient bundle. Denote the Thom space of the latter as

$$Z_N := Gr_n^+(\mathbb{R}^N)^{Q_{N-n}}.$$

Then the suspension ΣZ_N is the Thom space of $Q_{N-n} \oplus \underline{1} \to Gr_n^+(\mathbb{R}^N)$. But $Q_{N-n} \oplus \underline{1}$ is the pullback of $Q_{N+1-n} \to Gr_n^+(\mathbb{R}^{N+1})$ under the natural inclusion $Gr_n^+(\mathbb{R}^N) \hookrightarrow Gr_n^+(\mathbb{R}^{N+1})$, and in this manner we produce a map $\Sigma Z_N \to Z_{N+1}$. Whence the spectrum $MTSO_n = \{Z_N\}_{N\geq 0}$. The notation identifies $MTSO_n$ as an unstable version of the Thom spectrum MSO and also alludes to its appearance in the work of Madsen and Tillmann (2001). There are analogous spectra MTO_n, $MT\,\text{Spin}_n$, $MT\,\text{String}_n$, etc. If $F : S^N \to Z_N$ is transverse to the 0-section, then $X := F^{-1}(\text{0-section})$ is an n-manifold and the pullback of $Q_{N-n} - \underline{N}$ is stably isomorphic to $-TX$. Thus a map[2] $S \to MTSO_n$ classifies a map $M \to S$ of relative dimension n together with a rank n bundle $W \to M$ and a stable isomorphism[3] $T(M/S) \cong W$. An important theorem of Madsen and Weiss (2005) and Galatius *et al.* (2009) identifies $MTSO_n$ as a bordism theory of fiber bundles rather than a bordism theory of arbitrary maps.

If a *smooth manifold* M parametrizes a family of flat G-connections on X—that is, $P \to M \times X$ is a G-bundle with a partial flat connection along X—then there is a classifying map $M \to \mathcal{M}_X$ and the pullback of the stable tangent bundle of $\mathcal{M}_X$ to M is the index of the family of elliptic complexes (19.17). Note that if we replace the adjoint bundle $\mathfrak{g}_P$ in (19.17) by the trivial bundle of rank $\dim G$ then the resulting elliptic complex does not vary over M; its index is a trivializable bundle. Hence the reduced stable tangent bundle to $\mathcal{M}_X$ is computed by the de Rham complex coupled to the reduced adjoint bundle $\bar{\mathfrak{g}}_P = \mathfrak{g}_P - \underline{\dim G}$.

There is a corresponding reduced universal symbol (compare (19.20))

$$\bar{\sigma}_{\text{univ}} \in KO^0(MTSO_2 \wedge BG). \tag{19.21}$$

It induces a universal twisting in K-theory and a consistent orientation is constructed by trivializing this twisting.

Definition 19.1 *A* universal orientation *is a null homotopy of the composition*

$$MTSO_2 \wedge BG \xrightarrow{\bar{\sigma}_{\text{univ}}} ko \xrightarrow{\tau} pic_g K. \tag{19.22}$$

Two universal orientations are said to be equivalent if the null homotopies are homotopy equivalent.

The set of equivalence classes of universal orientations is a torsor for the abelian group

$$\mathcal{O}(G) := [\Sigma MTSO_2 \wedge BG, pic_g K]. \tag{19.23}$$

We prove in Theorem 19.1 below that universal orientations exist. In fact, there is a canonical universal orientation, so the torsor of universal orientations may be naturally identified with the abelian group (19.23). Definition 19.1 is designed to orient the moduli spaces attached to closed surfaces. In an equivalent form it leads to the pushforward maps we need for surfaces with boundary and to twisted K-theory of moduli spaces attached to the boundary (see the discussion preceding (19.31)).

Return now to the family of partial G-connections on $P \to M \times X$. The bundle $P \to M \times X$ is classified by a map $f : M \to MTSO_2 \times BG$ and the Atiyah–Singer index theorem (Atiyah and Singer 1971) implies that the index of the

[2] That is, a stable map from the suspension spectrum of S to $MTSO_n$.

[3] Thom bordism theories, such as MSO, retain the information of the stable normal bundle. Madsen–Tillmann theories, such as $MTSO_n$, track the stable *tangent* bundle, which is one more justification for the "T" in the notation.

family of operators (19.17) is $f^*\sigma_{\text{univ}}$. Therefore, a universal orientation pulls back to an orientation of the index of (19.17) (c.f. Remark 19.1). It follows that a universal orientation simultaneously orients $\mathcal{M}_X$ for all closed oriented two-manifolds X.

19.3.3 *Surfaces with boundary*

As a preliminary we observe two topological facts about $MTSO_n$.

Lemma 19.2

(1) *$MTSO_1 \simeq S^{-1}$, the desuspension of the sphere spectrum.*
(2) *There is a cofibration*

$$\Sigma^{-1}MTSO_{n-1} \xhookrightarrow{b} MTSO_n \twoheadrightarrow (BSO_n)_+. \tag{19.24}$$

Proof. For (1) simply observe

$$MTSO_1 \simeq BSO_1^{-V_1} \simeq \text{pt}^{-\mathbb{R}} \simeq S^{-1}. \tag{19.25}$$

For (2) begin with the cofibration built from the sphere and ball bundles of the universal bundle:

$$S(V_n)_+ \hookrightarrow B(V_n)_+ \twoheadrightarrow \big(B(V_n), S(V_n)\big)\ , \tag{19.26}$$

Then identify BSO_{n-1} as the unit sphere bundle $S(V_n)$ and write (19.26) in terms of Thom spaces:

$$BSO_{n-1}^{\underline{0}} \hookrightarrow BSO_n^{\underline{0}} \twoheadrightarrow BSO_n^{V_n}\ . \tag{19.27}$$

Here $\underline{0}$ is the vector bundle of rank zero. Now add $-V_n$ to each of the vector bundles in (19.27) and note that the restriction of V_n to BSO_{n-1} is $V_{n-1} \oplus \underline{1}$.

Consider the diagram

$$\begin{array}{ccccc} \Sigma^{-1}MTSO_1 \wedge BG & \xrightarrow{b} & MTSO_2 \wedge BG & \xrightarrow{q} & (MTSO_2, \Sigma^{-1}MTSO_1) \wedge BG \\ & & & \searrow^{\bar{\sigma}_{\text{univ}}} & \downarrow{\scriptstyle \bar{\sigma}'_{\text{univ}}} \\ & & & & ko \end{array} \tag{19.28}$$

The top row is a cofibration. From (19.24) we can replace $(MTSO_2, \Sigma^{-1}MTSO_1)$ with $(BSO_2)_+$. □

Lemma 19.3 *Define $\bar{\sigma}'_{\text{univ}}$ in (19.28) to be the map $(BSO_2)_+ \wedge BG \to ko$ induced by the standard representation of SO_2 smash with the reduced adjoint* *representation of G. Then the triangle in* (19.28) *commutes and the diagram gives a canonical null homotopy of the composite* $\bar{\sigma}_{\text{univ}} \circ b$.

Proof. Recalling the isomorphisms in (19.19), and using the fact that the universal symbol $\bar{\sigma}_{\text{univ}}$ is canonically associated to a representation of $SO_2 \times G$, we locate $\bar{\sigma}_{\text{univ}} \in KR^0_{SO_2\times G}(-\mathbb{R}^2)_c \cong KR^0_{SO_2\times G}(i\mathbb{R}^2)_c$. (Recall that the involution on $\mathbb{C}^2 \cong \mathbb{R}^2 \oplus i\mathbb{R}^2$ is complex conjugation and the subscript "c" denotes compact support.) Similarly, $\bar{\sigma}'_{\text{univ}} \in KR^0_{SO_2\times G}(\text{pt}) \cong KR^0_{SO_2}(\mathbb{C}^2)_c$. Using the Thom isomorphism we identify $\bar{\sigma}'_{\text{univ}}$ as the difference of classes represented by $\bigwedge^\bullet\mathbb{C}^2 \otimes \mathfrak{g}_{\mathbb{C}}$ and $\bigwedge^\bullet\mathbb{C}^2 \otimes \mathbb{C}^{\dim G}$, where in both summands $\theta \in \mathbb{C}^2$ acts as $\epsilon(\theta) \otimes \text{id}$. Exterior multiplication $\epsilon(\theta)$ is exact for $\theta \neq 0$, so this does represent a class with compact support. Also, as ϵ commutes with complex conjugation it is Real in the sense of Atiyah (1966). It remains to observe that its restriction under $KR^0_{SO_2\times G}(\mathbb{C}^2)_c \to KR^0_{SO_2\times G}(i\mathbb{R}^2)_c$ is the universal symbol $\bar{\sigma}_{\text{univ}}$ of the de Rham complex coupled to the reduced adjoint bundle. □

For any n a point of the 0-space of the pair of spectra $(MTSO_n, \Sigma^{-1}MTSO_{n-1})$ is represented by a map from (B^N, S^{N-1}) into the pair of Thom spaces attached to

$$\begin{array}{ccc} Q_{N-n} & \longrightarrow & Q_{N-n} \\ \downarrow & & \downarrow \\ Gr^+_{n-1}(\mathbb{R}^{N-1}) & \hookrightarrow & Gr^+_n(\mathbb{R}^N) \end{array}$$

for N sufficiently large. A map of (B^N, S^{N-1}) into this pair which is transverse to the 0-section gives a compact oriented n-manifold M with boundary embedded in (B^N, S^{N-1}), a rank n bundle $W \to M$ equipped with a stable isomorphism $TM \cong W$, and a splitting of $W\big|_{\partial M}$ as the direct sum of a rank $(n-1)$ bundle and a trivial line bundle. The composition with the boundary map

$$r : (MTSO_n, \Sigma^{-1}MTSO_{n-1}) \to MTSO_{n-1} \tag{19.29}$$

is the restriction of this data to ∂M.

Now let M be a smooth manifold, X a compact oriented two-manifold with boundary, and $P \to M \times X$ a principal G-bundle with partial flat connection along X. This data induces a classifying map $M \to \mathcal{M}_X$ and, forgetting the connection, a classifying map

$$f : M \longrightarrow (MTSO_2, \Sigma^{-1}MTSO_1) \wedge BG. \tag{19.30}$$

There are induced classifying maps $M \to \mathcal{M}_{\partial X}$ and

$$\dot{f} : M \longrightarrow MTSO_1 \wedge BG$$

for the boundary data; here $\dot{f} = r \circ f$. View X as a bordism $X : \emptyset \to \partial X$; later we incorporate incoming boundary components. The following key result relates the universal topology above to surfaces with boundary.

Lemma 19.4 *The map $\bar{\sigma}'_{\text{univ}} \circ f : M \to ko$ classifies the reduced tangent bundle of the restriction map $t : \mathcal{M}_X \to \mathcal{M}_{\partial X}$—the bundle $T\mathcal{M}_X - t^*T\mathcal{M}_{\partial X}$ reduced to rank zero—pulled back to M.*

Proof. At a point $A \in M$ there is a flat connection on $P\big|_{\{A\}\times X}$. The tangent space to $\mathcal{M}_X$ at that point is computed by the twisted de Rham complex (19.17), so is represented by the twisted de Rham cohomology $H^\bullet_A(X)$. Similarly, the tangent space to $\mathcal{M}_{\partial X}$ at the restriction $t(A)$ of A to the boundary is $H^\bullet_{t(A)}(\partial X)$. From the long exact sequence of the pair $(X, \partial X)$ we deduce that the difference $T\mathcal{M}_X - t^*T\mathcal{M}_{\partial X}$ at A is the twisted relative de Rham cohomology $H^\bullet_A(X, \partial X)$.

Now the twisted relative de Rham cohomology is the index of the deformation complex (19.17) with *relative boundary conditions* (Gilkey 1995, section 4.1). In other words, we consider the subcomplex of forms which vanish when restricted to ∂X. This is an example of a *local* elliptic boundary value problem. Atiyah and Bott (1964) interpret local boundary conditions in K-theory and prove an index formula. More precisely, the triple $\big(B(TX), \partial B(TX), S(TX)\big)$ leads to the exact sequence

$$KR^{-1}\big(B(TX)\big|_{\partial X}, S(TX)\big|_{\partial X}\big) \longrightarrow KR^0\big(B(TX), \partial B(TX)\big)$$
$$\longrightarrow KR^0\big(B(TX), S(TX)\big) \longrightarrow KR^0\big(B(TX)\big|_{\partial X}, S(TX)\big|_{\partial X}\big)$$

The symbol σ of an elliptic operator lives in the third group, and Atiyah and Bott construct a lift to the second group from a local boundary condition. (The image of σ in the last group is an obstruction to the existence of local boundary conditions; the image of the first group in the second measures differences of boundary conditions.) The relative boundary conditions on the twisted de Rham complex are *universal*, so the corresponding lift of the symbol occurs in the universal setting. Recall from the proof of Lemma 19.3 the exact sequence (19.28), now extended one step to the left:

$$KR^0_G(i\mathbb{R})_c \longrightarrow KR^0_{SO_2\times G}(\mathbb{C}^2)_c \longrightarrow KR^0_{SO_2\times G}(i\mathbb{R}^2)_c \longrightarrow KR^0_G(i\mathbb{R}^2)_c$$

The group G acts trivially in all cases. The Atiyah–Bott procedure applied to the relative boundary conditions gives a lift of $\bar{\sigma}_{\text{univ}} \in KR^0_{SO_2\times G}(i\mathbb{R}^2))_c$ to $KR^0_{SO_2\times G}(\mathbb{C}^2)_c$. Recall that $\bar{\sigma}'_{\text{univ}}$, constructed in the proof of Lemma 19.3, is also a lift of $\bar{\sigma}_{\text{univ}}$. But by periodicity we find $KR^0_G(i\mathbb{R})_c \cong KR^0_G(-\mathbb{R})_c \cong KO^0_G(-\mathbb{R})_c \cong KO^1_G(\text{pt})$ which vanishes by Anderson (1964). Thus the lift of $\bar{\sigma}_{\text{univ}}$ is unique and $\bar{\sigma}'_{\text{univ}}$ computes the relative twisted de Rham cohomology. This completes the proof. □

19.3.4 *The level*

A universal orientation induces a level, which is commonly used to identify the theory. One observation arising from this study is that the level is a derived quantity, and it is the universal orientation which determines the theory. We explain the relationship, and deduce the existence of universal orientations, in this subsection.

To begin we recast Definition 19.1 of a universal orientation in a form suited for surfaces with boundary. Consider the diagram

$$\begin{array}{ccccc} MTSO_2 \wedge BG & \xrightarrow{q} & (MTSO_2, \Sigma^{-1}MTSO_1) \wedge BG & \xrightarrow{r} & MTSO_1 \wedge BG \\ & \searrow^{\bar{\sigma}_{\mathrm{univ}}} & \downarrow{\scriptstyle \bar{\sigma}'_{\mathrm{univ}}} & & \vdots {\scriptstyle -\lambda} \\ & & ko & \xrightarrow{\tau} & pic_{\mathrm{g}}K \end{array} \tag{19.31}$$

The top row is a cofibration, the continuation of the top row of (19.28) in the Puppe sequence. Recall that a universal orientation is a null homotopy of $\tau \circ \bar{\sigma}_{\mathrm{univ}} = \tau \circ \bar{\sigma}'_{\mathrm{univ}} \circ q$.

Lemma 19.5 *A universal orientation is equivalent to a map* $-\lambda$ *in* (19.31) *and a homotopy from* $\tau \circ \bar{\sigma}'_{\mathrm{univ}}$ *to* $-\lambda \circ r$.

The proof is immediate. In view of (19.25) and adjunction we can write $\lambda : \Sigma^{\infty} BG \to \Sigma pic_{\mathrm{g}}K$ as a map from the suspension spectrum of BG to the spectrum $pic_{\mathrm{g}}K$, or equivalently as a map

$$\lambda : BG \longrightarrow B\,\mathrm{Pic}_{\mathrm{g}}\,K \tag{19.32}$$

on spaces.

Definition 19.2 *The map* (19.32) *is the* level *induced by a universal orientation.*

There is a map $K(\mathbb{Z}, 4) \to B\,\mathrm{Pic}_{\mathrm{g}}\,K$ (see (19.15)) and in some important cases, for example, if G is connected and simply connected, the level factors through $K(\mathbb{Z}, 4)$, that is, the level is a class $\lambda \in H^4(BG)$.

A universal orientation is more than a level: it is a map $-\lambda : MTSO_1 \wedge BG \to pic_{\mathrm{g}}K$ together with a homotopy of $-\lambda \circ r$ and $\tau \circ \bar{\sigma}'_{\mathrm{univ}}$ in (19.31). Our next result proves that universal orientations exist.

Theorem 19.1 *There is a canonical universal orientation* μ. *The corresponding level* h *is the negative of* $BG \to BO \to B\,\mathrm{Pic}_g\,K$, *where the first map is induced from the reduced adjoint representation* $\bar{\mathfrak{g}}$ *and the second is projection to a Postnikov section.*

Proof. Since complex vector spaces have a canonical K-theory orientation, the composite map $k \to ko \xrightarrow{\tau} pic_{\mathrm{g}}K$ is null, where k is the connective K-theory spectrum (see the text preceding (19.16)). Therefore, a universal orientation is given by filling in the left dotted arrow in the diagram:

$$\begin{array}{ccccc} MTSO_2 \wedge BG & \xrightarrow{q} & (MTSO_2, \Sigma^{-1}MTSO_1) \wedge BG & \xrightarrow{r} & MTSO_1 \wedge BG \\ \vdots & & \downarrow \bar{\sigma}'_{\mathrm{univ}} & & \vdots\ -\lambda \\ k & \longrightarrow & ko & \xrightarrow{\tau} & pic_{\mathrm{g}}K \end{array} \tag{19.33}$$

and specifying a homotopy which makes the left square commute. There is a natural choice: smash the K-theory Thom class $U : MTU_1 \simeq MTSO_2 \to k$ with the complexified reduced adjoint representation $\bar{\mathfrak{g}}_{\mathbb{C}}$. This is the universal rewriting of de Rham on a Riemann surface in terms of Dolbeault, at least on the symbolic level. In terms of the proof of Lemma 19.3, the map $\bar{\sigma}_{\mathrm{univ}}$, restricted to $MTSO_2$, is the exterior algebra complex $(\bigwedge^{\bullet}\mathbb{C}^2, \epsilon)$ in $KR^0_{SO_2}(i\mathbb{R}^2)_c$. Write $\mathbb{R}^2 = L_{\mathbb{R}}$ for the complex line $L = \mathbb{C}$ and $\mathbb{C}^2 \cong \mathbb{R}^2 \otimes \mathbb{C} \cong L \oplus \overline{L}$. Then the symbol complex at $\theta \in i\mathbb{R}^2$,

$$\mathbb{C} \xrightarrow{\epsilon(\theta)} L \oplus \overline{L} \xrightarrow{\epsilon(\theta)} L \otimes \overline{L}\ ,$$

is the realification of the complex

$$\mathbb{C} \xrightarrow{\epsilon(\theta)} L$$

which defines the K-theory Thom class. Tensor with the complexified reduced adjoint representation $\bar{\mathfrak{g}}_{\mathbb{C}}$ to complete the argument.

To compute the level of μ we factorize τ as $ko \xrightarrow{\eta} \Sigma^{-1}ko \to pic_{\mathrm{g}}K$, where the first map is multiplication by $\eta : S^0 \to S^{-1}$ and the second is projection to a Postnikov section. The map η fits into the Bott sequence $k \to ko \to \Sigma^{-1}ko$, and so we extend (19.33) to the left:

$$\begin{array}{cccccccc} & & \Sigma^{-1}MTSO_1 \wedge BG & \longrightarrow & MTSO_2 \wedge BG & \xrightarrow{q} & (MTSO_2, \Sigma^{-1}MTSO_1) \wedge BG \\ & & \vdots\ \alpha & & \downarrow U\wedge\bar{\mathfrak{g}}_{\mathbb{C}} & & \downarrow \bar{\sigma}'_{\mathrm{univ}} \\ \Sigma^{-1}ko & \longrightarrow & \Sigma^{-2}ko & \longrightarrow & k & \longrightarrow & ko \end{array} \tag{19.34}$$

The homotopy which expresses the commutativity of the right-hand square induces the map α in this diagram, and the map $-\lambda$ induced in (19.33) is the suspension of α. We claim that there is a unique α, up to homotopy, which makes the left square in (19.34) commute. For the difference of any two choices for α is a map $\Sigma^{-1}MTSO_1 \wedge BG \to \Sigma^{-1}ko$, and the homotopy classes of such maps form the group $KO^1(BG)$ which vanishes (Anderson 1964). It is easy to find a map α as follows. Since $\Sigma^{-1}MTSO_1 \simeq S^{-2}$ (Lemma 19.2(1)), the upper left map is the inclusion of the bottom cell of $MTSO_2 \wedge BG$. The composite $\Sigma^{-1}MTSO_1 \wedge BG \simeq \Sigma^{-2}BG \to k$ factors as $\Sigma^{-2}BG \to \Sigma^{-2}k \to k$, where the first map is the double desuspension of $\bar{\mathfrak{g}}_{\mathbb{C}}$ and the second Bott periodicity. Choose α to be the double desuspension of $\bar{\mathfrak{g}}$, the *real* reduced adjoint representation. This completes the proof. □

Since equivalence classes of universal orientations form a torsor for the group $\mathcal{O}(G)$ in (19.23), the canonical universal orientation identifies the torsor of universal orientations with $\mathcal{O}(G)$. Notice the natural map

$$\begin{aligned} \ell : \mathcal{O}(G) = [\Sigma MTSO_2 \wedge BG, pic_{\mathrm{g}}K] &\longrightarrow [MTSO_1 \wedge BG, pic_{\mathrm{g}}K] \\ &\cong [BG, B\,\mathrm{Pic}_{\mathrm{g}}\,K] \end{aligned} \tag{19.35}$$

from universal orientations to levels. If $g \in \mathcal{O}(G)$, then the level of $\mu + g$ is $\ell(g) - h$. If G is connected, simply connected, and simple, then $[BG, B\,\mathrm{Pic}_{\mathrm{g}}\,K] \cong H^4(BG;\mathbb{Z}) \cong \mathbb{Z}$ and h is the dual Coxeter number of G times a generator. Then $g \mapsto \ell(g) - h$ is a version of the ubiquitous "adjoint shift."

Remark 19.3 For any G the top homotopy group of $\mathrm{Map}(\Sigma MTSO_2, pic_{\mathrm{g}}K)$ and of $B\,\mathrm{Pic}_{\mathrm{g}}\,K$ is π_4, which is infinite cyclic. So there is a homomorphism of $H^4(BG;\mathbb{Z})$ into the domain and codomain of (19.35), and on these subspaces ℓ is an isomorphism. This means that we can *change* a consistent orientation by an element of $H^4(BG;\mathbb{Z})$, and the level changes by the same amount.

19.3.5 *Pushforward maps*

Suppose we have chosen a universal orientation with level λ. Let X be a compact-oriented two-manifold with boundary. We can work as before with a family of flat connections on X parametrized by a smooth manifold M, but instead for simplicity we work universally on $\mathcal{M}_X$. As in (19.30) fix a classifying map

$$f : \mathcal{M}_X \longrightarrow (MTSO_2, \Sigma^{-1}MTSO_1) \wedge BG; \tag{19.36}$$

then there is an induced classifying map

$$\dot{f} = r \circ f : \mathcal{M}_{\partial X} \longrightarrow MTSO_1 \wedge BG$$

for r the map (19.29). Set $\tau = \dot{f}^*(\lambda)$. Let $t : \mathcal{M}_X \to \mathcal{M}_{\partial X}$ be the restriction map on flat connections. According to Lemma 19.4 the composition $\tau \circ \bar{\sigma}'_{\mathrm{univ}} \circ f$ is the reduced twisting $\tau_t - (\dim \mathcal{M}_X - t^* \dim \mathcal{M}_{\partial X})$. The homotopy which expresses commutativity of the square in (19.31) gives an isomorphism

$$\tau_t - (\dim \mathcal{M}_X - t^* \dim \mathcal{M}_{\partial X}) \xrightarrow{\cong} -t^*\tau. \tag{19.37}$$

In principle, $\dim \mathcal{M}_X$ and $\dim \mathcal{M}_{\partial X}$ are locally constant functions which vary over the moduli space. However, the Euler characteristic of the deformation complex (19.17) is independent of the connection, and we easily deduce

$$\dim \mathcal{M}_X - t^* \dim \mathcal{M}_{\partial X} \equiv (\dim G)\, b_0(\partial X) \pmod 2,$$

where b_0 is the number of connected components. (We only track degrees in K-theory modulo 2 [see (19.10)].) According to the discussion in Section 19.2 (especially (19.14)), there then is an induced pushforward

$$t_* : K^0(\mathcal{M}_X) \longrightarrow K^{\tau + (\dim G)\, b_0(\partial X)}(\mathcal{M}_{\partial X}). \tag{19.38}$$

This is the pushforward (19.5) associated to the bordism $X : Y_0 \to Y_1$ with $Y_0 = \emptyset$ and $Y_1 = \partial X$. The invariant (19.5) is then $t_*(1)$.

Note the special case $\partial X = S^1$. Then $\mathcal{M}_{S^1} = G//G$ is the global quotient stack of the action of G on G by conjugation. The codomain of (19.38) is thus

$$K^{\tau + \dim G}(\mathcal{M}_{S^1}) \cong K^{\tau + \dim G}(G//G) = K_G^{\tau + \dim G}(G).$$

This is the basic space of the two-dimensional TQFT we construct (see Section 19.1).

Observe that the universal orientation, in the form described around (19.31), leads to the twist τ in the codomain of (19.38). This is the mechanism which was envisioned in (19.14) when we discussed twistings in general: we have constructed a pushforward from untwisted K-theory to twisted K-theory. The universal orientation neatly accounts for the construction of a Frobenius ring structure on twisted K-theory.

To treat an arbitrary bordism $X : Y_0 \to Y_1$ we note that the deformation complex at a flat connection a on a principal G-bundle $Q \to Y_0$ is

$$0 \longrightarrow \Omega^0_{Y_0}(\mathfrak{g}_Q) \xrightarrow{\ d_a\ } \Omega^1_{Y_0}(\mathfrak{g}_Q) \longrightarrow 0 \tag{19.39}$$

The operator d_a is skew-adjoint. Therefore, there is a canonical trivialization of the K-theory class of (19.39); for example, a canonical isomorphism $\ker d_a \cong \operatorname{coker} d_a$. Suppose a classifying map f is given as in (19.36) and let $\dot{f}_0, \dot{f}_1$ denote its restriction to the boundary connections on Y_0, Y_1. Set $\tau_i = \dot{f}_i^*(\lambda)$. Then (19.37) and the canonical trivialization of (19.39) on the incoming boundary lead to the desired push–pull map

$$\begin{aligned} K^{\tau_0 + (\dim G)\, b_0(Y_0)}(\mathcal{M}_{Y_0}) &\xrightarrow{\ s^*\ } K^{s^*\tau_0 + (\dim G)\, b_0(Y_0)}(\mathcal{M}_X) \\ &\xrightarrow{\ t_*\ } K^{\tau_1 + (\dim G)\, b_0(Y_1)}(\mathcal{M}_{Y_1}) \end{aligned} \tag{19.40}$$

This is the map (19.5) with the twistings induced from the universal orientation.

A universal orientation induces consistent orientations on the outgoing restriction maps of bordisms. In other words, if $X : Y_0 \to Y_1$ and $X' : Y_1 \to Y_2$ are bordisms, then the push–pull maps derived from the diagram

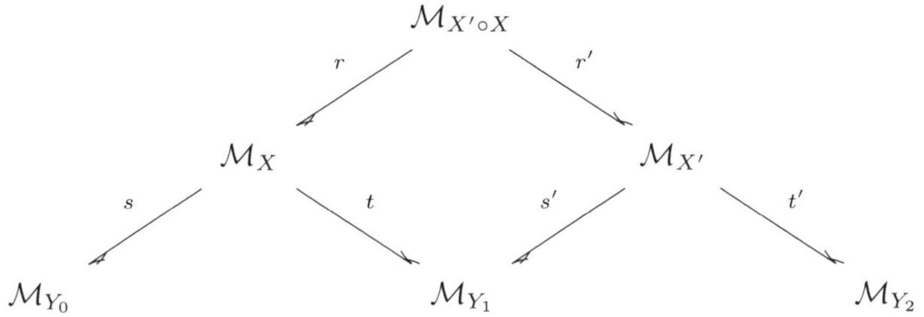

satisfy

$$(t'r')_* \circ (sr)^* = [t'_* \circ s'^*] \circ [t_* \circ s^*].$$

This follows from Lemma 19.1 and the "Fubini property"

$$(t'r')_* = t'_* r'_* \tag{19.41}$$

of pushforward. The orientation of t induces orientations of r' and $t'r'$, since the diamond is a fiber product. At stake in (19.41) is the consistency of the orientations, which is ensured by the use of a universal orientation. The details of this argument[4] will be given on another occasion.

One caveat: since $\mathcal{M}_X, \mathcal{M}_{\partial X}$ are stacks we can only pushforward along *representable* maps, and this forces every component of X to have a nonempty outgoing boundary. As mentioned at the end of section 19.1, the partial topological quantum field theory obtained from the push–pull construction extends to a full theory using the invertibility of the (co)pairing attached to the cylinder (Freed *et al.* 2003, section 17).

19.4 Families of surfaces, twistings, and anomalies

We begin with a general discussion of topological quantum field theories for families. Let F be an n-dimensional TQFT in the most naive sense. Thus F assigns a finite-dimensional complex vector space $F(Y)$ to a closed oriented $(n-1)$-manifold Y and a linear map $F(X) : F(Y_0) \to F(Y_1)$ to a bordism $X : Y_0 \to Y_1$. In particular, $F(X) \in \mathbb{C}$ if X is closed. Suppose that $\mathcal{Y} \to S$ is a fiber bundle with a closed oriented $(n-1)$-manifold. Then the vector spaces assigned to each fiber fit together into a complex vector bundle $F(\mathcal{Y}/S) \to S$. If $\gamma : [0,1] \to S$ is a path, then $\gamma^*\mathcal{Y} \to [0,1]$ is a bordism from the fiber $\mathcal{Y}_{\gamma(0)}$ to the fiber $\mathcal{Y}_{\gamma(1)}$. The topological invariance of F shows that $F(\gamma^*\mathcal{Y}) : F(\mathcal{Y}_{\gamma(0)}) \to F(\mathcal{Y}_{\gamma(1)})$ is unchanged by a homotopy of γ, and so $F(\mathcal{Y}/S) \to S$ carries a natural flat connection. Then a family of bordisms $\mathcal{X} \to S$ from $\mathcal{Y}_0 \to S$ to $\mathcal{Y}_1 \to S$ produces

[4] The main point is that consistent orientations themselves form an *invertible* topological field theory, and these field theories factor through the group completion of bordism, that is, the Madsen–Tillmann space.

a section $F(\mathcal{X}/S)$ of $\mathrm{Hom}\big(F(\mathcal{Y}_0/S), F(\mathcal{Y}_1/S)\big) \to S$; the topological invariance and gluing law of the TQFT imply that this section is flat. In other words, $F(\mathcal{X}/S) \in H^0\big(S; \mathrm{Hom}\big(F(\mathcal{Y}_0/S), F(\mathcal{Y}_1/S)\big)\big)$. It is natural, then, to postulate that a TQFT in families gives more, namely, classes of all degrees:

$$F(\mathcal{X}/S) \in H^\bullet\big(S; \mathrm{Hom}\big(F(\mathcal{Y}_0/S), F(\mathcal{Y}_1/S)\big)\big). \tag{19.42}$$

These classes are required to satisfy gluing laws and topological invariance as well as naturality under base change.

The idea of a TQFT in families—at least in two dimensions—was introduced in the mid-1990s. In two dimensions it is often formulated in a holomorphic language (e.g. Kontsevich and Manin 1994), and classes are required to extend to the Deligne–Mumford compactification of the moduli space of Riemann surfaces.

Our push–pull construction works for families of surfaces—with a twist. The purpose of this section is to alert the reader to the twist.[5]

Suppose $\mathcal{X} \to S$ is a family of bordisms from $\mathcal{Y}_0 \to S$ to $\mathcal{Y}_1 \to S$, where $\mathcal{Y}_i \to S$ are fiber bundles of oriented one-manifolds. Then the moduli stacks of flat connections form a correspondence diagram over S:

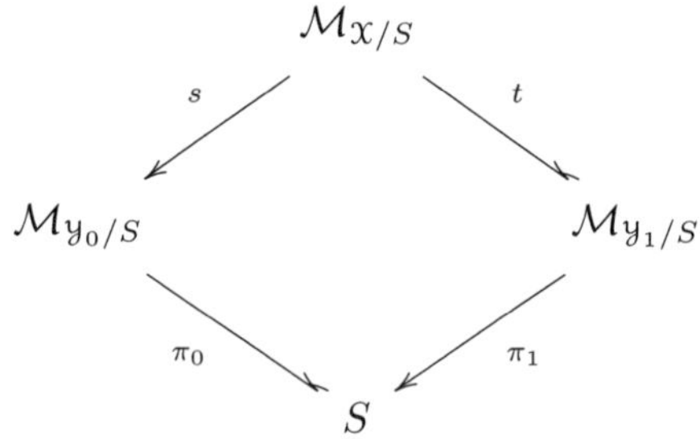

The push–pull constructs a map from twisted $K(\mathcal{M}_{\mathcal{Y}_0/S})$ to twisted $K(\mathcal{M}_{\mathcal{Y}_1/S})$. We can also work locally over S: the K-theory of the fibers of π_i form bundles of spectra over S and the push–pull gives a map of these spectra. But for our purposes the global push–pull suffices. This construction is a variation of (19.42): we use K-theory rather than cohomology.

The discussion of Section 19.3 goes through with one important change. It comes in the paragraph preceding (19.21). For simplicity suppose $\mathcal{Y}_0 = \emptyset$ so that the boundary $\partial\mathcal{X} = \mathcal{Y}_1$ is entirely outgoing. Fix $A \in \mathcal{M}_{\mathcal{X}_s}$ a flat connection on a principal G-bundle $P \to \mathcal{X}_s$. Then the KO-theory class of the de Rham complex coupled to the reduced adjoint bundle $\bar{\mathfrak{g}}_P = \mathfrak{g}_P - \underline{\dim G}$ computes the difference $T_A\mathcal{M}_{\mathcal{X}_s} - (\dim G)\, H^\bullet(\mathcal{X}_s)$, where $H^\bullet(\mathcal{X}_s)$ is the real cohomology of $\mathcal{X}_s$ viewed as a class in KO-theory. In Section 19.3 we treat $H^\bullet(\mathcal{X}_s)$ as a trivial vector space (there $S = \mathrm{pt}$), but now $H^\bullet(\mathcal{X}_s)$ varies with $s \in S$ and so can give rise

[5] We thank Veronique Godin for the perspicacious sign question which prompted this exposition.

to a non-trivial twisting. More precisely, $H^\bullet(\mathcal{X}_s)$ is the fiber at $s \in S$ of a flat vector bundle $\mathcal{H}^\bullet(\mathcal{X}/S) \to S$. Let $\tau_{\mathcal{X}/S}$ denote the twisting of complex K-theory attached to this bundle. This twisting replaces the degree shift in (19.37) and the pushforward (19.38) is modified to include that extra twist:

$$t_* : K^0(\mathcal{M}_{\mathcal{X}/S}) \longrightarrow K^{\tau + (\dim G)\, \pi_1^* \tau_{\mathcal{X}/S}}(\mathcal{M}_{\mathcal{Y}_1/S}). \tag{19.43}$$

The degree shift is now incorporated into the twist $\tau_{\mathcal{X}/S}$, and there may be non-trivial contributions to the twist from w_1 and W_3 of $\mathcal{H}^\bullet(\mathcal{X}/S) \to S$ as well. (The degree shift and twistings vanish canonically if $\dim G$ is even.)

Example 19.1 Consider the disjoint union X of two 2-disks. The boundary circles are outgoing, as above. Suppose that $\dim G$ is odd. For a single disk the pushforward $t_*(1)$ in (19.38) lands in $K_G^{\tau+1}(G)$ and is the unit $\mathbf{1}$ in the Verlinde ring. Thus for the disjoint union of two disks, $t_*(1)$ is the image of $\mathbf{1} \otimes \mathbf{1}$ under the external product $K_G^{\tau+1}(G) \otimes K_G^{\tau+1}(G) \to K_{G\times G}^{(\tau,\tau)}(G \times G)$. Now consider the family $\mathcal{X} \to S$ with fiber X and base $S = S^1$ in which the monodromy exchanges the two disks. The flat bundle $\mathcal{H}^\bullet(\mathcal{X}/S) \to S$ has rank 2 and non-trivial w_1. According to (19.43), then, $t_*(1)$ for the family lives in the twisted group $K^{(\tau,\tau)+\pi^*\tau_{\mathcal{X}/S}}(\mathcal{M}_{\partial\mathcal{X}/S})$. On each fiber of $\pi : \mathcal{M}_{\partial\mathcal{X}/S} \to S$ we recover the class $\mathbf{1} \otimes \mathbf{1}$ above. But upon circling the loop $S = S^1$ this class changes sign in the $\pi^*\tau_{\mathcal{X}/S}$-twisted K-group. Said differently, the diffeomorphism which exchanges the disks acts by a sign on $\mathbf{1} \otimes \mathbf{1}$. Of course, one might predict this from the sign rule in graded algebra: the Verlinde ring $K_G^{\tau+1}(G)$ is in odd degree, so upon exchanging the factors of $\mathbf{1} \otimes \mathbf{1}$ one picks up a sign. It shows up here as an extra twisting.

This extra twisting is a topological analog of what is usually called an *anomaly* in quantum field theory. In an anomalous theory in n dimensions the partition function on a closed n-manifold, rather than being a complex-valued function on a space of fields, is a section of a complex line bundle over that space of fields. Furthermore, there is a gerbe over the space of fields on a closed $(n-1)$-manifold, and for a bordism the partition function is a map of the gerbes attached to the boundary. In the homological version described at the beginning of this section, the parameter space S plays the role of the space of fields and for a family of closed n-manifolds the partition function in a non-anomalous theory is an element of $H^\bullet(S;\mathbb{R})$. An anomalous theory would assign a flat complex line bundle $L \to S$ to the family, and the partition function would live in the twisted cohomology $H^\bullet(S;L) = H^{L+\bullet}(S)$. In the two-dimensional TQFT we construct using push–pull on K-theory, the extra K-theory twist $\tau_{\mathcal{X}/S}$ is the anomaly (see (19.43)). Notice that there is no gerbe attached to a family of one-manifolds (better: it is canonically trivial). We remark that the anomaly is itself a particular example of an invertible topological quantum field theory.

Acknowledgments

The work of Daniel S. Freed, Michael J. Hopkins, and Constantin Teleman was supported by NSF grant DMS-0603964, DMS-0306519, and EPSRC GR/SO6165/01, respectively.

We thank Veronique Godin, Jacob Lurie, Ib Madsen, and Ulrike Tillmann for enlightening conversations.

References

Abrams, L. (1996). Two-dimensional topological quantum field theories and Frobenius algebras. *J. Knot Theory Ramifications* **5**(5), 569–87.

Anderson, D. W. (1964). The real K-theory of classifying spaces. *Proc. Nat. Acad. Sci.* **51**(4), 634–6.

Atiyah, M. F. (1966). K-theory and reality. *Q. J. Math. Oxford Ser. (2)* **17**, 367–86.

Atiyah, M. F. and Bott, R. (1964). The index problem for manifolds with boundary. *Differential Analysis, Bombay Colloquium, 1964*. Oxford University Press, London, pp. 175–86.

Atiyah, M. and Segal, G. (2006). *Twisted K-Theory and Cohomology*, Inspired by S. S. Chern, Nankai Tracts Mathematics, Vol. 11, World Scientific Publication, Hackensack, NJ, pp. 5–43, arXiv:math/0510674.

Atiyah, M. F. and Singer, I. M. (1971). The index of elliptic operators. V, *Ann. Math. (2)* **93**, 139–49.

Costello, K. (2007). Topological conformal field theories and Calabi-Yau categories. *Adv. Math.* **210**(1), 165–214, arXiv:math/0412149.

Donaldson, S. K. (1987). The orientation of Yang-Mills moduli spaces and 4-manifold topology. *J. Differential Geom.* **26**(3), 397–428.

Donaldson, S. K. and Kronheimer, P. B. (1990). The geometry of four-manifolds. *Oxford Mathematical Monographs.* Oxford Science Publications, New York.

Donovan, P. and Karoubi, M. (1970). Graded Brauer groups and K-theory with local coefficients. *Inst. Hautes Études Sci. Publ. Math.* No. 38, 5–25.

Freed, D. S. (2009). Remarks on Chern–Simons theory. *Bull. Amer. Math. Soc.* **46**, 221–254, arXiv:0808.2507.

Freed, D. S., Hopkins, M. J. and Teleman, C. Loop groups and twisted K-theory I, arXiv:0711.1906.

Freed, D. S., Hopkins, M. J. and Teleman, C. Loop groups and twisted K-theory II, arXiv:math/0511232.

Freed, D. S., Hopkins, M. J. and Teleman, C. Loop groups and twisted K-theory III, arXiv:math/0312155.

Freed, D. S., Hopkins, M. J. and Teleman, C. (2007*b*). Twisted equivariant K-theory with complex coefficients. *J. Topol.* **1**, arXiv:math/0206257.

Galatius, S., Madsen, I., Tillmann, U. and Weiss, M. (2009). The homotopy type of the cobordism category. *Acta Math.* **202**, 195–239, arXiv:math/0605249.

Gilkey, P. B. (1995). *Invariance Theory, the Heat Equation, and the Atiyah-Singer Index Theorem*, 2nd edn., *Studies in Advanced Mathematics.* CRC Press, Boca Raton, FL.

Hitchin, N. (2001). Lectures on special Lagrangian submanifolds. *Winter School on Mirror Symmetry, Vector Bundles and Lagrangian Submanifolds (Cambridge, MA,*

1999). AMS/IP Studies in Advanced Mathematics, Vol. 23, American Mathematical Society, Providence, RI, pp. 151–82, arXiv:math/9907034.

Hopkins, M. J. and Lurie, J. in preparation.

Kontsevich, M. and Manin, Yu. (1994). Gromov-Witten classes, quantum cohomology, and enumerative geometry. *Commun. Math. Phys.* **164**(3), 525–62, arXiv:hep-th/9402147.

Madsen, I. and Tillmann, U. (2001). The stable mapping class group and $Q(\mathbb{CP}^{\infty}_{+})$. *Invent. Math.* **145**(3), 509–44.

Madsen, I. and Weiss, M. (2005). The stable mapping class group and stable homotopy theory. *European Congress of Mathematics.* European Mathematical Society, Zürich, pp. 283–307.

Moore, G. W. and Segal, G. B. (2006). D-branes and K-theory in 2D topological field theory, arXiv:hep-th/0609042.

Murray, M. K. (1996). Bundle gerbes. *J. London Math. Soc. (2)* **54**(2), 403–16.

Segal, G. B. (1977). K-homology theory and algebraic K-theory. *K-theory and Operator Algebras (Proceedings of the Conference, University of Georgia, Athens, G.A, 1975).* Lectures Notes in Mathematics, Vol. 575, Springer, Berlin, Germany, pp. 113–27.

INDEX